BIOMEMBRANES

Volume 11

BIOMEMBRANES

A series edited by Lionel A. Manson
The Wistar Institute, Philadelphia, Pennsylvania

Recent Volumes in this Series

1972 • Biomembranes • Volume 3
Passive Permeability of Cell Membranes
Edited by F. Kreuzer and J. F. G. Slegers

1974 • Biomembranes • Volume 4A
Intestinal Absorption
Edited by D. H. Smyth

1974 • Biomembranes • Volume 4B
Intestinal Absorption
Edited by D. H. Smyth

1974 • Biomembranes • Volume 5
Articles by Richard W. Hendler, Stuart A. Kauffman, Dale L. Oxender, Henry C. Pitot, David L. Rosenstreich, Alan S. Rosenthal, Thomas K. Shires, and Donal F. Hoelzl Wallach

1975 • Biomembranes • Volume 6
Bacterial Membranes and the Respiratory Chain
By N. S. Gel'man, M. A. Lukoyanova, and D. N. Ostrovskii

1975 • Biomembranes • Volume 7
Aharon Katzir Memorial Volume
Edited by Henryk Eisenberg, Ephraim Katchalsi-Katzir, and Lionel A. Manson

1976 • Biomembranes • Volume 8
Articles by Robert W. Baldwin, William C. Davis, Paul H. DeFoor, Carl G. Gahmberg, Sen-itiroh Hakomori, Reinhard Kurth, Lionel A. Manson, Michael R. Price, and Howard E. Sandberg

1977 • Biomembranes • Volume 9
Membrane Transport—An Interdisciplinary Approach
By Arnost Kotyk and Karel Janácek

1979 • Biomembranes • Volume 10
Articles by Gloria Gronowicz, Eric Holtzman, Masayori Inouye, A. A. Jasaitis, Sandra K. Masur, Arthur Mercurio, D. H. Northcote, S. A. Ostroumov, V. D. Samuilov, M. Silverman, and R. J. Turner

1983 • Biomembranes • Volume 11
Pathological Membranes
Edited by Alois Nowotny

A Continuation Order Plan is available for this series. A continuation order will bring delivery of each new volume immediately upon publiction. Volumes are billed only upon actual shipment. For further information please contact the publisher.

BIOMEMBRANES, Volume 11

PATHOLOGICAL MEMBRANES

Edited by
Alois Nowotny
University of Pennsylvania
Philadelphia, Pennsylvania

PLENUM PRESS • NEW YORK AND LONDON

Library of Congress Cataloging in Publication Data

Main entry under title:

Pathological membranes.

(Biomembranes; v. 11)
Bibliography: p.
Includes index.
1. Pathology, Cellular. 2. Plasma membranes. 3. Cancer cells. I. Nowotny, A. (Alois), 1922- . II. Series: Biomembranes (Plenum Press); v. 11.
QH601.B53 vol. 11 [RB25] 574.87′5s 82-22343
ISBN 0-306-41065-6 [611′.01815

A Division of Plenum Publishing Corporation
233 Spring Street, New York, N.Y. 10013

Printed in the United States of America

Contributors

Leo G. Abood, Center for Brain Research and Department of Biochemistry, University of Rochester Medical Center, Rochester, New York
David J. Adams, Department of Medicine, University of Texas Health Science Center, San Antonio, Texas
S. R. Baker, The Wistar Institute, Philadelphia, Pennsylvania
R. W. Baldwin, Cancer Research Campaign Laboratories, University of Nottingham, Nottingham, United Kingdom
Jean M. Bidlack, Center for Brain Research and Department of Biochemistry, University of Rochester Medical Center, Rochester, New York
D. L. Blithe, The Wistar Institute, Philadelphia, Pennsylvania
Charles Eric Brown, Department of Biochemistry, The Medical College of Wisconsin, Milwaukee, Wisconsin
C. A. Buck, The Wistar Institute, Philadelphia, Pennsylvania
K. Chandrasekaran, National Cancer Institute, National Institutes of Health, Bethesda, Maryland
Margaret R. Clark, Department of Laboratory Medicine, School of Medicine, University of California, San Francisco, California
John F. Codington, Laboratory for Carbohydrate Research, Massachusetts General Hospital, and Departments of Biological Chemistry and Medicine, Harvard Medical School, Boston, Massachusetts
John S. Coon, Department of Pathology, Rush Medical College and Rush Presbyterian–St. Luke's Medical Center, Chicago, Illinois
Dean P. Edwards, Department of Medicine, University of Texas Health Science Center, San Antonio, Texas
M. J. Embleton, Cancer Research Campaign Laboratories, University of Nottingham, Nottingham, United Kingdom
S. Ferrone, Departments of Pathology and Surgery, College of Physicians and Surgeons of Columbia University, New York, New York

David M. Frim, Laboratory for Carbohydrate Research, Massachusetts General Hospital, and Departments of Biological Chemistry and Medicine, Harvard Medical School, Boston, Massachusetts

Ingegerd Hellström, Division of Tumor Immunology, Fred Hutchinson Cancer Research Center, Seattle, Washington

Karl Erik Hellström, Division of Tumor Immunology, Fred Hutchinson Cancer Research Center, Seattle, Washington

K. Imai, Departments of Pathology and Surgery, College of Physicians and Surgeons, Columbia University, New York, New York

F. Indiveri, Departments of Pathology and Surgery, College of Physicians and Surgeons, Columbia University, New York, New York

Marguerite M. B. Kay, Research and Medical Services, Olin E. Teague Veterans' Center, and Division of Geriatric Medicine, Texas A and M University, Temple, Texas

William L. McGuire, Department of Medicine, University of Texas Health Science Center, San Antonio, Texas

William C. Mentzer, Jr., Department of Pediatrics, School of Medicine, University of California, San Francisco, California

Peter T. Mora, National Cancer Institute, National Institutes of Health, Bethesda, Maryland

Karen Nelson, Division of Tumor Immunology, Fred Hutchinson Cancer Research Center, Seattle, Washington

A. K. Ng, Departments of Pathology and Surgery, College of Physicians and Surgeons, Columbia University, New York, New York

Garth L. Nicholson, Department of Tumor Biology, The University of Texas System Cancer Center, M. D. Anderson Hospital and Tumor Institute, Houston, Texas

A. Nowotny, School of Dental Medicine and School of Medicine, University of Pennsylvania, Philadelphia, Pennsylvania

M. A. Pellegrino, Departments of Pathology and Surgery, College of Physicians and Surgeons, Columbia University, New York, New York

George Poste, Smith Kline and French Laboratories, and Department of Pathology and Laboratory Medicine, University of Pennsylvania, Philadelphia, Pennsylvania

M. R. Price, Cancer Research Campaign Laboratories, University of Nottingham, Nottingham, United Kingdom

Ruta M. Radvany, Department of Surgery, Northwestern University Medical School, Chicago, Illinois

A. Vitiello, Departments of Pathology and Surgery, College of Physicians and Surgeons, Columbia University, New York, New York

L. Warren, The Wistar Institute, Philadelphia, Pennsylvania
Ronald S. Weinstein, Department of Pathology, Rush Medical College and Rush Presbyterian–St. Luke's Medical Center, Chicago, Illinois
B. S. Wilson, Departments of Pathology and Surgery, College of Physicians and Surgeons, Columbia University, New York, New York

Preface

A series of lectures entitled "Pathological Membranes," presented at the University of Pennsylvania School of Dental Medicine, provided the basis for this volume. These lectures were sponsored by the following pharmaceutical companies:

Bristol Laboratories
ICI United States, Inc., Stuart Pharmaceutical Division
McNeil Laboratories, Inc.
Merck Sharp & Dohme
Smith Kline & French Laboratories
Wyeth Laboratories

with minor contributions from additional sources. Therefore, I express my most sincere gratitude to these sponsors for their support.

Although the volume was to include coverage of as many as possible of the various diseases that are accompanied by membrane alterations, the paucity of sufficiently reliable information on certain membrane disorders limited the realization of this aim. Reviews on such disorders will be forthcoming when continued research brings a better understanding of the nature and significance of the membrane changes observed therein.

Accordingly, this volume reflects current knowledge of the pathology of membranes. Because most research has focused on membranes of malignant cells, many chapters are devoted to this topic. A few chapters report investigations on pathological alterations of erythrocyte membranes, including changes in these membranes during aging. The remaining chapters deal with more specialized topics, such as opiate receptors, physiochemical measurements of pathological changes, shedding of bacteria and eucaryotic cells under normal as well as pathological conditions, and the appearance of irregular immunogenic markers on some cell membranes.

Although the coverage of topics presented here cannot reflect the full range and scope of this field, it is hoped that this volume will convey the importance of pathological membranes, attract new investigators, and serve to intensify ongoing research.

Finally, I would like to thank my wife, Anne Nowotny, and Mrs. Dorothy Shanfeld, Miss Grace Nejman, and Dr. Eniko Kovats, for editorial assistance.

A. Nowotny

Philadelphia

Contents

Chapter 6

Alien Histocompatibility Antigens

Ruta M. Radvany

Chapter 7

Blood Group Antigens in Tumor Cell Membranes

John S. Coon and Ronald S. Weinstein

Chapter 8

Cell-Surface Macromolecular and Morphological Changes Related to Allotransplantability in the TA3 Tumor

John F. Codington and David M. Frim

Chapter 9

Simian Virus 40-Coded Antigens and the Detection of a 55K-Dalton Cellular Protein in Early Embryo Cells

Peter T. Mora and K. Chandrasekaran

Chapter 12

Experimental Systems for Analysis of the Surface Properties of Metastatic Tumor Cells

George Poste and Garth L. Nicolson

Chapter 13

Antigen-Specific Suppressor ("Blocking") Factors in Tumor Immunity

Karl Erik Hellström, Ingegerd Hellström, and Karen Nelson

Chapter 14

Estrogen Regulation of Specific Proteins as a Mode of Hormone Action in Human Breast Cancer

David J. Adams, Dean P. Edwards, and William L. McGuire

Chapter 15

Molecular Characteristics of Brain Opiate and Nicotine Receptors

Jean M. Bidlack and Leo G. Abood

Chapter 16

Investigation of Pathological Membranes with Nuclear Magnetic Resonance Spectroscopy

Charles Eric Brown

Chapter 1

Shedding Bacteria

A. Nowotny

School of Dental Medicine
and School of Medicine
University of Pennsylvania
Philadelphia, Pennsylvania

I. INTRODUCTION

The dynamic state of eucaryotic cell membranes is well established, and it is obvious that the phenomenon of active transport through cell envelopes, excitability, receptiveness, biosynthesis, and a number of other essential biological functions of the cell require a complex and highly functional boundary which not only keeps the subcellular organelles neatly together but actively participates in their dynamic functions. These observations and assumptions seem to be applicable, although in a more restricted fashion, to procaryotic cells.

The bacterial cell wall appears to be less dynamic and much more rigid than the membrane of eucaryotic cells. Its major role seems to be the maintenance of the shape of the bacterium (Salton, 1960). Under the cell wall lies the cytoplasmic membrane rich in enzymes, other proteins, and lipids. This plasma membrane resembles to some extent the membranes of eucaryotic cells, both structurally and functionally. This is the site of transport regulations, biosynthesis, and assembly of macromolecules, including components of the rigid cell wall.

In spite of the relative inertness of the bacterial cell wall, it would be fallacious to consider it as a stationary, passive container of the highly dynamic cellular apparatus. It has appendages with important functions. It has layers with great significance in disease and in health, and, above all, these and other components of the cell walls are released not only under pathological but also under apparently normal conditions, which may occur without interference with bacterial viability. This latter phenomenon, the shedding of bacteria, is the subject of this survey.

We should emphasize right at the onset the difficulties in distinguishing between products of shedding and products of cell decay. We have to realize that there are no isolation procedures which could yield a pure product completely free from the other. Neither are there any analytical or physicochemical methods which could help us in determining how much of the preparation is derived from the cell envelope by cell decay or shedding. The most important criteria of shedding we followed during the selection of research data for this review are the spontaneity of shedding and the survival of the cell during and after the release of surface components.

The first aim of this review is to substantiate the existence of the bacterial cell surface shedding phenomenon under normal and pathological conditions. The other aim is to elaborate on the potential consequences of this shedding. In attempting to survey this field, a few established facts will be briefly surveyed and a considerable portion of this chapter will deal with assumptions regarding the possible effects of shedding on the cell itself and particularly on the host exposed to shed cell surface components.

II. SHEDDING OF ENDOTOXIC LIPOPOLYSACCHARIDE (LPS)

One of the main distinctions between exo- and endotoxins one can find in textbooks of microbiology is that while exotoxins are produced and released by viable bacteria, endotoxins are not released without the disintegration of the cell itself. There is ample information which contradicts this and we present here only a few sets of data to prove that constant release of endotoxin may occur under certain environmental and nutritional conditions, without disintegration of the cell itself.

The discovery that bacterial contamination renders infusion solutions pyrogenic goes back to the nineteenth century. It was found that the higher the number of cultivable bacteria in such solution, the more pyrogenic it is. It was also established early that if a solution became pyrogenic, removal of bacteria by filtration did not eliminate the pyrogenicity because the "pyrogens" (Burdon-Sanderson, 1876), being much smaller than the bacterial cells, passed through the filters. It was also found that heat killing of bacteria did not inactivate the pyrogens. The fact that the product causing fever is also toxic if injected in large quantities was soon added to the reported biological activities. Pfeiffer coined the name *endotoxin* for this substance in 1892. It was Ecker (1917) who established that endotoxin can be obtained from the supernate of some but not all young gram-negative cultures. Branham (1925) showed that *Salmonella* bacteria can produce and release this toxic product into protein-free, synthetic media, proving that it is a product of the bacterium and not of the nutrients in the medium.

Tumor hemorrhage-inducing bacterial products were found in bacterial culture filtrates by Coley who used these preparations for therapeutic purposes and reported successful treatment of human lymphosarcomas around the turn of the century (1898, 1906). His pioneering work led to the development of the so-called "mixed bacterial toxin," one of the main components of which is bacterial endotoxin (for a review see Coley-Nauts *et al.*, 1946). Coley used *Serratia marcesens* and *Streptococcus pyogenes* culture filtrates; Gratia and Linz (1931), who revived the interest in these products, used an *Escherichia coli* culture supernate. Shear (1936) and Ikawa *et al.* (1952) also used *E. coli* and *S. marcescens* filtrates to study the chemical nature of the tumor hemorrhage-inducing component in such gram-negative culture supernatants. The active substance released by growing microorganisms showed some of the characteristic chemical and biological properties of endotoxins (Hartwell *et al.*, 1943).

When various extraction procedures became available to dislodge endotoxin from the cell surface (Boivin *et al.*, 1933; Morgan, 1937; Goebel *et al.*, 1945; Westphal and Lüderitz, 1954; Ribi *et al.*, 1959), attention became focused on these extracted materials. The interest in the spontaneously released endotoxin was somewhat revived by the results of Bishop and Work (1965), who reported that an extracellular glycolipid is released by *E. coli* cells under lysine-limiting conditions. Electron-microscopic pictures of the lysine-requiring mutant *E. coli* 12408 clearly showed not only a very convoluted cell surface but also numerous extracellular microvesicles shed from the growing cell (Knox *et al.*, 1966). These particles were isolated and subjected to thorough chemical and biological studies, the result of which established their identity with endotoxin (Taylor *et al.*, 1966).

Crutchley and co-workers (1968) designated such preparations as *free endotoxin*. It is important to note that the same authors also reported the presence of a nontoxic cell-wall component in the same preparation (Marsh and Crutchley, 1967). An excellent review on free endotoxins was written by Russell (1976). Several other publications followed and reported the release of endotoxin from a number of bacteria. Rothfield and Pearlman-Kothencz (1969) described the excretion of an LPS–protein–phospholipid complex by *E. coli* 12408 and *S. typhimurium*

LT2 and G30 cells and found that although the greatest amount was released after the exponential growth phase, a measurable quantity of labeled LPS was accumulating in the medium 10–20 min after the addition of [^{14}C]-galactose.

DeVoe and Gilchrist (1973) described the formation of blebs on the wall of three *Neisseria meningitidis* strains during their normal log phase growth. These protrusions later detach themselves and they can be separated from the cells. Chemical analysis of the preparation revealed the presence of 2-keto-3-deoxyoctonic acid (KDO). Jorgens and Smith (1974) measured Limulus lysate clotting activity to quantitate the amount of free and bound endotoxin produced by a clinical isolate *E. coli* strain. Filtration through a 0.22-μm membrane was used for the separation of the free endotoxin from the cells. The bound endotoxin was extracted from the cells with 45% phenol. These authors found that the two endotoxin preparations had similar activities in the lysate clotting test and that significant amounts of endotoxin were released during the early growth of this bacterium. Johnson and co-workers (1975) observed the release of LPS in eight nonpathogenic strains of *N. meningitidis*. They also observed that release of LPS can already occur in the early phase of growth. Chemical comparison of the free and bound LPS did not reveal any significant difference.

At this juncture, it is necessary to point out a few of the missing experiments in some of the above papers. To describe a preparation as LPS solely on the basis of chemcial analysis is not justified, particularly if the preparation was not subjected to any tests of homogeneity. As we showed a long time ago, even the so-called highly purified LPS preparations are heterogenous (Nowotny, 1966, 1971; Nowotny *et al.*, 1966; Chen *et al.*, 1973). If one has a preparation which contains a phospholipid and a polysaccharide component, chemical analytical results may be mistakenly interpreted as an indication of the presence of LPS. Determination of 3-OH-carboxylic acids or KDO is a better indicator, but still one has to realize that not only LPS contains such acids or carbohydrates. KDO determination detects all 2-keto-3-deoxy carbohydrates and related compounds, not only KDO (Nowotny, 1971). At the present time, there are no fully reliable chemical methods to detect LPS.

It is even more misleading to talk about the presence of endotoxin in a preparation if no characteristic biological assays for endotoxicity have been carried out. There are a few endotoxicity parameters which must be used, such as pyrogenicity, Shwartzman skin assay, and the more recently developed Limulus lysate clotting reaction. The use of one of these alone can also be misleading, since not only LPS is pyrogenic or positive in Shwartzman, and biological preparations left at room temperature can easily become contaminated with bacteria to the extent of displaying strong positivity in the Limulus test. Therefore, in addition to chemical analyses, it is most important to carry out at least two characteristic biological tests of endotoxicity in a quantitative fashion. Only if all chemical analytical and biological activity data so indicate, can one claim that the preparation is (or contains) endotoxic LPS.

What is quite evident from the above brief and by no means complete survey of shed bacterial products is that the distinction between gram-negative and gram-positive bacteria based on the release of their toxins cannot be accepted. Not only do gram-positives release their (exo)toxins without disintegration of the cell, but gram-negatives can do so also with their (endo)toxin, although in a more limited fashion. Statements found in several textbooks affirming that gram-negatives release them "only if the integrity of the cell is disturbed" (Davies *et al.*, 1973, p. 638) are inaccurate.

III. BACTERIAL APPENDAGES

Flagella and pili of bacteria can often be found in culture supernatants, particularly if vigorous aeration or stirring of the culture was applied. Capsular layers sometimes have a definite border, but frequently they are just diffusing into the medium as if they were dissolved in it. In both cases, antigens can often be detected in the cell-free supernatant by immunological methods, indicating that the capsular substance of some bacteria is released into the surrounding medium by living cells (Tomcsik, 1956).

The release of flagella from the bacterial cell requires rather drastic measures such as vigorous shaking or repeated freeze–thawing, but the cell itself will survive and will rapidly resynthesize the flagellum unless environmental conditions prevent this. Spontaneous release will occur mostly if the cells undergo autolysis, but they may detach and disintegrate either at low pH or elevated temperature (Smith and Koffler, 1971). Stocker and Campbell (1959) reported that there are differences in the resistance of individual flagella on a single cell at pH 2. Lacey (1961) reviewed nongenetic variations of bacterial antigens, among them changes of H (flagellar) antigens induced by raising the temperature from 18 to 37°C.

The effect of environment on the biosynthesis of some bacterial components has been studied by many laboratories, but really meaningful results were obtained only by those scientists who used the "chemostat," a continuous culture apparatus where all conditions of growth, but particularly the chemical composition of the nutrients, are precisely controlled (Monod, 1949, 1950). It was found that the quantity of vitally essential materials, such as proteins or nucleic acids, may undergo two- to three-fold changes due to nutritional conditions, but nonessential components such as polysaccharide or lipid storage may be increased or decreased by a factor of 10 or more. The environmental factors influencing the resynthesis of lost flagella were reviewed by Kerridge (1961). It was found that elevated temperature (44°C) leaves bacterial growth unaffected but the number of flagellated cells decreased rapidly, rendering the cell nonmotile (Quadling and Stocker, 1956).

Pili can be more easily removed from growing bacteria and can be found in large quantities in cell-free culture supernatants of some bacteria (Brinton, 1965; Every and Skerman, 1980). Often the yield of pili is higher from the culture than from the sedimented cells. These spontaneously released pili contained considerable quantities of LPS. Every (1979) assumed that their shedding is similar to the release of an LPS–protein–phospholipid complex described by Rothfield and Pearlman-Kothenz (1969). Similar LPS contamination was present in pili preparations obtained from *Neisseria gonorrhoea* (Novotny and Turner, 1975; Robertson *et al.*, 1977) and *Moraxella nonliquefaciens* (Frøholm and Sletten, 1977). Spontaneously detached fimbriae were seen by negative staining under electron microscopy in several Myxobacterial preparations (MacRae *et al.*, 1977). The same authors presented electron micrographs showing holes on bacterial surfaces which may have been the anchoring points of the released fimbriae.

These, as well as several other publications not quoted here, clearly indicate that several appendages and adhering capsular materials can be shed or released by viable, growing microorganisms.

IV. MICROVESICLES

Vibrio cholerae forms protrusions on the cell surface which separate later and are released into the medium (Chatterjee and Das, 1967). The ultrastructure of these microvesicles showed that they are closed sacs and their wall is indistinguishable from the wall of the cell itself, as already described above. DeVoe and Gilchrist (1973) reported the release of endotoxin by bleblike protrusions of *N. meningitidis* bacteria grown under normal conditions *in vitro*. Novotny *et al.* (1975) studied the morphology of naturally occurring and *in vitro*-maintained cultures of *N. gonorrhoeae* and observed that while naturally occurring bacteria had a smooth surface, the cultured cells showed a rough and disorganized structure. Bleb-like endotoxin-containing shed fragments were also isolated from cultured cells which were toxic in the chick embryo lethality test and pyrogenic in rabbits. KDO was present in this bleb-like endotoxin. Russell *et al.* (1975) also reported that *Neisseria sicca* released microvesicles containing LPS–protein complex.

Kahn *et al.* (1979) observed the release of vesicles with a diameter of 40–70 nm from a mutant strain of *Haemophilus parainfluenizae* which was defective in transformation. These vesicles were purified by centrifugation and showed a marked DNA binding activity which was resistant to DNase treatment. The same vesicles can also inhibit the transformation of competent *H. parainfluenzae* or *H. influenzae* cells because they successfully compete for transforming DNA with DNA receptors on competent cells. The authors assumed that the defect in transformation of this shedding mutant may be due to abnormal release of the vesicles which contain the DNA receptors. Wild-type cells do not release such microvesicles.

Bacteria treated with various antibiotics show extensive bleb formation (Nishino and Nakazawa, 1975; Ovcinnikov *et al.*, 1974). Chloramphenicol treatment induced inhibition of protein synthesis, caused profound surface changes on *E. coli* K12 (Klainer and Russell, 1974), and induced the liberation of "free endotoxin" into the medium (Russell, 1976).

Among the oral microorganisms which were isolated from the deep gingival pocket of periodontitis patients, several strains showed unusual electron microscopic morphology. One of the isolates recently investigated quite extensively is *Actinobacillus actinomycetemcomitans* (AaY4) which occurs in a significantly higher percentage in the periodontal lesions of rapidly progressing juvenile periodontitis cases than in nondestructive periodontitis or in the oral flora of healthy individuals (Slots *et al.*, 1980). Ultrathin sections of this microorganism were studied by electron microscopy and numerous microvesicles were found in the intercellular space. The vesicles appear as spherical structures, with a diameter varying from 45 to 60 nm. The wall of the vesicles is trilaminal, resembling the cell walls of AaY4. The first step in the isolation of vesicles was differential centrifugation. Cells were sedimented at 16,000*g* and the resulting supernatnat at 100,000*g*. The sediment containing the microvesicles was resuspended and centrifuged again at 100,000*g* over a 35% sucrose cushion. After repeated washings in saline at 100,000*g*, the pellet was studied under the electron microscope. It was found that these isolated vesicles are spherical or tubular; in the latter case their narrowest width is between 15 and 20 nm (Lai *et al.*, 1981).

The chemical analyses of the vesicles revealed, as expected, that they contain lipids, proteins, and carbohydrates. The biological measurement started with an estimation of endotoxicity (pyrogenicity, local Shwartzman, Limulus lysate clotting, and chick embryo lethality) which was followed by determination of the *in vitro* bone resorbing activity of the preparations in a quantitative assay. The results can be summarized as follows:

1. As expected, the microvesicles contained endotoxin. The percentage of endotoxin content calculated on dry weight basis is between 1 and 15%. The highest percentage was obtained by the local Shwartzman assay, the lowest by the Limulus lysate clotting assay.
2. In spite of the relatively low endotoxin content, the microvesicles were highly toxic in the chick embryo lethality test. The LD_{50} of the vesicles was 0.03 μg, while the AaY4 endotoxic LPS, isolated by the phenol–water procedure, had an LD_{50} = 0.06 μg. This toxin was sensitive to proteolysis, in contrast to LPS. This indicated that the vesicles contained a toxin not identical to and more potent than endotoxin.
3. The *in vitro* bone resorption studies revealed that in addition to endotoxin, there is another component (or components) present in the vesicles which can induce extensive bone resorption (as measured by the ^{45}Ca release assay). This component, just like the above-mentioned toxin, was sensitive to proteolysis but resistant to heat.
4. Independently from us, others found that the microvesicles contain a substance which is toxic to leukocytes. This toxin is sensitive to both heat and proteolysis.

Accordingly, it became clear that the microvesicles released by the AaY4 strain contain at least three but most probably more biologically active substances, which are all toxic and at least two of them are active in *in vitro* bone resorption. The potential role of these released vesicles in the pathogenesis of periodontitis is considerable. The vesicles are released abundantly and their release does not seem to cause disintegration of the cells. These small blebs contain not one but several highly active biological toxins and bone-resorption-inducing components. They are concentrated in the deep gingival pockets, and, being much smaller than the parents cells, they may be able to pass anatomical barriers not permeable to whole bacteria. A preliminary report was published (Nowotny *et al.*, 1981). Details of the findings were described recently (Nowotny *et al.*, 1982).

V. VARIOUS ANTIGENS AND OTHER MACROMOLECULES

Endotoxic LPS carries the O antigenic determinants of gram-negative bacteria (Kaufman *et al.*, 1961), flagella contain the powerful immunogens of the H antigens, and the capsular polymers consist of the K antigens. The release of these cell constitutents means the release of immunogens (or antigens) into the surrounding environment of the bacteria. Other antigenic components were also reported to be discharged by the cells. We will list only a very few examples of these here.

Erythrocyte sensitizing antigens were found in the supernate of tubercle bacilli (Middlebrook and Dubos, 1948; Middlebrook, 1950). In the presence of complement, antisera to tubercle bacilli could lyse the sensitized erythrocytes. The reaction was immunologically specific. Rantz *et al.* (1952) investigated the culture filtrate of 13 hemolytic streptococci and several other gram-positive and a few gram-negative strains, following the method of Middlebrook.

To obtain antisera, they used children with acute rheumatic fever as donors and healthy young adults as controls. Several of the sera were hemolytic to the sensitized erythrocytes in the presence of complement, but no correlations could be established between the health status of the donor and the titer of antibodies in their sera. As far as the presence of sensitizing antigens in the supernatant of the various strains investigated is concerned, they were present in several but not all cultures. The authors described the reactions they observed as immunologically nonspecific. Seal (1951) and Baker *et al.* (1951) isolated water-soluble, strain-specific and immunogenic components from *Pasteurella pestis*. They purified and characterized the major component by various physicochemical measurements and studied its immunogenicity.

Lipoteichoic acid (LTA), a component of gram-positive bacterial cell walls and quite similar to LPS in many biological assays, is released by oral strains of *Streptococci* and *Lactobacilli* (Markham *et al.*, 1975). The first observation made was the presence of an antigen in the culture fluid of the above stains, which could sensitize erythrocytes to antisera produced against a cross-reacting LTA. The amount of LTA in the ultrafiltered culture supernate was so high that it could be detected by the not-too-sensitive precipitin method. The authors also reported that the release appears to be spontaneous and does not require autolysis of the bacteria.

Evidence that actively growing gram-negative bacteria release surface antigens was obtained in our laboratories. Tripodi (1966) studied the effect of ethyl methane sulfonate-induced mutation on the O antigenic properties of *S. marcescens* 08. During these experiments, we observed that both the mutants as well as the parent strain release O antigens while growing on a nutrient agar. The antigen was detected by punching a hole in the agar near the colonies and filling this with hyperimmune rabbit O antiserum. Diffusion of these antibodies toward the colonies resulted in clear precipitin line formation as shown schematically in Fig. 1 (Nowotny, 1968).

The spontaneous release of somatic antigens from *E. coli* 055 was detected by Guckian and Perry (1966) through the use of the inhibition of passive hemolysis of erythrocytes coated with *E. coli* 055. The antigen was found in the 0.45-μm filtered culture supernatant of actively growing cells, and it already appeared after 2 to 3 hr of incubation. The authors assume that their somatic antigen is endotoxin, but no assays were carried out to substantiate it.

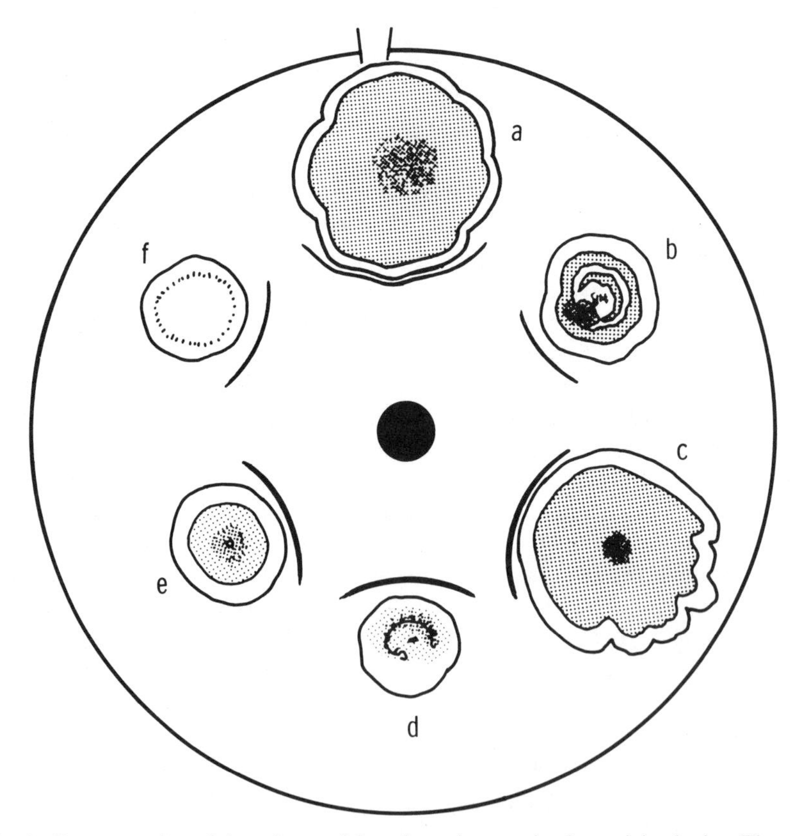

Fig. 1. Demonstration of the release of O antigens by growing bacterial colonies. The agar plate is a trypticase soy nutrient medium. The colonies were grown on the surface of the agar and they are various mutants of *Serratia marcescens* 08 strain. The center well is filled with hyperimmune rabbit antiserum to *S. marcescens* 08 O antigen.

Somatic antigens of *V. cholerae* were detected in the supernatant of two strains by using their capability to inhibit the vibriocidal activity of homologous antisera (Pike and Chandler, 1974). The antigens, which appear to be similar if not identical to those described by Chaterjee and Das (1967), were released readily during the exponential phase without detectable disintegration of the cells. Both LPS (O antigen) and alkaline phosphatase enzyme (a cell envelope) were released by *S. typhimurium* smooth and rough mutants (Lindsay *et al.*, 1972).

The above-listed publications are only a few selected from the literature which show clearly that constant shedding of bacterial cell wall components may occur and that the quantities shed depend upon the strain investigated and particularly upon the conditions of their growth.

VI. CONDITIONS OF RELEASE

As is already obvious, some bacterial strains shed their cell-wall components more readily than others. Ultrastructural studies of gram-negative bacteria revealed that while some strains appear to have a tightly adhering outermost layer, others have a loosely attached, highly convoluted structure surrounding the cell. The spontaneity of release may be influenced greatly by the number of attachment points of the outermost layer to the deeper peptidoglycan component of the wall, but this is apparently not the sole requirement. It has been known since the fundamental work of Toennies that nutritional factors may lead to irregularities in the biosynthesis and to lysis of gram-positive bacterial cell walls (Toennies and Gallant, 1948, 1949; Herbert, 1961).

The same seems to apply to gram-negatives as well. Careful studies were carried out by Collins (1964) to establish the effect of growth rate on the composition of the *Salmonella enteritidis* cell wall. The growth conditions were accurately controlled by the use of a chemostat. It was found that accelerated growth may lead to incomplete synthesis of O antigen LPS, thus resulting in a partial but reversible smooth to rough conversion. Adams (1971) and McDonald and Adams (1971) observed significant differences in the yield and chemical composition of LPS synthesized by *N. sicca* under high and low aeration. Tempest and Ellwood (1969) made similar observations, growing *Aerobacter aerogenes* in a chemostat under carbon limitation. Pearson and Ellwood (1974) grew bacteria with a significantly lower toxin content under sulfur limitation, while growth under glycerol limitation (Pearson and Ellwood, 1972) produced the most toxic cell walls.

The release of cell-wall components is similarly influenced by growth conditions. Municio *et al.* (1963) found a "mucopeptide" in the culture medium of a lysine-requiring *E. coli* mutant. Using the chemical method of Salton and Pavlik (1960) as well as immunochemical procedures, Municio and associates established that the mucopeptide is of cell-wall origin. A diaminopimelic acid-requiring mutant of *E. coli* was studied for growth characteristics by Meadow *et al.* (1957). These studies of Work and associates have been continued and the release of a glycolipid under lysine-limited growth condition was reported (Bishop and Work, 1965). This substance was described later as LPS; ultrastructure morphology showed extensive microvesicle formation on the lysine-starved cell-wall surface (Knox *et al.*, 1966).

What emerges from these selected references is that growth conditions and the supply of nutrients not only can induce variations in the chemical composition of the cell surface components, but they can also influence their toxicity. From our point of view the release of such bacterial products is most important and it appears that changes under the above conditions may lead to facilitated release or even to the lysis and disintegration of the cell body. Conditions pathological for the bacteria are particularly effective such as starvation, or quite obviously, low concentration of some antibiotics (Iida and Koike, 1974).

The growth conditions for individual bacteria in a micro-environment where a heterogenous flora exists should also be considered. One of the real-life examples is bacteria growing in the deep gingival pocket of periodontitis patients. In a chemostat, the growth condition can be regulated by the experimenter, and, for the most meaningful results, usually only one bacterium is cultivated in the chemostat. The micro-environment of a gingival pocket is quite different. The conditions of growth, such as aeration or nutrients, are beyond our control. The space is exceedingly crowded by many microorganisms, some of which could be discovered only by sampling the pockets with special anaerobic probes. These microorganisms fiercely compete for nutrients, thus creating a micro-environmental condition which may be acceptable for some strains, while much less than ideal or even bordering pathological for others. Under these conditions, some otherwise "harmless" microorganisms may start shedding and releasing components into this micro-environment, which may have great importance in the pathogenicity of periodontal diseases.

VII. CONSEQUENCES OF SHEDDING

A few experimentally substantiated and several nonproven assumptions will be offered in this chapter, which again is incomplete and will deal mostly with the aspects familiar to the author. The consequences are twofold: effects on the bacterium and effects on the environment of the growing bacterium. This latter obviously includes living organisms inhabited by the bacterium.

Consequences for the bacterium are in most cases short-lived and reversible, since the bacteria have a remarkable capacity not only to adapt to the new conditions, but also to resynthesize the lost component quite rapidly. Loss of immunogens or antigens due to shed LPS, capsule, or appendages does not seem to alter the viability or the virulence of the cells. The fact that they may become less recognizable to products of the immune response, such as antibodies or lymphocytes, may give them a transient escape advantage, but this is not proven and due to the rapid resynthesis of the lost component, this assumed advantage will be very short-lived indeed. Just to give a few examples on the rate of resynthesis, it is known that pure protein pili removed by mechanical breakage can be regenerated in 10–15 min in fresh medium at 37°C (Brinton, 1965). The complete regeneration of flagella takes only a few hours (Smith and Koffler, 1971). There are no firm data available concerning the rate of LPS or capsular material biosynthesis. Completion of the assembly of the polysaccharide moiety of LPS appears to take 1–2 hr. The rate of biosynthesis of the lipid moiety is unknown. Novotny *et al.* (1969) removed pili and flagella of bacteria by mixing them in a blender and observed the rate of resynthesis. Removal of the appendages did not interfere with the viability and growth rate of the cells. They found that 30% of the cells were flagellated again in 30 min. Resynthesis of F-type pili started within a few seconds and synthesis was complete in minutes. Synthesis of the I-type pili was somewhat slower, but still faster than the regeneration of the flagella. Rapid resynthesis of pili as well as flagella is possible from a pool of precursors. Intracellular flagellin was detected in *Proteus vulgaris* by Weinstein *et al.* (1960) and also in other strains by several other laboratories.

Accordingly, even if some functions of the bacterial cells are impaired by the removal of appendages, quick resynthesis will reduce these assumed, but not proven, consequences to a minimum. It is obvious, on the other hand, that nutritional or other inhibitions of the biosynthesis of essential cell-wall constituents will lead not only to microvesicle formation and extensive shedding but also to the death and disintegration of the cell, as shown by Toennies and Gallant as early as 1949.

The consequences of shedding for the environment are much more serious and these include reactions of the host as living environment.

The release of endotoxin locally or into the circulation will elicit all the known endotoxicity reactions. Small quantities will induce fever and other cardinal symptoms of inflammation. Increased vascular permeability, thrombocytopenia, disseminated intravascular coagulation, and a drop in blood pressure will follow immediately (Bennett and Beeson, 1950; Nowotny, 1969). The effects of LPS on the cells of the lymphatic system are quite significant. Macrophages are quickly activated by small quantities, but large amounts will reduce phagocytosis. B cells will undergo mitogenesis and, in a more direct way, LPS will also act on T-cell subclasses. The release of several mediators will be initiated or elevated, and several chain reactions will be set in motion. As a consequence the host's nonspecific natural defenses will be mobilized. The harmful effects of LPS are better known. Irreversible vascular collapse is the major and frequently lethal consequence of septicemia and endotoxic shock, which is elicited if greater quantities of gram-negative LPS reach the circulation. Local damage can also be caused by endotoxins, particularly by repeated application. The best example of this is the Sanarelli–Shwartzman phenomenon.

The bone resorbing activity of LPS is well documented by *in vitro* model systems (Raisz and Niemann, 1969; Hausman *et al.*, 1970; Robinson and Shapiro, 1975; Nowotny *et al.*, 1981). It is most probably induced via the activation of osteoclasts (Horton *et al.*, 1972). The inflammation and tissue destruction observed in periodontitis is caused by several etiological factors, and endotoxin may very well be the major causative agent, as we already discussed above. It has also been known for a long time that sensitivity to endotoxins can be greatly enhanced by previous infection with other microorganisms. Extensive tissue damage was reported in tubercle bacilli-infected guinea pigs or rabbits if they received iv cell-free culture filtrate of gram-negative bacteria. (Bordet, 1931; Freund, 1934). It was reported that infected mice are so sensitive to isolated endotoxins that they show an up to 1000-fold decrease in the LD_{50} (Suter, 1964).

LTA can exhibit some of the biological effects of LPS. It induces Shwartzman lesions, necrosis of the kidney, and bone resorption *in vitro*, and it fixes complement (Wicken and Knox, 1977), although to a lesser degree than LPS. Local accumulation of LTA shed under nutritionally limited conditions (Markham *et al.*, 1975) by otherwise nonpathogenic gram-positive bacteria may very well lead to localized tissue damage.

It is not quite clear whether the shed cell-wall components or appendages contribute significantly to the specific immune response of the host to the bacteria. LPS is a much less immunogenic bacterial product than flagellae. In order to obtain a high-titer antiserum to O antigens, it is best to use heat-killed whole bacterial cells instead of isolated and purified LPS. Depolymerized immunogens frequently lead to immune tolerance and this is especially true of flagellin (Nossal *et al.*, 1965; Shellam and Nossal, 1968). Extracted and isolated immunogens are frequently much less immunogenic than they were in their position on the cell membranes, and this is particularly apparent in the case of tumor-associated as well as normal histocompatibility antigens.

Therefore, it is difficult to determine whether the shed cell constituents differ in their immunogenicity from those which adhere to the cell surface, but there is no doubt that they are at least as antigenic. The fact that shed components have a high reactivity with antibodies and/or immunocytes has been documented in the past, not only for bacteria but also for mammalian cells (Abdelnoor *et al.*, 1972). If this is the case, and shed immunogens are less able to immunize but have a high reactivity with specific immune products, they may reduce the efficiency of the specific anti bacterial immune defense.

Although the shedding of bacterial cell-wall components does not seem to have any direct harmful effect on shedding bacteria, the fact that these shed particles like endotoxins can activate the reticuloendothelial system in a nonspecific fashion will have its consequences on the bacterium itself. Facilitated clearing of bacteria from the circulation by enhanced phagocytosis and accelerated intracellular killing of bacteria by activated lysozomal functions may be induced by LPS.

VIII. CONCLUSION

We have compiled some data showing that several bacterial cell-wall constituents can be shed into the environment. One of the biologically most active and therefore most important components is endotoxic LPS which forms the outermost layer in gram-negative bacteria.

Some of the shed components are constantly released by some cells and apparently spontaneously under "normal" growth conditions. Others will show microvesicle formation and/or shedding only under conditions pathological for the bacterium, such as starvation, the presence of antibiotics, or inhibition of the biosynthesis of essential constituents by drugs.

The consequences of shedding may be minimal for the bacterium itself, unless shedding reaches proportions of cell-wall disintegration. Repair or replacement of shed components appears to be rapid enough to reduce the consequences to a negligible degree. The consequences of shedding on the host can be quite considerable. While the release of small quantities may stimulate the nonspecific defenses of the host, continuous release of larger amounts may be responsible for local damage or for profound systemic disturbances.

IX. REFERENCES

Abdelnoor, A., Higgins, M., and Nowotny, A., 1972, Effect of cetyltrimethylammonium bromide on human erythrocyte membrane, *in* "Cellular Antigens," (A. Nowotny, ed.), Springer-Verlag, New York/Heidelberg.

Adams, G. A., 1971, The chemical composition of a cell-wall lipopolysaccharide from *Neisseria Sicca, Can. J. Biochem.* **49:**243.

Baker, E. E., Sommer, H., Foster, L. E., Meyer, E., and Meyer, K. F., 1951, Sutdies on immunization against plague. I. The isolation and characterization of soluble antigen of *Pasteurella pestris, J. Immunol.* **68:**131.

Bennett, I. L., and Beeson, P. B., 1950, The properties and biologic effects of bacterial pyrogens, *Medicine* **29:**365.

Bishop, D. G. and Work, E., 1965, An extracellular glycolipid produced by *Escherichia coli* grown under lysine-limiting conditions, *Biochem. J.* **96:**567.

Boivin, A., Mesrobeanu, J., and Mesrobeanu, L, 1933, Extraction d'un complexe toxique et antigenique à partir du bacille d'Aertrycke, *C. R. Soc. Biol. Paris* **114:**307.

Bordet, P., 1931, Contribution à l'étude de l'allergie non spécifique, *C. R. Soc. Biol. Paris* **106:**1251.

Branham, S. E., 1925, Toxic products of *Bacterium Enteritidis* and of related microorganisms, *J. Infect. Dis.* **37:**291.

Brinton, C. C., Jr., 1965, Contribution of pili to the specificity of the bacterial surface, and a unitary hypothesis of conjugal infectious heredity, *in "The Specificity of Cell Surfaces"* (B. D. Davies and L. Warren, eds.) Prentice–Hall, Englewood Cliffs, New Jersey.

Burdon-Sanderson, J., 1876, On the process of fever. III. Pyrexia, *Practitioner*, 417.

Chatterjee, S. N. and Das, J., 1967, Electron microscopic observations on the excretion of cell-wall material by Vibrio cholerae, *J. Gen. Microbiol.* **48:**1.

Chen, C. H., Johnson, A. G., Kasai, N., Key, B. A., Levin, J., and Nowotny, A., 1973, Heterogeneity and biological activity of endotoxic glycolipid from *Salmonella minnesota* R595, *J. Infect. Dis.* **128:**S43.

Coley, W. B., 1898, The treatment of inoperate sarcoma with the mixed toxins of *Erysipelas* and *Bacillus prodigiouses*: Immediate and final results in one hundred and forty cases, *J. Am. Med. Assoc.* **31:**389.

Coley, W. B., 1906, Late results of the treatment of inoperable sarcoma by the mixed toxins of *Erysipelas* and *Bacillus prodigiosus, Am. J. Med. Sci.* **131:**375.

Coley-Nauts, H., Swift, W. E., and Coley, B. L., 1946, The treatment of malignant tumors by bacterial toxins as developed by the late William B. Coley, M.D., Reviewed in the light of modern research, *Cancer Res.* **6:**205.

Collins, F. M., 1964, The effect of the growth rate on the composition of *S. enteritidis* cell walls, *Aust. J. Exp. Biol. Med. Sci.* **42:**255.

Crutchley, M. J., Marsh, D. G., and Cameron, J., 1968, Biological studies on free endotoxin and a non-toxic material from culture supernatant fluids of Escherichia coli 078 K 80, *J. Gen. Microbiol.* **50:**413.

DeVoe, I. W., and Gilchrist, J. E., 1973, Release of endotoxin in the form of cell wall blebs during in vitro growth of *Neisseria meningitidis, J. Exp. Med.* **138:**1156.

Ecker, E. E., 1917, The pathogenic effect and nature of a toxin produced by B. parathyphosus B., *J. Infect. Dis.* **21:**541.

Every, D., 1979, Purification of pili from *Bacteroides nodosus* and an examination of their chemical, physical and serological properties, *J. Gen. Microbiol.* **115:**309.

Every, D., and Skerman, T. M., 1980, Ultrastructure of the *Bacteroides nodosus* cell envelope layers and surface, *J. Bacteriol.* **141:**845.

Freund, J., 1934, Hemorrhages in tuberculous guinea pigs at the site of injection of irritants following intravascular injections of injurious substances (Shwartzman phenomenon), *J. Exp. Med.* **60:**669.

Frøholm, L. O., and Sletten, K., 1977, Purification and N-terminal sequence of a fimbrial protein from *Moraxella non-liquefaciens, FEBS Lett.* **73:**29.

Goebel, W. F., Binkley, F., and Perlman, E., 1945, Studies on the Flexner groups of dysentery bacilli. I. The specific antigen of *Shigella paradysenteriae* (Flexner), *J. Exp. Med.* **81:**315.

Gratia, A., and Linz, R., 1931, Le phénomène de Shwartzman dans le sarcome du cobaye, *C. R. Soc. Biol. Paris* **108:**427.

Guckian, J. C., and Perry, J. E., 1966, Release of somatic antigen into filtrates of an Escherichia coli broth culture, *Tex. Rep. Biol. Med.* **24:**432.

Hartwell, J. L., Shear, M. J., and Adams, J. R., 1943, Chemical treatment of tumors. VII. Nature of the hemorrhage-producing fraction from *Serratia marcescens* (*Bacillus prodigiosus*) culture filtrate, *J. Nat. Cancer Inst.* **4:**107.

Hausmann, E., Raisz, L. G., and Miller, W. A., 1970, Endotoxin: Stimulation of bone resorption in tissue culture, *Science* **168:**862.

Herbert, D., 1961, The chemcial composition of microorganisms as a function of environment, *Symp. Soc. Gen. Microbiol.* **11:**391.

Horton, J. E., Raisz, L. G., Simmons, H. A., Oppenheim, J. J., and Mergenhagew, S. E., 1972, Bone resorbing activity in supernatant fluid from cultured human peripheral blood leukocytes, *Science* **177:**793.

Iida, K., and Koike, M., 1974, Cell wall alterations of gram-negative bacteria by aminoglycoside antibiotics, *Antimicrob. Ag. Chemother.* **5:**95.

Ikawa, M., Koepfli, J. B., Mudd, S. G., and Niemann, C., 1952, An agent from *E. coli* causing hemorrhage and regression of an experimental mouse tumor. I. Isolation and properties, *J. Natl. Cancer. Inst.* **13:**157.

Johnson, K. G., Perry, M. B., McDonald, I. J., and Russell, R. R. B., 1975, Cellular and free lipopolysaccharides of some species of *Neisseria, Can. J. Microbiol.* **21:**1969.

Jorgensen, J. H., and Smith, R. F., 1974, Measurement of bound and free endotoxin by the limulus assay, *Proc. Soc. Exp. Biol. Med.* **146:**1024.

Kahn, M., Concino, M., Gromkova, R., and Goodgal, S., 1979, DNA binding activity of vesicles produced by competence deficient mutants of *Haemophilus, Biochem. Biophys. Res. Commun.* **87:**764.

Kauffmann, F., Krüger, L., Lüderitz, O., and Westphal, O., 1961, Zur Immunchemie der O-Antigene von *Enterobacteriaceae, Zentralbl. Bakteriol. Parasitenkd. Infektionskr. Hyg.* **182:**57.

Kerridge, D., 1961, The effect of environment on the formation of bacterial flagella, *Symp. Soc. Gen. Microbiol.* **11:**41.

Klainer, A. S., and Russell, R. R. B., 1974, Effect of the inhibition of protein synthesis on the Escherichia coli cell envelope, *Antimicrob. Ag. Chemother.* **6:**216.

Knox, K. W., Vesk, M., and Work, E., 1966, Relation between excreted lipopolysaccharide complexes and surface structures of a lysine-limited culture of Escherichia coli, *J. Bacteriol.* **92:**1206.

Lacey, B. W., 1961, Non-genetic variation of surface antigens in *Bordetella and other microorganisms, Symp. Soc. Gen. Microbiol.* **11:**343.

Lai, C. -H., Listgarten, M. A., and Hammond, B. F., 1981, Comparative ultrastructure of leukotoxic and non-leukotoxic strains of Actinobacillus actinomycetemcomitans, *J. Periodontol. Res.* **16**:379.

Lindsay, S. S., Wheeler, B., Sanderson, K. E., Costerton, J. W., and Cheng, K. -J., 1972, The release of alkaline phosphatase and of lipopolysaccharide during the growth of rough and smooth strains of *Salmonella typhimurium, Can. J. Microbiol.* **19**:335.

MacRae, T. H., Dobson, W. J., and McCurdy, H. D., 1977, Fimbriation in gliding bacteria, *Can. J. Microbiol.* **23**:1096.

Markham, J. L., Knox, K. W., Wicken, A. J., and Hewett, M. J., 1975, Formation of extracellular lipoteichoic acid by oral *Streptococci and Lactobacilli, Infect. Immun.* **12**:378.

Marsh, D. G., and Crutchley, M. J., 1967, Purification and physico-chemical analysis of fractions from the culture supernatant of *Escherichia coli 078K80: Free endotoxin and a non-toxic fraction, J. Gen. Microbiol.* **47**:405.

McDonald, I. J., and Adams, G. A., 1971, Influence of cultural conditions on the lipopolysaccharide composition of *Neisseria sicca, J. Gen. Microbiol.* **65**:201.

Meadow, P., Hoare, D. S., and Work, E., 1957, Interrelationships between lysine and αε-Diaminopimelic acid and their derivatives and analogues in mutants of *Escherichia coli, Biochem. J.* **66**:270.

Middlebrook, G., 1950, A hemolytic modification of the hemagglutination test for antibodies against tubercle bacillus antigens, *J. Clin. Invest.* **29**:1480.

Middlebrook, G., and Dubos, R. J., 1948, Specific serum agglutination of erythrocytes sensitized with extracts of tubercle bacilli, *J. Exp. Med.* **88**:521.

Morgan, W. T. J., 1937, Studies in immuno-chemistry. II. The isolation and properties of a specific antigenic substance from *B. dysentheriae* (Shiga), *Biochem. J.* **31**:2003.

Monod, J., 1949, The growth of bacterial cultures, *Annu. Rev. Microbiol.* **3**:371.

Monod, J., 1950, Le technique de culture continue, theorie et application, *Ann. Inst. Pasteur* **79**:390.

Municio, A. M., Diaz, T., and Martinez, A., 1963, The presence of a muco peptide in the media of an *E. coli* mutant and its relation to the cell wall, *Biochem. Biophys. Res. Commun.* **11**:195.

Nishino, T., and Nakazawa, S., 1975, Morphological alterations of *Pseudomonas aeruginosa* by aminoglycoside antibiotics, *J. Electron Microsc.* **24**:73.

Nossal, G. J. V., Ada, G. L., and Austin, C. M., 1965, Antigens in immunity. X. Induction of immunologic tolerance to *Salmonella adelaid* flagellin, *J. Immunol.* **95**:665.

Novotny, P., and Turner, W. H., 1975, Immunological heterogeneity of pili of *Neisseria gonorrhoeae, J. Gen. Microbiol.* **89**:87.

Novotny, C., Carnahan, J., and Brinton, C. C., Jr., 1969, Mechanical removal of F pili, Type I pili and flagella from Hfr and RTP donor cells and the kinetics of their reappearance, *J. Bacteriol.* **98**:1294.

Novotny, P., Short, J. A., and Walker, P. D., 1975, An electron microscope study of naturally occurring and cultured cells of Neisseria gonorrhoeae, *J. Med. Microbiol.* **8**:413.

Nowotny, A., 1966, Heterogeneity of endotoxic bacterial lipopolysaccharides revealed by ion-exchange column chromatography, *Nature* (*London*) **210**:278.

Nowotny, A., 1968, Recent problems in endotoxin research, *Ann. Immunol. Hung.* **11–12**:15.

Nowotny, A., 1969, Molecular aspects of endotoxic reactions, *Bacteriol. Rev.* **33**:72.

Nowotny, A., 1971, Chemical and biological heterogeneity of endotoxins, *in* "Microbial Toxins," Vol. IV, (G. Weinbaum, S. Kadis, and S. Ayl, eds.), 309 pp., Academic Press, New York.

Nowotny, A., Cundy, K., Neale, N., Nowotny, A. M., Radvany, R., Thomas, S., and

Tripodi, D., 1966, Relation of structure to function in bacterial O-antigens. IV. Fractionation of the components, *Ann. N.Y. Acad. Sci.* **133:**586.

Nowotny, A., Behling, U. H., Hammond, B., Lai, C-H., Listgarten, M., Pham, P. H., and Sanavi, F., 1981, Biological effects of membrane vesicles of *Actinobacillus actinomycetemcomitans* (Y4), *J. Dent. Res.* **60:**Abstract 552A.

Nowotny, A., Behling, U. H., Hammond, B., Lai, C.-H., Listgarten, M., Pham, P. H., and Sanavi, F., 1982, Release of toxic microvesicles by *Actinobacillus actinomycetemcomitans*, *Infect. Immun.* **37:**151.

Ovcinnikov, N. M., Delektorskij, V. V., and Afanas'ev, B. A., 1974, Electron microscopy of gonococci in the urethral sections of patients with gonorrhoeae treated with penicillin and erythromycin. *Br. J. Vener. Dis.* **50:**179.

Pearson, A. D. and Ellwood, D. C., 1972, The effect of growth conditions on the chemical composition and endotoxicity of walls of Aerobacter aerogenes N.C.T.C. 418, *Biochem. J.* **127:**72.

Pearson, A. D., and Ellwood, D.C., 1974, Growth environment and bacterial toxicity, *J. Med. Microbiol.* **7:**391.

Pike, R. M., and Chandler, C. H., 1974, The spontaneous release of somatic antigen from Vibrio cholerae, *J. Gen. Microbiol.* **81:**59.

Quadling, C. G., and Stocker, B. A. D., 1956, An experimentally induced transition from the flagellated to the non-flagellated state in *Salmonella*; the fate of parental flagella at cell division, *J. Gen. Microbiol.* **15:**i.

Raisz, L. G., and Niemann, I., 1969, Effect of phosphate, calcium and magnesium on bone resorption and hormonal responses in tissue culture, *Endocrinology* **85:**446.

Rantz, L. A., Zuckerman, A., and Randall, E., 1952, Hemolysis of red blood cells treated by bacterial filtrates in the presence of serum and complement, *J. Lab. Clin. Med.* **39:**443.

Ribi, E., Milner, K. C., and Perrine, T. D., 1959, Endotoxin and antigenic fractions from the cell wall of Salmonella enteritidis. Methods for separation and some biological activities, *J. Immunol.* **82:**75.

Robertson, J. N., Vincent, P., Ward, M. E., 1977, The preparation and properties of gonococcal pili, *J. Gen. Microbiol.* **102:**169.

Robinson, P. J., and Shapiro, I. M., 1975, The effect of endotoxin and human dental plaque of the respiratory rate of bone cells, *J. Periodontol. Res.* **10:**305.

Rothfield, L., and Pearlman-Kothencz, M., 1969, Synthesis and assembly of bacterial membrane components. A lipopolysaccharide-phospholipid-protein complex excreted by living bacteria, *J. Mol. Biol.* **44:**477.

Russell, R. R. B., 1976, Free endotoxin, *Microbios Lett.* **2:**125.

Russell, R. R. B., Johnson, K. G., and McDonald, I. J., 1975, Envelope proteins in Neisseria. *Can. J. Microbiol.* **21:**1519.

Salton, M. R. J., 1960, *Microbial Cell Wall,* Wiley, New York/London.

Salton, M. R. J., and Pavlik, J. G., 1960, Studies of the bacterial cell wall. VI. Wall composition and sensitivity to lysozyme, *Biochim. Biophys. Acta* **39:**398.

Seal, S. C., 1951, Studies on the specific soluble proteins of *Pasteurella pestis* and allied microorganisms. I. Isolation fractionation and certain physical chemical and serological properties, *J. Immunol.* **67:**93.

Shear, M. J., 1936, Chemical treatment of tumors. IV. Properties of hemorrhage-producing fraction from *B. coli* filtrate. *Proc. Soc. Exp. Biol. Med.* **34:**325.

Shellam, G. R., and Nossal, G. J. V., 1968, Mechanism of induction of immunological tolerance. IV. The effects of ultra-low doses of flagellin, *Immunology* **14:**273.

Slots, J., Reynolds, H. S., and Genco, R. J., 1980, Actinobacillus actinomycetemcomitans in human periodontal disease: A cross-sectional microbiological investigation. *Infect. Immun.* **29:**1013.

Smith, R. W., and Koffler, H., 1971, Bacterial flagella, *Adv. Microb. Physiol.* **6:**219.

Stocker, B. A. D., Campbell, J. C., 1959, The effect of non-lethal deflagellation on bacterial motility and observations on flagellar regeneration, *J. Gen. Microbiol.* **20:**670.

Suter, E., 1964, Hyperreactivity to endotoxin in infection, *in* "Bacterial Endotoxins" (M. Laudy and W. Braun, eds.), 435 pp., Rutgers Press, New Brunswick, New Jersey.

Taylor, A., Knox, K. W., and Work, E., 1966, Chemical and biological properties of an extracellular lipopolysaccharide from *E. coli* grown under lysine-limiting conditions, *Biochem. J.* **99:**53.

Tempest, D. W., and Ellwood, D. C., 1969, The influence of growth condition on the composition of some cell-wall components of *Aerobacter aerogenes, Biotechnol Bioeng.* **11:**775.

Toennies, G., and Gallannt, D. L., 1948, Bacterimetric studies. I. Factors affecting the precision of bacterial growth responses and their measurement, *J. Biol. Chem.* **174:**451.

Toennies, G., and Gallant, D. L., 1949, Bacterimetric studies. II. The role of lysine in bacterial maintenance, *J. Biol. Chem.* **177:**831.

Tomcsik, J., 1956, Bacterial capsules and their relation to the cell wall, *Symp. Soc. Gen. Microbiol.* **6:**41.

Tripodi, D., 1966, Active Sites of *Serratia marcescens* Endotoxin Preparations, Ph.D. Thesis, Temple University, Philadelphia, Pennsylvania.

Weinstein, D., Koffler, H., and Moskowitz, M., 1960, *Bacteriol Proc.* **63.**

Westphal, O., and O. Lüderitz, 1954, Chemische Erforschung von Lipopolysacchariden gram-negativer Bakterien, *Angew. Chem.* **66:**407.

Wicken, A. J., and Knox, K. W., 1977, Biological properties of lipoteichoic acids, *Microbiology* **1977:**360.

Chapter 2

Shedding Eucaryotic Cells

A. Nowotny

School of Medicine
and School of Dental Medicine
University of Pennsylvania
Philadelphia, Pennsylvania

I. INTRODUCTION

Attempts to distinguish shedding, secretion, and disintegration of cell envelopes face the same difficulties as those discussed in Chapter 1. High-molecular-weight compounds (such as IgM) are produced and released by certain cells, and some of these are believed to be assembled just before their release from the cell envelope. It is also known that dying or dead cells will autolyse and disintegrate, and the noncovalent forces building the membrane will no longer be sufficient to maintain integrity of this dynamic barrier. As a consequence, the breakdown products will accumulate in the medium. It is quite possible that some of the investigators who described cell membrane components in serum or medium detected products of cell decay and not shedding. True shedding should mean, therefore, that the components in question were parts of the surface structure, and their release does not result in the death of the cell. Finally, it should be stated here that isolation of shed components will almost certainly co-isolate autolysed cell-wall constitutents present in the same medium and, unless properly fractionated, will include some secreted products of the cell as well. Only the proportions of these three may vary, and this will obviously depend upon the starting material chosen and the procedures applied for isolation.

We attempted to keep the above points in mind in the selection of results to be included in this review. No attempt was made to be complete or to include every relevant publication. We omitted several valuable contributions partly because of the overwhelming number of papers written on this subject, and partly because excellent surveys of the field were published in recent years (Price and Baldwin, 1977; Black, 1980). The authors of these surveys attempted to be complete and they achieved this with varying success. One of our main efforts was to find and to give credit to the first publications on the subject matter we wished to discuss. From the several aspects of shedding, we chose to elaborate in more detail first on the accumulated evidence that shedding of normal as well as malignant cells occurs, second on the possible mechanisms of shedding, and last, but not least, on the consequences of the shedding for the cell and for the host.

II. SHEDDING OF NORMAL CELLS

Without going into a discussion of whether or not cells maintained *in vitro* can be considered *normal*, we use the term *normal cells* to describe cells originating from normal tissues.

The first reference we found on *in vitro* shedding of normal cells was published by Billingham and Sparrow (1955). These authors reported that cell-free supernates of epidermal cells contain a tissue-specific immunogen. Ben-Or and Doljanski (1960) found that not only single-cell suspensions of tissues but also tissue culture supernates, harvested after ½ to 2 hr of incubation, had antigenic properties as assessed by complement fixation. Lillie and Tyler (1962) reported that sea urchin eggs shed their surface components into the surrounding environment and these react with the proper sperm. Several other systems were described where the release of surface components could be clearly demonstrated, without any deliberate attempts by the investigators to dislodge these from the membrane by extraction, destruction, or any harsh treatment of the cell. Pellegrino *et al.* (1973) found HL-A antigens in good yield in the exhausted culture fluids of human lymphoid cells maintained *in vitro*. The antigens could be purified readily and they showed high specificity and antigenic reactivity. Nakamuro *et al.* (1973) isolated the common moiety of HL-A antigens, an 11,000-dalton component, in spent culture fluids of human cells. Among the more recent publications on *in vitro* shedding of normal cells we refer to the paper of Fernandez and Macsween (1977), who reported that human T cells lost their erythrocyte receptors if maintained in culture at 37°C, while this was negligible at 4°C, indicating that either an enzymatic process or a certain degree of membrane fluidity is one of the prerequisites for the release of these receptors. Emerson and Cone (1979b) studied the turnover and shedding of Ia antigens by mouse spleen cells *in vitro*. Isotope-labeled Ia antigen was rapidly released in the first 6 hr, but the release was slow in the next 20 hr. Shedding was inhibited by harsh iodination or other conditions which were pathological for the cells. The Ia antigen was replaced under "normal" *in vitro* conditions thus indicating that shedding is accompanied by biosynthesis. Thermodynamically, this turnover appears to be a process requiring high energy. Plesser and co-workers (1980) labeled the surface of chick embryo cells and reported that the rate of accumulation of exfoliated macromolecules in the culture medium was the mirror image of their disappearance from the cell membrane, thus indicating that no internalization of the macromolecules takes place. These authors found that the shed macromolecules are identical to those isolated from membranes, showing that shedding can occur without detectable degradation. Brennan *et al.* (1980) carried out similar experiments using porcine thyroid cells and confirmed that shedding is part of functioning cellular metabolism, and no peptide cleavage is required for the release.

In vivo shedding of normal cells was also reported. Zmijewski and co-workers (1967) were the first to report that serum contains histocompatibility antigens. They found that the amount of HL-A antigens in the circulation depends on exposure to heat or cold, overeating, or alcoholic overindulgence. Charlton and Zmijewski (1970) could isolate HL-A7 from a patient's serum who was isoimmunized by pregnancy. The antigen was associated with low-density α-lipoprotein. Similar results were reported by Van Rood *et al.* (1970). Brawn (1971) could desensitize sensitized lymphocytes by incubating them with sera of the mouse strain which provided the sensitizing histocompatibility antigen. The fact that such serum could neutralize the sensitized lymphocytes was interpreted as proof of the presence of histocompatibility antigens in the circulation of these normal, untreated mice.

Miyakawa and co-workers (1973) isolated soluble HL-A complexes from serum and used specific antisera to determine their antigenicity. Three major components, fractions of $2–8 \times 10^5$, 48,000, and 10,000 daltons, could be isolated. The first two had full HL-A specificity, but the smallest fragment behaved as the common portion (β_2 microglobulin) of the HL-A antigen complex. Billing and Terasaki (1974) reported similar results.

Nakamuro *et al.* (1973) found great similarity between the common portion of HL-A antigens and human β_2-microglobulin. They isolated such fragments from normal human sera as well as from urine. These studies support the concept that in the normal individual, self-tolerance is a function of antigen-type tolerance to soluble HL-A-type molecules. Pellegrino *et al.* (1975) found HL-A antigens in serum and reported that not all major HL-A specificities are released at the same rate, some being present in human sera in larger quantities than other types. Reisfeld *et al.* (1976) isolated and characterized HL-A histocompatibility antigens from serum and urine. They purified the HL-A antigen which was associated in the serum with high-density lipoprotein and β_2-microglobulin. The HL-A antigen preparations isolated from sera readily induced monospecific cytotoxic antibody production in rabbits. The HL-A antigen isolated from urine showed similar immunogenicity. The release of Ia antigens into normal serum in substantial amounts was shown by inhibition of cytotoxic anti-Ia sera with normal mouse serum (Parish *et al.*, 1976). Later, the same authors reported that mitogens, such as phytohemagglutinin, concanavalin A (Con A), and lipopolysaccharide (LPS), can cause a 125-fold increase in the serum Ia antigen level. The origin of the antigen was the most potent in inducing the release of Ia antigens from T cells (Parish and McKenzie, 1977). It is interesting that LPS, a know mitogen for B cells, was the most potent in inducing the release of Ia antigens from T cells. At the same time as Parish and associates, Callahan and co-workers (1976) reported that mouse Ia antigens in serum are associated with β_2-microglobulin and high-density lipoprotein. The experiments of Zeligs and Wollman (1977) are quite relevant to our subject. They found bleb formation followed by an apparent pinching-off process in intact rat tissues shown by thyroid epithelial cells which protruded into the follicular lumens *in vivo*.

All these reports indicate that membrane events which may lead to exfoliation, shedding, or other release processes are not restricted to *in vitro* or to pathological circumstances. Kapeller *et al.* (1976) and Doljanski and Kapeller (1976), who studied the turnover of membrane components in chick embryo cells extensively, reached the conclusion that cell surface shedding occurs under normal physiological conditions as part of the surface membrane turnover and this applies to both *in vitro* and *in vivo* conditions.

III. SHEDDING UNDER PATHOLOGICAL CONDITIONS

One of the earliest observations on *in vitro* release of infected cell envelope components was published by Marcus (1962) who inoculated cells in tissue culture with mixoviruses. When these cells were removed from the culture dish, cell surface residues (called *footprints*) could be observed on the plastic surface. Noninfected cells did not leave these components behind, indicating that viral infections might have induced the release of surface constituents. Winzler and associates (Molnar *et al.*, 1965a,b) studied the biosynthesis of glycoproteins released into the culture fluid of Ehrlich carcinoma cells maintained *in vitro*. They pulse-labeled the cells with [^{14}C]glucosamine and measured both the rate of uptake and the rate of appearance of high-molecular-weight labeled components in the culture fluid. Column chromatography on O-(diethylaminoethyl)-cellulose resolved these into several fractions. The authors assumed that the components in the culture supernatant were released by the tumor cells (Molnar *et al.*, 1965b). The glucosamine metabolism of HeLa cells was studied by Kornfeld and Ginsburg (1966). Their observations agreed with those made by Molnar *et al.*, since the HeLa cells also released labeled glucosamine in the form of macromolecules within 12 hr of incubation.

Herberman *et al.* (1973) found both cell surface and viral coat antigens in the sera of Gross virus-induced leukemia-bearing mice. By *in vitro* manipulation with the leukemic cells, such as repeated washings with saline or by autolysis, additional soluble antigens were obtained. Ben-Sasson *et al.* (1974) reported that cultured chick embryo cells infected with Rous sarcoma virus also release glucosamine- and choline-containing macromolecules. These products are selectively bound to spleen cells of Rous sarcoma-bearing chicken (sensitized?) lymphocytes but not to cells of normal chickens. Bolognesi and co-workers (1975) isolated virion glycoprotein from the supernatant of cell cultures infected with Friend leukemia virus.

The synthesis of the large external transformation-sensitive (LETS) surface gylcoprotein was studied in normal and transformed chick embryo cells by Critchley *et al.* (1976). The level of LETS concentration in the transformed cells was greatly reduced in spite of an apparently unchanged rate of synthesis. Rahman and associates (1977) measured similar parameters on cultured human melanoma cell lines, using [^{3}H]glucosamine as labeled precursor. They established that incorporation as well as shedding rates depend upon the medium. Bystryn (1977) also used human melanoma cultures but labeled the surface antigens using the lactoperoxidase technique. The release was considerably faster than cell death, since 98% viability could be maintained throughout the experiment. The melanoma-associated antigen was released as a complex which could be dissociated with nonionic detergent.

The literature is replete with references relevant to the shedding of tumor cells *in vivo* in spite of the fact that this is a rather new observation. Before going into the discussion of these, one should again turn to earlier literature, because there were a number of reports on phenomena where cell envelope exfoliation could have been a contributing factor. For example, Stern and Willheim (1943) gave a detailed description of the observation that cancer patients have a significantly higher protein-bound carbohydrate serum level than normal individuals. Seibert *et al.* (1947) reported the same phenomenon not only in cancer but also in other diseases; they attributed this to tissue destruction. Winzler and co-workers (1948, 1953) described high levels of serum mucoproteins in cancer and pneumonia, but they assumed that altered protein metabolism was the causative factor. Catchpole (1950) reported that sera of tumor-bearing mice have elevated glycoprotein levels. This was particularly evident in the case of large tumor masses. He suggested that the glycoprotein may originate from the ground substance of connective tissues at the site of tumor invasion.

Darcy (1957) used gel diffusion analysis to detect an abnormal serum component in tumorous rat sera. A very similar if not identical serum component was found during both pregnancy and the process of tissue regeneration. Accordingly, Darcy assumed that the appearance of the serum component is associated with rapid cell division. Miller and Bernfeld (1960) used rabbit antiserum to an isolated unknown serum component occurring in the sera of C3H mice bearing spontaneous mammary adenocarcinoma. Using this antiserum for diagnostic purposes, they could identify sera of tumor-bearing mice with better than 90% accuracy. Miller (1963) also isolated similar but not identical glycoproteins from the ascites fluid of Sarcoma 1-carrying A/J mice. Haughton and Davies (1962) reported the isolation of H2 antigens in the ascites fluid of tumor-bearing mice in a detailed study. They also detected the fact that tumor cell ghosts do not expose all their antigenic determinants on the cell surface. The existence of such cryptic receptors may have a profound effect on the reactivity of decaying cell membrane fragments, as we will discuss later.

The first paper in which the isolation of tumor-specific antigens from tumor-bearing murine sera was reported came from the laboratory of Stück *et al.* (1964). These authors used virally induced leukemias and found leukemia antigens in the sera if Rauscher or Friend virus was used for the induction of leukemia, but not if Gross virus was taken. Studies of unusual serum components of cancer patients were vigorously pursued by Hellström and Hellström, who published their first paper on this topic in 1969 in which they reported that cancerous serum contains a component which can block *in vitro* cell-mediated immune reaction to Moloney sarcoma. Several papers and reviews followed (Hellström *et al.*, 1970, 1971, 1973, 1974), and the first conclusion was that these sera, as well as sera of some cancer patients, contain a "blocking antibody," which is similar in its action to the "enhancing antibody" described by Kaliss (1967). It was assumed that these noncytotoxic blocking antibodies coat the surface of the target, thus protecting it from cytotoxic lymphocytes. This was later modified and evidence was offered that the "blocking factor" is an antibody–antigen complex (Sjögren *et al.*, 1971), and the antigen was derived from the tumor cell. The observation of the Hellströms was confirmed by several laboratories, both in human and in animal tumor studies. [For a review see Hellström and Hellström (1974), as well as Chapter 13 in this monograph, Heppner *et al.* (1973) , and Levy (1973)].

Of the several experimental tumor cell lines which shed surface components *in vivo* or *in vitro*, we will discuss the two TA3 types in more detail. These were both isolated as spontaneous mammary adenocarcinomas of A/J mice and adapted to the ascites form. One of the lines remained strain specific (called TA3-St); the other, after several hundred passages, lost specificity and now can grow across histocompatibility barriers and even in xenogeneic recipients. This line is called TA3-Ha.

The surface components of the TA3 cells were thoroughly investigated earlier. Gasic and Gasic (1962a) found that these cells are covered with an acidic, loosely adhering gel-like coat which they could stain by Hale's procedure. Enzymatic hydrolysis removed the coat, but regeneration was very rapid and a new layer was formed within a few hours. More significantly, Gasic and Gasic (1962b) reported that a neuramimidase-containing "receptor-destroying enzyme" from *Vibrio cholerae* could reduce the number of metastases formed by TA3-Ha injected iv if the enzyme was injected by the same route. Sanford (1967) removed the sialomucin coat by neuramimidase *in vitro* and thereby reduced the tumorigenicity by approximately 90%. The assumption was that the sialomucin coat covers histocompatibility markers and that is why TA3-Ha cells can grow across histocompatibility barriers. The strain-specific TA3-St cells which grew only in A/J mice did not have the same heavy coat, but some Hale-positive layers could still be detected on them. Codington *et al.* (1970) isolated this coating substance from TA3-Ha and did very thorough chemical structural studies on it. The results of this outstanding work are summarized in Chapter 8 of this monograph.

Beginning in 1970, we studied the mechanisms of escape of the TA3-Ha cells from immune recognition. One of our early observations (Grohsman and Nowotny, 1972) was that the ascites fluid of TA3-Ha-bearing mice contains a component which is not immunoglobulin and has a membranous appearance under electron microscopy. The same component, if injected ip together with a threshold dose of live TA3-Ha cells, facilitated tumor growth in allogeneic recipients. The enhancement of tumor growth was measured by determining the TD_{50} (number of viable TA3-

Ha cells needed to induce tumorous death in 50% of the inoculated mice). The TD_{50} significantly decreased if ascites fluid was injected together with the tumor cells. TD_{50} increased greatly if the allogeneic recipients were immunized by (1) irradiated TA3-Ha cells, (2) adaptive transfer of spleen cells from TA3-Ha tumor-bearing mice, (3) skin graft from $H\text{-}2^a$ donors, and (4) the rejection of TA3-St inocula indicating that the TA3-Ha line is immunogenic, carries $H2^a$ markers, but still can escape immune recognition. We also reported there for the first time that ip injection of bacterial endotoxin enhances the resistance of the allogeneic recipients to subsequent challenge with TA3-Ha cells (Grohsman and Nowotny, 1972).

Further isolation and characterization of this ascites fluid component was carried out (Nowotny *et al.*, 1974). Precipitation by ammonium sulfate gave a preparation (called P3) which did not go back into solution if dialyzed against saline. Biochemical analyses for membrane marker enzymes (alkaline α-glycerophosphatase and 5'-nucleotidase) gave positive results. Gas–liquid chromatographic analysis of the long-chain carboxylic acid of the isolated ascites fluid component and of the isolated tumor membrane revealed identical composition. Finally, no antibodies could be detected in P3. These results indicated that the ascites fluid of TA3-Ha cells contained membranous components shed from the tumor cells. Neonatal injection of P3 induced immune tolerance in allogeneic mice, thus completely abrogating their rather low but measurable innate resistance to TA3-Ha (Nowotny *et al.*, 1974). It could be shown that the same ascites fluid preparation could induce a specific immune resistance to viable TA3-Ha cells if injected intramuscularly 10 days before the challenge (Nowotny *et al.*, 1976). These results led to the conclusion that the exfoliated tumor membrane fragments can have a dual effect, tumor growth enhancement and immunogenicity. The possible mechanism of the tumor growth facilitation is discussed in Section V.

Human urine can contain tumor-derived antigens, and numerous papers dealt with such urine constituents, which were considered in the earlier publications to be the products of necrotizing tumor tissues or of altered cellular metabolism. Rudman *et al.* (1969) reported the identification of novel proteins and peptides in the urine of patients with advanced malignancies. Jehn *et al.* (1970) described a tumor-derived antigen in the urine of a melanoma patient. After the discovery of the carcino-embryonic antigen (CEA), several laboratories started to study its appearance in body fluids and excretions. Hall and co-workers (1972) were the first to detect CEA in the urine of patients with carcinomas of the urinary tract. Gozzo *et al.* (1974), Guinan *et al.* (1974), Nery *et al.* (1974), and Wu *et al.* (1974) confirmed this finding and elaborated procedures to measure CEA quantitatively and chromatographic methods to isolate and characterize it.

Detailed studies of non-CEA tumor antigens occuring in urine were initiated by Morton and associates (Gupta and Morton, 1975, 1976, 1979; Rote *et al.*, 1980a,b,c). These papers dealt with solid tumors, various sarcomas, carcinomas, or melanomas. These authors reported that if the tumor was removed surgically, the antigens were no longer excreted. They found that the detection of antigen in the urine is a very valuable diagnostic tool, since 87.4% of the patients were positive while only 6.9% of normal urine had immunologically detectable antigens. As far as the source of the antigens is concerned, Rote *et al.* (1980c) entertained several possibilities, including bacterial origin, but concluded that the antigen is most likely released by the tumor and is not identical with CEA. No explanation could be offered on how the large (mol. wt. 100,000) complex found its way into the urine without being filtered out by the kidney's glomeruli, as reported for B16 melanoma antigen complexes by Poskitt *et al.* (1974).

Several other laboratories reported non-CEA tumor antigens in the urine of patients, particularly in cases of progressing and/or chemotherapeutically treated melanomas (Carrel and Theilkaes, 1973, Bennett *et al.*, 1978; Volkers *et al.*, 1978), while Lopez and Thomson (1977) detected such antigens in the serum as well as in the urine of breast carcinoma patients. While the passage of these antigens from nonurinary tract malignancies into the urine remains to be investigated, the above-quoted papers and several others not included here show clearly that *in vivo* shedding of tumor components can occur and may appear not only in serum but also in urine.

In summary, there is ample evidence which proves that both normal and malignant cells may shed surface components. This can occur *in vitro* (which is anything but a normal condition for the cells, no matter how carefully the tissue culture medium is composed) but shedding could also be manifested *in vivo*. Their exfoliated products were found not only in tissue culture fluids but also in the sera, ascites exudate, or the urine of human patients as well as experimental animals.

The frequency of this occurrence seems to be quite difficult to determine. Certain cell types or cell surface components showed shedding, but these appear to be more the exception than the rule at this time. It is most likely, however, that with more extensive investigation and with the use of the latest membrane methodologies, more and more cell types, normal or pathological, *in vitro* or *in vivo*, will be shown to exfoliate membrane components without cellular death. It is to be expected that the major difference among them will be only in the quantity released.

IV. POSSIBLE MECHANISMS OF SHEDDING

Regarding the possible mechanism of shedding, one has to go back to the pioneering studies of Schoenheimer (1942) who in his classical work established that the constituents of living organisms are always in a dynamic state, and replacements for old components are constantly synthesized under normal conditions. Warren based his studies on this principle, and reviewed his results in 1968 and 1969. Part of these publications deals with the significance and role of turnover rates in the release of surface constituents. Warren argued convincingly that both dividing and nondividing cells synthesize cell membrane components at about equal rates. In the division phase, these will be used up for the formation of the needed additional cell envelope; therefore very little, if any, of these will be released unused into the medium. The release will be enhanced only in the nondividing phase, when they are not needed to enlarge the cell envelope. Accordingly, the release is governed by a supply-and-demand-type economy at the cellular level. Whether the rejected membrane components will be reutilized or just degraded in the medium is not known at this time. Warren (1969) also proposed a working hypothesis for the role of turnover rate in some of the characteristic properties of malignant cells.

Quite obviously, shedding can also be induced and/or accelerated. Krishan and Frey (1975) exposed leukemic cells to vinca alkaloids and observed the formation and release of membrane-lined vesicles from the cytoplasm. Skinnider and Ghadially (1977) described the formation of blisters and wormlike appendages on the cell surfaces of malignant histiocytes and a sarcoma, probably due to the action of vincristine therapy given to the patients. Emerson and Cone (1979a) studied the effect of colchicine- and cytochalasin-induced alterations on microfilaments and microtubules and the consequences of these on the release of murine B-cell IgM and IgD. Liepins and Hillman (1981) reported that mastocytoma cells formed and shed microvesicles if they were kept at 4°C for 1 hr and then slowly warmed up to 22°C. The cells remained viable during the process. Ferber *et al.* (1980) induced vesicle formation by a mild detergent which did not interfere with the activity of membrane-bound hydrolytic enzymes. Spontaneous release was observed by calf thymocytes maintained *in vitro* through enzyme activity measurement which included alkaline phosphatase, nucleotide pyrophosphatase, and γ-glutamyl transferase. This shedding was greatly accelerated by detergent action, which is an indication that no hydrolysis was required for the release and the authors reported that the cells remained intact. A comparison of the spontaneously shed and detergent-released fragments showed that they were identical in activity except for the γ-glutamyl transferase enzyme, which was lower in the spontaneously shed particles. The total cholesterol and phospholipid content was also identical, but the detergent-shed membrane flakes had a significantly higher percentage of polyunsaturated fatty acid, as if such areas on the membrane could be exfoliated more easily than the others. It is known that such areas have a greater membrane fluidity at body temperature (Inbar and Shinitsky, 1974a).

Sjögren *et al.* (1971) and Bansal *et al.* (1972) used low pH to dissociate membrane-bound immunoglobulin and the assumed blocking factor from tumor cell surfaces. Dissociation of membrane-bound antitumor antibodies was achieved by relatively mild procedures (Ran *et al.*, 1975). More recently, Rittenhouse *et al.* (1978) reported the spontaneous release of IgG and complement containing macromolecular aggregates from Ehrlich ascites cells if these were kept in cold isotonic buffer for 1 hr. Apparently this nonphysiological condition was sufficient to induce such rapid shedding. On the other hand, the immune components were quite firmly associated with their receptors on the membrane and they were released together. Hsu and associates (1980) maintained normal human lymphocytes conjugated with flouresceinated or radiolabeled antibodies and studied the mechanism of release. They concluded that the disappearance of the label from the cells (in 3 days) was not due to dissociation of antibodies from antigens but to the shedding of antigen–antibody complexes from the cells. If they prepared a covalently bound antigen–polyacrylamide artificial complex and reacted this with labeled antibody under identical conditions, no label release into the medium was noticed. These experiments indicate that the antigen–antibody complex is held together by forces which are greater than the avidity of some cellular antigens to the cell membrane.

What follows from this is that immunoglobulins can facilitate the release of surface macromolecules. Leonard (1973) was the first to report that antibodies to tumor antigens can cause "cap" formation on the cell surface which will later be pinched off and released into the incubation medium. Schroit *et al.* (1973) described a very similar phenomenon using mouse peritoneal macrophages and antimacrophage serum prepared in rabbits. They found that these antibodies removed their receptors from the cell membrane. The supernatant of such reaction mixtures contained antigen–antibody complexes, and the antigenic determinants in the complex was clearly established as of macrophagic origin. Calafat *et al.* (1976) induced redistribution and shedding of tumor virus antigens on GR mouse ascites leukemia cells by the use of antibodies to the virus. Nordquist *et al.* (1977) could show similar phenomena on epithelial human tumors. Leong *et al.* (1977) changed the distribution of membrane antigens of human melanoma cells by specific antibodies. All the above and several other experiments allow the assumption that antibodies, particularly those with high affinity, will not only react with antigens embedded in the fluid mosaic of the cell envelope but are quite capable of dislodging them, without introducing any visible harm to the cell itself. This concept, of course, has to be proven by proper thermodynamic studies and measurements.

Several investigators assumed that proteolytic processes are responsible for the shedding phenomenon. Membrane-bound enzymes are well known and they are used as markers of membranous preparations and fractions. Proteolytic activity on cell surfaces was demonstrated (Ossowski *et al.*, 1973) but it remains only a possibility that autolytic enzymes, located in the membrane itself, will cleave off and liberate cell surface components. Doljanski and Kapeller (1976) found that trypsin inhibitor obtained from the soybean reduced shedding by 30–50%. Trypsin treatment of the cells, on the other hand, will release surface components which are indistinguishable from spontaneously shed fragments. Pearlstein and Waterfield (1974) used such procedures and studied the rate of the repair process, which was found to be quite rapid.

Thrombin, a protease, converts fibrinogen to fibrin and activates blood clotting factor XIII. Exposure of human fibroblasts to this enzyme introduces losses of the cells' major surface glycoprotein component, fibronectin. Within 60 min of exposure, this glycoprotein accumulates in the cell culture supernatant, as reported by Mosher and Vaheri (1978). While these observations were made *in vitro*, there is no reason to believe that such events cannot take place during *in vivo* exposure, resulting in the release of identical or similar macromolecules from cell surfaces.

An enzyme, plasminogen activator (PA), occurs in both established normal cell lines and in transformed cells (Roblin *et al.*, 1975a,b). It is located on the cell membrane (Quigley, 1976). PA is released by both cell types and the conditions of the release were extensively studied (Chou *et al.*, 1977). According to Chou *et al.* (1979), Ca^{2+} at 4.3 mM concentration enhances the biosynthesis of PA and, as a consequence of this overproduction, membrane PA will be shed into the medium. The mechanism of the Ca^{2+} action is not clear, but the possibility that it may participate in enzyme activation (Allan and Mitchell, 1978) should not be overlooked.

It is feasible that certain proteolysis-sensitive materials (such as fibronectin) will be easily set free by an enzyme which cleaves either its attachment site on the cell surface or the part of the molecule which serves as an anchor to the cell. Baumann and Doyle (1978) studied the turnover of plasma membrane glycolipids in hepatoma cells and found that a trypsin-sensitive glycoprotein is rapidly degraded. They could isolate fragments of this in the serum-free culture supernatant. They assumed that either membrane or protoplasmic proteases are involved. Analogies for this possibility can be found in the literature (Rutishauser *et al.*, 1976, Roblin *et al.*, 1975a; Ran *et al.*, 1975), but convincing data are still lacking.

Although no experimental evidence provides definite proof, it is enticing to associate shedding with certain physicochemical properties of the cell envelope. One is inclined to assume that a highly dynamic membrane, with a rapid turnover rate but particularly with a high fluidity, will lose components more readily than a firm, tight, immobile envelope. The interesting experiments of Inbar and Shinitzky (1974a,b) and Shinitzky and Inbar (1974) dealt with the microviscosity measurements of living cell membranes. They used the instrument built by Shinitzky for this purpose, which measures the flourescence polarization of the membrane, due to the incorporation of flourescent probes in it. If the membrane has a low viscosity, the probe can rotate in the membrane, thus causing a measurable degree of depolarization of the polarized flourescent light passing through a thermostatically controlled cuvette which contains the live cell suspension. [The principle and the instrument were developed by Weber and Bablouzian (1966)]. The findings, published in several papers, indicated that malignant (leukemic) cells have a low viscosity as compared to isolated normal counterparts. They also showed that raising the viscosity of the leukemic cells by increasing the membrane's cholesterol content reduced their tumorigenicity. In a more recent work using the same methodology, van Blitterswijk and associates (1977) determined the lipid fluidity of isolated plasma membranes of normal and malignant leukocytes as well as of extracellular membranous microvesicles exfoliated by the cells and released into ascites fluid. The membranes of leukemic cells again had a significantly lower viscosity, but the shed microvesicle membranes had a high viscosity, as if shedding would occur on those parts of the cell envelope which are rigid but surrounded by a low-viscosity environment. Con A in high concentration is described as an inhibitor of lateral diffusion within the membrane (Yahara and Edelman, 1975; Schlessinger *et al.*, 1977). On the other hand, Jones (1973), Edelman

(1976), and Mosher *et al.* (1978) found increased shedding and no changes in receptor mobility if mitogenic quantities of Con A were used. Doetschman (1980) reported that stabilization of chick embryo cell membranes by Con A reduced the shedding of [^{3}H]fucose-containing glycoproteins from these cells, in apparent contradiction to the above findings. Doetschman used two- to fivefold higher concentrations and this may also explain the apparent descrepancies.

From the observations quoted above, as well as from simple considerations, it appears that certain specific locations on the membrane are more likely to become detached than others. Some cells contain microvilli, other spontaneously form uropods. Nonuniform distribution of certain receptors on lymphocytes was demonstrated by elegant experiments (Stackpole *et al.*, 1971; Orci *et al.*, 1979). Some of the receptors form patches on the relatively smooth cell surfaces, some seem to be concentrated on protruding appendages. Some, which were evenly distributed, will form caps if external forces pull the reactive surface components into a concentrated zone, as was observed by using hepatoma cells and antihepatoma antibodies (Leonard, 1973). Concentration of surface immunoglobulins on lymphocytes was shown quite convincingly by dePetris (1978). The use of monovalent anti-immunoglobulin labeled with ferritin allowed dePetris to locate the immunoglobulins on the microvilli. The immunoglobulins also formed dense patches on these appendages when spleen cells were treated with 10 or 20 mM Na-azide, which effectively induced the formation of microvilli. The production of microvilli was enhanced by adenosine triphosphate deprivation. Metabolically active lymphocytes showed enhanced microvilli formation, although to a lesser degree than the azide-induced cells. This latter phenomenon was particularly evident on lymphocytes activated by bacterial endotoxin (dePetris, 1978).

It is obvious that the composition of those areas where such accumulation can occur will be quite different from the rest of the cell body. For example, in the system studied by dePetris (1978), the lipid concentration will be lower and the number of protein-containing receptors will increase. This will introduce changes in membrane viscosity and permeability. Furthermore, one would expect that the cohesive forces such as hydrophobic interactions of the constituent molecules will also change in this area. It has already been pointed out above that shedding of cells does not appear to be a random phenomenon, but is restricted to certain domains of the membrane. Van Blitterswijk and co-workers (1977, 1979) showed that shed components have a higher viscosity and compared this phenomenon to virus budding since this selects the more rigid zones of the membrane. Petitou *et al.* (1978), from the same laboratory, continued this line of thought and suggested that an impaired exchange between serum and leukemic cell cholesterol content causes a lowering of microviscosity in the malignant cells. This discrepancy is further increased by selective exfoliation of leukemic cell membrane patches where cholesterol molecules accumulate. They found a high ratio of cholesterol to phospholipids in these shed fragments, which is responsible for the enhanced microviscosity, as mentioned above.

Regarding the mechanisms by which such extruded pseudopods or microvilli can be detached without impairing the viability of the parent cell they leave behind, studies of the mechanism of cell fusion may be of some help, at least on a theoretical basis. Black (1980) assumed that enhanced fusion may facilitate shedding. Although data are not available, one may consider that cell fusion at the base of microvesicles will seal the cell membrane and will cut the protruded parts adrift.

Interesting studies were carried out by Allan and Mitchell (1979) who studied the mechanisms of cellular secretions, and whose conclusions may also be applicable for the mechanism of shedding. The basic thesis of their treatise is that membrane fusion will occur unless the hydrophobic phases of two adjacent membranes are kept apart by their hydrophylic layers which seal both sides of the fluid hydrophobic phase. Some viruses, such as Sendai, bring these phases so close together through tight binding that the chances of fusion will greatly increase. Other processes will facilitate fusion by the removal of the charged functional groups from the hydrophylic layer which are responsible for the intermembrane repulsion forces. Phospholipase C, for example, can facilitate erythrocyte membrane fusion (Peretz *et al.*, 1974) and can either cause the formation of internal membrane vacuoles (Allan *et al.*, 1975) or the release of extracellular microvesicles (Allan and Mitchell, 1979). The role of Ca^{2+} in this enzymatic process has been well established. Basically the same assumptions led Lucy (1977) to the conclusion that protein depletion in some areas of the membrane will form foci of fusion, particularly if lysolecithin, a product of phospholipase action, is present in higher proportions in this zone of the membrane. Lucy also studied the "fusogenic" potentials of various lipids and found that while lysolecithin is the most potent, other lipids containing unsaturated fatty acids are also active. One of the roles of the fusogenic compounds is to introduce a local disorder in the lipid bilayer. It is obvious that there is more than one way to achieve fusion, and if fusion facilitates shedding, all of the events listed above may contribute to shedding. In the selected few articles we reported above, many of these events were indeed observed and reported.

In conclusion, the exact mechanism of shedding is not known. It seems likely that the dynamic fluidity of the membranes allows the formation of zones which will be chemically and thermodynamically sufficiently different from the rest of the cell to be severed. Induction of shedding can be achieved by a number of ways, including enhanced biosynthesis of the shed components, facilitated fusion and enforced pooling of some surface components into patches or pseudopods.

V. THE CONSEQUENCES OF SHEDDING

As far as the cells are concerned, there are several consequences one can consider and the very first is the survival of the shedding cells. It is clear that extensive shedding which can be induced by detergents or cytostatic drugs may reach a point where the repair mechanisms cannot supply replacements for lost components and the cell will disintegrate. It is important to emphasize here that during shedding the exfoliation of relatively large sheets of surface layers may occur without losing cell viability. There are, of course, instances where the cell will not disintegrate immediately, but will lose some of its functions, and as a consequence of this will be either eliminated sooner than its normal life span would predict or will lose viability in a protracted fashion.

It appears to be more important to discuss the consequences of losing recognizable surface markers and receptors. These events can result in "naked" cells, which are neither recognizable nor capable of interacting with each other or with intact cells. Cells with such asocial behavior may ignore molecular messengers (mediators) which require the presence of a specific receptor site, or a certain density of such receptors per square units of cell surface to be effective. Several papers were written on the possible routes of the spreading of malignancy [for a review see Nicholson (1979) and Poste and Nicholson in this volume]. So-called "antigenic modulation" is one of the mechanisms by which cells may lose recognizability. This was originally described by Boyse *et al.* (1967), who studied experimental leukemias and reported that the phenotypic expression of TL antigen on both normal and leukemic cells can be suppressed by TL antibody. This suppression leads to cells with reduced recognizability. Stackpole and co-workers (1978, 1980) studied "antigenic modulation" and its effect on the escape from immune recognition. It is assumed that nonimmunogenic tumor cells can sneak through immune surveillance and will be successful in establishing microfoci of less recognizable malignancies (Kim, 1970; Kim *et al.*, 1975; Davey *et al.*, 1976). This possibility gained support from the observations of several independent laboratories, which reported that metastatic tumors are much less immunogenic and more resistant to cytostatic drugs than the primary lesion in the same host. Dennis (1981) found that the rate of shedding is directly proportional to tumorigenicity. Others showed, by using isotope-labeled tumor cell inocula, that less than 0.1% of the cells can escape the numerous traps set by the natural defenses of the host. These cells must be significantly different from those which were eliminated.

Although is it attractive to consider that naked cells which shed their markers will have the greatest chances for survival in a hostile environment and form unnoticed metastatic foci, several experimental findings caution us against overemphasizing this possibility. Several laboratories have found that resynthesis of shed markers by the cell is very rapid. Some cells can perform this replacement within 1 hr, others take a longer period of time. This means that if shedding may give an advantage to escape recognition, it will only be a transcient one. Whether this short-lived condition is sufficient for the formation of a new colony in an area less accessible for the components of the defense system is not known at this time.

What effect the shed fragments may have on the host demands a more detailed discussion and, again, reading of the earlier literature. Flexner and Jobling (1907) worked with a transplantable rat sarcoma and observed that if a heat-killed cell emulsion of the same tumor was injected before the viable inoculum, growth of the tumor was promoted. If the emulsion of the cells was not heated, no growth enhancement could be seen. Haaland (1910) disintegrated mouse tumor cells by grinding them at liquid air temperature and injected the homogenate 15, 17, and 20 days prior to challenge with viable cells. Haaland observed that such pretreatment not only does not induce immunity to the challenge but "only seems to manure the soil for a subsequent growth of tumors," since he saw a greatly accelerated tumor development. Intact tumor cells given in subthreshold quantities induced immunity to the tumor. Casey (1934) minced Sarcoma 180 by grinding and sterilized the suspension by Berkefeld "V" filtration. The filtrate, 0.1 ml, could enhance the growth of fresh Sarcoma 180 challenge given 2 weeks later. The enhancement was statistically significant if the homogenate was of the same origin as the challenge. Using other experimental sarcomas and carcinomas, no cross-protection or enhancement could be observed.

What emerges from these data is that immunogen-containing preparations can enhance the growth of subthreshold inocula if the immunogen is well disintegrated, as if small fragments induce tolerance and intact cells of the same tumors cause immunity. The possible mechanisms of these early observations became the center of revived research efforts after inbred strains and syngeneic and allogenic tumors became available.

The results showed that shedding may provide fragments which will have the same growth-promoting effect as the preparation of Flexner and Joblin, Haaland, or Casey.

As we already described, we found that the TA3-Ha tumor shed spontaneously membranous fragments into the ascites fluid, particularly if grown in allogeneic recipients. Admixture of this membrane product to 1 or 1/10 TD_{50} of viable cells will enhance the tumor growth. In search of the possible mechanism, we found that the shed fragments are reacting with sensitized spleen cells (Grohsman and Nowotny, 1972). *In vitro* incubation of such spleen cells with the shed fragments in the presence of complement caused lysis of immune spleen cells, and those which remained viable lost their capacity to transfer *in vivo* resistance adoptively to normal recipients (Nowotny *et al.*, 1974). Similar results and conclusions were reached in the same year by Ben-Sasson *et al.* (1974).

While our experiments were among the first which showed that *in vivo*-shed tumor membrane components can have an *in vivo*-enhancing effect on tumor growth, several papers published previously or simultaneously showed that products exist with similar inhibitory effects on lymphocytic actions *in vitro*. The first among these were the results of Hellström and Hellström (1969), already discussed previously and summarized by them for this monograph. Several publications from various laboratories followed.

Currie and Basham (1972) found that patient sera contain components which can inhibit *in vitro* immune reactions to own tumor. Currie and Gage (1973) observed, while studying the evolution of cytotoxic rat lymphocytes against a spontaneously metastasizing sarcoma, that 7 days after tumor inoculation a serum component appeared which inhibited lymphocyte cytotoxicity. Baldwin and associates (Baldwin *et al.*, 1973a; Rees *et al.*, 1974) showed in numerous papers that chemically induced animal tumors can also release tumor-associated antigens into the circulation which can inhibit *in vitro* antitumor action of cytotoxic lymphocytes. [For reviews see Baldwin and Robins (1975), Price and Baldwin (1977), and also Baldwin *et al.* in this volume.] Baldwin *et al.* (1972), in an elegant experiment, produced such inhibitory complexes *in vitro* by extracting the antigen from the tumor with 3 M KC1 and combining it with specific antibodies. The Ag/Ab ratio was critical and the most extensive blocking could be obtained in antibody excess. Baldwin *et al.* (1973b) also showed that human tumor membrane digested with papain yielded a preparation which inhibited *in vitro* cytotoxicity of the patient's lymphocytes to colon carcinoma target cells.

Alexander and co-workers (Thomson *et al.*, 1973a,b) found both soluble antigen and soluble Ag/Ab complexes in the circulation of rats with chemically induced sarcomas. Currie and Gage (1973) found blocking factor to spontaneous fibrosarcoma. Ankerst (1971) and Skurzak *et al.* (1972) found that blocking factors may be more generally present than originally thought, and it was shown by Bansal *et al.* (1973) that such inhibitors can induce tolerance to allografts in rats. Along the same lines of thought, the theory of Alexander (1974) was quite important. He assumed that such shed blocking antigens prevent the rejection of both tumors and immunogenic embryos.

Unfortunately, there is a glaring deficiency in the avalanche of publications on the blocking factors. With the exception of few papers, none of the above reports described *in vivo* effects of these which would indicate that the *in vitro* inhibitions can also be manifested *in vivo* by enhanced tumor growth. We found only three papers where the results of such attempts were described. One of these was published by Pierce (1971), who took heat-inactivated serum from Moloney virus-induced, sarcoma-bearing mice and injected it into normal mice. Twenty-four hours later, he inoculated the mice with Moloney virus and observed the rate of development of the sarcoma. It was reported that the latent period was shortened and the tumor development accelerated. Pierce assumed that the serum contained blocking factors, similar to those described by the Hellströms, but unfortunately experimental confirmation of this was not carried out. Furthermore, Moloney sarcoma-bearing sera are highly viremic. As Pierce noted, 0.1 ml of unheated serum caused tumor development in 8 days. Although Pierce used heated serum (56°C for 30 min) for facilitating Moloney sarcoma development, one cannot exclude the possibility that this serum contained constituents of viral origin which promoted the tumor growth. Bansal *et al.* (1972) used serum of progressively growing polyoma-bearing rats and a low pH eluate of the same tumor. By injecting these into polyoma-inoculated rats, facilitated tumor growth was observed. Bansal did not use heat-inactivated serum. The growth promotion was highly significant only by tumor eluate. It was not significant if 0.3 ml serum was injected every other day. If 5 ml serum was injected, slight but only transient growth enhancement could be seen, which merged with normal growth rate after 24 days. Similar experiments were reported by Vaage (1972), who attempted but failed to influence the growth of chemically induced syngeneic sarcoma in C3Hf mice.

Considering all the above, it seems to be evident that the source of these serum components which inhibit lymphocyte cytotoxicity *in vitro* or of our ascites fluid component which promoted tumor growth *in vivo* is the tumor itself. The process of their release is shedding by viable cells and autolysis of the cytoskeleton of dead cells. Comparative studies of these two kinds of products with regard to their reactivity (receptor availability and receptor density) were not carried out. To discuss this point we have to use several assumptions and some sets of indirect evidence.

We already elaborated on the most likely general phenomenon that occurs due to lateral movement in the plan of the membrane regouping of certain functional components. Caps will be formed which ought to have different intermolecular forces in action than the rest of the cell surface. As a consequence of this, these caps or patches may be externalized and severed.

It follows that such shed surface components will have a chemical composition different from the cell surface they left behind. It is more important to realize that during cap formation the density of reactive sites will be greatly enhanced, since they were pulled together by external forces (such as antibodies or plant agglutinins or by recognizing lymphocytes) into a cap as was shown by Berke and Fishelson (1975) among others. It is only logical to assume that the reactivity of such shed surface fragments will be considerably greater (provided that the external cap-forming forces did not occupy all reactive sites irreversibly) than the cell they left behind or than those resting cells where the reactive sites are evenly distributed all around the surface. The experiments of dePetris (1978) showed convincingly that concentration of reactive sites can occur even without visible external forces. Scott (1976) induced shedding of microvesicles artificially and reported severalfold enrichment of cholesterol, sphingomyelin, and membranebound 5′-nucleotidase in these microvesicles. In determining the A and B blood group antigen reactivities on intact and mild detergent-treated human erythrocytes, our laboratory showed that many more isoagglutinin receptors are present on detergent-shed membrane fragments than on intact erythrocytes (Abdelnoor *et al.*, 1971). It is unlikely that mild detergent-caused dissociation of the membrane would produce such highly active fragments by cap formation or similar processes. As we assumed, and scanning electron microscopy supported, this treatment set free new A and B receptors not available for the immunoglobulins in the intact cells. These findings were similar to those reported by Haughton and Davies (1962) in measuring H2 antigens or murine cell ghosts. Although we found no evidence in the literature (but this may be entirely our fault), we believe that autolysis of decaying cell debris will similarly set free cryptic functional groups, thus making them available now for the initiation of reactions or for interaction with already available immune reaction products. Accordingly, there is a strong indication that the reactivity of shed (or released by other means) surface components is higher than that of the intact cell. This may apply particularly if we compare it with the remaining reactivity of the cells from which they were released. As we discussed in an earlier publication

(Nowotny *et al.*, 1974), these shed particles will very successfully compete for interaction. If the interacting agents are antibodies or cytotoxic lymphocytes, they will first meet the shed fragments which will greatly outnumber the cellular reactive sites, and, being much more reactive, will have much greater affinity then the cell itself. In the case of tumors, the most important consequence of shedding is the interception and neutralization of antitumor immunoglobulins and immunocytes by highly effective surface components released by the tumor itself. In support of this assumption, a few papers can be cited which show that such mechanisms may be at work *in vivo*.

The first observation about a local lymph node paralysis was published by Alexander and Hall (1969). They studied the function of lymphocytes obtained from distant nodes and from lymph nodes which drain the tumors. They found local node paralysis in the latter and assumed that this paralysis is caused by antigens of the tumor which react with lymphocytes. Flannery *et al.* (1973) described a local anergy of lymphocytes in the regional lymph nodes close to the tumor. They assumed that the tumors which invaded the region released tumor antigens which formed complexes with them. An *in vivo* blockade of antibodies was reported by Thomson *et al.* (1973b), who reported that immunoglobulins specific for tumor-associated antigens cannot be detected in rats unless the tumor is excised. The low Ig level in the tumor-bearing animals is not due to absorption by the tumor itself but to the release of tumor-specific transplantation antigens which rendered the lymphocytes in this node totally unresponsive.

Several laboratories published findings indicating that tumors may produce immunosuppressive compounds. Such activity was found in the supernatant of Ehrlich ascites cells, as reported by Biran *et al.* (1970), which could prolong the survival of a skin allograft if the component (or whole Ehrlich ascites cells) was injected into the skin donor before the transplantation. The mechanism of this unusual phenomenon remained obscure. We also observed a mild immunosuppression of anti-sheep red blood cell (SRBC) response in ICR mice, if we injected them simultaneously with TA3-Ha ascites fluid. At that time we attributed this observation to possible antigenic competition (Abdelnoor and Nowotny, unpublished). Our observation (Nowotny *et al.*, 1976) that while small quantities of the same ascites fluid component induced lymphoblast response in spleen cells from TA3-Ha immunized mice, large quantities inhibited the incorporation of [^{3}H]thymidine, lowering it far below the normal level, was more clear cut. Normal spleen cells showed no response in the same assay if they were exposed to ascites fluid or to the tumor growth-enhancing component isolated from it. Serum or ascites fluid of thymoma-bearing mice suppressed the number of antibody-forming cells (Chan and St. C. Sinclair, 1972). Ascites fluid of Ehrlich tumor was strongly immunosuppressive *in vitro* to SRBC as reported by Hršak and Marotti (1973). Kamo *et al.* (1975a) reported that mastocytoma cells (derived from mast cells of DBA/2 mice) suppressed the *in vitro* immune response of DBA/2 spleen cells to SRBC if the were cocultivated. It was also found that cell-free supernates of mastocytoma cells maintained *in vitro* can similarly suppress the anti-SRBC response measured *in vitro*. The same laboratory (Kamo *et al.*, 1975b) found that mastocytoma cells grown in ascites form release an immunosuppressive factor into the ascites fluid which passes 0.4-μm nucleopore filters but not dialysis membranes. The activity of this factor was abolished if it was heated for 30 min at 57°C. Another potentially immunosuppressive mechanism triggered by tumors was described by Pike and Snyderman (1976), who reported the existence of macrophage-inactivating factors produced by four different murine neoplasms. The activity of this low-molecular-weight, heat-stable compound could be demonstrated by depression of macrophage migration *in vivo* and of chemotactic response *in vitro*. Normal tissues and cells did not contain the inhibitor. The authors speculated that such inhibitors, which can suppress initial steps of the immune response, may aid the tumors to escape immune recognition and destruction. Following the same line of thought, the Hellströms described the activation of T suppressor (Ts) cells by tumor antigens (Hellström and Hellström, 1978). In the continuation of this experimental series, they removed Ts cells from tumor-bearing animals by low-dose irradiation and observed inhibition of tumor growth (Hellström *et al.*, 1978). They assume that tumors induce a Ts activator, and the consequence of this is suppression of antitumor immune response. Accordingly, their "blocking factor" is a Ts cell activator. The Hellströms have elaborated in detail on this in Chapter 13 of this monograph.

Immunization by shed fragments, particularly by those which have a higher density of immunodeterminants than the intact cells, is one of the expected consequences. It is only logical to assume that these shed membrane particles will be quickly taken up and processed by macrophages; therefore, their release by the tumor cells should be the most efficient way to induce immunity. It is quite puzzling that the opposite seems to be true. There are several observations which show that whole cells are more immunogenic than any immunogenic components isolated from them. For example, sheep erythrocytes are quite immunogenic if they are freshly taken; they lose immunogenicity if stored in a cold room and their immunogenicity will be almost completely abolished if the cells are lysed by the addition of a few drops of distilled water (Behling and Nowotny, unpublished). Bacterial antigens are another example. Isolated

O antigens (such as LPS) are much less immunogenic than heat-killed whole cells. If the LPS contains only 1 or 2% amino acids (phenol–water extraction), it is even less immunogenic than similar preparations containing 6 to 10% bound amino acids (trichloroacetic acid-extracted LPS), although the antibodies are directed primarily against the polysaccharide constituent which is the same in both LPS preparations, provided they were extracted from the same bacterial strain. The same applies to tumors. To achieve good immune resistance, the best procedure is to use whole cells, and the injection of a nontumorigenic dose of viable cells is particularly effective. The next best procedure is the injection of a low number of irradiated cells, and the worst is to attempt to immunize with cell membrances or their fragments. Quite frequently, such attempts to immunize with the improper preparation will result in reduced resistance of the host to the tumor in question. Such experimental reports can be found in newer as well as older literature (Flexner and Joblin, 1907; Casey, 1934; Révész, 1956; Hewitt *et al.*, 1973; Nowotny *et al.*, 1976). Quite obviously, induction of resistance and detectable immune response to the tumor are the end result of exceedingly complex events. Furthermore, they may not follow the same path and they do not need to be the logical consequences of each other. Immunization attempts can lead to Ts activation, which may allow tumor growth (Umiel and Trainin, 1974; Naor, 1979), or to "immune enhancement" (Kaliss, 1967), mediated by noncytotoxic antibodies which is an often-cited, although still controversial, phenomenon.

Returning to the possible role of shed fragments in the immunity to the tumor, one can envision that the reaction of the host to the first release of surface components is the production of antibodies, which will specifically seek and react with the tumor surface. If they can induce capping, they may be able to facilitate further release of antigen-dense flakes. As we speculated in an earlier publication (Nowotny *et al.*, 1974), initial immune response may only enhance the shedding, leading to the release of such quantities which will have a paralyzing effect on the antitumor immune response. The observation of Prehn (1971), which indicated that a weak immune response may be stimulatory to the tumor growth, makes such a hypothesis quite sustainable.

VI. CONCLUSION

The consequences of shedding may include the production of cells with reduced immunogenicity, and shedding may trigger initiation of an immune response as well as the induction of immune paralysis. The highly reactive fragments of the membrane surrounding the tumor or dispersed in the circulation may provide a smoke screen, a protective halo for the tumors, because they can intercept and effectively block or neutralize incoming cytotoxic lymphocytes and immunoglobulins. This may be a partial explanation of the events leading to the spread of malignant foci. Immunosuppression induced by tumor cell products may provide an alternate mechanism by which tumors learned how to escape immune destruction. At any rate, shedding by tumors appears to be one of the most efficient self-defense mechanisms of malignant cells.

VII. REFERENCES

Abdelnoor, A., Higgins, M., and Nowotny, A., Effect of cetyltrimethylammonium bromide on human erythrocyte membrane, *in* "Cellular Antigens" (A. Nowotny, ed.) p. 153, Springer-Verlag, New York Heidelberg Berlin.

Allan, D., and Michell, R. H., 1978, A calcium activated phosphoinositide phosphodiesterase in the plasma membrane of human and rabbit erythrocytes *Biochim. Biophys. Acta* **508:**277.

Allan, D., and Michell, R. H., 1979, The possible role of lipids in control of membrane fusion during secretion, *Symp. Soc. Exp. Biol.* **33:**323.

Allan, D., Low, M. G., Finean, J. B., and Michell, R. H., 1975, Changes in lipid metabolism and cell morphology following attack by phospholipase C on red cells and lymphocytes. *Biochim. Biophys. Acta* **413:**309.

Alexander, P., 1974, Escape from immune destruction by the host through shedding of surface antigens: Is this a characteristic shared by malignant and embryonic cells? *Cancer Res.* **34:**2077.

Alexander, P., and Hall, J. G., 1969, The role of immunoblasts in host resistance and immunotherapy of primary sarcomata, *Adv. Cancer Res.* **13:**1.

Ankerst, J., 1971, Demonstration and identification of cytotoxic antibodies and antibodies blocking the cell-mediated antitumor immunity against adenovirus type 12-induced tumors. *Cancer Res.* **31:**997.

Baldwin, R. W., and Robins, R. A., 1975, Humoral factors abrogating cell-mediated immunity in the tumor-bearing host, *Curr. Top. Microbiol. Immunol.* **72:**21.

Baldwin, R. W., Price, M. R., and Robins, R. A., 1972, Blocking of lymphocyte-mediated cytoxicity for rat hepatoma cells by tumour-specific antigen-antibody complexes, *Nature (London)* **238:**185.

Baldwin, R. W., Embleton, M. J., and Robins, R. A., 1973a, Cellular and humoral immunity to rat hepatoma-specific antigens correlated with tumour status, *Int. J. Cancer* **11:**1.

Baldwin, R. W., Embleton, M. J., and Price, M. R., 1973b, Inhibition of lymphocyte cytotoxicity for human colon carcinoma by treatment with solubilized tumor membrane fractions. *Int. J. Cancer* **12:**84.

Bansal, S. C., Hargreaves, R., and Sjögren, H. O., 1972, Facilitation of polyoma tumor growth in rats by blocking sera and tumor eluate, *Int. J. Cancer* **9:**97.

Bansal, S. C., Hellström, K. E., Hellström, I., and Sjögren, H. O., 1973, Cell-mediated immunity and blocking serum activity to tolerated allografts in rats, *J. Exp. Med.* **137:**590.

Baumann, H., and Doyle, D., 1978, Turnover of plasma membrane glycoproteins and glycolipids of hepatoma tissue culture cells, *J. Biol. Chem.* **253:**4408.

Bennett, C., Cooke, K. B., and Geck, P., 1978, Protein insolubilisation as an aid to the detection of melanoma antigen in human urine, *Prot. Biol. Fluids* **24:**667.

Ben-Or, S., and Doljanski, F., 1960, Single-cell suspensions as tissue antigens, *Exp. Cell Res.* **20:**641.

Ben-Sasson, Z., Weiss, D. W., and Doljanski. F., 1974, Specific binding of factor(s) released by Rous sarcoma virus-transformed cells to splenocytes of chickens with Rous sarcomas, *J. Natl. Cancer Inst.* **52:**405.

Berke, G., and Fishelson, Z., 1975, Localization of aggregagated cell surface antigens of target cells bound to cytotoxic T lymphocytes, *J. Exp. Med.* **142:**1011.

Billing, R. J., and Terasaki, P. I., 1974, Purification of HL-A antigens from normal serum, *J. Immunol.* **112:**1124.

Billingham, R. E., and Sparrow, E. M., 1955, The effect of prior intravenous injections of dissociated epidermal cells and blood on the survival of skin homografts in rabbits. *J. Embryol. Exp. Morphol.* **3:**265.

Biran, S., Ben-Hur, N., and Robinson, E., 1970, Factor influencing skin graft survival from donors with Ehrlich ascites tumor, *Oncology* **24:**344.

Black, P. H., 1980, Shedding from the cell surface of normal and cancer cells, *Adv. Cancer Res.* **32:**75.

Bolognesi, D. P., Langlois, A. J., and Schafer, W., 1975, Polypeptides of mammalian oncornaviruses. IV. Structural components of murine leukemia virus released as soluble antigens in cell culture, *Virology* **68:**550.

Boyse, E. A., Stockert, E., and Old, L. J., 1967, Modification of the antigenic structure of the cell membrane by thymus-leukemia (TL) antibody. *Proc. Natl. Acad. Sci. USA* **58:**954.

Brawn, R. J., 1971, *In vitro* desensitization of sensitized murine lymphocytes by a serum factor (soluble antigen?), *Proc. Natl. Acad. Sci. USA* **68:**1634.

Brennan, A., Povey, P. M., Smith, B. R., and Hall, R., 1980, Shedding of surface proteins by porcine thyroid cells. *J. Endocrinol.* **85:**245.

Bystryn, J. C., 1977, Release of cell-surface tumor-associated antigens by viable melanoma cells from humans *J. Natl. Cancer Inst.* **59:**325.

Calafat, J., Hilgers, J., Van Blitterswijk, W. J., Verbeet, M., and Hageman, P. C., 1976, Antibody-induced modulation and shedding of mammary tumor virus antigens on the surfaces of GR ascites leukemia cells as compared with normal antigens. *J. Natl. Cancer Inst.* **56:**1019.

Callahan, G. N., Ferrone, S., Poulik, M. D., Reisfeld, R. A., and Klein, J., 1976, Characterization of Ia antigens in mouse serum, *J. Immunol.* **117:**1351.

Carrel, S., and Theilkaes, L., 1973, Evidence for a tumour-associated antigen in human malignant melanoma. *Nature (London)* **242:**609.

Casey, A. E., 1934, Specificity of enhancing materials from mammalian tumors, *Proc. Soc. Exp. Biol. Med.* **31:**663.

Catchpole, H. R., 1950, Serum and tissue glycoproteins in mice bearing transplantable tumors, *Proc. Soc. Exp. Biol. Med.* **75:**221.

Chan, P. L., and St. C. Sinclair, N. R., 1972, Immunologic and virologic properties of chemically and α-irradiation-induced thymic lymphomas in mice, *J. Natl. Cancer Inst.* **48:**1629.

Charlton, R. K., and Zmijewski, C. M., 1970, Soluble HL-A7 antigen: Localization in the β-lipoprotein fraction of human serum, *Science* **170:**636.

Chou, I.-N., O'Donnell, S. P., Black, P. H., and Roblin, R. O., 1977, Cell-density dependent secretion of plasminogen activator by 3T3 cells, *J. Cell. Physiol.* **91:**31.

Chou, I.-N., Cox, R., and Black, P. H., 1979, Studies on the mechanism of Ca^{2+} stimulation of plasminogen activator synthesis/release by Swiss 3T3 cells, *J. Cell. Physiol.* **100:**457.

Codington, J. F., Sanford, B. H., and Jeanloz, R. W., 1970, Glycoprotein coat of the TA^{-3}cell. I. Removal of carbohydrate and protein material from viable cells. *J. Natl. Cancer Inst.* **45:**637.

Critchley, D. R., Wyke, J. A., and Hynes, R. O., 1976, Cell surface and metabolic labelling of the proteins of normal and transformed chicken cells, *Biochim. Biophys. Acta* **436:**335.

Currie, G. A., and Basham, C., 1972, Serum mediated inhibition of the immunological reactions of the patient to his own tumour: A possible role for circulating antigen, *Br. J. Cancer* **26:**427.

Currie, G. A., and Gage, J. O., 1973, Influence of tumour growth on the evolution of cytotoxic lymphoid cells in rats bearing a spontaneously metastisizing syngeneic fibrosarcoma, *Br. J. Cancer* **28:**136.

Darcy, D. A., 1957, Immunological demonstration of a substance in rat blood associated with tissue growth, *Br. J. Cancer* **11:**137.

Davey, G. C., Currie, G. A., and Alexander, P., 1976, Spontaneous shedding and antibody induced modulation of histocompatibility antigens on murine lymphomata: Correlation with metastatic capacity, *Br. J. Cancer* **33:**9.

Dennis, J. W., Donaghue, T. P., and Kerbel, R. S., 1981, Membrane-associated alterations detected in poorly tumorigenic lectin-resistant variant sublines of a highly malignant and metastatic murine tumor, *J. Natl. Cancer Inst.* **66:**129.

dePetris, S., 1978, Preferential distribution of surface immunoglobulins on microvilli, *Nature (London)* **272:**66.

Doetschman, T. C., 1980, The effects of ConA on cell surface shedding in cell cultures, *J. Cell Sci.* **46:**221.

Doljanski, F., and Kapeller, M., 1976, Cell surface shedding-The phenomenon and its possible significance, *J. Theor. Biol.* **62:**253.

Edelman, G. M., 1976, Surface modulation in cell recognition and cell growth, *Science* **192:**218.

Emerson, S. G., and Cone, R. E., 1979a, Differential effects of colchicine and cytochalasin on the shedding of murine B cell membrane IgM and IgD, *Proc. Nat. Acad. Sci. USA* **76:**6582.

Emerson, S. G., and Cone, R. E., 1979b, Turnover and shedding of Ia antigens by murine spleen cells in culture, *J. Immunol.* **122:**892.

Ferber, E., Schmidt, B., and Weltzien, H. U., 1980, Spontaneous and detergent-induced vesiculation of thymocyte plasma membranes, *Biochim. Biophys. Acta* **595:**244.

Fernandez, L. A., and Macsween, J. M., 1977, The spontaneous shedding of the lymphocyte receptor for sheep red blood cells, *Dev. Comp. Immunol.* **1:**385.

Flannery, G. R., Chalmers, P. J., Rolland, J. M., and Nairn, R. C., 1973, Immune response to a syngeneic rat tumour: Development of regional node lymphocyte anergy, *Br. J. Cancer* **28:**118.

Flexner, S., and Jobling, J. W., 1907, On the promoting influence of heated tumor emulsions on tumor growth, *Proc. Soc. Exp. Biol. Med.* **4:**156.

Gasic, G., and Gasic, T., 1962a, Removal and regeneration of the cell coating in tumor cells, *Nature (London)* **196:**170.

Gasic, G., and Gasic, T., 1962b, Removal of sialic acid from the cell coat in tumor cells and vascular endothelium, and its effects on metastasis, *Proc. Natl. Acad. Sci. USA* **48:**1172.

Gozzo, J. J., Schlesinger, R. M., and Monaco, A. P., 1974, Detection of bladder cancer-associated antigens in urine by microcomplement fixation, *Surg. Forum* **25:**117.

Grohsman, J., and Nowotny, A., 1972, The immune recognition of TA3 tumors, its facilitation by endotoxin and abrogation by ascites fluid, *J. Immunol.* **109:**1090.

Guinan, P., John, T., Sadoughi, N., Ablin, R. J., and Bush, I. 1974, Urinary carcinoembryonic-like antigen levels in patients with bladder carcinoma, *J. Urol.* **111:**350.

Gupta, R. K., and Morton, D. L., 1975, Presence of human tumor-associated antigen(s) in urine of patients with cancer, *Surg. Forum* **26:**158.

Gupta, R. K., and Morton, D. L., 1976, Tumor-associated antigens in urine of cancer patients, *Proc. Am. Assoc. Cancer Res.* **17:**92.

Gupta, R. K., and Morton, D. L., 1979, Detection of cancer-associated antigen(s) in urine of sarcoma patients. *J. Surg. Oncol.* **11:**65.

Haaland, M., 1910, The contrast in the reactions to the implantation of cancer after the inoculation of living and mechanically distintegrated cells, *Proc. R. Soc. London* **82:**293.

Hall, R. R., Laurence, D. J. R., Darcy, D., Stevens, U., James, R., Roberts, S., and Munro Neville, A., 1972, Carcinoembryonic antigen in the urine of patients with urothelial carcinoma, *Br. Med. J.* **3:**609.

Haughton, G., and Davies, D. A. L., 1962, Tissue cell antigens: Antigens of mouse tumor cell ghosts, *Br. J. Exp. Pathol.* **43:**488.

Hellström, I., and Hellström, K., 1969, Studies on cellular immunity and its serum mediated inhibition in Moloney virus-induced mouse sarcomas, *Int. J. Cancer* **4:**587.

Hellström, K. E., and Hellström, I., 1974, Lymphocyte-mediated cytotoxicity and blocking serum activity to tumor antigens, *Adv. Immun.* **18:**209.

Hellström, K. E., and Hellström, I., 1978, Evidence that tumor antigens enhance tumor growth in vivo by interacting with a radiosensitive (suppressor?) cell population, *Proc. Nat. Acad. Sci. USA* **75:**436.

Hellström, I., Hellström, K. E., and Sjögren, H. O., 1970, Serum mediated inhibition of cellular immunity to methylcholanthrene induced murine sarcomas, *Cell. Immunol.* **1:**18.

Hellström, I., Sjögren, H. O., Warner, G. A., and Hellström, K. E., 1971, Blocking of cell-mediated tumor immunity by sera from patients with growing neoplasms, *Int. J. Cancer* **7:**226.

Hellström, I., Warner, G. A., Hellström, K. E., and Sjögren, H. O., 1973, Sequential studies on cell-mediated tumor immunity and blocking serum activity in ten patients with malignant melanoma, *Int. J. Cancer* **11:**280.

Hellström, K. E., Hellström, I., Kant, J. A., and Tamerius, J. D., 1978, Regression and inhibition of sarcoma growth by interference with a radiosensitive T cell population, *J. Exp. Med.* **148:**799.

Heppner, G. H., Stolbach, L., Byrne, M., Cummings, F. J., McDonough, E., and Calabresi, P., 1973, Cell-mediated and serum blocking reactivity to tumor antigens in patients with malignant melanoma, *Int. J. Cancer* **11:**245.

Herberman, R. B., Aoki, T., and Nunn, M. E., 1973, Solubilization of G (Gross) antigens on the surface of G leukemia cells, *J. Natl. Cancer Inst.* **50:**481.

Hewitt, H. B., Blake, E., and Porter, E. H., 1973, The effect of lethally irradiated cells on the transplantability of murine tumours, *Br. J. Cancer* **28:**123.

Hršak, I., and Marotti, T., 1973, Immunosuppression mediated by Ehrlich ascites fluid, *Eur. J. Cancer* **9:**717.

Hsu, C. C. S., Wu, S. J.-Y., George, S., and Morgan, E. R., 1980, Assessment of shedding and reexpression of surface immunoglobulin and Ia-like antigen on human blood lymphocytes, *Cell. Immunol.* **52:**154.

Inbar, M., and Shinitzky, M., 1974a, Increase of cholesterol level in the surface membrane of lymphoma cells and its inhibitory effect on ascites tumor development, *Proc. Natl. Acad. Sci. USA* **71:**2128.

Inbar, M., and Shinitzky, M., 1974b, Cholesterol as a bioregulator in the development and inhibition of leukemia, *Proc. Natl. Acad. Sci. USA* **71**:4229.

Jehn, V. W., Nathanson, L., Schwartz, R. S., and Skinner, M., 1970, *In vitro* lymphocyte stimulation of a soluble antigen from malignant melanoma, *N. Engl. J. Med.* **283**:329.

Jones, G., 1973, Release of surface receptors from lymphocytes, *J. Immunol.* **110**:1526.

Kaliss, N., 1967, Immunological enhancement: Conditions for its expression and its relevance in grafts of normal tissues, *Ann. N.Y. Acad. Sci.* **129**:155.

Kamo, I., Kateley, J., and Friedman, H., 1975a, Mastocytoma-induced suppression of *in vitro* antibody formation, *Proc. Soc. Exp. Bio. Med.* **148**:883.

Kamo, I., Patel, C., Kateley, J., and Friedman, H., 1975b, Immunosuppression induced *in vitro* by mastocytoma tumor cells and cell-free extracts, *J. Immunol.* **114**:1749.

Kapeller, M., Plesser, Y. M., Kapeller, N., and Doljanski, F., 1976, Turnover and shedding of cell-surface constituents in normal and neoplastic chicken cells, *in* "Progress in Differentiation Research" (N. Muller-Berat ed.), pp. 397–405, North-Holland, Amsterdam.

Kim, U., 1970, Metastasizing mammary carcinomas in rat: Induction and study of their immunogenicity, *Science* **167**:72.

Kim, U., Baumler, A., Carruthers, C., and Bielat, K., 1975, Immunological escape mechanism in spontaneously metastasizing mammary tumors, *Proc. Nat. Acad. Sci. USA* **72**:1012.

Kornfeld, S., and Ginsburg, V., 1966, The metabolism of glucosamine by tissue culture cells, *Exp. Cell Res.* **41**:592.

Krishan, A., and Frei, E., III, 1975, Morphological basis for the cytolytic effect of vinblastine and vincristine on cultured human leukemic lymphoblasts, *Cancer Res.* **35**:497.

Leonard, E. J., 1973, Cell surface antigen movement: Induction in hepatoma cells by antitumor antibody, *J. Immunol.* **110**:1167.

Leong, S. P., Sutherland, C. M., and Krementz, E. T., 1977, Changes in distribution of human malignant melanoma membrane antigens in the presence of human antibody by immunofluorescence, *Cancer Res.* **37**:293.

Levy, N. L., 1973, Use of an *in vitro* microcytotoxycity test to assess human tumor-specific cell-mediated immunity and its serum-mediated abrogation, *Natl. Cancer Inst. Monograph* No. 37, 85.

Liepins, A., and Hillman, A. J., 1981, Shedding of tumor cell surface membranes, *Cell Biol. Int. Rep.* **5**:15.

Lopez, M. J., and Thomson, D. M. P., 1977, Isolation of breast cancer tumor antigen from serum and urine, *Int. J. Cancer* **20**:834.

Lucy, J. A., 1977, Cell fusion, *Trends Biochem. Sci.* **2**:17.

Marcus, P. I., 1962, Dynamics of surface modification in Myxovirus-infected cells, *Cold Spring Harbor Symp. Quant. Biol.* **27**:351.

Miller, E. E., 1963, Fractionation and characterization of sarcoma I ascites tumor fluid, *Growth* **27**:101.

Miller, E. E., and Bernfeld, P., 1960, Abnormal plasma components in C3H mice bearing spontaneous tumors, *Cancer Res.* **20**:1149.

Miyakawa, Y., Tanigaki, N., Kreiter, V. P., Moore, G. E., and Pressman, D., 1973, Characterization of soluble substances in the plasma carrying HL-A alloantigenic activity and HL-A common antigenic activity, *Transplantation* **15**:312.

Molnar, J., Lutes, R. A., and Winzler, R. J., 1965a, The biosynthesis of glycoproteins. V. Incorporation of glucosamine-1-^{14}C into macromolecules by Ehrlich ascites carcinoma cells, *Cancer Res.* **25**:1438.

Molnar, J., Teegarden, D. W., and Winzler, R. J., 1965b, The biosynthesis of glycoproteins. VI. Production of extracellular radioactive macromolecules by Ehrlich ascites carcinoma cells during incubation with glucosamine-^{14}C, *Cancer Res.* **25**:1860.

Mosher, D. F., and Vaheri, A., 1978, Thrombin stimulates the production and release of a major surface-associated glycoprotein (Fibronectin) in cultures of human fibroblasts, *Exp. Cell Res.* **112:**323.

Mosher, H., Schneider, D., and Falke, D., 1978, Influence of Concanavalin A on protein synthesis and protein release in BHK-21 cells, *Biochim. Biophys. Acta* **507:**445.

Nakamuro, K., Tanigaki, N., and Pressman, D., 1973, Multiple common properties of human β_2-microglobulin and the common portion fragment derived from HL-A antigen molecules, *Proc. Natl. Acad. Sci. USA* **70:**2863.

Naor, D., 1979, Suppressor cells: Permitters and promoters of malignancy? *Adv. Cancer Res.* **29:**45.

Nery, R., James, R., Barsoum, A. L., and Bullman, H., 1974, Isolation and partial characterization of macromolecular urinary aggregates containing carcinoembryonic antigen-like activity, *Br. J. Cancer* **29:**413.

Nicholson, G., 1979, Cancer metastasis, *Sci. Am.* March, 66.

Nordquist, R. E., Anglin, J. H., and Lerner, M. P., 1977, Antibody-induced antigen redistribution and shedding from human breast cancer cells, *Science* **197:**366.

Nowotny, A., Grohsman, J., Abdelnoor, A., Rote, N., Yang, C., and Waltersdorf, R., 1974, Escape of TA3 tumors from allogeneic immune rejection: Theory and experiments, *Eur. J. Immunol.* **4:**73.

Nowotny, A., Butler, R. C., Grohsman, J., and Keebler, C., 1976, Dual effect of tumor antigens: Induction of tumor resistance on tumor growth enhancement, *Ann. N.Y. Acad. Sci.* **276:**106.

Orci, L., Amherdt, M., Roth, J., and Perrelet, A., 1979, Inhomogeneity of surface labelling of B-cells at prospective sites of exocytosis, *Diabetologia* **16:**135.

Ossowski, L., Quigley, J. P., Kellerman, G. M., and Reich, E., 1973, Fibrinolysis associated with oncogenic transformation. Requirement of plasminogen for correlated changes in cellular morphology, colony formation in agar and cell migration, *J. Exp. Med.* **138:** 1056.

Parish, C. R., and McKenzie, I. F. C., 1977, Mitogens and T-independent antigens stimulate T lymphocytes to secrete Ia antigens, *Cell. Immunol.* **33:**134.

Parish, C. R., Chilcott, A. B., and McKenzie, I. F. C., 1976, Low molecular weight Ia antigens in normal mouse serum. I. Detection and production of a xenogeneic antiserum, *Immunogenetics* **3:**113.

Pearlstein, E., and Waterfield, M. D., 1974, Metabolic studies on ^{125}I-labeled baby hamster kidney cell plasma membrances, *Biochim. Biophys. Acta* **362:**1.

Pellegrino, M. A., Pellegrino, A., Ferrone, S., Kahan, B. D., and Reisfeld, R. A., 1973, Extraction and purification of soluble HL-A antigens from exhausted media of human lymphoid cell lines, *J. Immunol.* **111:**783.

Pellegrino, M. A., Ferrone, S., Pellegrino, A. G., Oh, S. K., and Reisfeld, R. A., 1975, Evaluation of two sources of soluble HL-A antigens: Platelets and serum, *Eur. J. Immunol.* **4:**250.

Peretz, H., Toister, Z., Laster, J., and Loyter, A., 1974, Fusion of intact human erythrocytes and erythrocyte ghosts, *J. Cell Biol.* **63:**1.

Petitou, M., Tuy, F., Rosenfeld, C., Mishal, Z., Paintrand, M., Jasnin, C., Mathe, G., and Inbar, M., 1978, Decreased microviscosity of membrane lipids in leukemic cells: Two possible mechanisms, *Proc. Natl. Acad. Sci. USA* **75:**2306.

Pierce, G. E., 1971, Enhanced growth of primary Moloney virus-induced sarcomas in mice, *Int. J. Cancer* **8:**22.

Pike, M. C., and Snyderman, R., 1976, Depression of macrophage function by a factor produced by neoplasms: A mechanism for abrogation of immune surveillance, *J. Immunol.* **117:**1243.

Plesser, Y. M., Weiss, D. W., and Doljanski, F., 1980, Cell-surface shedding by fibroblasts in culture, *Isr. J. Med. Sci.* **16:**519.

Poskitt, P. K. F., Poskitt, T. R., and Wallace, J., 1974, Renal deposition of soluble immune complexes in mice bearing B-16 melanoma, *J. Exp. Med.* **140:**410.

Prehn, T. R., 1971, Perspectives in oncogenesis: Does immunity stimulate or inhibit neoplasia? *J. Reticuloendothelial Soc.* **10:**1.

Price, M. R., and Baldwin, R. W., 1977, Shedding of tumor cell surface antigens, *Cell Surf. Rev.* **3:**423.

Quigley, J. P., 1976, Association of a protease (plasminogen activator) with a specific membrane fraction isolated from transformed cell, *J. Cell Biol.* **71:**472.

Rahman, A. F. R., Liao, S. K., and Dent, P. B., 1977, Characterization of human malignant melanoma cell lines. VII. Glycoprotein synthesis and shedding as revealed by {^{3}H} glucosamine labeling, *In Vitro* **13:**580.

Ran, M., Eshel, I., Witz, I. P., and Klein, G., 1975, Dynamic alterations in some surface properties of freshly explanted Moloney lymphoma cells, *J. Natl. Cancer Inst.* **55:**843.

Rees, R. C., Price, M. R., Baldwin, R. W., and Shah, L. P., 1974, Inhibition of rat lymph node cell cytotoxicity by hepatoma-associated embryonic antigen, *Nature* (*London*) **252:**751.

Reisfeld, R. A., Allison, J. P., Ferrone, S., Pellegrino, M. A., and Poulik, M. D., 1976, HL-A antigens in serum and urine: Isolation, characterization, and immunogenic properties, *Transplant. Proc.* **8:**173.

Révész, L., 1956, Effect of tumour cells killed by x-rays upon the growth of admixed viable cells, *Nature* (*London*) **178:**1391.

Rittenhouse, H. G., Ar, D., Lynn, M. D., and Denholm, D. K., 1978, The spontaneous release of a high-molecular-weight aggregate containing immunoglobulin G from the surface of Ehrlich ascites tumor cells, *J. Supramolec. Struct.* **9:**407.

Roblin, R. O., Chou, I.-N., and Black, P. H., 1975a, Proteolytic enzymes, cell surface changes and viral transformation, *Adv. Cancer Res.* **22:**203.

Roblin, R. O., Chou, I.-N., and Black, P. N., 1975b, Secretion of plasminogen activator by normal mouse cells *in vitro*, *J. Cell Biol.* **67:**366a (Abstract).

Rote, N. S., Gupta, R. K., and Morton, D. L., 1980a, Determination of incidence and partial characterization of tumor-associated antigens found in the urine of patients bearing solid tumors, *Int. J. Cancer* **26:**203.

Rote, N. S., Smith, W. G., and Johnson, G. H., 1980b, Urine excretion of tumor-associated antigens by patients with gynecologic neoplasms, *Gynecol. Oncol.* **9:**370.

Rote, N. S., Gupta, R. K., and Morton, D. L., 1980c, Tumor-associated antigens detected by autologous sera in urine of patients with solid neoplasms, *J. Surg. Res.* **29:**18.

Rudman, D., Del Rio, A., Akgun, S., and Frumin, E., 1969, Novel proteins and peptides in the urine of patients with advanced neoplastic disease, *Am. J. Med.* **46:**174.

Rutishauser, U., Thiery, J.-P., Brackenbury, R., Sela, B.-A., and Edelman, G. M., 1976, Mechanisms of adhesion among cells from neural tissues of the chick embryo, *Proc. Natl. Acad. Sci. USA* **73:**577.

Sanford, B. H., 1967, An alteration in tumor histocompatibility induced by neuraminidase, *Transplantation* **5:**1273.

Schlessinger, J., Elson, E. L., Webb, W. W., Yahara, I., Rutishauser, U., and Edelman, G. M., 1977, Receptor diffusion on cell surfaces modulated by locally bound Concanavalin A, *Proc. Natl. Acad. Sci. USA* **74:**1110.

Schoenheimer, R., 1942, "The Dynamic State of Body Constituents," Harvard Univ. Press, Cambridge, Massachusetts.

Schroit, A. J., Geiger, B., and Gallily, R., 1973, The capacity of macrophage components to inhibit anti-macrophage serum activity, *Eur. J. Immunol.* **3:**354.

Scott, R. E., 1976, Plasma membrane vesiculation: A new technique for isolation of plasma membranes, *Science* **194:**743.

Siebert, F. B., Seibert, M. V., Atno, A. J., and Campbell, H. W., 1947, Variation in protein and polysaccharide content of sera in the chronic diseases, tuberculosis, sarcoidosis and carcinoma, *J. Clin. Invest.* **26:**90.

Shinitsky, M., and Inbar, M., 1974, Differences in microviscosity induced by different cholesterol levels in the surface membrane lipid layer of normal lymphocytes and malignant lymphoma cells, *J. Mol. Biol.* **85:**603.

Sjögren, H. O., Hellström, I., Bansal, S. C., and Hellström, K. E., 1971, Suggestive evidence that the "blocking antibodies" of tumor bearing individuals may be antigen-antibody complexes, *Proc. Natl. Acad. Sci. USA* **68:**1372.

Skinnider, L. F., and Ghadially, F. N., 1977, Ultrastructure of cell surface abnormalities in neoplastic histiocytes, *Br. J. Cancer* **35:**657.

Skurzak, H. M., Klein, E., Yoshida, T. O., and Lamon, E. W., 1972, Synergistic or antagonistic effect of different antibody concentrations on *in vitro* lymphocyte cytotoxicity in the Moloney sarcoma virus system, *J. Exp. Med.* **135:**997.

Stackpole, C. W., and Jacobson, J. B., 1978, Antigenic modulation, *in* "The Handbook of Cancer Immunology," Vol. 2 (H. Waters, ed.), p. 55, Garland, New York.

Stackpole, C. W., Aouki, T., Boyse, E. A., Old, L. J., Lumney-Frank, J., and de Harven, E., 1971, Cell surface antigens: Serial sectioning of single cells as an approach to topographical analysis, *Science* **172:**472.

Stackpole, C. W., Cremona, P., Leonard, C., and Stremmel, P., 1980, Antigen modulation as a mechanism for tumor escape from immune destruction: Identification of modulation-positive and modulation-negative mouse lymphomas with xenoantisera to murine leukemia virus gp70. *J. Immunol.* **125:**1715.

Stern, K., and Willheim, R., 1943, "The Biochemistry of Malignant Tumors," Reference Press, Brooklyn, New York.

Stück, B., Old, L. J., and Boyse, E. A., 1964, Occurrence of soluble antigen in the plasma of mice with virus-induced leukemia *Proc. Natl. Acad. Sci. USA* **52:**950.

Thomson, D. M. P., Steele, K., and Alexander, P., 1973a, The presence of tumour-specific membrane antigen in the serum of rats with chemically induced sarcomata, *Br. J. Cancer* **27:**27.

Thomson, D. M. P., Eccles, S., and Alexander, P., 1973b, Antibodies and soluble tumour-specific antigens in blood and lymph of rates with chemically induced sarcomata, *Br. J. Cancer* **28:**6.

Umiel, T., and Trainin, N., 1974, Immunological enhancement of tumor growth by syngeneic thymus-derived lymphocytes, *Transplantation* **18:**244.

Vaage, J., 1972, Specific desensitization of resistance against a syngeneic methylcholanthrene-induced sarcoma in C3Hf mice, *Cancer Res.* **32:**193.

van Blitterswijk, W. J., Emmelot, P., Hilkmann, H. A. M., Oomenmeulemans, E. P. M., and Inbar, M., 1977, Differences in lipid fluidity among isolated plasma membranes of normal and leukemic lymphocytes and membranes exfoliated from their cell surface, *Biochim. Biophys. Acta* **467:**309.

van Blitterswijk, W. J., Emmelot, P., Hilkmann, H. A. M., Hilgers, J., and Feltkamp, C. A., 1979, Rigid plasma-membrane-derived vesicles, enriched in tumour-associated surface antigens (Mlr), occurring in the ascites fluid of a murine leukaemia (GRSL), *Int. J. Cancer* **23:**62.

van Rood, J. J., Van Leeuwen, A., and Van Santen, M. C. T., 1970, Anti HL-A2 inhibitor in normal human serum, *Nature* (*London*) **226:**366.

Volkers, C., Cooke, B., Bennett, C., Byrom, N., Campbell, M., Elliot, P., and Whitfield, P., 1978, The significance of urinary melanoma antigen excretion and the ability of thymosin to raise the level of depleted lymphocytes in vitro in malignant melanoma, *Aust. N. Z. J. Surg.* **48**:32.

Warren, L., 1969, The biological significance of turnover of the surface membrane of animal cells, *Curr. Top. Dev. Biol.* **4**:197.

Warren, L., and Glick, M. L., 1968, Membranes of animal cells. II. The metabolism and turnover of the surface membrane, *J. Cell Biol.* **37**:729.

Weber, G., and Bablouzian, B., 1966, Construction and performance of a fluorescence polarization spectrophotometer, *J. Biol. Chem.* **241**:2558.

Winzler, R. J., 1953, Plasma proteins in cancer, *Adv. Cancer Res.* **1**:503.

Winzler, R. J., Devor, A. W., Mehl, J. W., and Smyth, I. M., 1948, Studies on the mucoproteins of human plasma. I. Determination and isolation, *J. Clin. Invest.* **27**:609.

Wu, J. T., Madsen, A., and Bray, P. F., 1974, Quantitative measurement and chromatography of human urinary carcinoembryonic antigen activity, *J. Natl. Cancer Inst.* **53**:1589.

Yahara, I., and Edelman, G. M., 1975, Modulation of lymphocyte receptor mobility by locally bound Concanavalin A., *Proc. Natl. Acad. Sci. USA* **72**:1579.

Zeligs, J. D., and Wollman, S. H., 1977, Ultrastructure of blebbing and phagocytosis of blebs by hyperplastic thyroid epithelial cells *in vivo*, *J. Cell Biol.* **72**:584.

Zmijewski, C. M., McCloskey, R. V., and St. Pierre, R. L., 1967, The effect of environmental extremes on the detectability of leukocyte isoantigens in normal humans with inference to tissue typing of prospective donors, *Histocompat. Testing*, 397.

Chapter 3

The Role of Carbohydrates Bound to Proteins

L. Warren, S. R. Baker, D. L. Blithe, and
C. A. Buck

The Wistar Institute
Philadelphia, Pennsylvania

I. INTRODUCTION

In the past decade, there has been a growing interest in the surface membrane of the cell and its glycoproteins. During this time, our laboratory and others have been investigating changes that take place in the carbohydrates of glycoproteins when a cell becomes malignant. Although clear and consistent changes have been found, their significance is not known, because, it is felt, the function of bound carbohydrates is largely unknown. In this article some notions will be discussed on the form and function of bound carbohydrates both in normal and pathological processes that may help us to understand the meaning of the observed changes in malignancy. The argument will be made that protein- and lipid-bound carbohydrates may have played a special part in evolution and, in an unexpected way, may be involved in disease processes.

Central to our considerations of the role of bound carbohydrates is their apparently paradoxical nature. The bound carbohydrates of glycoproteins frequently appear to be dispensible. Whereas some processes involving glycoproteins are dependent on the precise structure of their carbohydrates (see Gottschalk, 1972; Cook and Stoddart, 1973; Rosenberg and Schengrund, 1976; Hughes, 1976; Warren *et al.*, 1978), these macromolecules appear to display an apparent imprecision and a shifting polymorphism that to some extent reflects the environmental state at the time of their synthesis. An hypothesis is presented that resolves contradictions and suggests lines of experimentation. Glycoproteins may endow the cell with a plastic, adaptable character that broadens the range of conditions within which the organism can operate effectively and survive—an environmental buffer. The bound carbohydrates may be a structural element in the cell that operates at the interface of genetic and environmental influences. They may constitute a mechanism established by genetic instruction that operates in response to environmental fluctuation.

II. FRACTIONATION OF GLYCOPEPTIDES

For several years, our laboratory has been comparing the glycopeptides derived by exhaustive digestion of glycoproteins from the surface and interior of control and malignant cells (Warren *et al.*, 1978). In the course of this work, methods have been developed to separate and purify glycopeptides (Buck *et al.*, 1970; Blithe *et al.*, 1980a), all of which come from cells metabolically labeled with [^{14}C]- or D-[^{3}H]-glucosamine or L-fucose. Isolated material is freed from as much polypeptide material as possible by digestion with the nonspecific protease(s), Pronase. The resulting labeled glycopeptides are first separated largely on the basis of size and, to some extent, shape on a column of Sephadex G-50 (Fig. 1).

The first peak in Fig. 1, representing material that is excluded from the column, consists largely (~70%) of glycosaminoglycans (GAG). Some of these are covalently attached to "hybrid" glycoproteins on the surface of the cell (Baker *et al.*, 1980). The peak on the right (tubes >80) is highly variable in size and composition and represents small dialyzable oligosaccharides of only a few sugars. The glucose- and fucose-containing

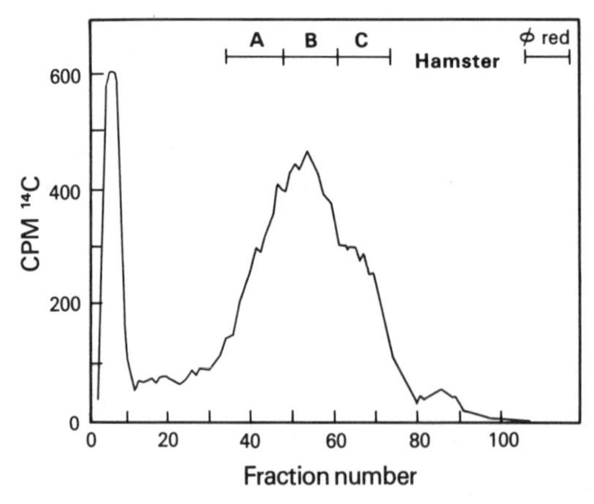

Fig. 1. Sephadex G-50 chromatography of glycopeptides. Pronase-digested, ^{14}C-labeled glycopeptides from the surface of hamster cells (BHK21/C13) were applied to a column of Sephadex G-50 (118 × 3 cm) and eluted with 0.01 M ammonium acetate in 20% ethanol. ΦR, Phenol red. Tubes in the indicated regions A, B, and C were pooled for further analysis (Buck *et al.*, 1970). Approximate molecular weight range: A, 4200–5500; B, 3000–4200; C, 1500–3000.

glycopeptides of Larriba *et al.* (1977) could be a component of this peak. Some glycopeptides are found in relatively small and variable amounts in the pre-A region (tubes 10–35). The carbohydrate structures of this size are especially prominent in embryonic teratocarcinoma cells (Muramatsu *et al.*, 1978), disappearing when the cell differentiates.

The A, B, and C region of the curve in Fig. 1 is quantitatively dominant (tubes 35–75). The group A (mol. wt. 4200–5500), group B (mol. wt. 3000–4200), and group C (mol. wt. 1500–3000) glycopeptides consist of *N*-acetyl-D-glucosamine, D-mannose, D-galactose, L-fucose, and sialic acid; some *N*-acetyl-D-galactosamine is also found (Blithe *et al.*, 1980a). The bulk of these glycopeptides are bound to the polypeptide chain by a relatively alkali-stable linkage to the amide-N of asparagine. Surprisingly, few glycopeptides are bound through the hydroxyl group of serine and/or threonine, unlike the situation in glycophorin of human erythrocytes where 15 out of 16 carbohydrate groups are bound by this linkage (Tomita and Marchesi, 1975).

These glycopeptides are further fractionated on a column of concanavalin A (Con A)-Sepharose into a (+) fraction that binds and is eluted with α-methyl-D-mannoside and into a (−) fraction that does not bind. Each of these fractions is further fractionated on a column of *O*-diethylaminoethyl (DEAE)-Sephadex, on the basis of molecular charge (Fig. 2). The few remaining amino acids of the glycopeptides have essentially no influence on their behavior on DEAE-Sephadex. Analytic work has revealed that the column separates glycopeptides according to the number of residues of sialic acid they bear (0,1,2,3,4 . . .), although some materials eluting at high salt concentrations are small chains of GAG (tubes 270–350 in Figs. 2A–D) (Blithe *et al.*, 1980a). If sialic acid residues are removed from glycopeptides, they do not adhere to the DEAE-Sephadex column under our conditions. By these column procedures, we are able to discern approximately 35 peaks from the trypsinate and/or pellet of vertebrate cells in tissue culture (Blithe *et al.*, 1980a) (Fig. 2). Recent work in progress using chromatography on thin-layer plates of silica gel 60 (Holmes and O'Brien, 1979) indicates that we can detect at least 50 peaks. Despite the great complexity, the glycopeptide patterns are remarkably reproducible.

III. GLYCOPEPTIDES FROM CELLS OF VARIOUS SPECIES

The fractionation techniques have been employed on isotopically labeled material from cells in culture of the human (WI-38), mouse (BALB/C 3T3), hamster (BHK21/C13), chicken embryo, Russel viper,

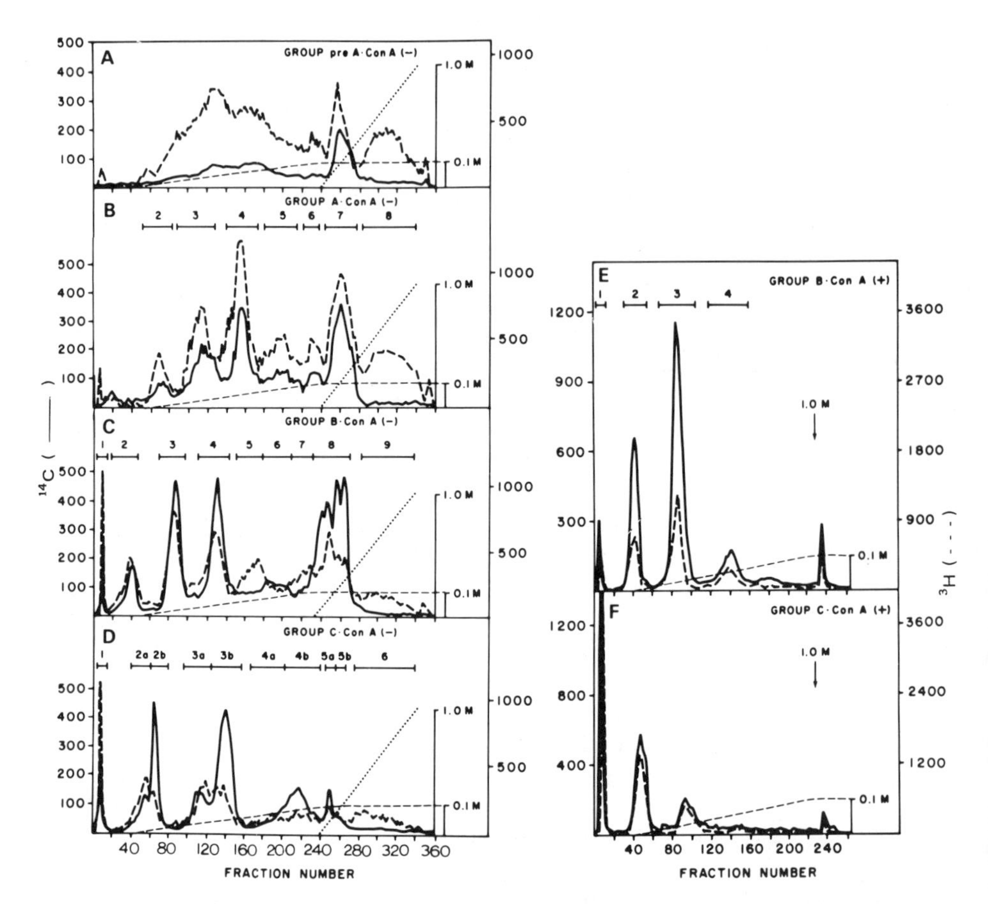

Fig. 2. DEAE-Sephadex chromatography. Glycopeptides from the surfaces of BHK21/C13 cells metabolically labeled with D-[^{14}C]glucosamine (——) and from C13/B4 transformed cells labeled with D-[^{3}H]glucosamine (– – –) were fractionated according to size (and shape) into groups pre-A, A, B, and C on a column of Sephadex G-50 (Fig. 1). Each fraction was applied to a column of Sepharose-Con A and divided into fractions that adhered (+) or did not adhere (−) to Con A. Those fractions with sufficient radioactivity were further fractionated according to charge on columns of DEAE-Sephadex. The glycopeptides were eluted first with a linear salt gradient from 0 to 0.1 M sodium borate in 0.01 M pyridine acetate, pH 4.5, and then with a second linear salt gradient from 0 to 1.0 M sodium acetate in 0.1 M sodium borate and 0.01 M pyridine acetate, pH 4.5. This was followed by a final wash with 1.5 M sodium acetate. (——) Glycopeptides from ^{14}C-labeled control cells; (– – –) glycopeptides from ^{3}H-labeled transformed cells. (– – –) Linear salt gradient based on the molarity of sodium borate; (···) linear salt gradient based on the molarity of sodium acetate (Blithe *et al.*, 1980a).

and the fathead minnow (*Pimephales promelas*). In Figs. 3A, B, and C are seen double-label elution patterns from columns of DEAE-Sephadex. The elution patterns of material from hamster, human (Fig. 3A), reptile (Fig. 3B), and fish (Fig. 3C) are very similar, although some differences can be seen. Glycopeptides metabolically labeled with [^{14}C]- or [^{3}H]-glucosamine were further fractionated on thin-layer plates of silica gel 60 and detected by autoradiography or by cutting and counting. Some peaks yielded as many as three spots. Almost all the glycopeptides detected on thin-layer plates were found in all of the vertebrate species analyzed.

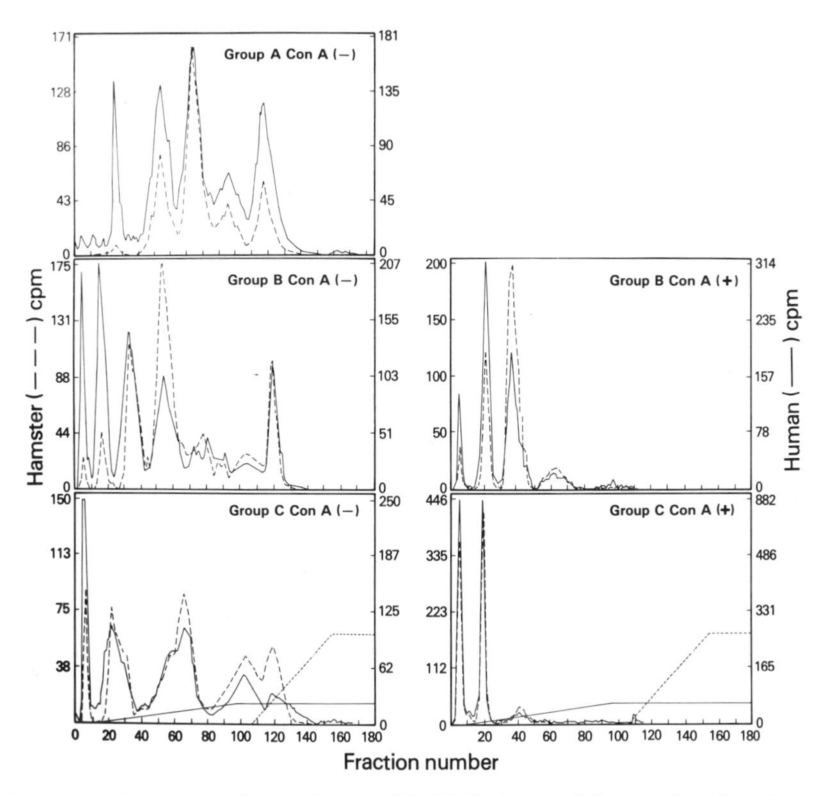

Fig. 3A. Elution patterns from columns of DEAE-Sephadex of glycopeptides from hamster and human. Double-label experiments. Hamster (BHK21/C13) (– – –) vs human (WI-38) (——). BHK21/C13 (hamster) cells were grown in the presence of D-[^{3}H]glucosamine and WI-38 (human) cells were grown in the presence of D-[^{14}C]glucosamine. Glycopeptides obtained by exhaustive digestion of cellular material were mixed and fractionated together on columns of Sephadex G-50 and then Sepharose-Con A. Fractions from the latter columns (+) for adhering material and (−) for material that passed through the column were fractionated on a column of DEAE-Sephadex as seen above.

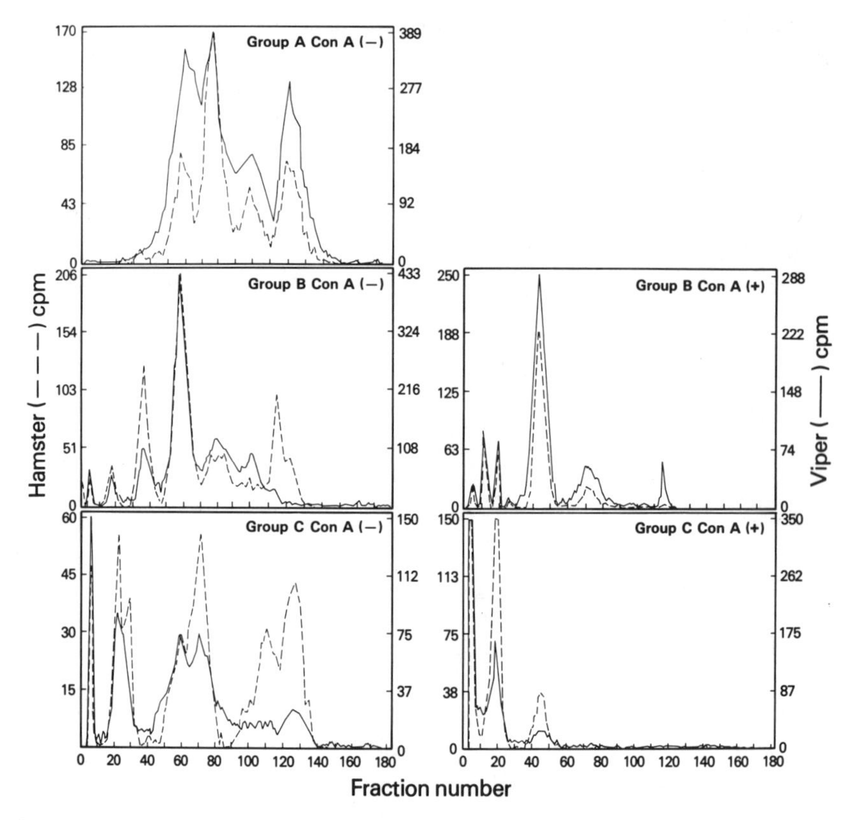

Fig. 3B. Elution patterns from columns of DEAE-Sephadex of glycopeptides from hamster and reptile. Double-label experiments. Hamster(BHK21/C13) (– – –) vs. Russel viper (RS-W) (——).

Future work will tell us whether co-eluting peaks and spots do, in fact, consist of identical material and whether they are homogeneous. At the present time, we can say that co-eluting peaks and spots consist of materials with the same size, affinity for Con A, charge, and mobility on plates of silica gel 60. All of these sialic acid-containing materials came from cells in tissue culture.

In summary, our preliminary data indicate striking similarities in the complex elution patterns of glycopeptides from cells of vertebrate classes labeled in culture or *in vivo*. These complex populations appear to have been stable and conserved over several hundred million years. Carbohydrate groups common to so many vertebrates may explain why, in general, they are of low immunogenicity. It is conceivable that they con-

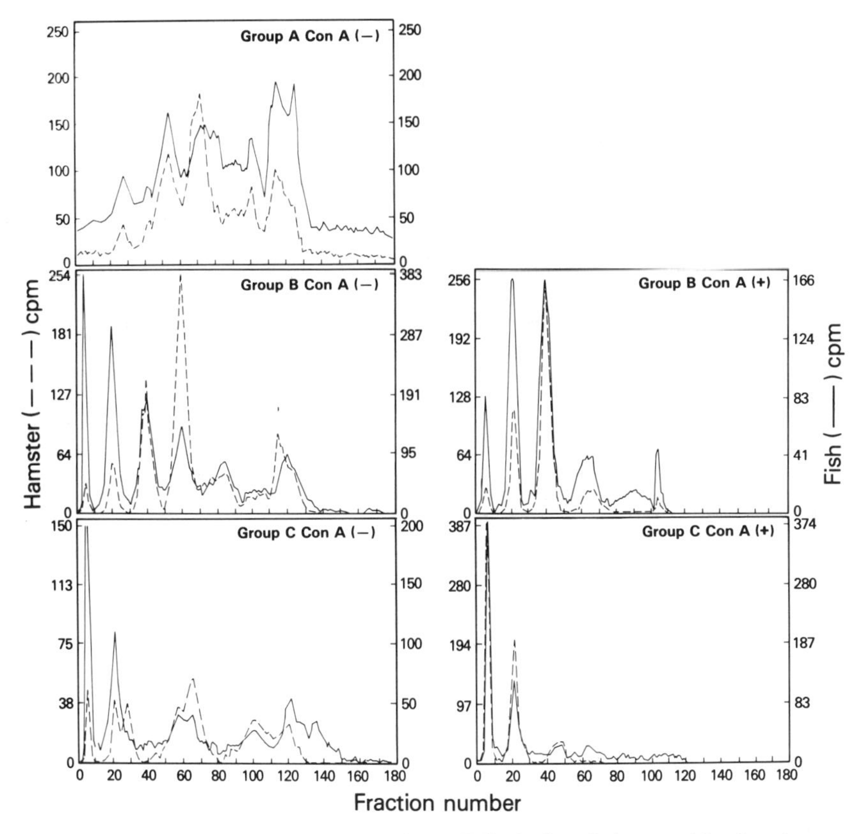

Fig. 3C. Elution patterns from columns of DEAE-Sephadex of glycopeptides from hamster and fish. Double-label experiments. Hamster (BHK21/C13) (– – –) vs. fathead minnow (FHM) (——).

stitute an alphabet capable of forming "words" whose meanings are unknown.

IV. MICROHETEROGENEITY: THE GLYCOPEPTIDES OF "HOMOGENEOUS" GLYCOPROTEINS

Individual membrane glycoproteins of hamster cells labeled with D-[^{14}C]glucosamine have been purified. These yield a single amino-terminal peptide upon fingerprint analysis (Baker *et al.*, 1980). After digestion with Pronase, routine glycopeptide fractionation has revealed at least 12 different glycopeptide populations (Fig. 4) in addition to molecules of GAG covalently attached to some of the glycoproteins. The patterns of glycopeptides derived from six purified glycoproteins, some from the cell surface, were found to be complex but were very similar and reproducible (Baker *et al.*, 1980). The molecular weights of two of these glycoproteins are 110,000 and 160,000 and from preliminary carbohydrate analyses, it has been determined that approximately 10% of their mass is carbohydrate. Thus the molecular weights of the carbohydrate components associated with the glycoproteins should be 11,000–16,000. However, column analysis of glycopeptides derived from these glycoproteins indicate the presence of at least 12 glycopeptides, each of mol. wt. 3000–5000, per glycoprotein. This does not include at least 1 mole-equiv of covalently bound GAG of mol. wt. 1200–6000 and several smaller carbohydrate groups of mol. wt. 800 linked to the hydroxyl groups of serine and/or threonine. It is estimated that the molecular weight of these groups is at least 60,000 which is clearly more than 10% of the total mass of the glycoprotein. The data and calculations suggest that every polypeptide chain of a "homogeneous" glycoprotein bears a variety of differing car-

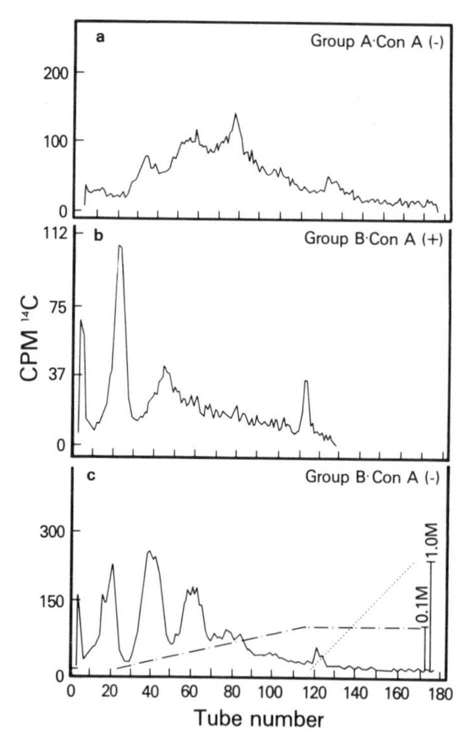

Fig. 4. DEAE-Sephadex chromatography of glycopeptides obtained from a single purified glycoprotein of hamster cells, BHK21/C13 (A1-3) (Baker *et al.*, 1980). The glycopeptides were subdivided on the basis of elution from a column of Sephadex G-50 and binding affinity for Con A into Group A·Con A(−), Group B·Con A(−), and Group C·Con A(+). There was not enough Group A·Con A(+) or Group C material for column chromatography. Each group was subjected to ion-exchange chromatography on DEAE-Sephadex A-25 as described in the legend to Fig. 2.

bohydrate groups since the presence of at least 12 carbohydrate groups has been revealed; yet each glycoprotein molecule probably bears not more than three or four oligosaccharides. It would appear that purification of glycoproteins has been dominated by the properties of the polypeptide chain leading to a product homogeneous with regard to that structure. However, the glycoprotein is heterogeneous with respect to the carbohydrate.

Microheterogeneity of the carbohydrate bound to protein has been known for many years, and the literature on the subject has been reviewed (see Montgomery, 1972; Spiro, 1973; Gallagher and Corfield, 1978). Extensive microheterogeneity of the carbohydrate component of purified glycoproteins is evident by examining the ratios of sugars to one another in glycopeptides purified after exhaustive digestion of the glycoprotein with Pronase. Rarely are the ratios of the various sugars integral, signifying heterogeneity. Fractions are invariably ignored, and numbers are usually rounded off to the nearest integer. If our proposal is correct, ignoring microheterogeneity in order to establish the structure of carbohydrate polymers not only reduces the validity of the work but, more importantly, may be conceptually in error when considering function. The special role of bound carbohydrates may lie in the array of structures on a polypeptide which differ only slightly from one another. To regard heterogeneity merely as an annoyance may be a repetition of an attitude held during early work on immunoglobulins where investigators dismissed variability because they had trouble enough establishing *the* structure of the molecule.

One criterion of heterogeneity is that there be more carbohydrate groups identified then there are sites of glycosylation on a polypeptide. This is certainly the case with ovalbumin where several discrete glycopeptides have been identified and only one glycosylation site is found (Cunningham *et al.*, 1957; Makimo and Yamashina, 1966; Cunningham and Ford, 1968; Marshall and Neuberger 1972). The same structures are found in material from one egg or in eggs from six different variants of chicken (Lush, 1961). Because of the constancy in amount of the various carbohydrate structures and the accounting for all carbohydrates during isolation, it is unlikely that they arose by random degradation or accidental, incomplete synthesis (see Montgomery, 1972).

There are numerous, documented examples of heterogeneity in the carbohydrate components of bovine and porcine ribonucleases, some forms of which are free of carbohydrates (see Plummer and Hirs, 1963; 1964; Jackson and Hirs, 1970a,b; Fukuda and Egami, 1971; Tarentino *et al.*, 1970; Beintema *et al.*, 1976). At least eight forms of ribonuclease can be identified in porcine pancreatic secretions, all different in their bound carbohydrate. In this instance, variations in carbohydrate structure attached to a polypeptide of only 125 amino acids (mol. wt. 14,040) is sufficient to effect separation of each form upon electrophoresis. Clear instances of heterogeneity are also found in thyroglobulin (Spiro, 1965; Arima *et al.*, 1972), thyroxine-binding globulin (Zinn *et al.*, 1978a,b), α-2 macroglobulin of human plasma (Dunn and Spiro, 1967), bronchial glycoproteins and mucins (Roussel *et al.*, 1975; Gallagher and Corfield, 1978), myeloma globulin (Dawson and Clamp, 1968), glycoprotein of vesicular stomatitis virus (Robertson *et al.*, 1976), and others. Both the anion (Yu and Steck, 1975) and sugar transporter (Gorga *et al.*, 1979) of human erythrocytes are glycoproteins that are heterogeneous in their sugar component. Probably, if one looks carefully enough, one can find heterogeneity in virtually all glycoproteins and much of this heterogeneity must go beyond variation in the number of terminal sialic acid and fucose residues. The application of nuclear magnetic resonance to the detection and study of microheterogeneity should prove invaluable (see Atkinson and Hakimi, 1980).

Heterogeneity of bound carbohydrates in the face of complex, conserved populations of carbohydrate oligomers signify that when asparagine, serine, or threonine residues of nascent polypeptide chains are glycosylated they may end up with a small number of oligomers (perhaps 1–5) from a total library of recurring carbohydrate groups (perhaps 50–100) that the cell is capable of providing. Certain sites may or may not be glycosylated. Our work suggests that the library of "completed" oligosaccharide forms is more or less common to cells of four classes of sialic acid-synthesizing vertebrates (Warren, 1963). We do not imply that all of the carbohydrate groups can be found in all glycoproteins. Clearly there is some control of the groups that different polypeptides bear (Rosner *et al.*, 1980), and various polypeptides may manifest different degrees of heterogeneity of their carbohydrate. There is also no known reason to exclude the existence of glycoproteins that are completely homogeneous in their carbohydrate structure.

V. ALTERATIONS IN THE PROTEIN-BOUND CARBOHYDRATES

Shifts in populations of carbohydrate groups of glycoproteins appear to be well tolerated. Much work has been done on changes in malignancy of the carbohydrate groups of glycoproteins (Meezan *et al.*, 1969; Buck *et al.*, 1970; Van Beek *et al.*, 1973, 1975; Emmelot, 1973; Glick, 1974; Glick *et al.*, 1973, 1974; Ogata *et al.*, 1976; Warren *et al.*, 1978). These groups are altered in over 80% of glycoproteins (Tuszynski *et al.*, 1979; Baker *et al.*, 1980) in every membrane system of the cell (Buck *et al.*, 1974; Van Nest and Grimes, 1977). In general, larger carbohydrate groups replace smaller groups on polypeptide chains in the malignant cell.

Further support for this type of change has been provided by Burridge (1976) who examined glycoproteins of control and transformed cells that had been separated by polyacrylamide gel electrophoresis. ^{125}I-Labeled lectins revealed that homologous bands of control and malignant cells often stained differently, and this was presumably due to different carbohydrate compositions of individual glycoproteins.

Shifts also occur with the growth state of the cell (Buck *et al.*, 1971; Glick and Buck, 1973; Ceccarini *et al.*, 1975; Muramatsu *et al.*, 1973, 1976; Gahmberg and Hakomori, 1974). Growth-dependent changes of glycolipids have also been observed (Robbins and Macpherson 1971; Gahmberg and Hakomori, 1974; Hirschberg *et al.*, 1974).

In Tables I and II are listed a variety of conditions and agents that appear to induce changes in the protein-bound carbohydrates. It is possible in some of these examples that the amounts of whole glycoprotein rather than the carbohydrate components alone are altered. The results of analysis of the carbohydrate components of a basement membrane glycoprotein produced by murine carcinoma cells growing in different media are presented in Table III. Clearly the bound carbohydrates of the glycoprotein differ when the cell is growing in different environments (Megaw and Johnson, 1979). In preliminary experiments using chromatography on columns of Sephadex G-50, we have found that the elution pattern of glycopeptides from hamster cells growing on glass surfaces is different from that derived from cells growing on plastic surfaces. We have also found differences in glycopeptides from fish cells in culture growing at 18 and 33°C.

The sialic acid content of epithelial tissue of the Prussian carp (*Carassius auratus*) growing in fresh water has been shown to be higher than from the same fish in seawater. This also holds for other sea- and freshwater fish (Hentschel and Müller, 1979). Elevated sialic acid content could be due to increased synthesis of sialic acid-rich glycoproteins rather than increased amounts of sialic acid per polypeptide.

The well-known microheterogeneity of the carbohydrate of serum proteins produced by the liver may not only be due to inherent mechanisms within the cell but also to the fact that these serum glycoproteins are the pooled products of hepatocytes operating close to or distant from an arterial supply, in a fasting or fed, resting, exercising, or sleeping individual. Each state produces a different metabolic situation in which, perhaps, the various sugars are produced, activated, and transferred to

Table I
Change in Elution Pattern of Glycopeptides from Sephadex G-50

System	Species	Agent or condition	Change (see Fig. 1)	Reference
AD6	Mouse	Mutant	Decreased Groups A and C	Blithe *et al.* (1980b)
PCC4; embryo	Mouse	Spontaneous differentiation	Pre-A group decreases sharply upon differentiation	Muramatsu *et al.* (1978; 1980)
Transformed and malignant	Many	Spontaneous, virus, carcinogens	Increase of Group A	Buck *et al.* (1970); Van Beek *et al.* (1973)
BHK21/C13	Hamster	pH 6.5 vs 7.2	Decrease of Groups A and B	Unpublished
BHK21/C13 and C13/B4	Hamster	Ethidium bromide	Marked increase in Group A	Soslau *et al.* (1974)
BHK21/C13	Hamster	2-Deoxyglucose	Decrease of Groups A and B	Unpublished
CHO	Hamster	cAMP	Decrease of Group A	Van Veen *et al.* (1976)
HeLa	Human	Na-butyrate	Decrease of Group A	Simmons *et al.* (1975); Via *et al.* (1980)
BHK21/C13	Hamster	Tunicamycin	Elution peak and Group D present; groups A, B, and C eliminated	Damsky *et al.* (1979)
Several cell lines	Several	Growth	Altered core structure; increase of Group A	Buck *et al.* (1971); Ceccarini *et al.* (1975)
3T12	Mouse	Retinoic acid deficiency	Increase of Group A	Sasak *et al.* (1980)

Table II
Changes in Carbohydrates Bound to Protein

System	Condition	Change	Reference
Rat (whole animal)	Retinoic acid deficiency	Change of carbohydrate in α_1 macroglobulin	Wolfe *et al.* (1979)
Murine carcinoma (cell culture)	Varying concentration of serum	Change of carbohydrate in a glycoprotein	Megaw and Johnson (1979)
Murine carcinoma (cell culture)	Buffers varied	Change of carbohydrate in a glycoprotein	Megaw and Johnson (1979)

the polypeptide leading to variation in the product. However, the overall balance of products formed in all these states may be quite constant under normal conditions.

VI. GLYCOLIPIDS

A large number of glycolipids have been isolated from various forms of life and have been precisely characterized (Hakamori, 1975). Gangliosides, cerebrosides, and other glycolipids have carbohydrate structure covalently bound to lipid that probably form stable, noncovalent associations with proteins and glycoproteins in membranes. A polypeptide associated with a glycolipid may in effect be the equivalent of a glycoprotein with the added feature that the location of the carbohydrate can be varied; the glycolipid can, perhaps, change its conformation or shift to several positions and thus prove a versatile agent. The role of lipid-bound carbohydrate could be essentially the same as that proposed for protein-bound carbohydrate.

VII. THE FUNCTION OF BOUND CARBOHYDRATE

Aside from certain involvements of protein-bound carbohydrate in specific processes to be discussed, the majority of studies reveals small or no changes in activity when the carbohydrates of functional glycoproteins are altered. This could mean that the carbohydrates function at a time earlier in cellular processes than that being studied or that they are of little importance and consequently silent. However, it is difficult to believe that a major, ubiquitous component of the cell covalently bound

Table III
Composition of Epithelial Basement Membrane Glycoprotein(s) of Murine Carcinoma Cells in Culture[a]

Sugar	μg sugar/mg protein				
	Medium; 10% FBS[b]	Medium; 1% FBS		Chemically defined medium; FBS, buffer $NaHCO_3$ + Hepes	Chemically defined medium; FBS, buffer $NaHCO_3$
		Days 1–15	Days 16–39		
L-Fucose	1.6	1.3	1.2	1.4	1.1
D-Mannose	18.0	12.2	1.1	3.1	0.7
D-Galactose	20.6	11.9	3.7	7.1	0
D-Glucose	10.0	4.0	2.7	10.1	0
N-Acetyl-D-galactose	41.5	3.7	3.4	27.7	0
N-Acetyl-D-glucosamine	40.3	16.4	3.9	22.7	1.8
Sialic acid	3.4	0.9	0.7	2.2	0
Total	135.6	50.4	16.7	74.3	3.6

[a] EBM glycoprotein isolated from medium and analyzed (mol. wt. 28,000–30,000). All glycoproteins were antigenically identical, and amino acid compositions were also the same. Data from Megaw and Johnson (1979).
[b] FBS, Fetal bovine serum; Hepes, *N*-2-hydroxyethylpiperazine-*N*′-2-ethanesulfonic acid.

to many proteins that arose early in living forms (Aberhalden and Heyns, 1933) and that appears in recurring, conserved combinations in a variety of classes of animals has only a small role to play.

We speculate that the well-documented, critical involvement of glycoprotein carbohydrates in certain, specific processes may have arisen during evolution when carbohydrates became structural components of mechanisms with direct participation in the processes. Removal or alterations of part or all of the oligosaccharide leads to a dramatic change in activity. The homing of glycoproteins to the liver after the removal of their sialic acid residues is a well-known example (Pricer and Ashwell, 1971; Ashwell and Morell, 1974). Dramatic changes in binding and activity of hormones can occur upon removal of bound sugars (see Hughes, 1976; Rosenberg and Schengrund, 1976). Bound carbohydrates can eliminate or decrease susceptibility of polypeptide to degradation by proteolytic enzymes (Olden *et al.*, 1979; Wang and Hirs, 1977) and can be important in the processing of viral components (Leavitt *et al.* 1977a). From a physiocochemical point of view, it is known that the presence of sugars bound to polypeptides affect their associations with other molecules, their solubility (Tarentino *et al.*, 1970; Leavitt *et al.*, 1977b; Cheng *et al.*, 1979), precipitibility (Gibson *et al.*, 1979), and viscosity in aqueous solutions and they can impart an antifreeze function to the molecules (Kömatsu *et al.*, 1970). The region of a polypeptide chain in the vicinity of associated carbohydrate will not sink into the lipid phase of membrane, just as floats of a fishing net keep the line that is near the float above water, the rest of the rope sinks. In other words, bound carbohydrates may play a critical role in the orientation of glycoproteins in membranes. They appear to be situated in an exposed position at the surface of the molecule in the region of β turns enriched in proline in the polypeptide chains (see Kornfeld and Kornfeld, 1980; Smith and Pease, 1980). For attachment of carbohydrates to the amide-N of asparagine, the sequence asparagine-X-serine or threonine in the polypeptide is necessary but not sufficient (Marshall, 1972). The carbohydrate polymers are hydrated structures that readily form hydrogen bonds and usually bear one or more highly exposed, charged sialic acid residues. In globular proteins, they protrude into the aqueous phase at the surface of a polypeptide component that is folded in on itself and is stabilized by hydrophobic bands. The group A, B, and C glycopeptides, usually the dominant groups of mol. wt. 1500–5000, seem to be ones of "optimal" size. However, shape and charge must also be important, especially in interactions between macromolecules.

The list of specific, dramatic consequences that follows from tampering with bound carbohydrates is by no means complete. Yet it seems that for every example of this type, there are many more tests, usually unpublished, where removal of sugars causes no more than minor changes of function. Disappointment with so many apparently negative results and the intuitive, subjective judgment that these oligomers must "do more" than what experimentation has revealed has induced us to re-evaluate their possible functioning and to establish a second category of "small changes" in function. The remaining highly speculative part of the chapter has evolved from this position.

In many instances where alteration of bound carbohydrate does not appear to result in changed activity, activity may in fact be altered but only to a small extent, e.g. the removal of carbohydrate from thyroglobulin binding protein (Cheng *et al.*, 1979). Few investigators deem as significant 5 or 20% changes or are willing to pursue or discuss them. It is possible that the bound carbohydrates have nothing to do with the particular function being assayed but might be involved in other processes in which the glycoprotein takes part. In many instances the assays used are not of sufficient precision or reliability to speak with conviction about small effects. Usually function is tested under one set of conditions (optimum), and one could argue that larger, significant differences might be found away from the optimum. We postulate that the modifying activity is due to the influence of the carbohydrate group on the mobility or conformation of the polypeptide chain. In contrast to this indirect effect, the extensive consequences of the first category discussed could result from carbohydrates participating directly in the mechanism being tested. Carbohydrate groups participating in activities of the second type are in effect multiple, covalently bound allosteric effectors where each site on the polypeptide chain may bear no group or a variety of carbohydrate oligomers in accordance with the known microheterogeneity. In this category, they are modifiers rather than critical effectors.

VIII. A POSSIBLE ROLE FOR BOUND CARBOHYDRATE GROUPS IN NATURE

Evidence has been provided that protein-bound carbohydrates exist as a complex array of structures that have been stable in vertebrates over a very long period of time. These characteristics leave little doubt that they are important. On the other hand, a firm belief in the critical participation of bound carbohydrates in cellular function is eroded by their shifting, nonunique character. It is difficult to ascribe an important, specific function to a molecule that can have many forms, that can vary with environmental conditions, and, usually, is not dramatically altered in activity upon modification of its structure. The following hypothesis may resolve the dilemna. Regardless of the activity of a glycoprotein (e.g., hormone, enzyme, receptor site, structural agent, clotting factor), it is not one form of a specific molecule that performs the function. The glycoprotein is polymorphic with a homogeneous polypeptide that is usually the effector bearing different combinations of a variety of groups each of which, we propose, modifies activity slightly. Whatever the environment in which the glycoproteins act, there are present polypeptides bearing combinations of carbohydrate groups that operate "most advantageously." The array of molecules provides a preadaptive advantage to the organism and a structural variation that could be the foundation of natural selection. The small differences in activity between the various forms may not look important to the experimental scientist (who insists on severalfold differences in effect to attract attention) but within an evolutionary time frame they suffice.

Several consequences follow from this notion:

1. If bound carbohydrates are agents that serve as environmental buffers enabling organisms to cope with a wider range of conditions, then they should be found where the cell has least control of the environment in which the polypeptide must operate. This is, in fact, the case. Sixty to seventy percent of glycoproteins of the cell are found in the cell surface, with the carbohydrate outside the lipid layer facing the intercellular world (see Nicolson and Singer, 1974; Steck, 1974; Warren, 1976; Rothman and Lenard, 1977). The cell probably has the most control in the nuclear compartment and this is where there is the least bound sugar. On the other hand, there appears to be an abundance of glycoprotein (and GAG) in purely extracellular locations (see Gottschalk 1972; Gallagher and Corfield, 1978). Few of the cellular, metabolic enzymes appear to be glycoproteins. However, lysosomal and other enzymes, which act either extracellularly or in compartments of internalized bits of the extracellular world of variable nature, are richly endowed with carbohydrates (Goldstone and Koenig, 1970; Neufeld *et al.*, 1975). On the other hand, while serum proteins and enzymes are glycoproteins, serum albumin found in abundance is not. This glaring exception to our generalization prompts us to ask what special function or property of this polypeptide chain permits the organism to dispense with glycosylation? Does the polypeptide chain of albumin have built into it a ratchet-like property that bound carbohydrates might be able to impart? For instance, is it possible that the albumin polypeptide inherently can exist in several discrete quasi-stable conformations that can interconvert depending on environmental conditions?

Consonant with the notion of a gradient (extracellular → nuclear) is the possibility that cells normally exposed to environments capable of rapid changes such as epithelial and endothelial cells will have a special, enhanced capability to synthesize glycoprotein of variable sugar composition. They might be contrasted in this regard with muscle or brain cells. Perhaps it is the stability and nature of the environment of a cell that to a large extent determines the type of sugar-containing macromolecules that are synthesized (e.g., globular glycoproteins, large mucin-like molecules, GAG).

2. If different carbohydrate groups or combinations of them modulate the activity of the polypeptide, and glycoproteins are extensively polymorphic in these structures, then a series of structures should, within limits, produce an array of activities, i.e. activities and their constants ascribable to glycoproteins should be described by Poisson distribution curves that can change in shape to some extent with environmental conditions.

3. If a glycoprotein bearing a set of carbohydrates has, in fact, a slightly different activity than the homologous glycoprotein with different carbohydrate groups or if alterations of the carbohydrate groups changes activity to a small extent, then the element of time becomes a crucial parameter in magnifying the consequences such as in evolution and perhaps also in long-term, chronic diseases and aging.

4. We have proposed that glycoproteins are heterogeneous in their carbohydrate and this may help the organism to cope with its environment. We have also provided some evidence that the array of carbohydrate groups can be induced to change and that this happens during the life of the cell. It follows that structural changes induced by a change in environment or a drug could assist the cell in more efficiently coping with the new environment. This constitutes a homeostatic mechanism, an adaptive response that is constantly in operation.

5. Based upon this reasoning, we can divide the polymeric constituents of the organism into two fundamental groups—those formed in the presence or in the absence of a template. DNA, RNA, and protein are formed on a template, relatively error and variation free. In some instances, there is a limited but controlled array of forms in a set of isozymes or their equivalents. The forms of these finely controlled molecules closely follow genetic dictates.

In the second group are placed the carbohydrate polymers and lipid membranes. The membrane is considered to be a polymer of lipid molecules bound by noncovalent linkages. It has been shown to change in composition and microviscosity depending on the state of growth of the cell or whether they are normal or malignant (Cheng and Levy, 1979).

Glycolipids could be of special importance because, through them, environment may have a double impact on cellular structure and function in the synthesis of their polymeric carbohydrate component and in the formation of the polymeric membrane.

The protein-bound carbohydrate groups under consideration are not made on a template and their modification of polypeptides is also not directly template determined. What has been said about the possible function of glycosylation holds for any post-translational (or post-template synthesis) modification, such as phosphorylation, acetylation, or methylation. All of these generate polymorphism of cellular components. In summary, the products of polymerization of carbohydrates or lipids are not identical but are polymorphic; their synthesis is influenced by the environment and the availability of substrates, and the ultimate product can vary significantly in composition within the time frame of cell division and turnover constants. Going one step further, it is postulated that the changed array of molecules can be beneficial in that it helps the animal to cope in the new environment.

6. It may be of use to arrange glycoproteins in a continuous series. At one end are those involved with "small effects" and environmental buffering. These would be associated with enhanced microheterogeneity. At the other end of the spectrum would be glycoproteins whose activity is radically altered by removal of one or more sugar residues. The carbohydrate of these macromolecules might be directly involved in the process and would be homogeneous in structure. This might apply to only one carbohydrate group of many in a glycoprotein.

The relationship of function to structure can also be classified in other ways. Does it take the removal of one, two, or three sugar residues to seriously impair the activity of a glycoprotein? Are one, two, or three carbohydrate antennas needed for an activity? Does the removal of one or more sugar residues cause the appearance of new activities? Can a carbohydrate group carry out a function essentially unrelated to the polypeptide chain? In this instance, the polypeptide would be a relatively inert rack bearing functionally active carbohydrate effectors.

IX. GLYCOPROTEINS IN PATHOLOGY

Our interest in the glycoproteins began with comparative studies of bound carbohydrates of normal and malignant cells. Although clear differences have been detected in numerous malignancies, the significance of the finding is unknown. Consistent with our speculations, we suggest that the changes observed in bound carbohydrates of over 80% of the glycoproteins of malignant cells may produce a scattering or cascade of functional changes that are tolerated by the malignant cell. One or more of these functional changes may lead to persistent cell division, the hallmark of malignancy. Others might lead to a more advantageous transport of metabolites, decreased adhesiveness, or other changes that assist the malignant cell to spread and to thrive.

In the previous discussion, we have stressed the small and nonacute consequences frequently following changes in bound carbohydrate. Since organisms can live with change over long periods of time the possibility exists that small, chronic changes in the carbohydrate groups of glycoproteins can result from altered environments and biosynthetic capabilities induced by arteriosclerosis, degenerative diseases, aging, or by metabolic diseases such as diabetes. These changes might be silent and unrecognized but ultimately after many years, lead to breakdown. However, even in acute disease, the glycoproteins could change. An inflammatory focus clearly creates a new environment for surrounding cells. Do these cells produce altered glycoproteins and, if so, what are the consequences of their presence? Experiments to test these possibilities are now in progress.

X. CONCLUSIONS

We have speculated at length as to what glycoproteins do and why they exist, and we will attempt now to evaluate the arguments. These are based upon (1) the belief in the pervasiveness of microheterogeneity of glycoproteins; (2) the ability of the bound carbohydrates to change with environmental and other conditions; (3) the existence of a category of carbohydrate effects that are relatively small but significant; (4) the proposals that polypeptides with different sets of carbohydrate groups have slightly different activities; and (5) the activities of specific molecules can vary with environmental conditions.

From this, it follows that organisms are armed with an array of glycoprotein effectors that enhance their ability to cope with environmental change. On the other hand, environmental conditions have some control of the final form of the bound carbohydrates. This shifting polymorphism is proposed as the basis for a homeostatic mechanism, an adpative response which may have been the original function of bound carbohydrates. The functions that are acutely dependent on the integrity of protein-bound carbohydrates cannot be predicted and may have developed later in evolution when carbohydrate groups that happened to be present became, by chance, components of specific mechanisms.

What has been ascribed to carbohydrate polymers holds for any nontemplate determined process of polymer synthesis or embellishment such as any post-translational modification of protein or the formation of lipid membranes. The special significance of bound carbohydrate may be that they operate at the interface of genetic and environmental influences. Within certain genetically determined limits, environmental forces may be capable of altering structure at the molecular level and this process may be ongoing.

There is good experimental support for some parts of this story, such as the existence of extensive microheterogeneity and the change of carbohydrate structure with environmental conditions. The proposed small functional changes with change in protein-bound carbohydrates require more substantiation. The most speculative part of the argument resides in items 4 and 5 above, but these are testable. It is, in fact, possible that small or no changes in function with modification of carbohydrate means just that. In these negative experiments, where removal of sugars have no apparent effect, insufficient amounts or types of sugars may have been removed. It is very likely that the bound carbohydrates have functions yet to be discovered; it is doubtful whether the functions known at the present time constitute the bulk of their activities. Intuitively, we feel that the bound carbohydrates with their peculiar distribution are too ancient, too ubiquitous, and too controlled in their complex, polymorphic structure to justify their existence only as agents for the functions known at the present time. However, the intuition of others may lead to different conclusions. Although it is difficult to clearly demonstrate the validity of the ideas outlined in this lecture they do suggest many experiments and relevant data can be gathered without difficulty.

ACKNOWLEDGMENTS. The work described in this paper was carried out with the support of USPHS Grants CA-10815, CA-19144, and CA-19130 and ACS Grant BC-275.

We wish to thank Dr. F. Clark and Ms. N. Parks of the Wistar Institute for cells of *Pimephales promelas* and Russel viper. The notions presented here have been presented in a briefer form in *Clinical Biochemistry* **13** (5), 191–197, 1980, by L. Warren and C. A. Buck, and L. Warren, "Fundamental Mechanism in Human Cancer Immunology," Chapter 7; University of Texas Press, Austin, 1980.

XI. REFERENCES

Aberhalden, E., and Heyns, K., 1933, Demonstration of chitin in the wing remains of coleoptera of the upper middle Eocene, *Biochem. Z.* **259:**320.

Armia, T., Spiro, M. J., and Spiro, R. G., 1972, Studies on the carbohydrate units of thyroglobulin. Evaluation of their microheterogeneity in the human and calf proteins, *J. Biol. Chem.* **247:**1825.

Ashwell, G., and Morell, A. G., 1974, The role of surface carbohydrates in the hepatic recognition and transport of circulating glycoproteins, *in* Advances in Enzymology. (A. Meister and J. Wiley, eds.), Vol. 41, Academic Press, NY, p. 99.

Atkinson, P. H., and Hakimi, J., 1980, Alterations in glycoproteins of the cell surface, *in* "The Biochemistry of Glycoproteins and Proteoglycans " (W. J. Lennarz, ed.), p. 209, Plenum Press, NY.

Baker, S. R., Blithe, D. L., Buck, C. A., and Warren, L., 1980, Glycosaminoglycans and other carbohydrate groups bound to proteins of control and transformed cells, *J. Biol. Chem.* **255:**8719.

Beintema, J. J., Gastra, W., Scheffer, A. J., and Wetting, G. W., 1976, Carbohydrate in pancreatic ribonucleases, *Eur. J. Biochem.* **63:**441.

Blithe, D. L., Buck, C. A., and Warren, L., 1980a, Comparison of glycopeptides from control and virus-transformed baby hamster kidney fibroblasts, *Biochemistry* **19:**3386.

Blithe, D. L., Pastan, I., Buck, C. A., and Warren, L., 1980b, Carbohydrate groups of glycoproteins in a glycosylation-defective, low-adherent mutant mouse cell, *Biochem. Int.* **1:**71.

Buck, C. A., Glick, M. C., and Warren, L., 1970, Comparative study of glycoproteins from the surface of control and Rous sarcoma virus-transformed hamster cells, *Biochemistry* **9:**4567.

Buck, C. A., Glick, M. C., and Warren, L., 1971, Effect of growth on the glycoproteins from the surface of control and Rous sarcoma virus-transformed hamster cells, *Biochemistry* **10:**2176.

Buck, C. A., Fuhrer, J. P., Soslau, G., and Warren, L., 1974, Membrane glycopeptides from subcellular fractions of control and virus-transformed cells, *J. Biol. Chem.* **249:**1541.

Burridge, K., 1976, Changes in cellular glycoproteins after transformation: Identification of specific glycoproteins and antigens in sodium dodecyl sulfate gels, *Proc. Natl. Acad. Sci. USA* **73:**4457.

Ceccarini, C., Muramatsu, T., Tsang, J., and Atkinson, P. H., 1975, Growth-dependent alterations in oligomannosyl cores of glycopeptides, *Proc. Natl. Acad. Sci. USA* **72:**3139.

Cheng, S., and Levy, D., 1979, The effect of cell proliferation on the lipid composition and fluidity of hepatocyte plasma membranes, *Arch. Biochem. Biophys.* **196:**424.

Cheng. S., Morrone, S., and Robbins, J., 1979, Effect of deglycosylation on the binding and immunoreactivity of human thyroxine-binding globulin, *J. Biol. Chem.* **254:**8830.

Cook, G. M. W., and Stoddart, R. W., 1973, "Surface Carbohydrates of the Eukaryotic Cell," Academic Press, New York.

Cunningham, L. W., and Ford, J. D., 1968, A comparison of glycopeptides derived from soluble and insoluble collagens, *J. Biol. Chem.* **243:**2390.

Cunningham, L. W., Neunke, B. J., and Neunke, R. H., 1957, Preparation of glycopeptides from ovalbumin, *Biochim. Biophys. Acta* **26:**660.

Damsky, C. H., Levy-Benshimol, A., Buck, C. A., and Warren, L., 1979, Effect of tuni-

camycin on the synthesis intracellular transport and shedding of membrane glycoproteins in BHK cells, *Exp. Cell Res.* **119:**1.
Dawson, G., and Clamp, J. R., 1968, Investigations on the oligosaccharide units of an A myeloma globulin, *Biochem. J.* **107**;341.
Dunn, J. T., and Spiro, R. G., 1967, The α_2-macroglobulin of human plasma, *J. Biol. Chem.* **242:**5556.
Emmelot, P., 1973, Biochemical properties of normal and neoplastic cell surfaces; a review, *Eur. J. Cancer* **9:**319.
Fukuda, M., and Egami, F., 1971, The Structure of a glycopeptide purified from porcine thyroglobulin, *Biochem. J.* **123:**415.
Gahmberg, C. G., and Hakomori, S., 1974, Organization of glycolipids and glycoproteins in surface membranes: Dependency on cell cycle and on transformation, *Biochem. Biophys. Res. Commun.* **59:**283.
Gallagher, J. T., and Corfield, A. P., 1978, Mucin-type glycoproteins. New perspective on their structure and synthesis, **3:**38.
Gibson, R., Schlesinger, S., and Kornfeld, S., 1979, The nonglycosylated glycoproteins of VSV is temperature-sensitive and undergoes intracellular aggregation at elevated temperatures, *J. Biol. Chem.* **254:**3600.
Glick, M. C., 1974, Chemical components of surface membranes related to biological properties, *Miami Winter Symposium* **7:**213.
Glick, M. C., and Buck, C. A., 1973, Glycoproteins from the surface of metaphase cells, *Biochemistry* **12:**85.
Glick, M. C., Rabinowitz, Z., and Sachs, L., 1973, Surface membrane glycopeptides correlated with tumorigenesis, *Biochemistry* **12:**4864.
Glick, M. C., Rabinowitz, Z., and Sachs, L., 1974, Surface membrane glycopeptides which coincide with virus transformation and tumorigenesis, *J. Virol.* **13:**967.
Goldstone, A., and Koenig, H., 1970, Lysosomal hydrolases as glycoproteins, Part II, *Life Sci.* **9:**1341.
Gorga, F. R., Baldwin, S. A., and Lienhard, G. E., 1979, The monosaccharide transporter from human erythrocytes is heterogeneously glycosylated, *Biochem. Biophys. Res. Commun.* **91:**955.
Gottschalk, A., 1972, "Glycoproteins, Their Composition, Structure and Function," Elsevier, Amsterdam.
Hakomori, S., 1975, Structures and organization of cell surface glycolipids dependency on cell growth and malignant transformation, *Biochim. Biophys. Acta* **417:**55.
Hentschel, H., and Müller, M., 1979, Sialic acids in epithelial tissues of *Carassius auratus gibelio* (Bloch) (prussian carp), *Cottus gobio* L. (bullhead) and *Myoxocephalus scorpius* (1.) (bull rout), *Comp. Biochem. Physiol.* 64A:585.
Hirschberg, C. B., Wolf, B. H. and Robbins, P. W., 1974, Synthesis of glycolipids and phospholipids in hamster cells: Dependence on cell density and the cell cycle, *J. Cell. Physiol.* **85:**31.
Holmes, E. W., and O'Brien, J. S., 1979, Separation of glycoprotein-derived oligosaccharides by thin-layer chromatography, *Anal. Biochem.* **93:**167.
Hughes, R. C., 1976, "Membrane Glycoproteins," Butterworths, London.
Jackson, R. L., and Hirs, C. H. W., 1970a, The primary structure of porcine pancreatic ribonuclease. I. The distribution and sites of carbohydrate attachment, *J. Biol. Chem.* **245:**624.
Jackson, R. L., and Hirs, C. H. W., 1970b, The primary structure of porcine pancreatic ribonuclease. II. The amino acid sequence of the reduced S-aminoethylated protein. *J. Biol. Chem.* **245:**637.

Kömatsu, S. K., DeVries, A. L., and Feeney, R. E., 1970, Studies of the structure of freezing point-depressing glycoproteins from an Antarctic fish, *J. Biol. Chem.* **245:**2909.

Kornfeld, R., and Kornfeld, S., 1980, Structure of glycoproteins and their oligosaccharide units, *in* "The Biochemistry of Glycoproteins and Proteoglycans" (W. J. Lennarz, ed.), Plenum Press, New York.

Larriba, G., Klinger, M., Sramek, S., and Steiner, S., 1977, Novel fucose-containing components of rat tissues, *Biochem. Biophys. Res. Commun.* **77:**79.

Leavitt, R., Schlesinger, S., and Kornfeld, S., 1977a, Tunicamycin inhibits glycosylation and multiplication of sindbis and vesicular stomatitis virus, *J. Virol.* **21:**375.

Leavitt, R., Schlesinger, S., and Kornfeld, S., 1977b, Impaired intracellular migration and altered solubility of nonglycosylated glycoproteins of vesicular stomatitis virus and sindbis virus, *J. Biol. Chem.* **252:**9018.

Lush, I. E., 1961, Genetic polymorphisms in the egg albumen porteins of the domestic fowl, *Nature* (*London*) **189:**981.

Makino, M., and Yamashina, I., 1966, Periodate oxidation of glycopeptides from ovalbumin, *J. Biochem.* (*Tokyo*) **60:**262.

Marshall, R. D., 1972, *Glycoproteins*, *Annu. Rev. Biochem.* **41:**673.

Marshall, R. D., and Neuberger, A., 1972, Hen's egg albumin, *in* "Glycoproteins," (A. Gottschalk, ed.), Part B, pp. 732–761, Elsevier, Amsterdam.

Meezan, E., Wu, H. C., Black, P. H. and Robbins, P. W. 1969, Comparative studies on the carbohydrate-containing membrane components of normal and virus-transformed mouse fibroblasts. II. Separation of glycoproteins and glycopeptides by Sephadex chromatography, *Biochemistry* **8:**2518.

Megaw, J. M., and Johnson, L. D., 1979, Glycoprotein synthesized by cultured cells: Effects of serum concentrations and buffers on sugar content, *Proc. Soc. Exp. Biol. Med.* **161:**60.

Montgomery, R., 1972, Heterogeneity of the carbohydrate groups of glycoproteins, *in* "Glycoproteins,' (A. Gottschalk, ed.), Part A, pp. 518–527, Elsevier, Amsterdam.

Montgomery, R., Wu, Y. C., and Lee, Y. C., 1965, Periodate oxidation of glycopeptides from ovalbumin, *Biochemistry* **4:**578.

Muramatsu, T., Atkinson, P. H., Nathenson, S. G., and Ceccarini, G., 1973, Cell surface glycopeptides: Growth-dependent changes in the carbohydrate-peptide linkage region, *J. Mol. Biol.* **80:**781.

Muramatsu, T., Koide, N., Ceccarini, C., and Atkinson, A., 1976, Characterization of mannose-labeled glycopeptides from human diploid cells and their gorwth-dependent alterations, *J. Biol. Chem.* **251:**4673.

Muramatsu, T., Gachelin, G., Nicolas, J. F., Condamine, H., Jakob, H. and Jacob, F., 1978, Carbohydrate structure and cell differentiation. Unique properties of fucosyl-glycopeptides isolated from embryonal carcinoma cells, *Proc. Natl. Acad. Sci. USA* **75:**2315.

Muramatsu, T., Condamine, H., Gachelin, G., and Jacob, F., 1980, Changes in fucosyl-glycopeptides during early post-implantation embryogenesis in the mouse, *J. Embryol. Exp. Morph.* **57:**25.

Neufeld, E. F., Lim, T. W., and Shapiro, L. J., 1975, Inherited disorders of lysosomal metabolism, *Annu. Rev. Biochem.* **44:**357.

Nicolson, G. L., and Singer, S. J., 1974, The distribution and asymmetry of mammalian cell surface saccharides utilizing ferritin conjugated plant agglutinins as specific saccharide stains, *J. Cell Biol.* **60:**236.

Ogata, S. I., Muramatsu, T., and Kobata, A., 1976, New structural characteristics of the large glycopeptides from transformed cells, *Nature* (*London*) **259:**580.

Olden, K., Pratt, R. M., and Yamada, K. M., 1979, Role of carbohydrate in biological function of the adhesive glycoprotein fibronectin, *Proc. Natl. Acad. Sci. USA* **76**:3343.

Plummer, T. H., Jr., and Hirs, C. H. W., 1963, The isolation of ribonuclease B, a glycoprotein from bovine pancreatic juice, *J. Biol. Chem.* **238**:1396.

Plummer, T. H., Jr., and Hirs, C. H. W., 1964, On the structure of bovine pancreatic ribonuclease B, *J. Biol. Chem.* **239**:2530.

Pricer, W. E., and Ashwell, G., 1971, The binding of desialylated glycoproteins by plasma membranes of rat liver, *J. Biol. Chem.* **246**:4825.

Robbins, P. W., and Macpherson, I. A., 1971, Control of glycolipid synthesis in a cultured hamster cell line, *Nature (London)* **229**:569.

Robertson, J. S., Etchison, J. R., and Summers, D. F., 1976, Glycosylation sites of vesicular stomatitis virus glycoprotein, *J. Virol.* **19**:871.

Rosenberg, A., and Schengrund, C. L., 1976, *in* "Biological Roles of Sialic Acid" (Rosenberg, A., and Schengrund, C. L., eds.), Plenum Press, New York.

Rosner, M. R., Grina, L. S., and Robbins, P. W., 1980, Differences in glycosylation patterns of closely related murine leukemia viruses, *Proc. Natl. Acad. Sci. USA* **77**:67.

Rothman, J. E., and Lenard, J., 1977, Membrane asymmetry, *Science* **195**:743.

Roussel, P., Lamblin, G., Degard, P., Walker-Nasier, E., and Jeanloz, R. W., 1975, Heterogeneity of the carbohydrate chains of sulfated bronchial glycoproteins isolated from a patient suffering from cystic fibrosis, *J. Biol. Chem.* **250**:2114.

Sasak, W., Deluca, L. M., Dion, L. D., and Silverman-Jones, C. S., 1980, Effect of retinoic acid on cell surface glycopeptides of cultured spontaneously transformed mouse fibroblasts (Balb/C 3T12-3 cells), *Cancer Res.* **40**:1944.

Simmons, J. F., Fishman, P. H., Freese, L., and Brody, R. O., 1975, Morphological alterations and ganglioside sialytransferase activity induced by small fatty acids in HeLa cells, *J. Cell Biol.* **66**:414.

Smith, J. A., and Pease, L. G., 1980, Reverse terms in peptides and proteins, *Crit. Rev. Biochem.* **8**:315.

Soslau, G., Fuhrer, J. P., Nass, M. M. K., and Warren, L., 1974, The effect of ethidium bromide on the membrane glycopeptides in control and virus-transformed cells, *J. Biol. Chem.* **249**:3014.

Spiro, R. G., 1965, The carbohydrate units of thyroglobulin, *J. Biol. Chem.* **240**:1603.

Spiro, R. G., 1973, Glycoproteins, *in* "Advances in Protein Chemistry," Vol. 27, pp. 349–467, Academic Press, New York.

Steck, T. L., 1974, Organization of proteins in the human red blood cell membrane, *J. Cell Biol.* **62**:1.

Tarentino, A., Plummer, T. H., Jr., and Maley, F., 1970, Studies on the oligosaccharide sequence of ribonuclease B, *J. Biol. Chem.* **245**:4150.

Tomita, M., and Marchesi, V. T., 1975, Amino acid sequence and oligosaccharide attachment sites of human erythrocyte glycophorin, *Proc. Natl. Acad. Sci. USA* **72**:2964.

Tuszynski, G. P., Baker, S. R., Fuhrer, J. P., Buck, C. A., and Warren, L., 1979, Glycopeptides derived from individual membrane glycoproteins from control and Rous sarcoma virus-transformed hamster fibroblasts, *J. Biol. Chem.* **253**:6092.

Van Beek, W. P., Smets, L. A., and Emmelot, P., 1973, Increased sialic acid density in surface glycoproteins of transformed and malignant cells. A general phenomenon? *Cancer Res.* **33**:2913.

Van Beek, W. P., Smets, L. A., and Emmelot, P., 1975, Changes in surface glycoprotein as a marker of malignancy in human leukemic cells, *Nature (London)* **253**:457.

Van Nest, G. A., and Grimes, W. J., 1977, A comparison of membrane components of normal and transformed Balb/C cells, *Biochemistry* **16**:2902.

Van Veen, J., Noonan, K. D., and Roberts, R. M., 1976, A correlation between membrane glycopeptide composition and losses in concanavalin A. Agglutinability induced by db-cAMP in Chinese hamster overy cells, *Exp. Cell Res.* **103:**405.

Via, D. P., Sramek, S., Larriba, G., and Steiner, S., 1980, Effect of sodium butyrate on the membrane glycoconjugates of murine sarcoma virus-transformed rat cells, *J. Cell Biol.* **84:**225.

Wang, F. C., and Hirs, C. H. W., 1977, Influence of the heterosaccharides in porcine pancreatic ribonuclease on the conformation and stability of the protein, *J. Biol. Chem.* **252:**8358.

Warren, L., 1963, Distribution of sialic acids in nature, *Comp. Biochem. Physiol.* **10:**153.

Warren, L., 1976, The distribution of sialic acid within the eucaryotic cell, *in* "Biological Roles of Sialic Acid " (A. Rosenberg and C. L. Schengrund, eds.), pp. 103–121, Plenum Press, New York.

Warren, L. Buck, C. A., and Tuszynski, G. P., 1978, Glycopeptide changes and malignant transformation. A possible role for carbohydrate in malignant behavior, *Biochim. Biophys. Acta* **516:**97.

Wolf, G., Kiorpes, T. C., Masushige, S., Schrieber, J. B., Smith, M. J., and Anderson, R. S., 1979, Recent evidence for the participation of vitamin A in glycoprotein synthesis, *Fed. Proc.* **38:**2540.

Yu, J., and Steck, T. L., 1975, Isolation and characterization of Band 3, the predominant polypeptide of the human erythrocyte membrane, *J. Biol. Chem.* **250:**9170.

Zinn, A. B., Marshall, J. S., and Carlson, D. M., 1978a, Preparation of glycopeptides and oligosaccharides from thyroxine-binding globulin, *J. Biol. Chem.* **253:**6761.

Zinn, A. B., Marshall, J. S., and Carlson, D. M., 1978b, Carbohydrate structures of thyroxine-binding globulin and their effects on hepatocyte membrane binding, *J. Biol. Chem.* **253:**6768.

Chapter 4

Disorders of Erythrocyte Cation Permeability and Water Content Associated with Hemolytic Anemia

William C. Mentzer, Jr.

Department of Pediatrics
School of Medicine
University of California
San Francisco, California

and

Margaret R. Clark

Department of Laboratory Medicine
School of Medicine
University of California
San Francisco, California

I. INTRODUCTION

One of the important functions of the red blood cell membrane is to regulate cell volume. A constant cell volume is critical to the maintenance of a stable intracellular milieu and to the preservation of the capacity to undergo extreme deformation. This ability permits the cell, whose resting diameter is approximately 8 μm, to pass through even the smallest capillaries of the microcirculation, whose diameters are only 3 μm. Loss of cell deformability causes a reduction in the normal 120-day survival of the red cell (Mohandas *et al.*, 1979), at least partly because of the particularly stringent requirements for passage through the spleen.

Whole cell deformability, defined as the capacity of a single cell to distort in response to mechanical stress, is influenced by three cellular properties: intrinsic membrane flexibility, the viscosity of the cellular contents, and the relative amount of unfilled membrane surface area to accommodate redistribution of the cell contents. While a change in red cell water content and volume has no effect on the first of these properties, it can have profound effects on intracellular viscosity and the amount of cell surface relative to cell volume (surface area-to-volume ratio; *S/V*). The viscosity of hemoglobin solutions increases extremely rapidly with increasing hemoglobin concentrations (Chien, 1977). Hence, loss of cell water and elevation of mean cell hemoglobin concentration (MCHC) substantially above the normal 35 g/100 ml results in a marked decrease in cell deformability because of increased intracellular viscosity. A substantial increase in cell water, on the other hand, causes a reduction in cell deformability as the cell becomes more spherical and has less redundant surface area relative to its volume. Measurement of red cell deformability as a function of suspending medium osmolality has shown that normal cell deformability is maximal in isotonic medium, decreasing when the cell either loses or gains water in response to an osmotic gradient (Mohandas *et al.*, 1980). Thus, with respect to cell deformability, the normal water content of the red cell represents an optimal balance between the effects of cell water on *S/V* and intracellular viscosity.

Ultimately, water regulation in red cells depends on regulation of the cellular content of monovalent cations, because they constitute the major osmotically active species inside the cell for which membrane permeability is low. The low, but finite, permeability of the membrane to sodium (Na) and potassium (K) is one of two major components of cell ion and water regulation. The other major component is the ability of the membrane to induce active transport of Na into the cell and K out of the cell, thus establishing transmembrane gradients for these ions. A very stable steady-state balance between passive flow of ions along the concentration gradients and active transport against the gradients provides the remarkably close regulation of ion and water content that is characteristic of red cells for virtually all of their circulatory lifetimes.

The transmembrane passive permeability routes for Na and K have not yet been completely identified, although some characterization has been achieved (Passow 1969; Knauf and Rothstein 1971; Beaugé and Lew 1977). However, it is not yet clear whether all of the channels in normal cells are specific, nor is it known whether pathologic increases in permeability occur through modification of the normal channels or creation of new ones.

The Na/K pump, responsible for active transport, has been well characterized, as recently reviewed by Dunham and Hoffman (1978). Several aspects of the pump are relevant to our present discussion. First, the pump is stimulated by an increase in intracellular Na concentrations, thus providing a means of compensating for increased Na permeability and influx. Second, the pump transports Na and K in a fixed ratio, exporting three Na ions for every two K ions it imports. We have recently shown that at very high levels of intracellular Na, prolonged pump stimulation brings out this net dehydrating effect of the pump, causing measurable water loss (Clark *et al.*, 1981a). This desalting effect works together with the Na-mediated stimulation to help compensate for increased Na and water influx into cells whose Na permeability is greatly increased. On the other hand, there is no defined mechanism in human red cells to compensate for a net loss of total ions and water (Glader and Sullivan, 1979; Clark *et al.*, 1981a).

A final important feature of the pump is that it requires ATP. Metabolic depletion, or inhibition of the pump by any other means (e.g., ouabain, low temperature), eliminates the pump component of volume regulation.

Other transport routes for Na and K which function in the red cell include Na/Na exchange (Lubowitz and Whittam, 1969), and Na/K cotransport (Wiley and Cooper, 1974). A defect in Na/K cotransport has been described for certain individuals suffering from hypertension (Garay *et al.*, 1980), and Na/Na exchange is abnormal in hereditary xerocytosis (Dutcher *et al.*, 1975). However, the physiologic role for these transport functions is unknown, and no hemolytic disorders have been found that appear to result directly from their malfunction. For those reasons, they will not be discussed in detail here.

One other transport function which has more relevance, however, is that of Ca transport. The red cell maintains a very low intracellular Ca concentration through its ATP-dependent Ca pump (Lew and Ferriera, 1978). An increase in red cell Ca content has many effects, but two are important with respect to water regulation. At very low levels, Ca causes a specific increase in K permeability (Gardos 1966; Lew and Ferriera, 1978), known as the Gardos effect. The permeability increase is specific for K, resulting in a net loss of ions and water, with disastrous effects on cell deformability (Clark *et al.*, 1981b). At higher levels, Ca can inhibit the Na/K pump as well (Dunn, 1974).

Through their osmotic influence, the absolute quantities of Na and K inside each cell at a given time determine the water content and cell volume. If the total ion content increases, the cell swells; if it decreases, the cell shrinks. Study of pathologic cells and manipulation of normal cells has helped to define the consequences of alterations in the various components of the pump/leak complex. A large increase in Na permeability leads to excessive Na influx and cell swelling; a specific increase in K permeability leads to excessive K efflux and cell shrinkage. A nonselective increase in both Na and K permeability appears generally to result in an increase in total cation and water content, presumably because of the large extracellular reservoir of Na and a membrane potential that favors cation influx (Cook, 1965). Inhibition of the pump also results in an increase in total cation content and in cell swelling, presumably for the same reasons. The changes in cell water content cause reductions in cell deformability, which would presumably jeopardize *in vivo* cell survival (Mohandas *et al.*, 1979).

A little more than a decade ago, Nathan and Shohet (1970) reviewed a group of hemolytic anemias in which hemolysis appeared to result from abnormalities in cell water regulation. They distinguished two classes of disorders, "hydrocytosis" and "desiccytosis," depending upon whether the cells contained too much or too little water. (The term "desiccytosis" has been replaced by "xerocytosis" in current terminology, to maintain consistent use of Greek roots.) Wiley and his colleagues (1975) believe that both overhydrated and underhydrated red cells may result from very similar underlying membrane abnormalities. However, we feel that the differences between overhydrated and dehydrated cells are sufficiently great to warrant separate classification. Therefore, we will use the same division originally employed by Nathan and Shohet (1970) as a framework for discussion of new information pertaining to hemolytic anemias in which abnormal water regulation may contribute to cell destruction. Our discussion will draw heavily upon experience gained in our own laboratories with these conditions. Other reviews that discuss these disorders are also available to the interested reader (Wiley 1977; Glader and Sullivan, 1979).

II. HYDROCYTES

As the name implies, hydrocytes are erythrocytes with excess cell water. Accumulation of water causes swelling and imparts a characteristic appearance to the cells. On peripheral blood smears the customary central area of pallor is reduced to a narrow slit, somewhat reminiscent of a mouth, giving rise to the term stomatocyte often used to describe hydrocytes. Since stomatocytes can be seen even in the absence of hemolytic anemia or abnormalities of cell wate or cation composition, the term stomatocytosis lacks precision and we prefer to refer to such cells as hydrocytes. *In vitro*, hydrocytes exhibit increased osmotic fragility and vulnerability to metabolic depletion. Similar factors are likely to be involved in their increased susceptibility to hemolysis *in vivo*.

A. Hereditary Hydrocytosis

1. Clinical and Hematologic Features

Lock and his co-workers (1961) described a mother and daughter with hereditary hemolytic anemia whose red cells were greatly swollen. On peripheral blood smears many stomatocytes were evident. A similar case was described by Meadow (1967). In these early reports, no mention was made of intracellular cation and water content, but in all eight subsequent cases the cell water was increased and striking abnormailities in monovalent cation composition were noted (Table I). Although hereditary hydrocytosis is an exceedingly rare disorder, as more cases have been identified a certain degree of heterogeneity has emerged. The majority of cases have been clinically severely afflicted. However, in one family, anemia was mild, and in two of the three affected family members was not discovered until adult life (Oski *et al.*, 1969).

Anemia has usually presented in early infancy, often accompanied by neonatal hyperbilirubinemia which has occasionally been severe enough to require exchange transfusion. The chronic hemolytic anemia is of variable severity with presplenectomy hemoglobin levels ranging from 3.6 to 10 g/100 ml and reticulocyte counts as high as 47%. Blood transfusions are usually required on an intermittent rather than regular basis, particularly following anemic crises which, in some patients are accompanied by jaundice and appear to be related to infections. Isotope studies have demonstrated a marked shortening of the red cell life span with sequestration of affected red cells in both the spleen and liver. In a patient we studied (W. D.), the mean red cell life span, measured with [^{32}P]-diisopropyl-fluorophosphate was only 7.2 days (normal >100 days) (Mentzer *et al.*, 1975). Equivalent results have been obtained in other patients using ^{51}Cr-labeled autologous red cells (Lock *et al.*, 1961; Lo *et al.*, 1970; Bienzle *et al.*, 1975). Splenectomy has usually diminished the rate of hemolysis but, in contrast to hereditary spherocytosis, hemolytic anemia persists after splenectomy. In patient W. D., we used ^{59}Fe to quanitate the effect of splenectomy on erythrokinetics. Before splenectomy, 23.5 g of new hemoglobin was synthesized per liter of blood daily (normal = 1.3 ± 0.4 g). Hemoglobin synthesis fell to 10.6 g/liter of blood per day post splenectomy. The ^{59}Fe-labeled mean erythron life span increased from 3.7 to 9.2 days (normal = 117 ± 15 days). These changes were accompanied by a fall in the reticulocyte count from 20–40% to approximately 10% and a rise in the hemoglobin level to 10–11 g/dl. Such blood counts are representative of those obtained in other patients sub-

Table I
Properties of Hydrocytes

Author	Stomatocytes (%)	MCV (fl)	MCHC (%)	Mean osmotic fragility (% saline)	RBC cations (mequiv/liter cells)		
					Na	K	Na+K
Severe							
Lock *et al.* (1961)	10–30	—	—	0.57	—	—	—
Meadow (1967)	30–40	—	—	0.75	—	—	—
Zarkowsky *et al.* (1968)	Many	118–151	23–27	>0.75	87	38	125
Lo *et al.* (1970, 1971)	24–30	128	28	>0.5	65	39	104
Schröter and Ungefehr (1976); Schröter *et al.* (1981)	40	108–115	28–30	0.55	95–103	16–22	~120
Bienzle *et al.* (1975)	5–30	120	28	0.66	86–94	20–25	—
Mentzer *et al.* (1975)	10–75	115–122	29–31	0.53	61–73	37–55	110–116
Mild							
Oski *et al.* (1969)	Many	95–112	—	0.43–0.48	44–59	72–73	—
Normal range		80–94	32–36	0.40–0.45	6–12	90–103	100–108

jected to splenectomy. All reported cases of hereditary hydrocytosis have been of European origin. No clear pattern of inheritance has emerged. In one family, a mother and daughter were affected (Lock *et al.*, 1961), while in another, afflicted males were found in three generations (Oski *et al.*, 1969). The parents and siblings of all other patients have been clinically and hematologically normal.

2. *Cellular Properties of Hydrocytes*

The excess cell water characteristic of hydrocytes is responsible for their swollen, abnormal morphology. Hydrocytes appear as stomatocytes on peripheral blood smears and as cup-shaped cells on wet preparations when examined by phase contrast microscopy. The mean cell volume is increased, and the mean cell hemoglobin concentration is correspondingly decreased. Monovalent cation composition is abnormal. Unlike normal human erythrocytes, which contain an abundance of K but relatively little Na, hydrocytes are high in Na and low in K. More importantly, the total monovalent cation content of hydrocytes is increased, requiring an equivalent increase in cell water to maintain osmotic equilibrium. Water constitutes 69–77% of the total weight of hydrocytes (normal 67–69%). Such cells are closer to the critical hemolytic volume than are normal erythrocytes and exhibit increased osmotic fragility.

a. Permeability. The abnormal monovalent cation content of hydrocytes results from greatly increased passive movements of both Na and K across the cell membrane. The abnormality in Na permeability is more striking than that for K and has been estimated to be between 25 and 50 times normal (Zarkowsky *et al.*, 1968; Mentzer *et al.*, 1975; Schröter and Ungefehr, 1976). The permeability of other monovalent cations (cesium, rubidium) is also increased while that of divalent cations, sulfate anion, sugars, nucleosides, and water appears to be normal (Zarkowsky *et al.*, 1968; Bienzle *et al.*, 1975; Wiley *et al.*, 1979). The normal reflection coefficients of uncharged, non-lipid-soluble small molecules such as urea, thiourea, and maleimide found *in vitro* have been taken to indicate normal membrane pore size (Zarkowsky *et al.*, 1968). Thus, the data accumulated to date indicate that the permeability defect is limited to monovalent cations.

b. Active Cation Transport. The striking accumulation of intracellular Na in hydrocytes does not reflect a failure of Mg-dependent Na/K-ATP ase-mediated active cation transport. In fact, the ouabain-inhibitable net flux of Na and K is usually greater than 10 times normal (Zarkowsky *et al.*, 1968; Mentzer *et al.*, 1975; Bienzle *et al.*, 1975). Measurement of Na pump-associated ATPase has shown a three- to fourfold increase in activity (Bienzle *et al.*, 1975). In one patient, the number of binding sites for [^{3}H]-ouabain, a measure of Na pump sites, was 12 times normal (Wiley, 1976). The Na : K flux ratio has usually been normal (Zarkowsky *et al.*, 1968; Mentzer *et al.*, 1975), although exceptions have been reported (Bienzle *et al.*, 1975; Schroter and Ungfehr, 1976). Not surprisingly, hydrocytes consume glucose at a greatly increased rate in order to support their extraordinary monovalent transport requirements. Ouabain-inhibitable lactate productio has ranged from 3.2 to 23.2 mmole/liter of cells per hour (normal 0.33–0.5 mmole/liter) (Zarkowsky *et al.*, 1968; Mentzer *et al.*, 1975; Schröter *et al.*, 1981). No primary abnormalities in glycolysis have been recognized in hydrocytes. 2, 3-diphosphoglycerate (2,3-DPG) levels have been normal or low instead of elevated, as would be expected in a young red cell population obtained from an anemic patient. Wiley has suggested that in hydrocytes a portion of the 1,3-DPG normally utilized for 2,3-DPG synthesis may be diverted through phosphoglycerate kinase in order to provide ATP for cation transport (Wiley *et al.*, 1979). On the other hand, Schröter and his colleagues (1981) reported that the rate of 2,3-DPG synthesis in hydrocytes was normal.

Erythrocyte ATP levels are normal in freshly obtained hydrocytes but may fall rapidly during 1-2 hr of storage or laboratory manipulation unless the cells are maintained at 37°-C and provided with adequate glucose. Under such optimal conditions, however, normal ATP levels can be maintained by such cells, at least during brief (1–3 hr) *in vitro* incubations (Zarkowsky *et al.*, 1968; Mentzer *et al.*, 1975). The addition of ouabain may allow the preservation of ATP at even higher levels since inhibition of the Na pump reduces consumption of ATP (Menzter *et al.*, 1975; Schröter *et al.*, 1981). However, several authors have noted a paradoxical and unexplained precipitous fall in ATP levels if hydrocytes treated with ouabain are incubated for 3 hr or longer (Zarkowsky *et al.*, 1968; Mentzer *et al.*, 1976). Such changes presumably result from the marked alterations in cell cation and water content that occur in this setting.

It has been shown *in vitro* that glucose deprivation of hydrocytes leads to rapid ATP depletion (Mentzer *et al.*, 1975). Ouabain prevents this loss (Schröter *et al.*, 1981) indicating that the ATP is consumed to support active cation transport. Unlike metabolically depleted normal cells where K loss occurs in excess of Na gain, glucose-deprived hydrocytes gain Na in excess of their concomitant loss of K. As a consequence, they swell and presumably are eventually the victims of osmotic lysis. This is reflected in the strikingly abnormal autohemolysis test results characteristic of such cells, where hemolysis may exceed 50% after 48 hr incubation in saline. The increased hemolysis can largely be prevented by incubation with glucose.

Metabolic depletion and osmotic lysis are likely to be the mode of hemolysis following entrapment of hydrocytes in the spleen since the acidic and hypoglycemic splenic environment would powerfully reduce the ability of hydrocytes to generate ATP. Initial detainment of hydrocytes by the spleen may be due to lack of deformability. In our patient W. D., even fresh red cells could not be filtered through 3-μ Nuclepore filters and passed only with difficulty through 8-μ Millipore filters (Mentzer *et al.*, 1975; 1976). Fresh hydrocytes from another patient filtered normally through 5-μ Nuclepore filters, but brief incubation or exposure to low pH reduce the flow rate substantially (Schröter *et al.*, 1981).

3. Membrane Composition

a. Lipids. The total lipid phosphorus in hydrocytes expressed on a per cell basis is increased by as much as 35% compared to normal red cells (Mentzer *et al.*, 1975; Wiley, 1976; Schröter *et al.*, 1981). However, such changes are characteristic of postsplenectomy reticulocyte-rich blood and are not specific to hydrocytes. The cholesterol : phospholipid ratio and the distribution of phospholipids have been normal, as has the incorporation of [^{14}C]-palmitic acid into membrane phospholipids. Thus, no evidence of a defect in lipid composition has yet been discovered.

b. Membrane Proteins. Hydrocyte membrane protein composition has been analyzed in three instances. In one, separation of erythrocyte membrane proteins by isoelectric focusing in one dimension followed by sodium dodecyl sulfate–polyacrylamide gel electrophoresis (SDS–PAGE) in the second revealed a small protein (mol. wt. 25,000 daltons) whose first-dimension migration was abnormal (Bienzle *et al.*, 1977). This protein was not seen in erythrocyte membranes from normal individuals or individuals with other hemolytic anemias. In the second patient, one-dimensional SDS–PAGE demonstrated an apparent lack of a high-molecular-weight membrane protein migrating to the position between bands 2.1 and 3 (Schröter *et al.*, 1981). We have more extensively analyzed erythrocyte membrane proteins from a third patient, W. D. (Mentzer *et al.*, 1978; Lande *et al.*, 1980). One-dimensional SDS–PAGE has consistently revealed a reduction in the amount of protein in the band 7 region in the absence of other evident abnormalities. SDS–PAGE as described by Laemmli (1970) resolves band 7 into three distinct bands designated 7.1 (mol. wt. 30,000), 7.2 (mol. wt. 28,000), and 7.3 (mol. wt. 26,000). Normally, reticulocyte-rich red cells demonstrate an increased staining intensity of band 7.2. In contrast, the reticulocyte-rich hydrocytes from W. D. demonstrated only a faint band 7.2. This abnormality was further analyzed by two-dimensional electrophoretic techniques that employed a nonequilibrium pH gradient in the first dimension and SDS–PAGE (NEPHGE) in the second (O'Farrell *et al.*, 1977). This approach was used because bands 7.2 and 7.3 of normal membranes were excluded from the standard first-dimension isoelectric focusing gel because of their basic isoelectric points. NEPHGE separates both acidic and basic proteins with higher resolution. In our laboratory, all three band 7 components of normal erythrocyte membranes are detected by NEPHGE. In contrast, the protein corresponding to band 7.2 was completely absent from hereditary hydrocytes from W. D. Reticulocyte-rich control red cells showed an increased band 7.2, while relatives of patient W. D. had a normal protein pattern. The function of band 7.2 is not currently known. It appears to be an integral membrane protein since it is not extracted by Triton X-100, strong alkaline solutions, or low ionic strength buffers. Its location as an integral membrane protein suggests that it may participate in the formation of a transmembrane monovalent cation channel, perhaps in concert with other membrane proteins. The absence of band 7.2 in hydrocytes could then result in formation of a defective channel with increased cation permeability. This possibility remains purely hypothetical at present in the absence of definitive information regarding the topography and function of band 7.2.

The two-dimensional electrophoretic technique is less satisfactory for the analysis of high-molecular-weight membrane proteins since appreciable amounts of such large proteins fail to enter the gel even in NEPHGE. It remains possible that abnormalities of such proteins are also present in hydrocytosis as suggested by the report of Schröter and his co-workers (1981). In patient W. D., the phosphorylation of spectrin *in vitro* was considerably less than normal in either short- (5 min) or long- (1 hr) term incubations. In contrast, phosphorylation of casein, an exogenous protein, by W. D. ghost preparations was normal, implying a defect in the spectrin molecule itself rather than in the phosphorylation enzymes (Mentzer *et al.*, 1978). The red cells of individuals whose hemolytic anemia is thought to be due to an intrinsic abnormality of the spectrin molecule often exhibit a susceptibility to budding and fragmentation when heated to temperatures between 45 and 49°C. No such susceptibility was evident in hydrocytes from W. D. which fragmented at the same temperature as normal erythrocytes. More detailed studies of spectrin are needed in W. D. and other hydrocyte patients in order to define a possible role for abnormalities of this protein in the pathogenesis of hemolytic anemia.

4. *Correction of the Permeability Defect in Hydrocytes with Imidoesters*

Incubation of hydrocytes from patient W. D. *in vitro* with imidoesters corrected the abnormal monovalent cation permeability of these cells

(Mentzer *et al.*, 1976, 1978; Mentzer and Lubin, 1979). When the cells were subsequently incubated at 37°C in the presence of glucose, active cation transport restored normal cation gradients and total cation content within 5–6 hr. Simultaneously, cell water, cell size, and intracellular hemoglobin concentration also returned to normal, and the cells regained a normal appearance under both light and scanning electron microscopy. The effect was not due to an acceleration of active transport since ouabain-inhibitable glycolysis was actually lower in imidoester-treated hydrocytes than in untreated hydrocytes. Interestingly, the precipitous fall in intracellular ATP content associated with ouabain (mentioned previously) was not seen in imidoester-treated hydrocytes where, in fact, the converse was true. The reduced deformability of hydrocytes, assessed by filtration, was also restored to normal in imidoester-treated cells. The process of endocytosis, which was subnormal in red ghosts prepared from hydrocytes, became normal after treatment with imidoesters as well. As a result of all these favorable changes, the survival of imidoester-treated hydrocytes *in vivo*, measured in a rat model, was prolonged into the normal range. However, the diminished phosphorylation of spectrin noted in hydrocyte was not restored to normal by imidoesters. The effect did not appear to be modulated by alterations in divalent cation transport, since calcium uptake and efflux in hydrocytes and hydrocyte ghosts, respectively, was normal and was not altered by imidoesters.

Correction of the permeability defect by imidoesters, which are amino-reactive reagents, establishes that membrane amino groups are likely to be involved in the primary defect. Attempts have been made to use the imidoesters as chemical probes to localize the site of the putative defect within the membrane. For the most part, these studies have employed dimethyladipimidate (DMA), the most effective of the imidoesters in correcting the hydrocyte defect. DMA reacts with membrane proteins and aminophospholipids. At higher concentration (>1 mM) evidence of protein–protein or lipid–lipid cross-linking is present, but at lower concentrations there is little detectable cross-linking. The reaction of imidoesters with membrane proteins is nonspecific. When [^{14}C]-DMA was reacted with hydrocytes, virtually all membrane proteins became labeled. Such experiments establish the ubiquity of free amino groups in the membrane but fail to specify which groups are actually responsible for the hydrocyte defect.

Employment of bifunctional imidoesters of varying chain length established that those whose maximal crosslinking dimension was less than 5 Å had no effect on the hydrocyte permeability defect (Table II). At a concentration of 5mM, reagents whose maximal crosslinking dimension ranges from 5 to 11 Å had a favorable effect on hydrocyte cation content. However, at a lower concentration (0.1 mM) DMA, whose maximal cross-

Table II
Correction of the Abnormal Cation Content of Hydrocytes by Bifunctional Imidoesters

Reagent	Maximum crosslink dimension (Å)	Concentration (mM)	RBC cation content (mequiv/liter RBC)	
			K^+	Na^+
None	—	—	28.9	99
Dimethylmalonimidate	3.7–5.0	5	28.9	83.9
Dimethylsuccinimidate	5.0–7.3	0.1	33.6	82
Dimethylsuccinimidate	5.0–7.3	5	74.5	21.7
Dimethyladipimidate	7.3–9	0.1	86.1	8.6
Dimethyladipimidate	7.3–9	5	87.2	14.4
Dimethylsuberimidate	9.7–11	0.1	11.1	68.2
Dimethylsuberimidate	9.7–11	5	78.1	15.5
Normal values	—	—	90–110	6–12

Note. Red blood cells (RBC) were incubated with indicated reagent for 6 hr in Krebs–Henseleit buffer pH 7.4, T 37°, HCT 5%.

linking dimensions are 7.3 to 9 Å, was clearly the most effective agent. Small monofunctional agents, such as methylacetimidate, were without effect. Larger monofunctional agents, such as methylbuterimidate, were effective at a concentration (20 mM) approximately 400 times the minimal effective concentration of DMA. Monofunctional reagents are capable of crosslinking as well as amidination but are less efficient than bifunctional reagents. Thus, it is likely that their effect on hydrocyte cation permeability is in fact due to crosslinking rather than amidination.

Efforts have been made to employ monofunctional imidoesters, which themselves do not influence the permeability defect, to block certain reactive amino groups within the membrane prior to reaction with DMA. It was anticipated that this might increase the specificity of the DMA reaction with the amino groups responsible for the hydrocyte permeability defect. Isothionyl acetimidate, which reacts only with amino groups exposed on the external surface of the membrane, had no effect on the monovalent cation permeability of hydrocytes. In addition, the ability of DMA to repair the hydrocyte permeability defect was not impeded by prior reaction of such cells with 15 mM isothionyl acetimidate, suggesting that the important reactive groups are not exposed on the external surface of the membrane but lie deeper within the membrane where they are inaccessible to nonpenetrating reagents. Methylacetimidate, which readily penetrates the erythrocyte membrane, also had no effect upon the hydrocyte permeability defect. Treatment of hydrocytes with 30 mM methylacetimidate blocked 83% of all amino-reactive sites from subsequent reaction with dimethyladipimidate but, as with isothionyl acetimidate, did not prevent correction of the permeability defect by

DMA. It is to be hoped that employment of several imidoesters in concert will enhance their usefulness as chemical probes to define the site of the hydrocyte defect.

B. Cryohydrocytes

1. Clinical and Hematologic Features

Miller and his co-workers (1965) described a boy with the clinical and hematologic features of hydrocytosis whose erythrocytes exhibited striking hemolysis on exposure to cold *in vitro*. The osmotic fragility of fresh red cells was normal but, after incubation at 37°C, the cells became abnormally fragile. The increase in osmotic fragility was even greater if the incubation was carried out at 4°C. Autohemolysis after 48 hr incubation in saline was greatly increased at both 4 and 37°C. The increased hemolysis could be corrected in part at 37°C by prior addition of glucose, but at 4°C, glucose had no effect. However, if the erythrocytes were incubated in ACD anticoagulant at 4°C, no hemolysis occurred. Although red cell glutathione levels were low in this patient, no metabolic disturbance or other etiology for the hemolytic anemia was discovered. Both parents and one sister were clinically and hematologically normal, but a son of the proband was affected, suggesting an autosomal dominant mode of inheritance (Townes and Miller, 1980).

We have encountered a similar patient (K. G.), a 9-year-old female of Irish ancestry who was anemic at birth (cord hematocrit, 42%) with normoblasts evident on the peripheral blood smear (Lande *et al.*, 1979, 1980; Mentzer and Lande, 1980). Subsequently, her hematocrit has ranged from 22 to 36%, and she has exhibited reticulocytosis (3–30%) and splenomegaly. Occasional stomatocytes have usually been evident on her peripheral blood smear. Her parents and three siblings are clinically and hematologically normal. Like the previous case reported by Miller, the red cells of this patient have subnormal glutathione levels (51 mg/dl). Heparnized blood stored at 4°C completely hemolyzed in 24 hr; hemolysis was less (21% in 48 hr) in EDTA anticoagulant and did not occur in ACD. When suspended in normal plasma, K. G. cryohydrocytes completely hemolyzed at 4°C, whereas normal red cells suspended in K. G. plasma did not lyse.

2. Cellular Properties of Cryohydrocytes

The cation content of fresh cryohydrocytes from K. G. was abnormal (Na = 41 and K = 63 mequiv/liter of red cells) and resembled that of hydrocytes. However, the total monovalent cation content (Na + K) was normal, as were the MCV and MCHC (34 g/100 ml). Thus, unlike hydrocytes, the hydration status of freshly obtained cryohydrocytes appeared to be normal. However, under the stress imposed by incubation *in vitro* (and presumably also in unfavorable environments *in vivo*), the cation content and cell water of these cells changed dramatically. In Table III, the effect of incubation at 4°C on monovalent cation content and cell size is shown. In heparin anticoagulant, cryohydrocytes rapidly gained Na and lost K. Na gain exceeded K loss, leading to a net accumulation of intracellular cations. As a consequence, cell water content increased, leading to cell swelling as reflected by an increase in MCV. Within 3 hr hemolysis was noted. Neither normal cells nor hereditary hydrocytes exhibited changes in cation or water content in equivalent circumstances. The cation content of cryohydrocytes suspended in ACD anticoagulant also changed during incubation, but the rate of change was less than that observed in heparin. Na and K changes were nearly balanced, leading to little net change in total cation content and little or no cell swelling.

These contrasting results in heparin and ACD appeared to be due to differences in incubation pH associated with the two anticoagulants. When heparinized blood with a pH of 7.4 at 37°C is cooled to 4°C, the pH rises to 8, whereas the pH of ACD anticoagulated blood at 4°C rises to about 7.6. Incubation experiments at both 37 and 4°C in *N*-2-hydroxyethylpiperazine-N^1-2-ethanesulfonic acid (Hepes) buffer clearly demonstrated a marked pH dependence of cation permeability of cryohydrocytes, particularly evident at alkaline pH. In contrast to hydrocytes, cryohydrocytes failed to show improvement in cation permeability upon treatment with imidoesters (at 37°C).

Table III
Effect of Incubation at 4°C on Cryohydrocytes

Anticoagulant	Incubation time (hr)	RBC cations (mequiv/10^3 RBC)			MCV (fl)	Hemolysis
		Na^+	K^+	$Na^+ + K^+$		
Heparin	0	51	71	122	100	0
	1	84	43	127	100	0
	2	108	29	137	100	0
	3	133	29	162	111	+
ACD	0	50	71	121	99	0
	1	56	62	118	99	0
	2	76	56	132	99	0
	3	91	42	133	101	0

The observation that EDTA anticoagulated blood hemolyzed more slowly than did heparinized blood suggested a possible role for membrane calcium in the permeability abnormality. However, when cryohydrocytes were incubated in Hepes buffer at 4°C, there was little difference between calcium-supplemented flasks and EDTA-containing flasks at pH 7.4 or 7.8. At pH 8.2, however, a modest reduction in permeability of approximately 25% was noted in EDTA-containing flasks. Therefore, the effect of EDTA or ACD on cryohydrocyte permeability may in part be due to chelation of divalent cations. Red cell ATP levels were normal (1.8 mmole/liter of red cells) in fresh cryohydrocytes and remained stable for up to 4 hr of incubation at 37°C but fell by 23% after 2 hr incubation at 4°C and by 66% after 5 hr. In such ATP-depleted cells, divalent cation chelation may assume greater importance.

It should be emphasized that hemolysis at 4°C is purely an *in vitro* phenomenon, of value in diagnosing cryohydrocytosis, but of no physiologic relevance. However, the cation permeability of cryohydrocytes is considerably increased even at 37°C, pH 7.4. In a representative experiment the net flux of K was 8.6 mequiv/liter cells/hr and of Na 9.1 mequiv/liter cells/hr yielding a Na : K flux ratio of 1.06 (normal 1.1–1.8). Ouabain-inhibitable lactate production was 4.74 mequiv/liter cells/hr (normal 0.3–0.5). As is the case with hydrocytes, such metabolically active cells would be vulnerable to detainment in the spleen, where low pH and relative lack of glucose constrain erythrocyte glycolysis. Unlike hydrocytes, the deformability of fresh cryohydrocytes, measured in the ecktacytometer, was normal. As would be predicted from the changes shown in Table 3, deformability rapidly decreased only after exposure to low temperature. Although it is not clear why cryohydrocytes should be selectively detained by the reticuloendothelial system, the favorable response to splenectomy noted in the patient described by Miller certainly suggests that the splenic environment is an unfavorable one and that the spleen contributes to hemolysis in this disorder. However, the continuing hemolytic anemia noted postplenectomy indicates that other nonsplenic factors contribute to hemolysis as well.

3. *Membrane Composition*

The molecular basis for the cryohydrocyte permeability abnormality in K. G. has not yet been established. Membrane lipid composition of the cells was normal. However, one-dimensional SDS–PAGE of cryohydrocyte membrane proteins demonstrated the same striking diminutation in band 7.2 noted in hydrocytes. Two-dimensional electrophoresis (EPHGE) confirmed that a protein present in normal erythrocyte membranes with a molecular weight of 28,000 was missing in cryohydrocyte membranes. The abnormality was not present in family members or patients with hemolytic anemia and reticulocytosis from a variety of other causes. The relationship between this membrane protein abnormality and the permeability defect is unclear at present. Although hydrocytes and cryohydrocytes appear to share a similar abnormality of membrane proteins, there clearly must be additional differences between the two disorders at the molecular level to explain their different responses to low temperature *in vitro* and to imidoesters.

C. Other Related Conditions

1. Stomatocytosis in the Dog

Hereditary hydrocytosis associated with hemolytic anemia has been described in Alaskan Malamute dogs (Pinkerton *et al.*, 1974). The condition resembles human hydrocytosis in that cell water, size, and total monovalent cation content are increased. However, it differs in several important ways from the human disease. First, there is no improvement in the hemolytic anemia following splenectomy in the dog. Second, short-legged dwarfism is an invariable associated finding in dogs but not in man. Most importantly, the red cells of normal dogs, which lack an ouabain-inhibitable Na/K-ATPase, are high-Na, low-K cells. Hydrocytic dog red cells have even more Na than do normal dog cells and they also have slightly more K than normal. Studies of the membranes of these cells have not yet been reported in detail but apparently membrane protein kinase activity and SDS–PAGE of membrane proteins are normal (Smith, 1981). Red cell-reduced glutathione levels are low, but glutathione stability is normal, as are the enzymes of glutathione synthesis (Smith, 1981).

The relationship between low glutathione levels and the permeability defect of dog hydrocytes, if any, is not clear at present. It is of interest that similar reductions in glutathione level have also been noted in some cases of human hydrocytosis (Miller *et al.*, 1965; Lo *et al.*, 1971; Mentzer and Lande, 1980).

2. Rh_{null} Disease

Individuals who are Rh_{null} completely lack the Rh antigens normally present on the erythrocyte membrane. The condition is inherited and is invariably associated with mild to moderate hemolytic anemia (Schmidt, 1979). It is mentioned here because the erythrocytes have the appearance of stomatocytes and exhibit increased osmotic fragility (Sturgeon, 1970). Although these observations suggest abnormal cell volume regulation, cell cations, size, and water content were normal in the only patient in whom they have so far been determined (Lauf and Joiner, 1976). Thus, these cells cannot be classified as true hydrocytes. They do, however, exhibit increased ouabain-insensitive K permeability and, although not measured, it is likely that Na permeability is also increased. Ouabain-inhibitable active K transport is also greater than normal. Further characterization of these cells will be of considerable interest.

III. XEROCYTES

In this section we will discuss several hematologic disorders in which red cell dehydration is found and possibly contributes to shortened red cell survival. We hypothesize that the reduced water content of the cells in these disorders has its origin in a derangement of mechanisms that regulate intracellular cation content and that potassium (K) loss is the central pathologic problem. Further, we suggest that the cells are subject to premature hemolysis because of their limited deformability, a property imposed by the profound dependence of hemoglobin viscosity on hemoglobin concentration. Within this context, the central problem for each disorder is to define the mechanism that causes the loss of intracellular K and water. So far, such attempts have had limited success, and the major progress of the last decade has been in the recognition of additional disorders in which dehydration occurs and in the suggestion of new possible mechanisms to be tested.

A. Hereditary Xerocytosis

1. Clinical and Hematologic Features

Hereditary xerocytosis is a red cell disorder associated with a variable degree of hemolysis, from very mild to moderate (Miller *et al.*, 1971; Glader *et al.*, 1974; Wiley *et al.*, 1975; Clark *et al.*, 1978a; Snyder *et al.*, 1978). Splenomegaly or hepatosplenomegaly and jaundice are frequently found but not always. Splenectomy, performed in a few patients, has not been of clear benefit in contrast to the generally favorable response in hereditary stomatocytosis. The consanguineous family studied by Miller *et al.* (1971) provided evidence for an autosomal recessive mode of inheritance, producing anemia only in the homozygotes. However, in other kindreds described subsequently, even probable heterozygotes have been anemic. It may be that more than one disorder is represented by these patients. Reported reticulocyte counts in xerocytosis have varied from 3 to 27%, and hematocrits from 27 to 43%. Where measured, osmotic fragility has been decreased and MCHC generally somewhat elevated, even in the presence of increased numbers of reticulocytes.

2. *Cellular Properties of Xerocytes*

The high MCHC and decreased osmotic fragility apparently reflect the reduced water content of the red cells, which appears to derive from increased membrane permeability to monovalent cations. In contrast to hydrocytosis, the characteristic finding has been that red cell K is significantly reduced and Na only moderately increased, producing an overall reduction in total monovalent cation content. Measurements of cation fluxes have confirmed a moderate increase in membrane permeability to both Na and K (Glader *et al.*, 1974), but it appears that under physiologic conditions, K loss predominates, leading to cell dehydration rather than overhydration. Glader *et al.* (1974) observed this dehydration process during *in vitro* incubations without ouabain. Interestingly, incubation of hereditary xerocytes with ouabain caused the cells to maintain a constant water content, suggesting that the $3Na_{out}:2K_{in}$ stoichiometry of the stimulated Na/K pump combined with the increase in K permeability to cause cell dehydration. A further finding of interest is that although the next passive flux of Na across the membrane is only modestly elevated, isotopic flux measurements have consistently shown a large increase in unidirectional Na influx, up to six times the normal rate (Dutcher *et al.*, 1975; Wiley *et al.*, 1975; Clark *et al.*, 1978a). This apparently reflects an increase in Na/Na exchange (Dutcher *et al.*, 1975). Active transport of Na and K is also increased, but this is likely to be an effect secondary to the elevation of intracellular Na concentrations and the increased proportions of reticulocytes.

The patients whose cells have the lowest K content are those with the highest reticulocyte counts, even though the K levels in reticulocyte-rich cell subpopulations are not as depressed as those in mature cell subpopulations (Clark *et al.*, 1978a). This suggests that K is lost progressively throughout the lifetime of the cell and that the probability of cell destruction increases with increasing K and water loss. We consider it likely that cells are removed from the circulation when their deformability falls below a certain level as a result of the increase in intracellular viscosity that follows K and water loss (Williams and Morris, 1980). Our experiments showed that the least dense subpopulation of xerocytes, whose K and water content are only moderately reduced, deformed normally, whereas the most dense, low-K cells were virtually undeformable in isotonic medium (Clark *et al.*, 1978a). Suspension of the undeformable cells in hypotonic medium permitted an increase in water content and restored normal deformability, indicating that their initially low water content and high MCHC provided the major limitation to cell deformation. In addition to its probable contribution to steady-state hemolysis, the high MCHC of xerocytes appears to make the cells susceptible to stress-induced hemolysis in special circumstances. Platt *et al.* (1981) described a xerocytosis patient who had little or no anemia and was able to engage in competitive swimming. However, during intense training sessions, he had episodes of hemolysis that were proposed to result from shear-induced fragmentation of his relatively undeformable cells. *In vitro* experiments showed that, indeed, cells were vulnerable to hemolysis under high shear stress. Moreover, artificial rehydration of xerocytes reduced their susceptibility to shear-induced hemolysis. This phenomenon was thought to be similar to the *in vitro* shear-induced fragmentation of other high MCHC-cells described by MacCallum *et al.* (1975).

So far, the molecular defect of hereditary xerocytosis has eluded definition. Analysis of membrane proteins by SDS PAGE failed to detect any specific abnormalities in membrane protein composition (Sauberman *et al.*, 1979). The same group of investigators found that cell fragmentation during *in vitro* depletion of ATP was more rapid in xerocytes than in normal cells or those from various other hematologic disorders (Snyder *et al.*, 1978). However, this phenomenon seems likely to have been a nonspecific, secondary effect associated with the increased ATP consumption of the Na/K pump and perhaps with the availability of excess membrane surface that could readily bud off into myelin forms. Another, as yet unexplained, abnormality of hereditary xerocytes was their increased susceptibility to peroxidation by exogenous H_2O_2 (Snyder *et al.*, 1981). This effect was also found in other osmotically resistant cells, such as hereditary spherocytes and sickle cells, but not to the degree seen for xerocytes. Red cell Ca levels were also normal (Glader *et al.*, 1974; Wiley *et al.*, 1977), suggesting that K loss reflects a primary defect in K permeability, rather than a Gárdos effect secondary to increased Ca permeability. Measurement of 2,3-DPG has consistently shown a reduction in this intracellular constituent, but it appears to be a secondary phenomenon, perhaps associated with accelerated activity of the Na/K pump (Wiley *et al.*, 1979). The apparently high level of Na/Na exchange may bear some mechanistic relationship to the K leak, but this question has not yet been examined directly. The overhydrated hereditary hydrocytes also appear to have increased Na/Na exchange (Schröter and Ungefehr, 1976), but the molecular defect in those cells is probably different from that in the xerocytes since DMA, which corrects the permeability defect of hydrocytes (Mentzer *et al.*, 1978), had no effect on that of xerocytes (Glader and Sullivan, 1979). Finally, except for the patients studied by Miller *et al.* (1971), a consistent elevation in erythrocyte phosphatidylcholine has been found in membrane samples from several patients. The relationship of this observation to the permeability defect is not clear since such a lipid abnormality was found in patients affected by a different disorder, who show reduced red cell K concentrations but no reduction of cell water (Jaffé and Gottfried, 1968; Shohet *et al.*, 1971, 1973; Godin *et al.*, 1980; Godin and Herrins, 1981). Albala *et al.* (1978) have described yet another patient whose red cells, like hereditary xerocytes, contained reduced amounts of total monovalent cations and reduced levels of 2,3-DPG. However, those cells had a normal water content and were also considered to constitute a separate disorder. It may be that the lipid changes in hereditary xerocytes, similar to those seen in β-thalassemia minor (Kalofoutis *et al.*, 1980), are a secondary consequence of ongoing membrane peroxidation, as suggested by their increased susceptibility to peroxidation *in vitro* (Snyder *et al.*, 1981).

From this discussion it is clear that much remains to be learned about hereditary xerocytosis and the mechanisms that control the normal membrane permeability barrier to K. Perhaps a concerted study of the relationships between K permeability and Na/Na exchange in the xerocytes may provide clues for the identification of the molecular defect in this disorder.

B. Sickle Cell Disease

1. Clinical and Hematologic Features

In sickle cell disease, it is possible to divide the clinical manifestations into two components, a hemolytic component and a vaso-occlusive component. The severity of both components varies greatly, not only across the spectrum of disorders that involve double heterozygosity for hemoglobin S with that for other abnormal hemoglobins but even among individuals who are homozygous for hemoglobin S. Mentzer and Wang (1980) have recently reviewed the complexities of the clinical expression of this disorder. The present discussion will focus on the role of cell dehydration, which appears to relate primarily to the hemolytic component of the disease. Several years ago, it was shown by Serjeant *et al.* (1969) that the extent of hemolysis correlated with the percentage of circulating irreversibly sickled cells (ISC), cells that were subsequently found to be severely dehydrated (Glader *et al.*, 1978). The frequency or severity of painful sickling crises, a major feature of the vaso-occlusive component of the disease, did not correlate with ISC counts, although ISC may contribute to major organ damage (Serjeant *et al.*, 1978).

2. *Properties of ISC*

ISC are defined by their retention of an enlongated shape even under fully oxygenated conditions, when their hemoglobin has been completely depolymerized. The cells are presumed to have been formed during prolonged periods of sickling. The proportion of ISC in peripheral blood samples has varied from 1–2% up to more than 40% in samples from a recently studied group of patients in our laboratory. While the most obvious abnormality of the cells is their shape, their reduced water content is striking and provides a useful means of isolating the cells by density gradient centrifugation. Studies employing this technique have shown that the dehydrated state of these cells is associated with a profound depletion of intracellular K content, together with only a moderate increase in Na content (Clark *et al.*, 1978b; Glader *et al.*, 1978). The consequence is a marked reduction in total monovalent cation content and a MCHC that sometimes exceeds 45 g/100 ml. It should be noted that the presence of these cells is not generally detectable through an increase of whole blood MCHC, because their high MCHC is usually offset by the low MCHC of reticulocytes which are also numerous (Oda *et al.*, 1978). In contrast to hereditary xerocytes, the reduced K content of ISC is accompanied by reduced Na/K pump activity, in spite of increased levels of Na that would ordinarily be expected to stimulate the pump (Clark *et al.*, 1978c). Other membrane abnormalities that may be related to cell dehydration have been reported. ISC contain excess amounts of Ca (Palek, 1977), and there is substantial evidence that their ability to pump Ca out of the cell is severely impaired (Gopinath and Vincenzi, 1979; Bookchin and Lew, 1980; Litosch and Lee, 1980; Dixon and Winslow, 1981). The elevation of intracellular Ca led Glader and Nathan (1978) to suggest that cell K and water loss might be the consequence of a Ca-mediated Gardos effect. However, recent studies have shown that sickle cells cannot fully activate the Gardos channel (Lew and Bookchin, 1980), so the relationship between their high Ca and low K is not yet clear. Other membrane abnormalities that may or may not be related to dehydration include the tenacious binding of denatured hemoglobin S to the ISC membrane (Asakura *et al.*, 1977; Lessin *et al.*, 1978), and an increased tendency of the more dehydrated sickle cells to adhere to cultured endothelial cells (Hebbel *et al.*, 1980a). This latter property has been shown to correlate to a striking degree with clinical severity (Hebbel *et al.*, 1980b).

The primary consequence of cell dehydration in ISC appears to be a profound reduction in cell deformability. The impaired deformability of ISC was for some time believed to derive from a stiffening of the cell membrane, possibly caused by excess intracellular Ca (Eaton *et al.*, 1979; Lorand *et al.*, 1979). However, experiments in our laboratory have provided evidence that, even though the ISC membrane may not be quite as deformable as normal membranes, the extremely high intracellular viscosity that results from severe dehydration is the major determinant of reduced deformability (Clark *et al.*, 1980a,b). As suggested recently, some ISC may have less deformable membranes than others (Smith *et al.*, 1981). Nevertheless, our studies of the deformation of ISC in hypotonic medium support the concept that any impairment of whole cell deformability that originates in the membrane is small compared to that caused by water loss (Clark *et al.*, 1980b).

The mechanism that leads to dehydration in ISC has not yet been defined. Ultimately, it probably derives from a nonspecific permeability increase that occurs during acute sickling. First described by Tosteson *et al.* (1955) as a nonselective leak for monovalent cations, it was subsequently found that permeability to Ca also increased during sickling (Palek, 1977). Glader and Nathan (1978) reported that sickling-induced changes in ion permeability occurred in two stages. The first involved a nonselective balance leakage of Na and K in opposite directions across the membrane, and the second involved a relative augmentation in K loss, resulting in dehydration. The proposition that this second stage reflected a net accumulation of Ca and its stimulation of a Gardos leak was attractive and was consistent with the observed elevation of Ca in ISC. However, more recent findings have raised the possibility that another process, that of membrane peroxidation, may also contribute to cellular dehydration in sickle cells. Evidence has been found for increased susceptibility of ISC and deoxygenated sickle cells to peroxidation both *in vitro* (Chiu *et al.*, 1979) and *in vivo* (Chiu and Lubin, 1979; Das and Nair, 1980; Jain and Shohet, 1981a). A recent paper (Zimmerman and Natta, 1981) suggests that one piece of this evidence, the elevation of glutathione peroxidase activity , is not specific to sickle cell disease. Nevertheless, the other hemolytic conditions in which the activity of the enzyme was increased were hemoglobin abnormalities that could also result in increased membrane peroxidation. Membrane peroxidation, like Ca accumulation, could explain most of the abnormalities of water regulation in sickle cells. Both Ca accumulation and membrane peroxidation cause a specific increase in membrane permeability to K and result in cell dehydration (Gárdos, 1966; Ham *et al.*, 1973; Orringer and Parker, 1977), and both processes can also inhibit the Na/K pump (Dunn, 1974; Kesner *et al.*, 1979; Koontz and Heath, 1979). In fact, it is possible, or even likely, that both processes reinforce one another, since Ca appears to increase the vulnerability of cells to peroxidative damage (Jain and Shohet, 1981b), and peroxidation increases the permeability of red cell membranes to Ca (Gardos *et al.*, 1976).

A third factor that may contribute to dehydration of sickle cells is the binding of hemoglobin S to the membrane. There is abundant evidence that extensive binding occurs (Schneider *et al.*, 1972; Asakura *et al.*, 1977; Jones, 1979; Lau *et al.*, 1979; Kim *et al.*, 1980; Sayare and Schuster, 1980). As suggested by Rachmilewitz (1976) for thalassemic cells, this may possibly provide a source for active oxidizing species in the vicinity of the membrane, largely bypassing the oxidative defenses of the cell. From another perspective, it is also possible that membrane peroxidation contributes to hemoglobin binding, since Goldstein *et al.* (1980) have suggested that malonaldehyde, a product of unsaturated lipid peroxidation, is capable of linking hemoglobin to the membrane.

3. *Therapies Based on Membrane Abnormalities of Sickle Cells*

It should be obvious that careful studies are required to dissect the possible interactions of Ca accumulation, membrane peroxidation, and hemoglobin–membrane binding and their potential role in causing K and water loss during prolonged sickling. Nevertheless, the evidence suggesting a role for one or another of these processes has led to the exploration of possible therapeutic approaches based on their blockage or reversal. A limited trial was conducted in which vitamin E was administered to patients in an attempt to protect the red cells against oxidative damage (Natta *et al.*, 1980). The authors of the study reported a decrease in the percentage of ISC during the administration of the vitamin, but the number of patients was too small to determine whether there was amelioration of clinical problems. A second approach has been to give patients zinc in an effort to limit the effects of Ca accumulation during sickling (Brewer and Kruckeberg, 1979). A large-scale study has not yet been completed, but preliminary results suggest that zinc can reduce the percentage of circulating ISC and, presumably, the proportion of dehydrated cells

(Brewer *et al.*, 1977). Finally, it has been proposed that if one could increase the salt or water content of sickle cells, it might be possible not only to ameliorate sickling-induced dehydration but also to reduce the probability of initial sickling because of the strong dependence of hemoglobin S polymerization on hemoglobin S concentration (Orringer *et al.*, 1980). A clinical study was undertaken in which anti-diuretic hormone was administered in an effort to reduce extracellular Na concentrations and plasma osmolality (Rosa *et al.*, 1980). This was expected to produce a corresponding increase in intracellular water content and reduction in MCHC. The initial trial involving three patients provided encouragement for this approach, but others have reported difficulty in achieving good results (Leary and Abramson, 1981; Charache, 1981). Another agent which influences cell water content is cetiedil, a drug already used in Africa to treat sickling crises. *In vitro* studies (Asakura *et al.*, 1980; Benjamin *et al.*, 1980; Berkowitz and Orringer, 1981) showed that the drug both increased N permeability, causing cell swelling, and blocked the Gardos effect in Ca-loaded cells. It remains to be seen whether these actions occur *in vivo* or whether they provide the basis for its reported efficacy in ameliorating crises (Cabannes, 1977).

4. Properties of Non-ISC

Water regulation in non-ISC, the mature discoid sickle cells with normal morphology, appears to be relatively normal. Even if a cell has increased membrane permeability during sickling, it appears to be able to restore normal intracellular cation concentrations after reoxygenation if sickling is not too prolonged and no dehydration occurs (Glader and Nathan, 1978). When whole sickle bood is analyzed by density gradient centrifugation, one frequently sees a bimodal distribution of cells, such that most of the mature discoid cells equilibrate at normal cell density, and most of the ISC equilibrate at considerably higher density (Fig. 1). At this time, it is not clear whether the sickling of initially discoid cells results from their entrapment in areas of low oxygen tension behind rheologically sluggish dehydrated cells or whether as yet undefined factors precipitate sickling in cells with normal water content without contribution from previously dehydrated cells. The expectation has been that any reduction in water content of a sickle cell would increase the probability that it would sickle. This vulnerability could result from a decrease in the delay time for hemoglobin S polymerization, which is extremely sensitive to hemoglobin concentration (Eaton *et al.*, 1976). Alternatively, it could result from an increase in the amount of hemoglobin S polymer present in the cell even at oxygen tensions above those required to produce

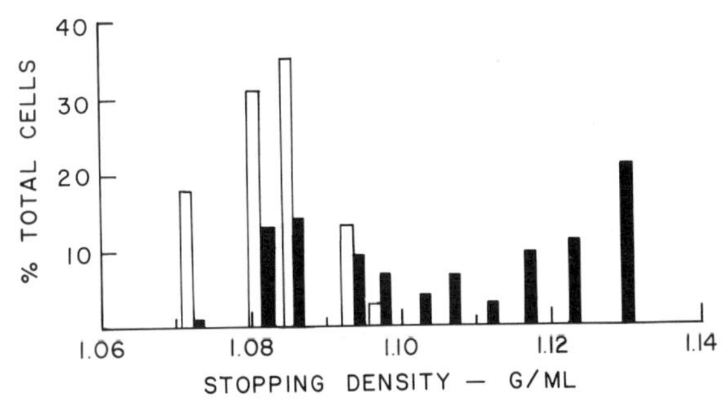

Fig. 1. Density distribution of sickle and normal red cells. Samples of homozygous SS and normal blood were centrifuged on discontinuous Stractan gradients, whose density ranged from 1.06 to 1.13 g/ml in approximately 0.004 g/ml steps. Cells that collected at the layer interfaces were removed, and their relative number was estimated by hemoglobin determination. The percentage of total cells at each interface is plotted as a function of the density of the Stractan layers on which they rested. Black bars, sickle cells; white bars, normal cells. Note the bimodal distribution of sickle cell densities. Most, but not all, of the high-density cells were ISC.

morphologic sickling (Noguchi and Schechter, 1981). In this latter situation, the presence of polymer might both provide nuclei for rapid hemoglobin polymerization upon further deoxygenation and also increase the intracellular viscosity sufficiently to impede rapid circulation of the cell through areas of dangerously low oxygen tension. Whether either of these possibilities has any relevance to physiologic processes of sickling remains to be clarified.

C. Hemoglobin CC Disease

Elsewhere we have recently discussed the apparent effects of several abnormal hemoglobins on cell water regulation (Clark and Shohet, 1981). Here we will focus on hemoglobin CC disease, because that is the only example other than sickle cell disease for which unequivocal evidence for cellular dehydration is available.

1. Clinical and Hematologic Features

Hemoglobin CC disease is a relatively mild disorder, and perhaps infrequent encounters between patient and investigator have limited the information available. Splenomegaly is always found, but there are no debilitating hemolytic episodes, and cell survival is only moderately shortened (Murphy, 1968). Microscopic examination of the blood shows large numbers of target cells, some of which may contain crystals of hemoglobin C if the patient has been splenectomized (Charache *et al.*, 1967; Fabry *et al.*, 1981). Microspherocytes are also frequently observed (Charache *et al.*, 1967; Murphy, 1968). The MCHC of the whole blood is commonly elevated, and the MCV is correspondingly reduced, thus accounting for the reduced osmotic fragility (Charcche *et al.*, 1967; Murphy, 1968).

2. *Properties of Hemoglobin CC Cells*

These changes in the whole blood reflect the fact that virtually all of the cells have reduced water content (Mohandas *et al.*, 1980) and that there are few reticulocytes to obscure the effect (Murphy, 1968). The intracellular content of K was found by Murphy (1968) to be reduced. Perhaps surprisingly, the Na content was also somewhat lower than normal, regardless of the units in which it was expressed. In our laboratory, we have measured the passive K efflux rate for red cells from one hemoglobin CC patient. This preliminary study gave a first-order rate constant for K efflux into Tris–HCl-buffered NaCl (pH 7.4) of 0.089 hr^{-1}, compared to a control value of 0.026 hr^{-1}. Bookchin recently measured the Ca content of hemoglobin CC cells by atomic absorption spectroscopy and found that it was elevated to approximately the same degree as in unseparated sickle cells (R. M. Bookchin, unpublished data). The only defined defect in the intracellular milieu is a reduction in the intracellular pH, an effect that may be responsible for a decrease in the oxygen affinity of the cells. This, in turn, could explain the failure of the bone marrow to maintain a normal hematocrit despite only a moderate shortening of red cell survival (Murphy, 1976). It is possible that the decrease in pH could be related to the low cell water content, since concentration of the impermeant intracellular anions, including hemoglobin, would tend to depress the concentration of intracellular chloride ions and correspondingly increase the concentration of hydrogen ions, through its effect on the membrane potential.

The deleterious effect of high MCHC on the deformability of hemoglobin CC cells has been recognized for a long time. Murphy (1968), then Charache *et al.* (1976), showed that the filterability and bulk viscosity of hemoglobin CC blood indicated poor deformability in isotonic medium that was largely reversible in hypotonic medium. Moreover, Murphy (1968) and Self *et al.* (1977) have noted that the subnormal hematocrit maintained by the patients is probably beneficial, since it reduces the blood viscosity and improves oxygen delivery. Thus, the major effect of poor cell deformability is upon cell survival (Murphy, 1968). More recently, Mohandas *et al.* (1980) and Fabry *et al.* (1981) have re-examined the rheological properties of hemoglobin CC cells, using density gradient centrifugation to isolate various subpopulations of cells for study. Both groups of investigators found that all populations of cells were dehydrated, even the least dense, reticulocyte-rich populations. Fabry *et al.* found that crystals of hemoglobin C, seen only in the one splenectomized patient, were in the most dense cells. Finally, these same authors performed studies of blood flow in an artificially perfused rat mesoappendix, and they concluded that the reduced deformability of hemoglobin CC cells had little effect on their flow through capillaries, suggesting that only the spleen represents a circulatory challenge to those cells.

The molecular mechanism underlying dehydration of cells containing only hemoglobin C is not known and has not been extensively investigated. Since cells from splenectomized individuals are just as dehydrated as those from patients who have spleens, it is evident that the spleen has no role in promoting dehydration (Fabry *et al.*, 1981). One might speculate that increased binding of hemoglobin C to the red cell membrane (Natta and Muir, 1980), perhaps associated with the presence of intracellular inclusions similar to Heinz bodies (Kim *et al.*, 1980), could cause abnormal hemoglobin–membrane interactions that either promote membrane oxidative damage or directly increase membrane permeability. The water loss itself could be associated with increased intracellular Ca concentration (R. M. Bookchin, unpublished data), in turn caused by increased Ca permeability. Experiments designed to address these questions might help define the membrane defect in these cells and deepen our understanding of hemoglobin–membrane interactions.

D. Hereditary Spherocytosis

Hereditary spherocytosis (HS) is a hemolytic disorder whose pathophysiology probably derives from membrane instability (Lux and Wolfe, 1980), rather than defective ion transport. Nevertheless, there is a subpopulation of dehydrated cells whose destruction may be accelerated by their high intracellular viscosity (Erslev and Atwater, 1963), and the disease thus merits a brief discussion here.

1. Clinical and Hematologic Features

HS is a disease of generally moderate severity, and splenectomy usually corrects any anemia. Lux and Wolfe (1980) have recently reviewed the clinical and hematologic features. The "spherocytes" seen on stained blood films as round cells with no central pallor are not actually spherical cells but usually have one concavity, like a stomatocyte (Bessis, 1977). In contrast to stomatocytes, the hemoglobin inside these HS cells is concentrated, giving the dried cell a uniform dark color. Very often, the whole blood MCHC is elevated despite reticulocytosis, and this abnormality reflects the presence of a severely dehydrated population of cells (Table IV). However, the osmotic fragility of HS blood, if abnormal, is typically increased rather than decreased. This has been attributed to a decrease in S/V presumably the result of membrane loss *in vivo* (Cooper and Jandl, 1969).

2. *Properties of HS Cells*

Abnormalities of cation transport in HS are well documented (Jacob and Jandl, 1964; Johnson and Salminen, 1980), but the major interest has been focused on increased Na permeability. Initially it was proposed that increased Na permeability of cells during splenic stasis resulted in a sequence of metabolic depletion followed by cell swelling and hemolysis (Jacob and Jandl, 1964). However, studies by Wiley (1970) showed that there was no correlation between *in vitro* Na permeability and *in vivo* hemolysis. A correlation between hemolysis and increased osmotic fragility was found, but the increased osmotic fragility reflected a reduction in cell surface area rather than an increase in cell volume and was thus not related to cell water alterations. The K content of peripheral red cells was found to be decreased and that of cells obtained from spleens at the time of splenomegaly was even lower (Mayman and Zipursky, 1974). However, K and ^{86}Rb permeability were found to be normal or only slightly increased (Jacob and Jandl, 1964: Johnsson and Salminen 1980), and the processes leading to water loss and increase of MCHC in some of the cells therefore remain somewhat enigmatic.

Table IV
MCHC of Density Gradient Subpopulations of Hereditary Spherocytes

Subpopulation	MCHC (g/100 ml)
1 (top)	28.0
2	32.5
3	35.7
4	37.7
5 (bottom)	47.7

Note. Cells were separated on a continuous gradient of Stractan (Clark *et al.*, 1978b) with a density range of 1.074–1.124 g/ml, and approximately equal fractions were isolated.

Deformability of HS cells is generally reduced in isotonic medium (Allard *et al.*, 1977; Mohandas *et al.*, 1980), and reduced whole cell deformability is thought to be a major factor in splenic destruction of the cells (Lux and Wolfe, 1980). The major cellular property responsible for impaired deformability appears to be reduced cell surface area, as suggested by the correlation between increased osmotic fragility and shortened cell survival (Wiley, 1970). A more elegant demonstration of the primary effect of membrane loss was provided by the observations of Cooper and Jandl (1969) that HS cells transfused into patients with obstructive jaundice improved their survival, apparently as a result of lipid acquisition and membrane expansion. Additional experiments were performed by Mohandas *et al.* (1980) to define the mechanism for reduced whole cell deformability in gradient separated subpopulations of HS cells. Abnormally large decreases in deformability in hypotonic medium identified the deficiency in membrane area as the limiting factor in the deformability of the least dense subpopulations. Moreover, even the two most dense populations, which showed an increase in deformability between 290 and 200 mosm/kg, showed a rapid decrease at lower tonicities, as a result of limited membrane surface area.

The physiologic consequence of water loss in the high MCHC subpopulation of HS cells has not been clarified. Although a high MCHC would ordinarily be expected to imapir cell flow properties, volume reduction could provide an advantage for surface-deficient HS cells that might partially offset the effect of increased intracellular viscosity. Godal and Refsum (1979) reported that athletes who had mild HS had increased hemolysis during periods of intensive training. This is reminiscent of the swimmer with hereditary xerocytosis described by Platt *et al.* (1981), suggesting that the high MCHC cells in HS might also have an increased sensitivity to shear stress as did the xerocytes. It is also consistent with the reported increase in mechanical fragility of HS cells (Griggs *et al.*, 1960). However, both MacCallum *et al.* (1975) and Williams *et al.* (1977) found that HS cells were less likely to fragment than normal cells under *in vitro* fluid shear stress. The discrepancy among these findings could reflect differences in the type of mechanical stress applied; alternatively, hemolysis could have resulted from an alteration of membrane properties in the cells of some patients.

3. *Membrane Proteins*

Recent progress has been made in defining the molecular defect in HS. Goodman *et al.* (1981) and Lux (1982) separately found evidence for a defect in red cell spectrin in several HS patients. The defect appeared to cause reduced binding of spectrin to another protein of the red cell membrane skeletal complex, band 4.1, and thereby provided a reasonable explanation for membrane instability in HS. However, other HS patients did not show the defect (Lux, 1982) and it is likely that several different molecular defects may be found in variants of this disorder. At this point, no information is available to explain the relationship between this primary abnormality and the accompanying alterations in cation permeability of HS membranes.

E. Pyruvate Kinase Deficiency

Our final example of physiologically significant red cell dehydration is pyruvate kinase (PK) deficiency. Mentzer (1981) and Glader and Nathan (1975) have discussed the clinical and hematologic features of PK deficiency. The point of interest for the present discussion is the original observation by Mentzer *et al.* (1971) that PK-deficient reticulocytes are especially susceptible to destruction in the spleen. That initial study showed that when PK-deficient blood was incubated with cyanide or other inhibitors of mitochondrial respiration, the reticulocytes developed an increase in K permeability, losing water and becoming more dense than the mature cells. Similar results were later reported by Glader and Sullivan (1979). This phenomenon was proposed to result from the greater ATP demands of the young cells, causing rapid ATP depletion and cell shrinkage. Recently, Koller *et al.* (1979) showed that PK-deficient cells took up ^{45}Ca during incubation with cyanide and that the increase in K permeability could be inhibited by quinine. Since quinine is known to block the Ca-mediated Gardos leak, this observation provides strong support for the porposal that Ca accumulation is the immediate cause of K and water loss in PK-deficient reticulocytes. Because this would appear to be a general result of ATP depletion, it might be expected that other glycolytic enzyme deficiencies would show a similar process. However, this has not yet been demonstrated, and, in fact, the immediate cause of hemolysis in the various glycolytic enzymopathies remains to be clarified (Valentine and Paglia, 1980). One other point of clinical interest is Glader's (1976a) demonstration that salicylate, an uncoupler of mitochondrial oxidative phosphorylation, can cause ATP depletion, K loss, and dehydration of PK cells *in vitro* similar to that caused by cyanide. This raises the possibility that salicylates could increase hemolysis in patients whose PK deficiency was severe although there is some question as to whether physiologically attainable levels of salicylate would constitute a hazard (Henry, 1976; Glader, 1976b).

The initial studies of cyanide-induced dehydration of PK-deficient cells failed to note any deformability abnormality of freshly drawn cells although incubation with cyanide decreased cell filterability (Mentzer *et al.*, 1971). Recently, Leblond *et al.* (1978b) measured deformability of PK cells, using more sensitive methods that employed smaller pore (3 μ versus 8 μ) filters and micropipets. They found that the deformability of fresh cells was in fact decreased, partly because of the presence of large, multilobulated, very young reticulocytes, and partly because of highly viscous, apparently dehydrated (Leblond *et al.*, 1978a,b), spiculated cells. The authors suggest than when the spleen is present, it effectively removes both the young reticulocytes and the dehydrated echinocytes because of their poor deformability. Thus splenectomy not only improves cell survival, but also results in an increase in the number of morphologically abnormal cells.

IV. CONCLUSION

Virtually all of the abnormalities of regulation of cell water and monovalent cation content we have discussed are due to alterations in passive permeability. Active cation transport mechanisms are unaffected and in most instances exhibit appropriately increased activity in an attempt to restore homeostasis. Hydrocytosis seems to result from a primary defect in the erythrocyte membrane while, with the exception of hereditary xerocytes, dehydration appears to be the consequence of membrane injury secondary to a defect elsewhere within the cell. For example, in sickle cell disease or hemoglobin CC disease, hemoglobin–membrane interactions, accumulation of membrane Ca, or peroxidation of membrane lipids are pathologic processes, related to the hemoglobinopathy, that may alter cation permeability. In PK deficiency, metabolic depletion results in membrane Ca accumulation and selective loss of potassium (Gardos effect). Only in hereditary xerocytosis itself is it likely that a primary membrane permeability defect exists.

The presumed linkage between abnormalities of cell water and hemolysis is abnormal cell deformability. Both hydrocytes and xerocytes are abnormal in this regard due either to an altered *S*/*V* (hydrocytes) or to increased internal viscosity (xerocytes). The greater severity of hemolytic anemia usually associated with hydrocytosis suggests that a reduced

S/V, particularly when accompanied by an increase in cell size, may compromise cell survival more than does increased internal viscosity. However, the evidence is somewhat contradictory on this point since the family with hydrocytosis described by Oski and his colleagues (1969) had severe derangement in cell water and cation content but only mild hemolysis.

The ultimate cause of *in vivo* hemolysis in these disorders is not known. Several processes may contribute, separately or in concert. First, the impaired deformability of either swollen, stomatocytic cells or shrunken, viscous cells would be expected to result in detention, at times prolonged, within the hostile environment of the spleen and other reticuloendothelial organs. Prolonged residence by cells whose increased pump rate requires increased ATP generation, or for those whose ability to produce ATP is impaired, may lead to ATP depletion, cation pump cessation, cell swelling, and lysis. In addition, xerocytes in particular may exhibit increased mechanical fragility, at least under the stress imposed by extreme exertion.

The molecular basis for inherited abnormalities in red cell cation permeability remains unknown. This is perhaps not surprising since the normal route for passage of cations across the erythrocyte membrane has not yet been described in molecular terms. In fact it is to be hoped that more detailed analysis of these rare but fascinating experiments of nature will identify normal as well as pathologic transmembrane Na and K channels and perhaps, in addition, provide other new insights into the regulation of cation and water content in normal red cells.

V. REFERENCES

Albala, M. M., Fortier, N. L., and Glader, B. E., 1978, Physiologic features of hemolysis associated with altered cation and 2,3-diphosphoglycerate content, *Blood* **52:**135.

Allard, C., Mohandas, N., and Bessis, M., 1977, Red cell deformability changes in hemolytic anemias estimated by diffractometric methods (ektacytometry), *Blood Cells* **3:**209.

Asakura, T., Minakata, K., Adachi, K., Russell, M. O., and Schwartz, E., 1977, Denatured hemoglobin in sickle erythrocytes, *J. Clin. Invest.* **59:**633.

Asakura, T., Ohnishi, S. T., Adachi, K., Ozguc, M., Hashimoto, K., Singer, M., Russel, M. O., and Schwartz, E., 1980, Effect of cetiedil on erythrocyte sickling: New type of antisickling agent that may affect erythrocyte membranes, *Proc. Nat. Acad. Sci. USA* **77:**2955.

Beaugé, L., and Lew, V. L., 1977, Passive fluxes of sodium and potassium across red cell membranes, *in* "Membrane Transport in Red Cells" (J. C. Ellory and V. L. Lews, eds.), pp. 39–51, Academic Press, New York.

Benjamin, L. J., Kakkini, G., and Peterson, C. M., 1980, Cetiedil: Its potential usefulness in sickle cell disease, *Blood* **55:**265.

Berkowitz, O. R., and Orringer, E. P., 1981, Effects of cetiedil on monovalant cation permeability in the RBC: An explanation for the efficacy of cetiedil in the treatment of sickle cell anemia (SCA), *Clin. Res.* **29:**330A (Abstract).

Bessis, M., 1977, "Blood Smears Reinterpreted," p. 96, Springer-Verlag, Berlin.

Bienzle, U., Niethammer, D., Kleeberg, U., Ungefehr, K., Kohne, E., and Kleihauer, E., 1975, Congenital stomatocytosis and chronic haemolytic anaemia, *Scand. J. Haematol.* **15:**339.

Bienzle, U., Bhadki, S., Knüfermann, H., Niethammer, D., and Kleihauer, E., 1977, Abnormality of erythrocyte membrane protein in a case of congenital stomatocytosis, *Klin. Wochenschr.* **55:**569.

Bookchin, R. M., and Lew, V. L., 1980, Progressive inhibition of the Ca pump and Ca: Ca exchange in sickle red cells, *Nature* (*London*) **284:**561.

Brewer, G. J., and Kruckeberg, W. C., 1979, The anti calcium and erythrocyte membrane effects of zinc, and their potential value in the treatment of sickle cell anemia, *in* "Development of Therapeutic Agents for Sickle Cell Disease" (J. Rosa, Y. Beuzard, and J. Hercules, eds.), pp. 195–204, Elsevier/North-Holland, Amsterdam.

Brewer, G. J., Brewer, L. F., and Prasad, A. S., 1977, Suppression of irreversibly sickled erythrocytes by zinc therapy in sickle cell anemia, *J. Lab. Clin. Med.* **90:**549.

Cabannes, R., 1977, Preliminary Studies on the Effects of Cetiedil in Acute Episodes of Sickle Cell Anemia, presented at Sickle Cell Conference, Washington, D. C., November.

Charache, S., 1981, Failure of desmopressin to maintain low serum sodium or prevent crises in sicklers, *Clin. Res.* **29:**572A (Abstract).

Charache, S., Conley, C. L., Waugh, D. F., Ugoretz, R. J., and Spurrell, J. R., 1967, Pathogenesis of hemolytic anemia in homozygous hemoglobin C disease, *J. Clin. Invest.* **46:**1795.

Chien, S., 1977, Rheology of sickle cells and erythrocyte content, *Blood Cells* **3:**279.

Chiu, D., and Lubin, B., 1979, Abnormal vitamin E and glutathione peroxidase levels in sickle cell anemia. Evidence for increased susceptibility to lipid peroxidation in vivo, *J. Lab. Clin. Med.* **94:**542.

Chiu, D., Lubin, B., and Shohet, S. B., 1979, Erythrocyte membrane lipid reorganization during the sickling process, *Br. J. Haematol.* **41:**223.

Clark, M. R., and Shohet, S. B., 1981, Abnormal hemoglobins and cell hydration, *Tex. Rep. Biol. Med.* **40:**417.

Clark, M. R., Mohandas, N., Caggiano, V., and Shohet, S. B., 1978a, Effects of abnormal cation transport on deformability of desiccytes, *J. Supramol. Struct.* **8:**521.

Clark, M. P., Unger, R. C., and Shohet, S. B., 1978b, Monovalent cation composition and ATP and lipid content of irreversibly sickled cells, *Blood* **51:**1169.

Clark, M. R., Morrison, C. E., and Shohet, S. B., 1978c, Monovalent cation transport in irreversibly sickled cells, *J. Clin. Invest.* **62:**329.

Clark, M. R., Guatelli, J. C., and Mohandas, N., 1980a, Influence of red cell water content on the morpholody of sickling, *Blood* **55:**823.

Clark, M. R., Mohandas, N., and Shohet, S. B., 1980b, Deformability of oxygenated irreversibly sickled cells, *J. Clin. Invest.* **65:**189.

Clark, M. R., Guatelli, J. C., White, A. T., and Shohet, S. B., 1981a, Study on the dehydrating effect of the red cell Na^+/K^+-pump in Nystatin-treated red cells with varying Na^+ and water contents, *Biochim. Biophys. Acta* **646:**422.

Clark, M. R., Mohandas, N., Feo, C., Jacobs, M. S., and Shohet, S. B., 1981b, The separate mechanisms of deformability loss in ATP-depleted and Ca-loaded erythrocytes. *J. Clin. Invest.* **67**:531.

Cook, J. S., 1965, The quantitative interrelationships between ion fluxes, cell swelling, and radiation dose in ultraviolet hemolysis, *J. Gen. Physiol.* **48**:719.

Cooper, R. A., and Jandl, J. H., 1969, The role of membrane lipids in the survival of red cells in hereditary spherocytosis, *J. Clin. Invest.* **48**:736.

Das, S. K., and Nair, R. C., 1980, Superoxide dismutase, glutathione peroxidase, catalase and lipid peroxidation of normal and sickled erythocytes, *Br. J. Haematol.* **44**:87.

Dixon, E., and Winslow, R. M., 1981, The interaction between ($Ca^{2+} + Mg^{2+}$)-ATPase and the soluble activator (calmodulin) in erythrocytes containing hemoglobin S, *Br. J. Haematol.* **47**:391.

Dunham, P. B., and Hoffman, J. F., 1978, Na and K transport in red blood cells, *in* "Physiology of Membrane Disorders" (T. E. Andreoli, J. F. Hoffman, and D. D. Fanestil, eds.), pp. 255–272, Plenum, New York.

Dunn, M. J., 1974, Red blood cell calcium and magnesium. Effects upon sodium and potassium transport and cellular morphology, *Biochim. Biophys. Acta* **352**:97.

Dutcher, P. O., Segal, G. B., Feig, S. A., Miller, D. R., and Klemperer, M. R., 1975, Cation transport and its altered regulation in human stomatocytic erythrocytes, *Pediatr. Res.* **9**:924.

Eaton, W. A., Hofrichter, J., and Ross, P. D., 1976, Delay time of gelation: A possible determinant of clinical severity in sickle cell disease, *Blood* **47**:621.

Eaton, J. W., Jacobs, H. S., and White, J. G., 1979, Membrane abnormalities of irreversibly sickled cells, *Semin. Hematol.* **16**:52.

Erslev, A. J., and Atwater, J., 1963, Effect of mean corpuscular hemoglobin concentration on viscosity, *J. Lab. Clin. Med.* **62**:401.

Fabry, M. E., Kaul, D. K., Raventos, C., Baez, S., Rieder, R., and Nagel, R. L., 1981, Some aspects of the pathophysiology of homozyous Hb CC erythrocytes, *J. Clin. Invest.* **67**:1284.

Garay, R. P., Dagher, G., Pernollet, M.-G., Devynck, M.-A., and Meyer, P., 1980, Inherited defect in a Na^+, K^+-co-transport system in erythrocytes from essential hypertensive patients, *Nature* (*London*) **284**:281.

Gárdos, G., 1966, The mechanism of ion transport in human erthrocytes, *Acta Biochim. Biophys. Acad. Sci. Hung.* **1**:139.

Gárdos, G., Szász, I., and Hollán, S. R., 1976, Potassium and calcium permeability changes in normal and pathological red cells, *in* "Membranes and Disease" (L. Bolis, J. F. Hoffman, and Leaf, eds.), pp. 105–107, Raven Press, New York.

Glader, B. E., 1976a, Salicylate-induced injury of pyruvate-kinase-deficient erythrocytes, *N. Eng. J. Med.* **294**:916.

Glader, B. E., 1976b, Salicylates and PK deficient reticulocyte, *N. Eng. J. Med.* **295**:230 (response to letter).

Glader, B. E., and Nathan, D. G., 1975, Haemolysis due to pyruvate kinase deficiency and other glycolytic enzymopathies, *Clin. Haematol.* **4(1)**:123.

Glader, B. E., and Nathan, D. G., 1978, Cation permeability alterations during sickling: Relationship to cation composition and cellular hydration of irreversibly sickled cells, *Blood* **51**:983.

Glader, B. E., and Sullivan, D. W., 1979, Erythrocyte disorders leading to K loss and dehydration, *Prog. Clin. Biol. Res.* **30**:503.

Glader, B. E., Fortier, N., Albala, M. M., and Nathan, D. G., 1974, Congenital hemolytic anemia associated with dehydrated erythrocytes and increased potassium loss, *N. Engl. J. Med.* **291**:491.

Glader, B. E., Lux, S. E., and Muller-Soyano, A., 1978, Energy reserve and cation composition of irreversibly sickled cells (ISC's) *in vivo*, *Br. J. Haematol.* **40**:527.

Godal, H. C., and Refsum, H. E., 1979, Haemolysis in athletes due to hereditary spherocytosis, *Scand. J. Haematol.* **22**:83.

Godin, D. V., and Herring, F. G., 1981, Spin label studies of erythrocytes with abnormal lipid composition: Comparison of red cells in a hereditary hemolytic syndrome and lecithin: Cholesterol acyltransferase deficiency, *J. Supramol. Struct. Cell. Biochem.* **15**:213.

Godin, D. V., Gray, G. R., and Frohlich, J., 1980, Study of erythrocytes in a hereditary hemolytic syndrome (HHS): Comparison with erythrocytes in lecithin: cholesterol acyltransferase (LCAT) deficiency, *Scand. J. Haematol.* **24**:122.

Goldstein, B. D., Rozen, M. G., and Kunis, R. L., 1980, Role of red cell membrane lipid peroxidation in hemolysis due to phenylhydrazine, *Biochem. Pharmacol.* **29**:1355.

Goodman, S. A., Kesselring, J. J., Weidner, S. A., and Eyester, E. M., 1981, The molecular alteration in the cytoskeleton of hereditary spherocytes, *J. Supramol. Struct.* Suppl. 5:131.

Gopinath, R. M., and Vincenzi, F. F., 1979, (Ca^{2+} + Mg^{2+})-ATPase activity of sickle cell membranes: Decreased activation by red blood cell eytoplasmic activator, *Am. J. Hematol.* **7**:303.

Griggs, R. C., Weisman, R., and Harris, J. W., 1960, Alterations in osmotic and mechanical fragility related to *in vivo* erythrocyte aging and splenic sequestration in hereditary spherocytosis, *J. Clin. Invest.* **39**:89.

Ham, T. H., Grauel, J. A., Dunn, R. A., Murphy, J. A., White, J. G., and Kellermeyer, R. W., 1973, Physical properties of red cells as related to effects in vivo. IV. Oxidant drugs producing abnormal intracellular concentration of hemoglobin (eccentrocytes) with a rigid-red-cell hemolytic syndrome, *J. Lab. Clin. Med.* **82**:898.

Hebbel, R. P., Yamada, O., Moldou, C. F., Jacob, H. S., White, J. G., and Eaton, J. W., 1980a, Abnormal adherence of sickle erythrocytes to cultured vascular endothelium. Possible mechanism for microvascular occlusion in sickle cell disease, *J. Clin. Invest.* **65**:154.

Hebbel, R. P., Boogaerts, M. A. B., Eaton, J. W., and Steinberg, M. H., 1980b, Erythrocyte adherence to endothelium in sickle cell anemia. A possible determinant of disease severity, *N. Engl. J. Med.* **302**:992.

Henry, R., 1976, Salicylates and PK deficient reticulocytes, *N. Engl. J. Med.* **295**:229 (letter).

Jacob, H. S., and Jandl, J. H., 1964, Increased cell membrane permeability in the pathogenesis of hereditary spherocytosis, *J. Clin. Invest.* **43**:1704.

Jaffé, E. R., and Gottfried, E. L., 1968, Hereditary nonspherocytic hemolytic disease associated with an altered phospholipid composition of the erythrocytes, *J. Clin. Invest.* **47**:1375.

Jain, S. K., and Shohet, S. B., 1981a, A novel phospholipid in irreversibly sickled erythrocytes: Evidence for a possible role of *in vivo* peroxidation for membrane damage, *Clin. Res.* **29**:519A (Abstract).

Jain, S. K., and Shohet, S. B., 1981b, Calcium potentiates the peroxidation of erythrocyte membrane lipids, *Biochim. Biophys. Acta* **642**:46.

Johnsson, R., and Salminen, S., 1980, Effect of ouabain on osmotic resistance and monovalent cation transport of red cells in hereditary spherocytosis, *Scand. J. Haematol.* **25**:323.

Jones, G. L., 1979, Spin label study of hemoglobin membrane interactions in normal and sickle erythrocytes, *Proc. West. Pharmacol. Soc.* **22**:79.

Kalofoutis, A., Diskakis, E., Stratakis, N. J., and Papademetriou, A., 1980, Changes of red cell phospholipids in β-thalassemia minor, *Biomed. Med.* **23**:1.

Kesner, L., Kindya, R. J., and Chan, P. C., 1979, Inhibition of erythrocyte membrane $(Na^+ + K^+)$-activated ATPase by ozone-treated phospholipids, *J. Biol. Chem.* **254**:2705.

Kim. H. C. Friedman, S., and Asakura, T., 1980, Inclusions in red blood cells containing Hb S or Hb C, *Br. J. Haematol.* **44**:547.

Knauf, P. A., and Rothstein, A., 1971, Chemical modification of membranes. I. Effects of sulfhydryl and amino reactive reagents on anion and cation permeability of the human red blood cell, *J. Gen. Physiol.* **58**:190.

Koller, C. A., Orringer, E. P., and Parker, J. C., 1979, Quinine protects pyruvate-kinase deficient red cells from dehydration, *Am. J. Hematol.* **7**:193.

Koontz, A. E., and Heath, R. L., 1979, Ozone alteration of transport of cations and the Na^+/K^+ ATPase in human erythrocytes, *Arch. Biochem. Biophys.* **198**:493.

Laemmli, U. K., 1970, Cleavage of structural proteins during the assembly of the head of bacteriophage, *Nature* (*London*) **227**:680.

Lande, W., Cerrone, K., and Mentzer, W., 1979, Congenital hemolytic anemia with abnormal cation permeability and cold hemolysis in vitro, *Blood* **54(1)**:29a.

Lande, W. M., Cerrone, K. L., and Mentzer, W. C., 1980, Abnormal RBC Membrane Protein in Two Patients with High Na^+, Low K^+ RBC, Proceedings—18th Congress of the International Society of Hematology, Montreal, Canada, August.

Lau, P.-W., Hung, C., Minakata, K., Schwartz, E., and Asakura, T., 1979, Spin-label studies of membrane-associated denatured hemoglobin is normal and sickle cells. *Biochim. Biophys. Acta* **552**:499.

Lauf, P. K., and Joiner, C. H., 1976, Increased potassium transport and ouabain binding in human Rh_{null} red blood cells, *Blood* **48(3)**:457.

Leary, M., and Abramson, N., 1981, Induced hyponatremia for sickle cell crisis, *N. Engl. J. Med.* **304**:844 (Letter).

Leblond, P. F., Lyonnais, J., and Delage, J-M., 1978a, Erythrocyte populations in pyruvate kinase deficiency anemia following splenectomy. I. Cell morphology, *Br. J. Haematol.* **39**:55.

Leblond, P. F., Coulombe, L., and Lyonnais, J., 1978b, Erythrocyte populations in pyruvate kinase deficiency anemia following splenectomy. II. Cell deformability, *Br. J. Haematol.* **39**:63.

Lessin, L. S., Kurantsin-Mills, J., and Wallas, C., 1978, Membrane alterations in irreversibly sickled cells: Hemoglobin-membrane interaction, *J. Supramol. Struct.* **9**:537.

Lew, V. L., and Bookchin, R. M., 1980, A Ca^{2+}-refractory state of the Ca-sensitive K^+ permeability mechanism in sickle cell anemia red cells, *Biochim. Biophys. Acta* **602**:196.

Lew, V. L., and Ferriera, H. G., 1978, Calcium transport and the properties of a calcium-activated channel in red cell membranes, *in* "Current Topics in Membranes and Transport" (F. Bronner and A. Kleinzeller, eds.), pp. 217–277, Academic Press, New York.

Litosch, I., and Lee, K. S., 1980, Sickle red cell calcium metabolism: Studies on Ca^{2+}-Mg^{2+} ATPase and Ca-binding properties of sickle red cell membranes, *Am. J. Hematol.* **8**:377.

Lo, S. S., Hitzig, W. H., and Marti, H. R., 1970, Stomatozytose, *Schweiz. Med. Wochenschr.* **100(46)**:1977.

Lo, S. S., Marti, H. R., and Hitzig, W. H., 1971, Hemolytic anemia associated with decreased concentration of reduced glutathione in red cells, *Acta Haematol* **46**:14.

Lock, S. P., Sephton Smith, R., and Hardisty, R. M., 1961, Stomatocytosis: A hereditary red cell anomaly associated with haemolytic anemia, *Br. J. Haematol.* **7**:303.

Lorand, L., Siefring, G. E., and Lowe-Krentz, L., 1979, Enzymatic basis of membrane stiffening in human erythrocytes, *Semin. Hematol.* **16:**65.

Lubowitz, H., and Whittam, R., 1969, Ion movements in human red cells independent of the sodium pump, *J. Physiol.* (*London*) **202:**111.

Lux, S. E., 1982, Report of Workshop on the Membrane Skeleton of Abnormal Red Blood Cells, *in* "Differentiation and Function of Hemopoietic Cell Surfaces" (V. P. Marchesi, R. Gallo, P. Majerus, and F. Fox, eds.), Liss, New York, in press.

Lux, S. E., and Wolfe, L. C., 1980, Inherited disorders of the red cell membrane skeleton, *Pediatr. Clin. No. Am.* **27:**463.

MacCallum, R. N., Lynch, E. C., Hellums, J. D., and Alfrey, C. P., 1975, Fragility of abnormal erythrocytes evaluated by response to shear stress, *J. Lab. Clin. Med.* **85:**67.

Mayman, D., and Zipursky, A., 1974, Hereditary spherocytosis: The metabolism of erythrocytes in the peripheral blood and in the splenic pulp, *Br. J. Haematol.* **27:**201.

Meadow, S. R., 1967, Stomatocytosis, *Proc. R. Soc. Med.* **60:**13.

Mentzer, W. C., 1981, Pyruvate kinase deficiency and disorders of glycolysis, *in* "Hematology of Infancy and Childhood" (D. G. Nathan and F. A. Oski, eds.), pp. 566–607, Saunders, Philadelphia, Pennsylvania.

Mentzer, W. C., Jr., and Lubin, B. H., 1979, The effect of crosslinking reagents on red-cell shape, *Sem in. Hematol.* **16(2):**115.

Mentzer, W. C., Jr., and Lande, W. M., 1980, Hemolytic anemia resulting from abnormal red cell membrane cation permeability-hydrocytosis and cryohydrocytosis, *in* "Red Blood Cell and Lens Metabolism" (S. Srivastava, ed.), pp. 311–314, Elsevier/North Holland, Amsterdam.

Mentzer, W. C., and Wang, W. C., 1980, Sickle cell disease: Pathophysiology and diagnosis, *Pediatr. Ann.* **9(8):**10.

Mentzer, W. C., Baehner, R. L., Schmidt-Schoenbein, H., Robinson, S. H., and Nathan, D. G., 1971, Selective reticulocyte destruction in erythrocyte pyruvate kinase deficiency, *J. Clin. Invest.* **50:**688.

Mentzer, W. C., Jr., Smith, W. B., Goldstone, J., and Shohet, S. B., 1975, Hereditary stomatoctosis: Membrane and metabolism studies, *Blood* **46(5):**659.

Mentzer, W. C., Lubin, B. H., and Emmons, S., 1976, Correction of the permeability defect in hereditary stomatocytosis by dimethyl adipimidate, *N. Engl. J. Med.* **294:**1200.

Mentzer, W. C., Lam, G. K. H., Lubin, B. H., Greenquist, A., Schrier, S. L., and Lande, W., 1978, Membrane effects of imidoesters in hereditary stomatocytosis, *J. Supramol. Struct.* **9:**275.

Miller, D. R., Rickles, F. R., Lichtman, M. A., LaCelle, P. L. Bates, J., and Weed, R. I., 1971, A new variant of hereditary hemolytic anemia with stomatocytosis and erythrocyte cation abnormality, *Blood* **38:**184.

Miller, G., Townes, P. L., and MacWhinney, J. B., 1965, A new congenital hemolytic anemia with deformed erythrocytes (? "Stomatocytes") and remarkable susceptibility of erythrocytes to cold hemolysis in vitro. I. Clinical and hematologic studies, *Pediatrics* **35:**906.

Mohandas, N., Phillips, W. M., and Bessis, M., 1979, Red blood cell deformability and hemolytic anemias, *Semin. Hematol.* **16:**954.

Mohandas, N., Clark, M. R., Jacobs, M. S., and Shohet, S. B., 1980, Analysis of factors regulating erythrocyte deformability, *J. Clin. Invest.* **66:**563.

Murphy, J. R., 1968, Hemoglobin CC disease: Rheological properties of erythrocytes and abnormalities in cell water. *J. Clin. Invest.* **47:**1483.

Murphy, J. R., 1976, Hemoglobin CC erythrocytes: Decreased intracellular pH and decreased O_2 affinity-anemia, *Semin. Hematol.* **13:**177.

Nathan, D. G., and Shohet, S. B., 1970, Erythrocyte ion transport defects and hemolytic anemia "Hydrocytosis" and "desiccytosis," *Semin. Hematol.* **7**:381.

Natta, C., and Machlin, L., 1979, Plasma levels of tocopherol in sickle cell anemia subjects, *Amr. J. Clin. Nutr.* **32**:1359.

Natta, C., and Muir, M., 1980, Preferential binding of β^C relative to β^S globin to stroma in hemoglobin SC disease, *Hemoglobin* **4**:157.

Natta, C. L., Machlin, L. J., and Brin, M., 1980, A decrease in irreversibly sickled erythrocytes in sickle cell anemia patients given Vitamin E, *Amer. J. Clin. Nutrit.* **33**:968.

Noguchi, C. T., and Schechter, A. N., 1981, The intracellular polymerization of sickle hemoglobin and its relevance to sickle cell disease, *Blood* **58**:1057.

Oda, S., Oda, E., and Tanaka, K. R., 1978, Relationship of density distribution and pyruvate kinase electrophoretic pattern of erythrocytes in sickle cell disease and other disorders, *Acta Haematol.* **60**:201.

O'Farrell, P. Z., Goodman, H. M., and O'Farrell, P. H., 1977, High resolution two-dimensional electrophoresis of basic as well as acidic proteins, *Cell* **12**:1133.

Orringer, E. P., and Parker, J. C., 1977, Selective increase of potassium permeability in red blood cells exposed to acetylphenylhydrazine, *Blood* **50**:1013.

Orringer, E. P., Roer, M. E. S., and Parker, J. C., 1980, Cell density profile as a measure of erythrocyte hydration: Therapeutic alteration of salt and water content in normal and SS red blood cells, *Blood Cells* **6**:345.

Oski, F. A., Naiman, J. L., Blum, S. F., Zarkowsky, H. S., Whaun, J., Shohet, S. B., Green, A., and Nathan, D. G., 1969, Congenital hemolytic anemia with high-sodium low-potassium red cells, *New Engl. J. Med.* **280(17)**:909.

Palek, J., 1977, Red cell calcium content and transmembrane calcium movements in sickle cell anemia, *J. Lab. Clin. Med.* **89**:1365.

Passow, H., 1969, Passive ion permeability of the erythrocyte membrane, *Prog. Biophys. Mol. Biol.* **19**:425.

Pinkerton, P. H., Fletch, S. M., Brueckner, P. J., and Miller, D. R., 1974, Hereditary stomatocytosis with hemolytic anemia in the dog, *Blood* **44**:557.

Platt, O. S., Lux, S. E., and Nathan, D. G., 1981, Exercise-induced hemolysis in xerocytosis; Red cell dehydration and shear sensitivity, *J. Clin. Invest.* **68**:631.

Rachmilewitz, E. A., 1976, The role of intracellular hemoglobin precipitation, low MCHC, and iron overload on red blood cell membrane peroxidation in thalassemia, *Birth Defects* **12(8)**:123.

Rosa, R. M., Bierer, B. E., Thomas, R., Stoff, J. S., Kruskal, M., Robinson, S., Bunn, H. F., and Epstein, F. H., 1980, A study of induced hyponatremia in the prevention and treatment of sickle cell crisis, *N. Engl. J. Med.* **303**:1138.

Sauberman, N., Fortier, N. L., Fairbanks, G., O'Connor, R. J., and Snyder, L. M., 1979, Red cell membrane in hemolytic disease. Studies on variables affecting electrophoretic analysis, *Biochim. Biophys. Acta* **556**:292.

Sayare, M., and Schuster, T. M., 1980, Association of hemoglobins A and S with cytoplasmic surface of the erythrocyte membrane, *Fed. Proc.* **39**:1916 (Abstract).

Schmidt P. J., 1979, Hereditary hemolytic anemias and the null blood types, *Arch. Int. Med.* **139**:570.

Schneider, R. G., Takeda, I., Gustavson, L. P., and Alperin, J. B., 1972, Intraerythrocytic precipitations of hemoglobins S and C, *Nature (London) New Biol.* **235**:88.

Schröter, W., and Ungefehr, K., 1976, Studies on the cation transport in high sodium and low potassium red cells in hereditary hemolytic anemia associated with stomatocytosis, *in* "Membranes and Disease" (L. Bolis, J. F. Hoffman, and A. Leaf, eds.), pp. 95–98, Raven Press, New York.

Schröter, W., Ungefehr, K., and Tilmann, W., 1981, Role of the spleen in congenital stomatocytosis associated with high sodium low potassium erythrocytes, *Klin. Wochenschr.* **59:**173.

Self, F., McIntire, L. V., and Zanger, B., 1977, Rheological evaluation of hemoglobin S and hemoglobin C hemoglobinopathies, *J. Lab. Clin. Med.* **89:**488.

Serjeant, G. R., Serjeant, B. E., and Milner, P. F., 1969, The irreversibly sickled cell; a determinant of haemolysis in sickle cell anemia, *Br. J. Haematol.* **17:**527.

Serjeant, G. R., Serjeant, B. E., Desai, P., Mason, K. P., Swewell, A., and England, J. M., 1978, The determinants of irreversibly sickled cells in homozygous sickle cell disease, *Br. J. Haematol.* **40:**431.

Shohet, S. B., Livermore, B. M., Nathan, D. G., and Jaffe, E. R., 1971, Hereditary hemolytic anemia associated with abnormal membrane lipids. I. Mechanism of accumulation of phosphatidyl choline, *Blood* **38:**445.

Shohet, S. B., Nathan, D. G., Livermore, B. M., Feig, S. A., and Jaffé, E. R., 1973, Hereditary hemolytic anemia associated with abnormal membrane lipid. II. Ion permeability and transport abnormalities, *Blood* **42:**1.

Smith, J. E., 1981, Animal models of human erythrocyte metabolic abnormalities, *Clin. Haematol.* **10(1):**239.

Smith, C. M., Kuettner, J. F., Tukey, D. P., Burris, S. M., and White, J. G., 1981, Variable deformability of irreversibly sickled erythrocytes, *Blood* **58:**71.

Snyder, L. M., Lutz, H. U., Sauberman, N., Jacobs, J., and Fortier, N. L., 1978, Fragmentation and myelin formation in hereditary xerocytosis and other hemolytic anemias, *Blood* **52:**750.

Snyder, L. M., Sauberman, N., Condara, H., Dolan, J., Jacobs, J., Szymanski, I., and Fortier, N. L., 1981, Red cell membrane response to hydrogen peroxide. Sensitivity in hereditary xerocytosis and other abnormal red cells, *Br. J. Haematol.* **48:**435.

Sturgeon, P., 1970, Hematological observations on the anemia associated with blood type Rh null, *Blood* **36(3):**310.

Tosteson, D. C., 1955, The effects of sickling on ion transport II. The effect of sickling on sodium and cesium transport, *J. Gen. Physiol.* **39:**55.

Tosteson, D. C., Carlsen, E., and Dunham, E. T., 1955, The effects of sickling on ion transport I. Effect of sickling on potassium transport, *J. Gen. Physiol.* **39:**31.

Townes, P. L., and Miller, G., 1980, Further studies of cold-sensitive variant of stomatocytosis, *Am. J. Human Gen.* **32:**57A.

Valentine, W. N., and Paglia, D. E., 1980, The primary cause of hemolysis in enzymopathies of anaerobic glycolysis: A viewpoint, *Blood Cells* **6:**819, with a commentary by E. Beutler, pp. 827–829.

Wiley, J. S., 1970, Red cell survival studies in hereditary spherocytosis, *J. Clin. Invest.* **49:**666.

Wiley, J. S., 1976, Hereditary stomatocytosis: A disease of cell water regulation, *in* "Membranes and Disease" (L. Bolis, J. F. Hoffman, and A. Leaf, eds.), pp. 89–94, Raven Press, New York.

Wiley, J. S., 1977, Genetic abnormalities of cation transport in the human erythrocyte, *in* "Membrane Transport in Red Cells" (J. C. Ellory and V. L. Lew, eds.), pp. 337–361, Academic Press, New York.

Wiley, J. S., and Cooper, R. A., 1974, A furosemide-sensitive co-transport of sodium plus potassium in the human red cell, *J. Clin. Invest.* **53:**745.

Wiley, J. S., Ellory, J. C., Shuman, M. A., Shaller, C. C., and Cooper, R. A., 1975, Characteristics of the membrane defect in the hereditary stomatocytosis syndrome, *Blood* **46:**337.

Wiley, J. S., Cooper, R. A., Adachi, K., and Asakura, T., 1979, Hereditary stomatocytosis: Association of low 2,3-diphosphoglycerate with increased cation pumping by the red cell, *Br. J. Haematol.* **41:**133.

Williams, A. R., and Morris, D. R., 1980, The internal viscosity of the human erythrocyte may determine its lifespan in vivo, *Scand. J. Haematol.* **24:**57.

Williams, A. R., Escoffery, C. T., and Gorst, D. W., 1977, The fragility of normal and abnormal erythrocytes in a controlled hydrodynamic shear field, *Br. J. Haematol.* **37:**379.

Zarkowsky, H. S., Oski, F. A., Sha'afi, R., Shohet, S. B., and Nathan, D. G., 1968, Congenital hemolytic anemia with high sodium, low potassium red cells I. Studies of membrane permeability, *N. Engl. J. Med.* **278(11):**573.

Zimmerman, C. P., and Natta, C., 1981, Glutathione peroxidase activity in whole blood of patients with sickle cell anemia, *Scand. J. Haematol.* **26:**177.

Chapter 5

Appearance of a Terminal Differentiation Antigen on Senescent and Damaged Cells and Its Implications for Physiologic Autoantibodies

Marguerite M. B. Kay

Research and Medical Services
Olin E. Teague Veterans' Center
and Division of Geriatric Medicine
Texas A and M University
Temple, Texas

I. INTRODUCTION

One of the homeostatic activities vital to the evolution of vertebrates and to the development and survival of individuals is the ability to remove cells programmed for death at the end of their useful life span. During metamorphosis, for example, selective tissues are "reabsorbed" by macrophages in order to prepare the organism for the next stage of development (Saunders, 1966). A satisfactory explanation of the specific mechanism which enables macrophages to ingest all cells distal to a definite site while sparing all cells proximal to that location has not been found (Weber, 1962; Tata, 1966). During embryogenesis in all species, cells in a circumscribed location are destroyed during tissue remodeling, yet both their successors and cells adjacent to that area are spared (Zwilling, 1964; Saunders and Fallon, 1966). Daily maintenance of homeostasis in adults requires removal of senescent and damaged cells and tissue repair following trauma. For example, approximately 360 billion senescent red cells are removed daily in humans. An understanding of this basic homeostatic mechanism is essential to both development and aging and to the concept of recognition and specificity which permeates all disciplines in biomedicine.

As an initial approach to this problem, the mechanism by which senescent and damaged red blood cells (RBC) are removed from the body were investigated. Red cells were used as a model for studying terminally differentiated cells because they are removed at the end of their 120-day life span, and they are an ideal experimental system in many respects. Large quantities of red cells are available, and they can be separated into populations of different ages. In addition, we know more about the structure and function of red cell membrane proteins than we do about those of any other cell. Although the initial studies were performed on red cells, the studies have been extended to include other somatic cells such as white cells and hepatic cells.

Results of the studies which are described herein reveal that a terminal differentiation antigen appears on senescent and damaged cells. An immunoglobulin (Ig) G autoantibody in serum recognizes and binds to the senescent cell antigen and initiates the removal of cells and cellular debris by macrophages. Thus, it appears that cells code for their own death.

II. GENERAL METHODS

A. RBC Separation

RBC were separated into young, middle-aged, and old populations on the basis of density (Bennett and Kay, 1981; Kay, 1975a,1978b; Murphy, 1973). Nine volumes of Percoll (Pharmacia) was diluted with 1 vol of 10 times concentrated (10X) phosphate-buffered saline (PBS) and then diluted to a final density of 1.106 with alpha minimum essential media (AMEM). Freshly prepared sterile solutions and sterile procedures were used throughout the studies. Peripheral blood was layered on Percoll in sterile, Sorvall polycarbonate tubes and centrifuged at 27,000*g* for 60 min (Sorvall RC-5). Platelets and white cells band at the top and were removed (Bennett and Kay, 1981). Young RBC are in the least dense fraction (ρ 1.090) and the old RBC are found in the most dense fraction (ρ 1.120) as determined by ^{59}Fe labeling *in situ* (Bennett and Kay, 1981).

B. Isolation of IgG from Senescent RBC

Blood was obtained from 250 healthy individuals between the ages of 21 and 35 years, and was processed the same day that it was obtained. The blood was centrifuged for 15 min at 2000*g* and the serum and white cells were removed. RBC were separated into populations of different ages by centrifugation (Bennett and Kay, 1981), and young RBC were removed. The remaining RBC were washed three times with 50–100 vol of Dulbecco's PBS, pH 7.4. IgG was eluted as described previously (Kay, 1978b; Kochwa and Rosenfield, 1964). Briefly, RBC membranes (ghosts) were prepared by digitonin lysis, washed three times with PBS, and IgG was eluted with 0.1 M glycine–HCl buffer, pH 2.3. Sodium azide (0.02%) was added. Eluates were neutralized with 1 N NaOH and concentrated to 0.25–1.00 ml using an Amicon Diaflow with a PM 10 filter. Solid material was removed by centrifugation at 27,000*g*. Samples were stored at −80°C until use. They were then thawed, pooled, and incubated with protein A-Sepharose 4B for 30–60 min at 37°C and 24–48 hr at 4°C. The protein A-Sepharose was poured into a 12-ml column prepared from a syringe and washed with 100 vol of PBS, then 100 vol of 10X PBS followed by 100 vol of PBS. IgG was eluted with 2 vol of glycine–HCl buffer, pH 2.3, neutralized with 1 N NaOH, and dialyzed against distilled water.

C. Senescent Cell IgG Affinity Columns

IgG eluted from senescent RBC and purified by chromatography on protein A-Sepharose (Pharmacia) was coupled to cyanogen bromide (CNBr)-activated Sepharose 4B (Pharmacia). Columns were prepared and washed with 15 vol of glycine–HCl buffer, pH 2.3, to break the bond between the insolubilized IgG and its senescent cell reporter. After neutralization with NaOH, preparations of pure RBC membrane sialoglycoproteins, obtained by extraction with lithium diiodosalicyclate (LIS) and partition in a phenol–water mixture according to the procedure of Marchesi and Andrews (1971), were processed with the senescent cell IgG column (see below). After extensive washing, the material on the column was eluted with glycine–HCl buffer, pH 2.3, neutralized, dialyzed against deionized, distilled water at 4°C, lyophilized, and analyzed by gel electrophoresis.

D. Sialoglycoprotein Preparations

Blood was obtained from healthy donors 21–30 years old. Donor cells were typed for 27 known RBC blood group antigens, including ABO, Rh, MN, Ss̄, Duffy, Kell, and Kidd. Donors did not have detectable autoantibodies to RBC or antibodies to any known blood group antigen, as determined by the Immunohematology Laboratory, Blood Services,

American Red Cross, nor did they have antibodies to nuclei, DNA, mitochondria, parietal cells, or smooth muscle (Hollister *et al.*, 1981). All blood was processed the same day that it was obtained. RBC from 500 ml of human blood were depleted of white cells and young RBC by centrifugation on self-forming Percoll gradients (Bennett and Kay, 1981). RBC were washed three times in 50–100 vol of PBS with 75 μg/ml phenylmethysulfonyl fluoride (PMSF) at 4°C. Right-side-out RBC membranes were prepared by hypotonic lysis in 5 mM sodium phosphate buffer, pH 7.5, containing 50 μg/ml PMSF (Marchesi and Andrews, 1971). After five washes in 100 vol of the same buffer, ghosts were resuspended in a solution of 0.3 M LIS (Marchesi and Andrews, 1971) or 1.0% Triton X-100 (TX-100), and partitioned in a phenol–water mixture according to the procedure of Marchesi and Andrews (1971). The water phase was dialyzed extensively against distilled water at 4°C and then lyophilized. The lyophilized material extracted with LIS was resuspended in 5 mM phosphate buffer and extracted with chloroform–methanol (2:1) (Silverberg *et al.*, 1976). The aqueous phase was dialyzed against distilled water and processed with the senescent cell IgG affinity column. TX-100 extracts were processed with the affinity column without chloroform–methanol treatment.

E. Polyacrylamide Gel Electrophoresis

Electrophoresis was performed using gradient gels containing from 2–16 or 4–30% acrylamide and the buffer system of Fairbanks *et al.* (1971) except that 0.02% sodium dodecyl sulfate (SDS) was used. The sample buffer was 10 mM Tris–HCl, pH 8.0, containing 1 mM Na-EDTA, 40 mM dithiothreitol, 5% glycerol, 0.02% SDS, and bromphenol blue. Gels were stained for proteins with Coomassie blue and for glycoproteins with dansyl hydrazine following the method of Eckhardt *et al.* (1976) with the modifications of Lutz *et al.* (1979). Molecular weights of 250,000 for band 1, 100,000 for band 3, 82,000 for band 4.1, 72,000 for band 4.2, and 45,000 for band 5 were used for calculating the mol. wt. of the IgG binding receptor and its "parent" molecule.

F. Erythrophagocytosis-Inhibition Assay

IgG was isolated by ion-exchange chromatography (Kay, 1975a, 1978b), and absorbed with the peptide eluted from the senescent cell IgG affinity column or the sialoglycoprotein mixture from which the peptide was isolated. The absorbed IgG was then incubated with RBC stored for 1 week in AMEM with 10% fetal calf serum. Stored RBC were used because they have the exposed receptor to which IgG isolated from senescent RBC aged *in situ* binds (Kay, 1975a, 1979). Senescent RBC could not be used because they already have IgG bound to the receptor. RBC were washed three times in AMEM and incubated with autologous macrophage cultures (Kay, 1975a, 1978b) (Figure 1).

Macrophages were isolated by centrifugation on performed Percoll gradients (Cochrum *et al.*, 1978; Bennett and Kay, 1981). After three washes, 5×10^5 macrophages in AMEM with 10% fetal calf serum were pipeted into sterile glass tubes (Kay, 1975a, 1979; Bennett and Kay, 1981). The culture medium was changed daily. These cultures contain 99–100% macrophages, as determined by esterase staining of cells recovered from the tubes at the end of the experiment and parallel cultures performed on coverslips in Leighton tubes. After 3–7 days, the medium was changed, and 10 RBC per macrophage were added. RBC were incubated with macrophages for 16–25 hr, and the percentage of RBC phagocytized was determined as described previously (Kay, 1975a, 1978b).

Lymphocytes, neutrophils, and platelets were isolated by centrifugation on Percoll gradients (Bennett and Kay, 1981). They were washed and stored at 4°C for 48 hr. The specific IgG autoantibody eluted from senescent cells (6 μg) was added to 1-ml suspensions of $\simeq 2 \times 10^7$ platelets and incubated for 30 min at 37°C and 30 min at 24°C (1,6,7,12). Positive controls consisted of autoantibody incubated with media only. Cells were removed by centrifugation and the supernate containing the remaining antibody was added to 5×10^8 stored RBC. RBC were indubated for 15 min at 37°C and 1.5 hr at 24°C. Aliquots of each sample (0.1 ml) containing 5×10^7 RBC were added to macrophage cultures.

The amount of IgG or antibody used for absorption and the number of cells used was based on preliminary dose–response experiments. The phagocytosis-inducing ability of the senescent cell IgG could not be absorbed with freshly isolated young RBC (43 ± 9% phagocytosis) but was abolished by absorption with stored RBC (0% phagocytosis).

Human adult liver cells (Chang) and primary cultures of human embryonic kidney cells (Flow 4000) were obtained from Flow Laboratories. Cells were stored in liquid nitrogen, thawed, and grown in tissue culture 7 days prior to use. Media was changed 2 days after initiation of culture. The less viable nonadherent cells were harvested by centrifugation and resuspended in 1 ml of AMEM containing 10% fetal calf serum.

IgG eluted from senescent cells (3 μg) was incubated with 2.1×10^6 liver cells (57% did not exclude trypan blue) or 5×10^6 kidney cells (82% did not exclude trypan blue) for 20 min at 37°C and 2 hr at 24°C. Cells

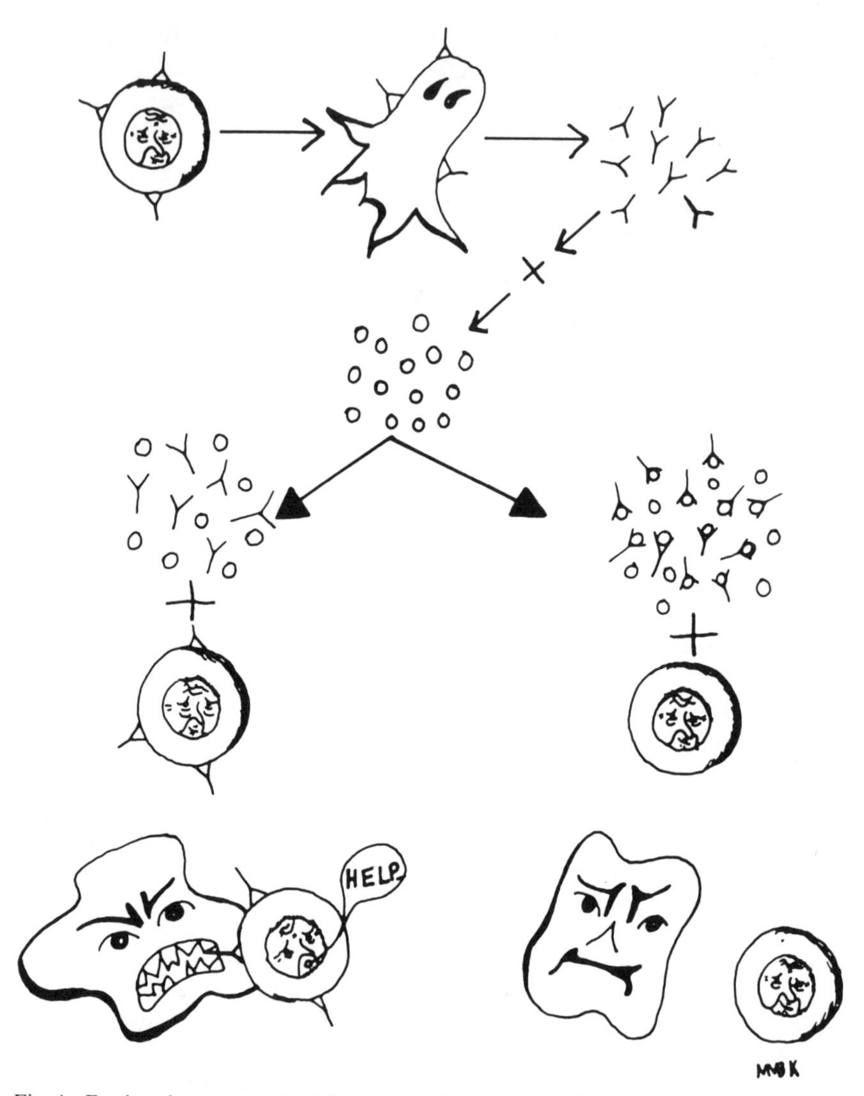

Fig. 1. Erythrophagocytosis-inhibition assay. Senescent RBC which carry IgG (Y) on their surface are washed, and ghosts are prepared by digitonin lysis. IgG is eluted from senescent RBC with glycine–HCl buffer, pH 2.3. IgG is then incubated with a molecule (○) or cell to determine whether it carries the antigenic determinants recognized by senescent cell IgG. If the molecule is the senescent cell antigen, then IgG binds via its Fab region to the antigen (right column). The Fab region of IgG is blocked and thus is not free to bind stored RBC which are mixed with it. Because the stored RBC lack IgG, They are not phagocytized when they are incubated with macrophages. If, on the other hand, the molecule which is mixed with the senescent cell IgG is not the senescent cell antigen, then IgG does not bind to it (left column). The Fab region of IgG is free to bind to stored RBC which are phagocytized when incubated with macrophages.

were removed by centrifugation, and the supernate was incubated with stored RBC. The phagocytosis-inhibition assay was performed as described above.

III. RECOGNITION AND REMOVAL OF SENESCENT CELLS

Investigation of the mechanism by which senescent cells are removed from the body was approached by postulating that Ig in normal serum attaches to the surface of senescent human red cells until a critical level is reached, which results in removal of these cells by macrophages (Kay, 1974, 1975a). Testing of this hypothesis was initiated by asking whether immunoglobulin alone could initiate phagocytosis of unaltered, freshly drawn RBC by macrophages.

A. Requirement for Ig

Freshly drawn human RBC were incubated with the IgG fraction of rabbit anti-human RBC antibody for 30 min at 37°C, washed, and added to macrophage cultures. Controls consisted of RBC incubated in medium or Ig-depleted serum. Sixty-three percent of the RBC (standard error of mean, 19%) treated with rabbit anti-human RBC reagent were phagocytized, whereas only 3–7% phagocytosis was observed in the controls. Rabbit IgG was demonstrated on the surface of the RBC with scanning immunoelectron microscopy (Kay, 1975a). These results indicate that IgG can initiate phagocytosis and might be required. However, the RBC were deliberately coated with foreign IgG in the form of a specific antibody. Therefore, it could not be determined from this procedure whether normal circulating Ig would attach to RBC. In order to test this, stored human RBC were incubated in medium, autologous Ig, autologous IgG, another individual's allogeneic Ig, or pooled, normal human (PNH)-IgG, PNH-IgM, PNH-IgA, washed, and incubated with the individual's macrophages. The results of these experiments (Fig. 2) demonstrated that (1) the percentage phagocytosis of stored RBC ("0" RBC) incubated in IgM and IgA is only slightly more than that in medium alone (<10%); (2) the percentage phagocytosis of stored RBC incubated in autologous IgG was essentially the same as that of stored RBC incubated in either autologous whole serum or the Ig fraction (42–49%); and (3) allogeneic Ig and PNH-IgG (≃30% phagocytosis) were not as effective as autologous whole serum, Ig, or IgG in promoting phagocytosis, although they had not been tested for blood group compatibility. These results, which confirm the

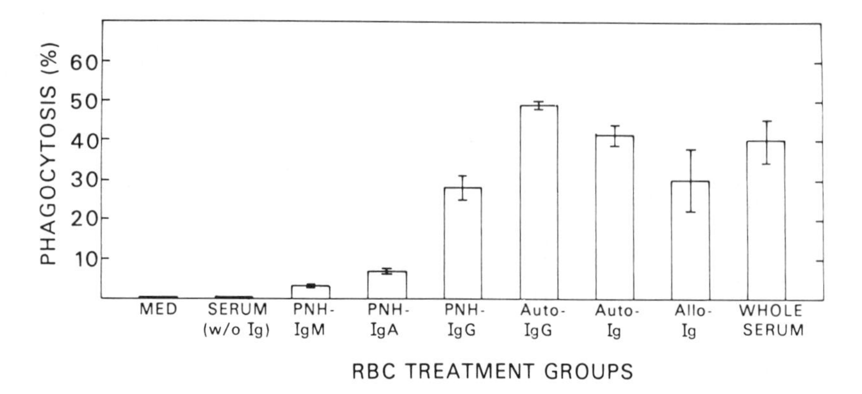

Fig. 2. Phagocytosis of stored RBC. Stored RBC were washed and incubated in Medium 199 (Med), autologous Ig-depleted or agamma serum (Serum w/o Ig), pooled normal human IgM, IgA, or IgG (PNH-Ig), autologous IgG (Auto-IgG), autologous Ig (Auto-Ig), allogeneic Ig (Allo-Ig), or autologous, fresh whole serum. RBC were then washed and incubated with macrophages. Vertical bars indicate one standard error of the mean. (From Kay, 1975a.)

requirement for Ig, indicate that normal circulating Ig can attach to RBC, and suggest that the Ig that attaches is IgG. These experiments support but do not prove the initial working hypothesis which states that immunoglobulins attach to the surface of aging RBC until a threshold level is reached, at which time macrophages phagocytize the cell.

B. Phagocytosis of RBC Aged *in Situ* and Detection of IgG on Their Surface

In an attempt to test this hypothesis directly, and to determine whether the Ig which attached *in situ* was IgG, the following experiments were performed.

Freshly drawn human RBC were separated into young and old populations according to their different densities. Aliquots from each population were incubated with scanning immunoelectron microscopy marker conjugates and prepared for scanning electron microscopy (Kay, 1975a, 1977, 1978a,b). At the same time, each population was incubated with macrophages in medium, Ig-depleted serum, or whole serum. Old RBC were phagocytized regardless of whether the final incubations were performed in medium *without* serum, in autologous Ig-*depleted* serum, or whole serum containing Ig (Fig. 3). This suggested that Ig was attached *in situ* to the RBC and that phagocytic recognition was not inhibited by other serum components. Scanning immunoelectron microscopy of the two populations revealed that young RBC were essentially unlabeled, whereas senescent RBC were labeled with SV40 anti-human IgG but not with T2 anti-human IgA or with KLH anti-human IgM (Kay, 1975a). The number of IgG molecules on senescent RBC varied between 40 and 100 per half cell. On the basis of these findings, it can be concluded that IgG attaches *in situ* to senescent human RBC.

These results were confirmed *in vivo* using mice which were bred and maintained in a Type I Maximum Security Barrier. RBC labeled *in situ* with ^{59}Fe were separated on Percoll gradients into young and old populations (Figs. 4 and 5) and injected into separate groups of syngeneic mice (Fig. 6). Kinetic studies revealed that 90% of the ^{59}Fe-labeled young RBC were removed from the circulation within 45 days. In contrast, 90% of the ^{59}Fe-labeled old RBC were removed within 20 days (Fig. 5). The difference in the rate of removal of young and old RBC was statistically significant ($P \leq 0.001$). Kinetic studies on density-separated spleen cell populations revealed that the radioactivity decreased in the RBC fraction concomitantly with an increase in radioactivity in the splenic macrophage fraction (Fig. 7). The radioactivity was found to be inside macrophages (Bennett and Kay, 1981).

Studies performed *in vitro* with mouse splenic macrophages and autologous young and old RBC revealed that mouse macrophages phagocytized senescent but not young RBC ($P \leq 0.001$; Fig. 8). The phagocytosis of middle-aged RBC ($\simeq$23%) was intermediate between that of young RBC ($\simeq$5%) and old RBC ($\simeq$50%). This suggests that the appear-

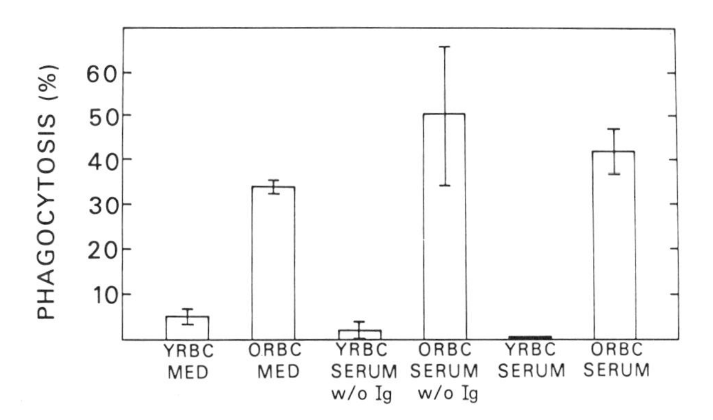

Fig. 3. Phagocytosis of RBC aged *in situ*. Freshly drawn RBC were separated by density into young and old RBC popultions, washed, resuspended in Medium 199 (Med), autologous gamma serum (Serum w/o Ig), or autologous fresh whole serum (Serum), and incubated with autologous macrophages. Vertical bars indicate 1 S.E.M. (From Kay, 1975a.)

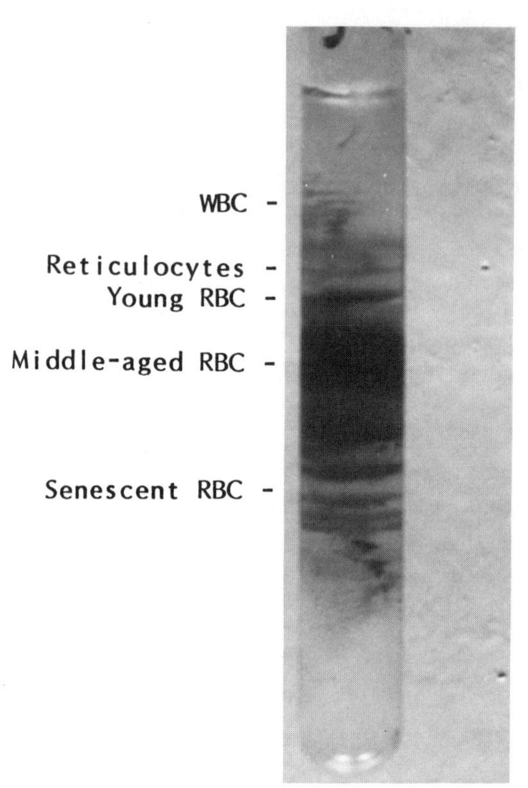

Fig. 4. RBC separated into populations of different ages on a Percoll density gradient. The density of each fraction was determined using a parallel gradient run with density markers beads (Pharmacia).

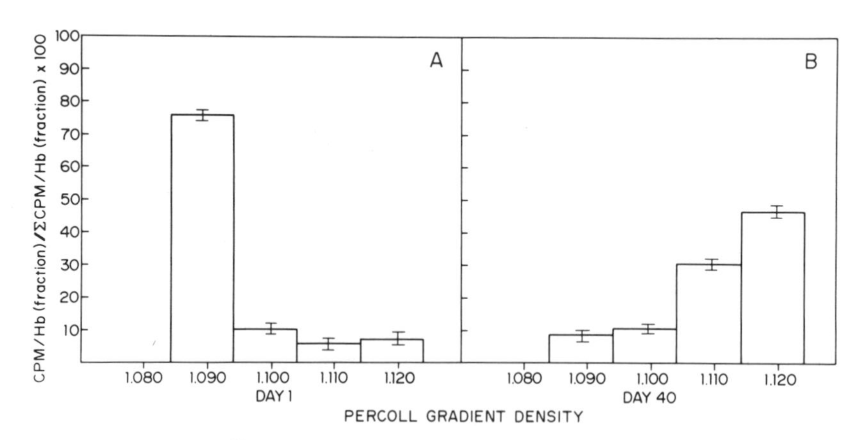

Fig. 5. Distribution of ^{59}Fe-labeled murine RBC on a Percoll density gradient on Days 1 and 40 after ^{59}Fe pulse-labeling. Seventy-six percent of the radioactivity is located in the least dense RBC fraction on Day 1 (A). The radioactivity shifts to the most dense fractions by Day 40 (B). (From Bennett and Kay, 1981.)

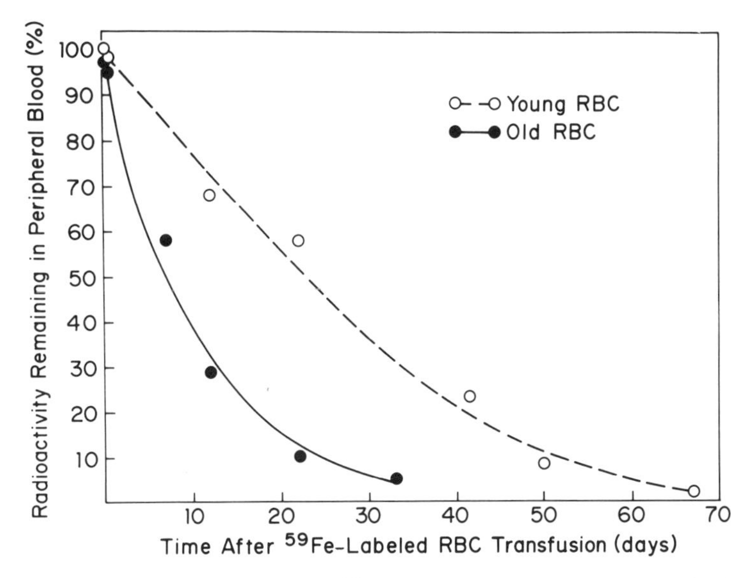

Fig. 6. Decrease with time of circulating ^{59}Fe-labeled young and senescent RBC after transfusion into mice. Senescent RBC are cleared significantly faster than young RBC ($P \leq 0.001$). Standard deviations for each point are less than one. (From Bennett and Kay, 1981.)

ance of the antigen to which autologous IgG binds, and, thus, molecular aging of membranes may be a cumulative process.

The results presented thus far suggest that macrophages distinguish senescent from mature RBC both *in vitro* and *in vivo* on the basis of selective attachment of autologous IgG to the membrane of senescent RBC. However, the presence of IgG on senescent cells does not indicate that an immunological binding has occurred. The IgG could be attached by nonspecific absorption or by RBC Fc receptors. For example, it has been shown that gamma globulin attaches to RBC in low ionic strength sucrose solutions (Fidalgo *et al.*, 1967; Najjar, 1974). However, the gamma globulin which attaches under these conditions is readily eluted by isotonic saline (Fidalgo *et al.*, 1967; Najjar, 1974). Since physiological salt solutions were used throughout the experiments described here, it is unlikely that the antibody being investigated was such a cytophilic antibody. Nonetheless, the IgG–red cell binding reaction was examined to determine whether the binding was immunological and to determine whether the antibody involved was an autoantibody. Definitive evidence for the role of autoantibodies in the selective removal of senescent cells can be obtained by first dissociating the antibodies from senescent cells, and then by demonstrating their specific immunologic reattachment via the Fab portion of the IgG molecule to homologous senescent, but not mature, cells.

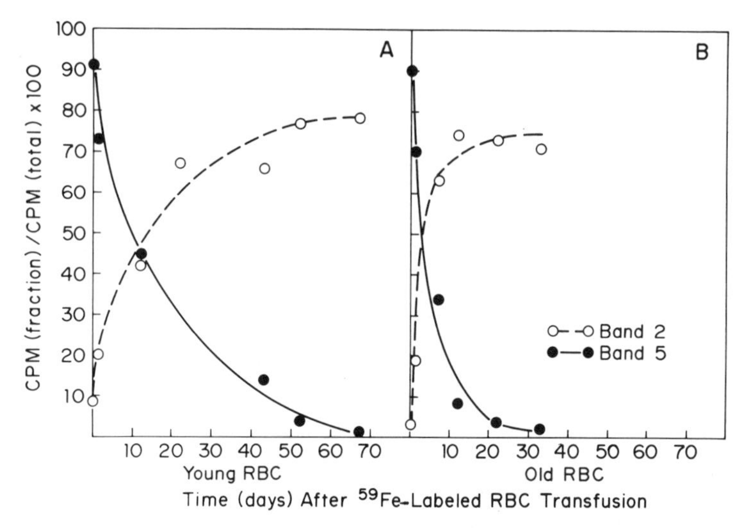

Fig. 7. Appearance of ^{59}Fe label in the splenic macrophage fraction of Percoll gradients (band 2) relative to the disappearance of ^{59}Fe label from the RBC fraction (band 5). ^{59}Fe-labeled young or old RBC were injected intravenously into BALB/C mice. At varying time intervals thereafter, the spleens were removed. Spleen cell suspensions were layered on Percoll gradients (formed by centrifugation at 27,000 × *g* for 30 min) and centrifuged at 450 × *g* for 10 min. Macrophages were isolated from the least dense fraction and RBC from the densest fraction. The rate of clearance of radioactivity in the RBC fraction (band 5) of gradients from mice receiving old RBC was more rapid than that of mice receiving young RBC ($P \leq 0.01$). Likewise, the appearance of radioactivity in the splenic macrophage fraction (band 2) of gradients from mice receiving old RBC was more rapid than that of mice receiving young RBC ($P \leq 0.01$). (From Bennett and Kay, 1981.)

C. Identification and Characterization of Senescent RBC Ig

Ig was eluted from RBC aged *in situ*. It was shown to be an IgG containing kappa and lambda light chains by immunodiffusion and immunoelectrophoresis (Kay, 1978b). Other Ig's were not detected by immunodiffusion (ID), immunoelectrophoresis (IEP), or polyacrylamide gel electrophoresis. Thus, the antibody attached to senescent cells is an IgG which is polyclonal with respect to light chains.

D. IgG Binding and Specificity

In order to determine whether the IgG eluted from old RBC aged *in situ* would reattach to homologous cells, the IgG was incubated with autologous or allogeneic young, stored RBC. These RBC were then washed and incubated with autologous macrophages. The percentage phagocytosis of stored RBC incubated with autologous IgG and then with autologous macrophages (46 ± 0%) was essentially the same as that of RBC aged *in situ* (50 ± 4%). Macrophages phagocytized autologous stored RBC incubated with autologous IgG (27 ± 0% phagocytosis), as well as allogeneic stored RBC incubated with autologous IgG (56 ± 4% phagocytosis). However, they did not phagocytize allogeneic cells which had not been incubated with IgG (0% phagocytosis) nor did they phagocytize young allogeneic cells which were incubated with allogeneic IgG

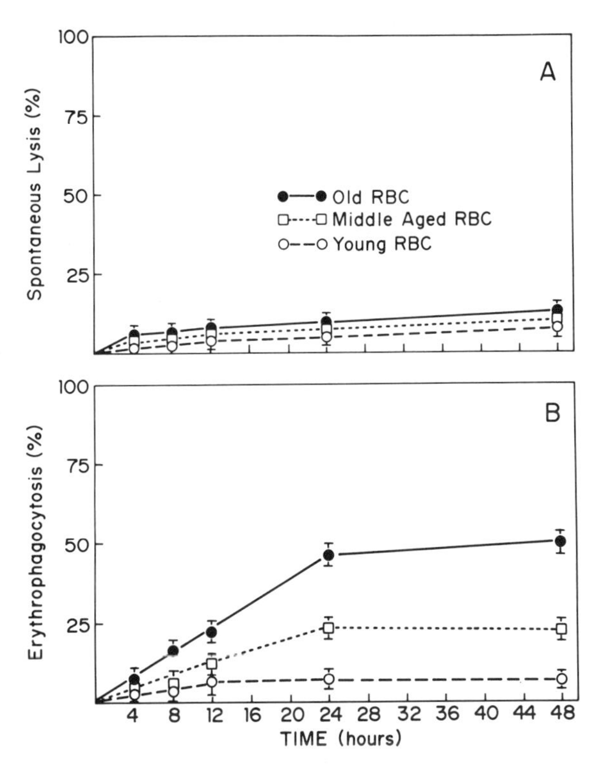

Fig. 8. (A) Kinetics of spontaneous lysis (%) of young, middle-aged, and senescent RBC cultured in triplicate without macrophages. Cultures were incubated at 37°C in humidified air containing 5% CO_2 for 48 hr. RBC were counted in 10-μl aliquots at the time points indicated. (B) Kinetics of phagocytosis (%) of autologous young, middle-aged, and senescent RBC by isolated splenic macrophages *in vitro*. Macrophages were cultured with RBC at 37°C in humidified air containing 5% CO_2 for 48 hr. RBC were counted in 10-μl aliquots taken at the time points indicated. The percentage phagocytosis was determined as described in the text. Macrophages phagocytized significantly more senescent RBC than middle-aged or young RBC ($P \leq 0.001$). (From Bennett and Kay, 1981.)

(5 ± 3% phagocytosis). IgG was demonstrated on the surface of stored RBC incubated with IgG eluted from autologous or allogeneic cells with scanning immunoelectron microscopy. Absorption of the eluted IgG with stored RBC, but not with freshly isolated young RBC, abolished its phagocytosis-inducing ability (Table I).

These binding and specificity experiments indicate that (1) IgG is required for the phagocytosis of stored autologous and allogeneic RBC; (2) nonspecific binding of IgG does not play a major role in these experiments because absorption of both PNH-IgG and IgG eluted from senescent cells with stored RBC abolishes its phagocytosis-inducing activity; (3) IgG eluted from senescent RBC is reactive against stored, but not young, cells; and (4) IgG eluted from senescent RBC cannot discriminate between autologous and allogeneic cells. The last two findings suggest that the antigen appearing on the surface of cells aged *in situ* and that appearing on stored cells is the same, or closely related, for all individuals.

To determine whether the Fab or Fc portions of IgG attaches to RBC, antigen blockade studies were performed. The results are summarized in Fig. 9. Stored RBC were incubated for 30 min with either PNH-IgG, its Fab, or its Fc fragment. The RBC were washed, and all three groups were incubated with IgG for another 30 min. RBC were washed again and incubated with autologous macrophages (Fig. 9A). For the control cultures, stored RBC were treated with IgG, Fab, or Fc before they were exposed to mononuclear phagocytes (Fig. 9B). Control results show that phagocytosis was achieved by exposing the RBC to IgG. Treatment with Fab or Fc did not promote phagocytosis. Experimental results show that treatment of RBC with Fab prior to incubation with IgG reduced the phagocytosis to essentially zero, whereas pretreatment with Fc did not inhibit phagocytosis (Fig. 9A). Thus, these blockade studies dem-

Table I
Phagocytosis of Stored RBC ("O" RBC) Incubated with IgG Eluted from Senescent Cells before and after Absorption with Stored RBC or Freshly Isolated Young RBC (YRBC)

		Phagocytosis (% ± S.E.M.)[b]		
Experiment No.	Quantity[a]	Before absorption	After absorption with YRBC	After absorption with "O" RBC
1	3	49 ± 2	43 ± 9	0
2	3	35 ± 1	34 ± 7	0
3	3	43 ± 11	46 ± 16	0

[a] Quantity (μg) of IgG added to 1.5 × 10^8 RBC in 1 ml of Medium 199.
[b] S.E.M., standard error of the mean of triplicate or quadriplicate cultures.

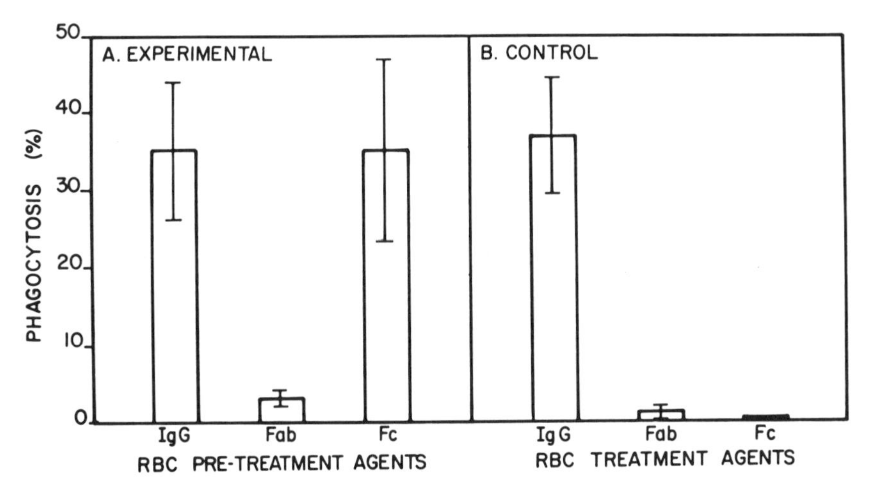

Fig. 9. (A) Susceptibility to phagocytosis of aged RBC as influenced by their pretreatment with either IgG, Fab, or FC before exposure to IgG. Stored RBC were incubated with either IgG, Fab, or Fc for 30 min, washed, and incubated with IgG for 30 min. The RBC were washed and incubated with mononuclear phagocytes for 3 hr. (B) Susceptibility to phagocytosis of aged RBC as influenced by their exposure to either IgG, Fab, or Fc. Stored RBC were incubated with either IgG, Fab, or Fc for 30 min, washed, and then incubated with mononuclear phagocytes for 3 hr. Bars indicate standard error of the mean; sample size, 6. (From Kay, 1978b.)

onstrate that IgG binds to old RBC via its Fab region. Binding of Fab but not Fc was also demonstrated with scanning electron microscopy. Binding of the IgG molecule to aged RBC via the Fab region is consistent with an immunological binding of IgG with a surface antigen. The Fc portion of IgG would then be available to the macrophage Fc receptor.

These experiments suggest that the IgG eluted from senescent RBC is an autoantibody, as it specifically reattaches to homologous RBC via its Fab region and initiates their selective destruction by macrophages.

Other investigators have confirmed the presence of IgG on senescent, damaged, and stored red cells (Tannert, 1978; Durocher, 1978; Wegner *et al.*, 1980; Halbhuber *et al.*, 1980). Alderman *et al.* (1980) found immunoglobulin on the surface of density-fractionated, stored and fresh crenated red cells washed and maintained in saline *without* calcium and magnesium. Both calcium and magnesium are required for membrane integrity. Since they detected the immunoglobulin by immunofluorescence, they were dealing with much larger quantities of Ig than is observed on freshly isolated senescent cells.

IgG has been demonstrated on the surface of senescent and stored red cells (Tannert, 1978; Wegner *et al.*, 1980) using ^{125}I-labeled anti-IgG and IgG. The amount of IgG on the surface of cells increases with storage (Wegner *et al.*, 1980; Kay, unpublished; Halbhuber *et al.*, 1980); Halbhuber (personal communication) demonstrated IgG on the surface of red cells stored by the blood bank. Wegner *et al.* (1980) have calculated that IgG bound to freshly washed RBC has an association constant of 1×10^{14} M^{-1} indicating a high affinity of IgG for the antigen. In addition, they found that ATP depletion caused a seven-fold increase in IgG binding after a delay of $\simeq$18 hr (Wegner *et al.*, 1980; Tannert, personal communication). The finding that there is no correlation between erythrocyte life span and ATP level *in situ* in humans with aplastic anemias that are not due to RBC alterations (Syllm-Rapoport *et al.*, 1969) suggests that ATP depletion is not the primary or causative event in the appearance of a senescent cell antigen.

IV. SENESCENT CELL ANTIGEN ISOLATION

The experiments described thus far indicate that macrophages distinguish between mature and senescent cells on the basis of selective IgG binding to the latter. Binding of an IgG autoantibody to senescent RBC through immunologic mechanisms indicates that antigenic determinants recognized by these IgG autoantibodies emerge on the surface membrane as RBC senesce. However, these experiments do not provide insight into membrane changes during cellular aging nor do they reveal the mechanism responsible for appearance of a "neo-antigen." In order to obtain this information, it is necessary to isolate and identify the antigen to which IgG binds on senescent RBC.

A. Use of Vesicles to Distinguish between a Cryptic and a Neo-Antigen

Previous studies on neuraminidase-treated young RBC, which indicated increased IgG binding to RBC (Kay, 1975a, 1978b; Kay and Baker, 1980; Jancik and Schauer, 1974) and a rapid clearance of these RBC from the circulation (Jancik and Schauer, 1974; Durocher *et al.*, 1975; Aminoff *et al.*, 1976, 1977; Jancik *et al.*, 1978), suggested that the senescent cell antigen may be a sialoglycoprotein. For this reason, sialoglycoproteins were extracted from membranes of RBC using TX-100 (Lutz *et al.*, 1979). Vesicles were reconstituted from extracts enriched in sialoglycoproteins in order to remove the TX-100.

Vesicles reconstituted from glycophorin-enriched extracts obtained from human RBC contained sialoglycoprotein bands PAS_1 through PAS_4

and trace contaminants of other minor proteins, some of which were not sialoglycoproteins (Lutz and Kay, 1981).

IgG binding to the vesicles from unseparated RBC could be demonstrated by binding of ^{125}I-labeled anti-human IgG and ^{125}I-labeled protein A (Fig. 10). Binding of protein A to IgG attached to vesicles supports the earlier finding that the Fab portion of IgG binds to the senescent cell antigen, since protein A binds to the Fc region of IgG. The phagocytosis-inducing ability of IgG was absorbed by vesicles as determined by the erythrophagocytosis-inhibition assay (Fig. 10).

Vesicles also absorbed the erythrophagocytosis-inducing ability of IgG eluted from senescent RBC. When IgG (10 μg) eluted from senescent RBC is absorbed with 50 μg of vesicle–protein prior to incubation with RBC, 7 ± 4% of the RBC were phagocytized; whereas 45 ± 10% were phagocytized when eluted IgG that was *not* absorbed was added to RBC. These findings indicated that the vesicles contain the antigen present on senescent RBC to which antibodies bind.

If the appearance of an IgG binding antigen on senescent RBC were due to exposure of a hidden or cryptic antigen, solubilization of the membranes of both young and old cells with a detergent should render the antigen accessible to IgG. On the other hand, if the antigen were an altered membrane component, then the antigen should be extracted primarily from senescent cells by detergent treatment.

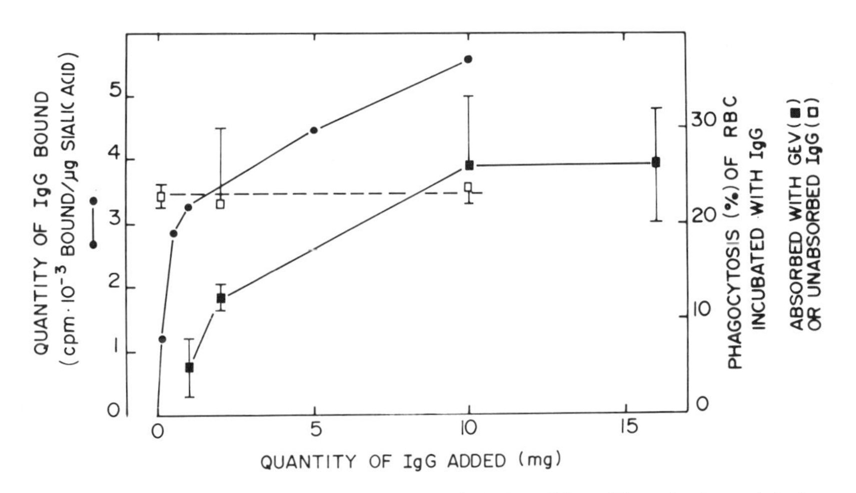

Fig. 10. IgG binding capacity of glycophorin-enriched vesicles. Glycophorin-enriched vesicles were isolated from unseparated RBC and incubated with IgG. IgG binding was assessed by binding of ^{125}I-labeled protein A and the erythrophagocytosis-inhibition assay. The data are given for 400 μg vesicle–protein. (From Lutz and Kay, 1981.)

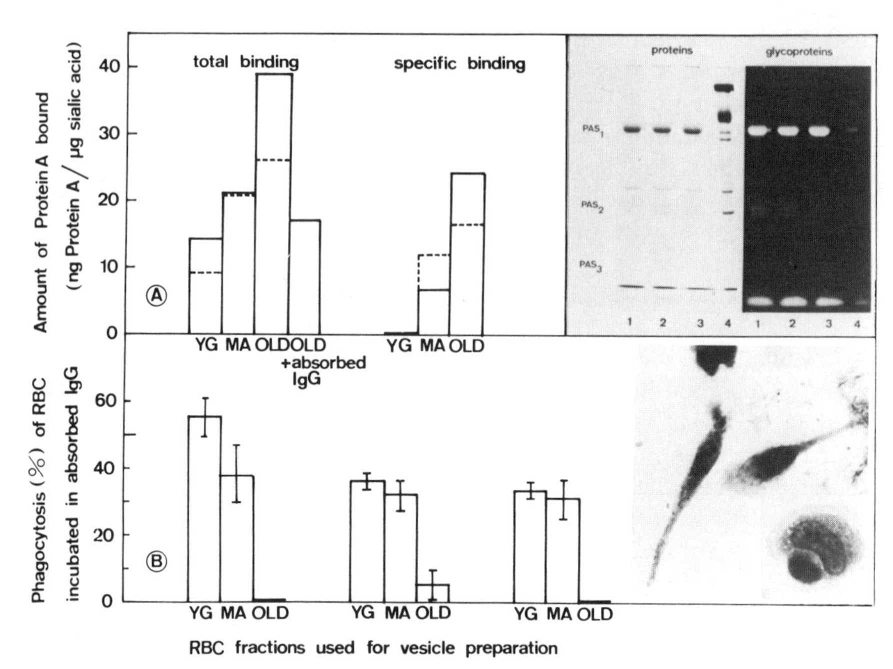

Fig. 11. (A) Binding of IgG to glycophorin-enriched vesicles from young (YG), middle-aged (MA), and old RBC. Glycophorin-enriched vesicles were prepared from membranes of young, middle-aged, and old human RBC. Solid line: 2.3 mg IgG was added per 400 μg vesicle–protein. Following washing, 6 μg protein A (5.53×10^5 cpm/μg) was added to each tube. The creatine contents of the cells were 204 (young), 53 (middle-aged), 38 (old) μg/10^{10} cells. Dashed lines: 13.3 mg IgG was added per 400 μg vesicle–protein. Following washing, 3 μg protein A (1.37×10^6 cpm/μg) was added to each tube. The creatine contents were 139 (young), 78 (middle-aged), and 51 (old). Similar results were obtained with ^{125}I-labeled anti-human IgG: the binding was 97 and 54 ng of anti-human IgG for vesicles of old and young cells, respectively. The creatine contents were 38 (old) and 87 (young) μg per 10^{10} cells. Photograph shows glycoprotein and protein composition of vesicles. Each gel was run with 15 μg protein. 1, Vesicles from senescent RBC; 2, vesicles from middle-aged RBC; 3, vesicles from young RBC; 4, RBC membranes. The samples shown here correspond to those used for IgG binding studies (solid line experiment in A). Gels contained 10% acrylamide. (From Lutz and Kay, 1981.) (B) Phagocytosis-inducing ability of IgG absorbed with vesicles from young (YG), middle-aged (MA), or old fractions. IgG (125 μg) isolated by column chromatography (1, 2) was absorbed with 50 μg of glycophorin-enriched vesicle–protein from young, middle-aged, or old cell populations for 24–48 hr at 4°C. The absorbed IgG was incubated with stored RBC which were then washed and incubated with macrophages. After a 16 to 24-hr incubation, the percentage of RBC phagocytized was determined. The ratios of the creatine contents of young to old RBC fractions from which vesicles were prepared were 3.2, 1.9, and 2.3 (left to right). (From Lutz and Kay, 1981.) Photograph shows macrophage cultures stained with esterase prior to addition of red cells. Inset shows a cell partially engulfed by a macrophage at the end of a phagocytosis experiment.

The reconstituted glycophorin-enriched vesicles from TX-100 extracts of young, middle-aged, and old RBC populations had the same protein composition and the same sialic acid content (Lutz and Kay, 1981). Binding of serum IgG to these vesicles was measured by the amount of bound ^{125}I-protein A. The results from two independent experiments showed that total binding of protein A to vesicles from senescent RBC is three- to four-fold higher than to vesicles from young cells (Fig. 11A).

Erythrophagocytosis-inhibition experiments were carried out with IgG absorbed with vesicles from young, middle-aged, and old fractions. The results showed that absorption of IgG with vesicles prepared from senescent RBC abolished the phagocytosis-inducing ability of serum IgG, whereas absorption with vesicles prepared from young cells did not (Fig. 11B).

Addition of lipids to the vesicles did not alter IgG binding (Lutz and Kay, 1981). Absorption of IgG with liposomes prepared from lipid extracts of young and old cells did not alter its phagocytosis-inducing ability (Lutz and Kay, 1981). Results of the above experiments indicate that the antigen to which IgG binds is an age-specific cell antigen, because it is extracted primarily from senescent RBC. Thus, the data suggest that the antigen may be an altered membrane component or "neo-antigen" rather than a cryptic antigen. Approximately 10–30% as much antigen can be extracted from middle-aged as from senescent RBC, suggesting that appearance of the senescent cell antigen is a cumulative process. This implies that cells "remember" environmental insults.

B. Isolation of the Senescent Cell Antigen from Sialoglycoprotein Extracts

Siologlycoprotein preparations were obtained by extraction of washed, right-side-out ghosts with 0.3 M LIS or 1.0% TX-100 and partitioned in a phenol–water mixture (Marchesi and Andrews, 1971). IgG was eluted by the method of Kochwa and Rosenfield (1964; Kay, 1978b) and isolated from eluates with protein A-Sepharose 4B. IgG eluted from senescent RBC was coupled to CNBr-activated Sepharose 4B.

Sialoglycoprotein mixtures were processed with the senescent cell IgG affinity columns. The material bound to the column was eluted with glycine–HCl buffer. Both dansyl hydrazine and Coomassie blue stains of gels of the eluted material revealed a single band migrating at a relative molecular weight of ≃62,000 which appears below band 4.2 in Fig. 12. When this experiment was repeated using 1% TX-100 extracts of RBC membranes, the same results were obtained. These experiments suggest

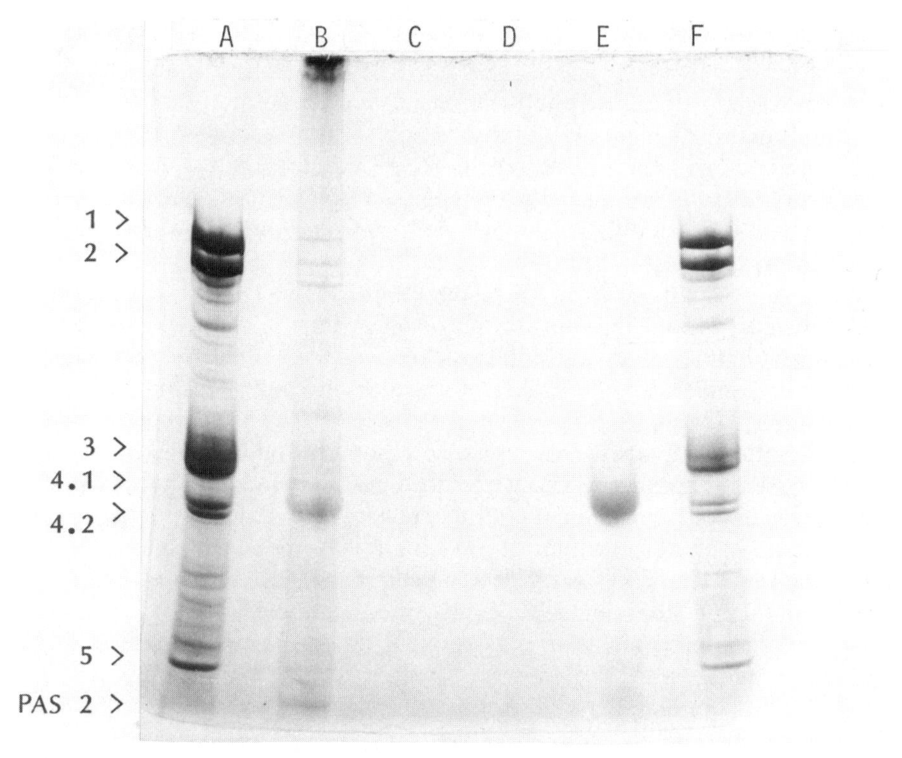

Fig. 12. Isolation of the senescent cell antigen from human RBC sialoglycoproteins with affinity columns prepared from the IgG eluted from senescent cells. A, RBC ghost; B, TX-100 extract of ghosts; C, ≃62,000-dalton peptide isolated with the senescent cell IgG affinity column from 1% TX-100 extracts of ghosts; D, ≃62,000-dalton peptide isolated from LIS extract of ghosts; E, LIS extract of ghost; F, RBC ghost. Gels were stained with Coomassie blue. (From Kay, 1981b.)

that the ≃62,000-dalton glycopeptide carries the antigenic determinants recognized by IgG obtained from freshly isolated senescent cells.

To confirm this finding, an erythrophagocytosis-inhibition assay was performed. IgG was absorbed with either the 62,000-dalton peptide eluted from senescent cell IgG affinity columns or with the sialoglycoprotein mixture that had not bound to these affinity columns. The absorbed IgG was then incubated with stored RBC. The phagocytosis-inhibition assay was performed as described previously (Kay, 1975a, 1978b, 1981a, 1981b). RBC were washed and incubated with autologous macrophage cultures. RBC incubated with the IgG absorbed with excess 62,000-dalton peptide were not phagocytized (0% phagocytosis), indicating that no free antigen-binding IgG remained in solution. In contrast, 52 ± 3% (mean ± S. D.) of the RBC incubated with IgG absorbed with the eluate that was not bound by senescent cell affinity columns were phagocytized, indicating the presence of free antigen-binding IgG. The same amount of phagocytosis was observed when stored RBC were incubated with IgG that was *not* absorbed (55 ± 6%). Thus, the 62,000-dalton peptide, but not the remaining sialoglycoprotein mixture from which it was isolated, abolishes the phagocytosis-inducing ability of IgG. This indicates that the 62,000-dalton peptide is the antigen which emerges on the membrane of cells as they senesce.

Other somatic cells were then examined for the presence of the antigen which appears on senescent RBC. Macrophages, lymphocytes, platelets, and neutrophils were isolated by centrifugation on Percoll density gradients (Bennett and Kay, 1981). Cells were stored at 4°C for 48 hr. They were then washed and incubated with the autoantibody eluted from senescent cells. Cells were removed by centrifugation, and the supernates containing any remaining IgG were incubated with stored RBC. Stored RBC were washed and incubated with macrophages. The percentage phagocytosis observed in control cultures containing IgG autoantibody that was not absorbed was 37±4 (mean ± 1 S. D.). The phagocytosis-inducing ability of the autoantibody eluted from senescent cells was absorbed by stored lymphocytes (0% phagocytosis), neutrophils (0%), and platelets (2±2%).

Cultured human adult liver cells and primary cultures of human embryonic kidney cells were also examined for presence of the antigen. The less viable cells which were no longer adherent were harvested from culture supernates by centrifugation (Kay, 1981a,b). Phagocytosis was 38±6% in control cultures. Absorption of the IgG eluted from senescent RBC with both liver and embryonic kidney cells abolished its phagocytosis-inducing ability (0 and 7±7%, respectively).

Since the results of the phagocytosis-inhibition-assay experiments indicated that the antigen that appears on senescent RBC also emerges on other somatic cells, isolation of the antigen from lymphocytes was attempted. Sialoglycoprotein mixtures were prepared from lymphocyte membranes in the same manner as described for RBC. The lymphocyte sialoglycoprotein mixture was processed with the senescent RBC IgG affinity column. Get electrophoresis of the material eluted from the column with glycine–HCl buffer revealed a band migrating at a mol. wt. of ≃62,000, at the same position as the antigen isolated from senescent RBC (Fig. 13). This finding confirms the results obtained with the phagocytosis-inhibition assay, indicating that the antigen which appears on senescent RBC also appears on other somatic cells. Emergence of the ≃62,000-dalton antigen on RBC initiates binding of IgG autoantibodies *in situ* and phagocytosis of senescent cells by macrophages. The antigen is present on stored lymphocytes, platelets, and neutrophils and on cultured liver

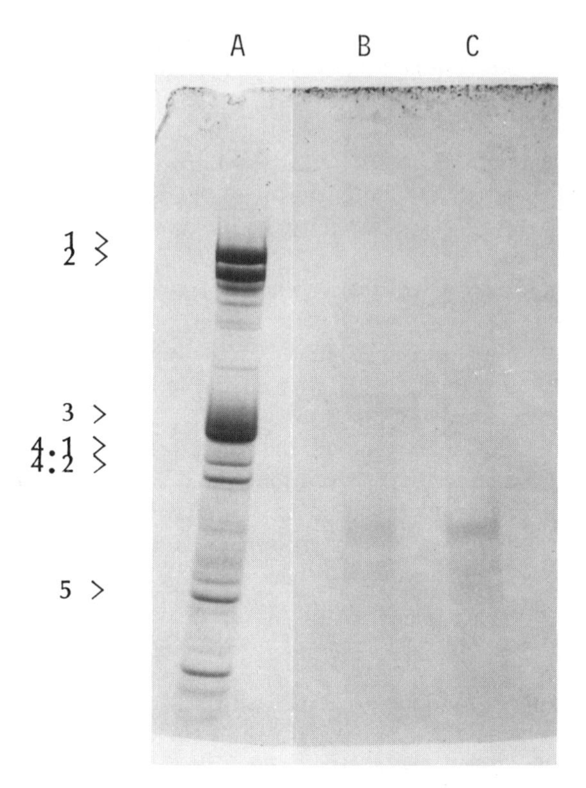

Fig. 13. Isolation of the antigen for IgG autoantibodies from human lymphocyte sialoglycoproteins with affinity columns prepared from the IgG eluted from senescent RBC. A, RBC ghost; B, 62,000-dalton peptide isolated with the senescent cell IgG affinity column from LIS extract of RBC ghost; C, 62,000-dalton peptide isolated with the senescent RBC IgG column from LIS extract of lymphocyte membranes. Lymphocytes (8×10^8) were isolated from 500 ml of blood by centrifugation on Percoll gradients (Bennet and Kay, 1981). After three washes, membranes were prepared by hypotonic lysis in 5 mM $NaPO_4$ and 1 mM EDTA, pH 7.6 (Dodge *et al.*, 1963). Membrane suspensions were centrifuged at 248*g* to remove large particles. The supernatant containing the membranes was centrifuged at 27,000*g*. Membranes were washed three times in 100 vol of the same buffer, extracted with LIS, and partitioned in a phenol–water mixture. The aqueous phase was dialyzed against distilled water and processed with the senescent RBC IgG affinity column. (From Kay, 1981b.)

and kidney cells. Therefore, the immunological mechanism for removing senescent and damaged red cells (Kay, 1975a) may be a general physiological process for removing cells programmed for death in mammals and, possibly, other vertebrates.

The identity of the 62,000-dalton glycoprotein was investigated. The sialoglycoproteins PAS_{1-4} can probably be excluded from further consideration, since they did not absorb the phagocytosis-inducing ability of IgG eluted from senescent cells in the experiment described earlier. Furthermore, desialylation of glycophorin A (PAS_1) changes its relative mobility from 83,500 to 71,000, *not* 62,000 (Fairbanks *et al.*, 1971). Spectrin had to be considered as a possible candidate for the origin of the senescent cell antigen because serum IgG from healthy 20- to 23-year olds contains antibodies to human spectrin (Fig. 14). TX-100 extracts (1%) of RBC

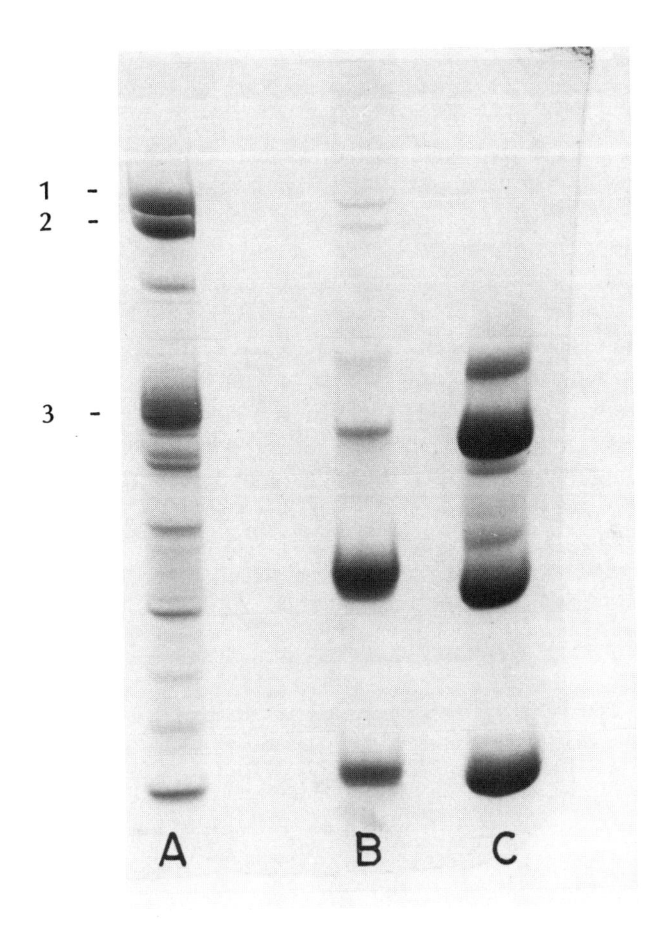

Fig. 14. Binding of normal human IgG to spectrin (bands 1 and 2). A, RBC ghost; B, spectrin eluted with glycine–HCl, pH 2.3, from IgG–CNBr-Sepharose; C, IgG isolated by ion-exchange chromatography. Spectrin was isolated by low ionic strength extraction at 37°C (Tyler *et al.*, 1979) and incubated with IgG conjugated to CNBr-activated Sepharose (52 mg of IgG/ml Sepharose). After extensive washing, bound material was eluted from the IgG–Sepharose immunoabsorbent with glycine–HCl, pH 2.3. IgG was also eluted with the antigen, probably because of the high IgG-to-Sepharose ratio during conjugation.

ghosts which contain spectrin (see Fig. 12) did not absorb the phagocytosis-inducing ability of senescent cell IgG. Specific rabbit anti-human spectrin antisera did not react with the senescent cell antigen in ID or IEP, although it reacted with purified spectrin in the same assays (results not shown). The immunoglobulin fraction of anti-human RBC antisera did not react with the senescent cell antigen, although it gives a strong precipitin reaction with RBC proteins (but not glycoproteins). Finally, anti-spectrin antisera did not react with white blood cell (WBC) membrane proteins, as determined by ID and IEP, although WBC contain the senescent cell antigen (results not shown). Anti-human RBC antisera reacted with two proteins in WBC membranes, one of which was actin. The identity of the other is not yet known. For these reasons, it is unlikely that the senescent cell antigen is derived from spectrin. The senescent cell antigen may be component 4.5 since it migrates in the same position electrophoretically as the upper band of 4.5, and/or it might be derived from carbohydrates of band 3.

It is interesting to speculate that a terminal differentiation antigen similar to the one described here may be the signal for the removal of cells during tissue remodeling in morphogenesis and embryogenesis. Neoplastic cells should be investigated to determine if part of their autonomy is attributable to lack of a terminal differentiation antigen. It is hoped that investigations into the mechanisms responsible for appearance of a terminal differentiation antigen will reveal methods for manipulating the antigen in order to prolong or terminate the life of cells physiologically. It is anticipated that such a therapeutic modality would be useful in the treatment of numerous diseases, including neoplasias, particularly those of the lymphoid system, aplastic anemias, and, possibly, birth defects.

V. AUTOANTIBODIES AS REGULATORS OF HOMEOSTASIS

The concept of self-tolerance is a basic tenet of immunology. Erhlich used the term *horror autotoxicus* to describe autoimmune disease, which was felt to result from a breakdown in internal regulation that prevented reactions against "self." The demonstrated existence of physiologic autoantibodies requires a revision of this view. Indeed, it is quite probable that immune cells recognize "self" rather than "non-self." Thus, for example, histocompatibility antigens present on cell surfaces identify a "self" cell to the immune cells which carry the same antigens. Cooperation of cells in an immune response often requires the interaction of cells with the same histocompatibility antigen haplotype. Nonself would simply be any cell or organism which did not carry self antigens. Investigations into the etiology of "pathologic" "autoimmune disease" will probably reveal that most cases represent an appropriate immune response to microbial antigens, such as viral or mycoplasmal antigens, that have been incorporated into the histocompatibility antigens on the surface of infected cells (see Kay, 1980b).

The results of the experiments described here indicate that as cells age a new antigen appears on their surface membrane, enabling pre-existing IgG autoantibodies to attach immunologically to these antigens. This provides the necessary signal for macrophages to selectively phagocytize senescent and damaged cells or their membranes. Thus, it would appear that certain autoantibodies are contributing to the maintenance of immunologic homeostasis by removing senescent and damaged cells. Autoantibodies of this type have been identified as "physiologic" (Kay, 1975a, 1978b; Kay and Makinodan, 1978). The extent to which this physiologic autoantibody protects against the formation of pathologic autoantibodies that could arise as a consequence of death and decay of senescent cells within the organism is not yet known.

Besides the physiologic autoantibody described here, for which a homeostatic function has been demonstrated and a mechanism of action has, in part, been elucidated, autoantibodies are being found with increasing frequency in normal, healthy individuals (Martin and Martin, 1975; Roder *et al.*, 1978; Dresser and Popham, 1976; Dresser, 1978).

Several cellular autoantibodies have been described. Normal mouse serum contains antibodies that react with autologous thymus cells (Martin and Martin, 1975). Autoantibodies against mouse RBC treated with bromelain have been described, and plaque-forming cells producing autoantibodies against bromelain-treated mouse RBC have been demonstrated (Cunningham, 1976; Steele and Cunningham, 1978). The "new" antigens that are exposed by bromelain treatment are not the same as those that are present on senescent RBC aged *in situ* or RBC aged by storage *in vitro* (Cox and Cunliffe, 1979). Thus, the function or significance of cells capable of responding to bromelain-treated RBC is unclear at this time because bromelain hydrolyzes proteins, peptides, amides, and esters of amino acids and peptides and because bromelain treatment represents a nonphysiologic situation. The implication of bromelain treatment studies is that mice have a high proportion ($\geq$50%) of antibody plaque-forming cells that recognize endogenous antigens. One problem with these studies is that bromelain may generate "new" antigens by altering many molecules. Another problem is that bromelain treatment renders RBC more susceptible to lysis. Further, alteration of a self-recognition antigen, such as one of the H-2 or HLA antigens, could result in recognition of the molecule as "nonself" or "altered self" (Shearer, 1974), just as the addition of a single sugar (β-*N*-acetylgalactosamine) to a glycophingolipid produced by adenocarcinoma of the stomach and colon leads to a "non-self" antigen passing as a "self" antigen (Levine, 1978).

Antisperm autoantibodies have been detected in >60% of normal, healthy males (Tung, 1975). Following vasectomy, the incidence of antisperm antibodies increases (Tung, 1975). The significance of these autoantibodies is not known, but they may be involved in the removal of sperm in vasectomized males. Autoantibodies to other antigens such as nuclei, mitochondria, and smooth muscle do not develop after vasectomy (Tung, 1975).

Autoantibodies against endogenous biological molecules have been described. For example, normal mice possess autoantigen-sensitive cells capable of producing anti-single-stranded DNA (Roder *et al.*, 1978). Autoantibodies to albumin have been reported in human sera (Mihāescu, 1981). Essentially 100% of the young healthy individuals tested have low levels of antibodies to thyroglobulin (Kay, unpublished).

B cells producing autoantibodies to human IgG and thyroglobulin have been demonstrated in umbilical cord blood and both young and elderly adults (Fong *et al.*, 1981). Studies on the effect of age on these autoreactive B cells revealed that those to IgG increased in number between birth and young adulthood while those reacting to thyroglobulin increased between young adulthood and old age (Fong *et al.*, 1981). The anti-IgG autoantibody production is T-cell dependent and is regulated by suppressor T cells (Tsoukas *et al.*, 1980). Injection of the polyclonal activator, lipopolysaccharide, into mice results in production of IgG antibodies which are specific for mouse IgG (Dresser, 1978). Plaque-forming cells producing IgM anti-Ig autoantibody have been demonstrated (Dresser, 1978).

An IgG autoantibody to sperimine, a polyamine, has been found in normal rabbits (Bartos *et al.*, 1980; Furuichi *et al.*, 1980). Autoantibodies to spectrin, a major structural protein of RBC cells present on the cytoplasmic side of the membrane, have been detected in the IgG fractions of autologous human serum (see Fig. 14 and Kay, unpublished). IgG autoantibodies to lactate dehydrogenase and IgG–lactate dehydrogenase complexes have been found in normal human sera (Grubb, 1975). An interaction between guinea pig Ig and F-actin has been observed (Fechheimer *et al.*, 1979); however, immunological binding by the Fab region of Ig has not been demonstrated. The function of these autoreactive Ig's has not been elucidated. They might be involved in the regulation of molecular synthesis or removal of macromolecules from the circulation. Autoantibodies against cellular molecules may be involved in the removal of cellular debris following tissue damage or necrosis caused by trauma (e.g., burns, bruises).

Auto-anti-idiotype antibodies are another example of a physiologic autoantibody. Idiotypic determinants of Ig are located on the Fab portion of the IgG molecule. Idiotypic determinants elicit specific anti-idiotypic antibodies in the same individual that synthesized the idiotype (Rodkey, 1974; Brown and Rodkey, 1979), as well as in isologous and heterologous species.

Jerne (1974) hypothesized that the quantitative expression of antibodies or idiotypes may be regulated by the production of anti-idiotype antibodies within the same individual. Regulation of antibody synthesis by autologous anti-idiotype responses has since been demonstrated. For example, a decrease in the plaque-forming cell response to Pneumococcus R36A vaccine and a subsequent increase in the number of anti-idiotype-specific plaque-forming cells has been demonstrated (Kluskens and Kohler, 1974; Cosenza, 1976). Auto-anti-idiotypic antibody present in human sera can block secretion of anti-TNP antibody by plaque-forming cells *in vitro* (Schrater *et al.*, 1979). Thus, it appears that auto-anti-idiotypic antibody regulates antibody secretion, at least *in vitro*. In this regard, it is interesting that antibodies to the Fab fragments of human IgG are found in human sera (Vos, 1977). Xenogeneic anti-Fab antibodies can induce lymphocyte DNA synthesis and cell division *in vitro* (Scribner *et al.*, 1978; Kay, 1980a). Thus, physiologic autoantibodies are capable of initiating, modulating, and terminating cellular and metabolic processes.

Swartzendruber (1964) described removal of plasma cells by macrophages in the spleen of mice between 3 and 6 days after primary immunization, as determined by electron microscopy. Phagocytosis of plasma cells was most extensive on Day 6. Jerne *et al.* (1963), utilizing a plaque assay, reported a 90% decrease in the number of antibody-producing cells in the mouse spleen between 4 and 7 days after primary immunization with the same antigen. Schooley (1961), using autoradiography, estimated the mean life span of a plasma cell at 8–12 hr after the last division in a stimulated lymph node. In addition, the phagocytosis of plasma cells observed by Swartzendruber correlates with the cessation of log phase of antibody appearance in the circulation (Congdon and Makinodan, 1961; Makinodan and Albright, 1962). In view of the preceding results, it is tempting to speculate that a clone(s) of plasma cells, and, thus, antibody production to a specific antigen, is terminated at the end of log phase of antibody appearance *in situ* by phagocytic macrophages. It is possible that an autoantibody similar to that involved in the removal of senescent cells is responsible for initiating the phagocytosis of terminally differentiated plasma cells.

One of the implications of such speculation is that multiple myeloma might result from a defect in macrophages or a regulatory autoantibody allowing plasma cells to function autonomously. Freed of their normal mechanism of destruction, plasma cells could then accumulate and the plasmacytosis of multiple myeloma would result.

The existence of physiologic autoantibodies indicates that there are B cells which produce autoantibody with regulatory functions that are fundamental to an individual's survival. Hence, these B-cell clones cannot be "forbidden" but must be part of the normal immune response. Regulatory theories which explain both facultatively pathologic and pathologic antibodies in terms of a decrease in the number or activity of regulator cells that suppress B-cell response to autoantigens are inadequate because they do not satisfactorily explain the existence of physiologic autoantibodies. Furthermore, there is an increase with age in the number and activity of suppressor cells and auto-anti-idiotypic antibodies in humans and long-lived mice (for a review, see Kay, 1981c). Thus, the observed age-related increase in autoimmune manifestations can not be explained by a decrease in these regulatory parameters.

It is interesting to speculate that physiologic autoantibodies similar to the one described here are responsible for tissue remodeling during embryogenesis and morphogenesis, tissue repair during wound healing following burns or trauma, and intercellular communication networks. As physiologic regulators, autoantibodies initiate, modulate, and terminate cellular and metabolic processes (Kay, 1980a, 1981c). The effect of age on physiologic autoantibodies and the role physiological autoantibodies may play in protecting against facultatively pathologic or pathologic autoimmune manifestations, and, possibly, neoplasia, have yet to be investigated. These autoantibodies might provide the missing immunoregulatory link between the immune system and the neuroendocrine system (see Kay, 1981d). Further, antibodies are amenable to external manipulation and could be used to restore immune function or to regulate other systems and cells. The role of these antibodies in diseases such as infection, wound healing, neoplasia, and dementia needs to be delineated.

ACKNOWLEDGMENTS. This work was supported by NIH Grant 22671 and the Veterans Administration Research Service.

VI. REFERENCES

Alderman, E. M., Fudenberg, H. H., and Lovins, R. E., 1980, Binding of immunoglobulin classes to subpopulations of human red blood cells separated by density-gradient centrifugation, *Blood* **55:**817.

Aminoff, D., Bell, W. C., Fulton, I., and Ingebrigsten, N., 1976, Effect of sialidase on the viability of erythrocytes in circulation, *Am. J. Hematol.* **1**:419.

Aminoff, D., Bruegge, W. F. W., Bell, W. C., Sarpolis, K., and Williams, R., 1977, Role of sialic acid in survival of erythrocytes in the circulation: Interaction of neuraminidase-treated and untreated erythrocytes and spleen and liver at the cellular level, *Proc. Natl. Acad. Sci. USA* **74**:1521.

Bartos, D., Bartos, F., Campbell, R. A., and Grettie, D. P., 1980, Antibody to spermine: A natural biological constituent, *Science* **208**:1178.

Bennett, G., and Kay, M. M. B., 1981, Homeostatic removal of senescent murine erythrocytes by splenic macrophages, *Exp. Hematol.* **9**:295.

Brown, J. C., and Rodkey, L. S., 1979, Autoregulation of an antibody response via network-induced auto-anti-idiotype, *J. Exp. Med.* **150**:67.

Cochrum, K., Hanes, D., Fagan, G., Speybroeck, J. V., and Sturtevant, F. K., 1978, A simple and rapid method for the isolation of human monocytes, *Transplant. Proc.* **10**:867.

Congdon, C. C., and Makinodan, T., 1961, Splenic white pulp after antigen injection: Relation of time of serum antibody production, *Am. J. Pathol* **39**:697.

Cosenza, H., 1976, Detection of anti-idiotype reactive cells in the response to phosphorylcholine, *Eur. J. Immunol.* **6**:114.

Cox, K. O., and Cunliffe, D. A., 1979, Density-dependent fractionation of murine RBC: Antigenic relationships between young and old RBC and bromelain-treated RBC, *Clin. Immunol. Immunopathol.* **13**:394.

Cunningham, A. J., 1976, Self-tolerance maintained by active suppressor mechanisms, *Transplant. Rev.* **31**:23.

Dodge, J. T., Mitchell, C., and Hanahan, D. J., 1963, The preparation and chemical characteristics of hemoglobin-free ghosts of human erythrocytes, *Arch. Biochem. Biophys.* **100**:119.

Dresser, D. W., 1978, Most IgM-producing cells in the mouse secrete auto-antibodies (rheumatoid factor), *Nature* (*London*) **274**:480.

Dresser, D. W., and Popham, A. M., 1976, Induction of an IgM anti-(bovine)-IgG response in mice by bacterial lipopolysaccharide, *Nature* (*London*) **264**:552.

Durocher, J. R., 1978, Membrane glycoproteins in erythrocyte survival, *J. Supramol. Struct. Suppl.* **2**:199.

Eckhardt, A. E., Hayer, C. E., and Goldstein, I. J., 1976, A sensitive fluorescent method for the detection of glycoproteins in polyacrylamide gels, *Anal. Biochem.* **73**:192.

Fairbanks, G., Steck, T. L., and Wallach, D. F. H., 1971, Electrophoretic analysis of the major polypeptides of the human erythrocyte membrane, *Biochemistry* **10**:2606.

Fechheimer, M., Daiss, J. L., and Cebra, J., 1979, Interaction of immunoglobulin with actin, *Mol. Immunol.* **16**:881.

Fidalgo, B. V., Katayama, Y., and Najjar, V. A., 1967, The physiological role of the lumphoid system. V. The binding of autologous (erythrophilic) gamma-globulin to human red blood cells, *Biochemistry* **6**:3378.

Fong, S., Tsoukas, C., Frincke, L., Lawrence, S., Holbrook, T., Vaughan, J., and Carson, D., 1981, Age-associated changes in Esptein-Barr virus induced human lymphocyte autoantibody responses, *J. Immunol.* **126**:910.

Furuichi, K., Ezoe, H., Obara, T., and Oka, T., 1980, Evidence for a naturally occurring anti-spermine antibody in normal rabbit serum, *Proc. Natl. Acad. Sci. USA* **77**:2904.

Grubb, A. O., 1975, Demonstration of circulating IgG-lactate dehydrogenase immune complexes by crossed immunoelectrophoresis, *Scand. J. Immunol.* **4**(Suppl. 2):53.

Halbhuber, K.-T., Stibenz, D., Feuerstein, H., Linss, W., Meyer, H.-W., Frober, R., Rumpel, E., and Geyer, G., 1980, Defined rearrangement of the membrane of banked erythrocytes, *in* "Abstracts, IXth International Symposium on Structure and Function of Erythoid Cells, Berlin, German Democratic Republic," p. 66.

Hollister, P. J., Bennett, G. D., Onari, K., Kay, M. M. B., and Makinodan, T., 1981, The influence of late radiation effects on immunologic aging, Proceedings, Western Gerontological Society Meeting, Submitted for publication.

Jancik, J., and Schauer, R., 1974, Sialic acid—A determinant of the lifetime of erythrocytes, *Hoppe-Seyler's Z. Physiol. Chem.* **355:**395.

Jancik, J. M., Schauer, R., Andres, K. H., and von Düring, M., 1978, Sequestration of neuraminidase-treated erythrocytes, *Cell Tissue Res.* **186:**209.

Jerne, N. K., 1974, Towards a network theory of the immune system, *Ann. Immunol.* (*Paris*) **125c:**373.

Jerne, N. K., Nordin, A. A., and Henry, C., 1963, The agar plaque technique for recognizing antibody-producing cells, *in* "Cell-Bound Antibodies" (D. B. Amos and H. Koprowski, eds.), pp. 109–125, Wistar, Philadelphia.

Kay, M. M. B., 1974, Mechanism of macrophage recognition of senescent red cells, *Gerontologist* **14:**33.

Kay, M. M. B., 1975a, Mechanism of removal of senescent cells by human macrophages *in situ*, *Proc. Natl. Acad. Sci. USA* **72:**3521.

Kay, M. M. B., 1975b, Multiple labeling technique used for studies of activated human B lymphocytes, *Nature* (*London*) **254:**424.

Kay, M. M. B., 1977, High resolution scanning electron microscopy and its application to research on immunity and aging, *in* "Immunity and Aging" (T. Makinodan and E. Yunis, eds.), pp. 135–150, Plenum Press, New York.

Kay, M. M. B., 1978a, Multiple labeling technique for scanning electron microscopy, *in* "Principles and Techniques of Scanning Electron Microscopy" (M. A. Hayat, ed.), pp. 338–357, Van Nostrand–Reinhold, New York.

Kay, M. M. B., 1978b, Role of physiological autoantibody in the removal of senescent human red cells, *J. Supramol. Struct.* **9:**555.

Kay, M. M. B., 1979, Effect of age on human immunological parameters including T and B cell colony formation, *in* "Recent Advances in Gerontology" (H. Orimo, K. Shimada, M. Iriki, and D. Maeda, eds.), pp. 442–443, Excerpta Medica, Amsterdam/Oxford/Princeton.

Kay, M. M. B., 1980a, Cells, signals and receptors, *in* "Aging Phenomena: Relationships among Different Levels of Organization" (K. Oota, T. Makinodan, M. Iriki, and L. S. Baker, eds.), pp. 171–200, Plenum Press, New York.

Kay, M. M. B., 1980b, Immunological aspects of aging, *in* "Aging, Immunity and Arthritic Disease," Vol. 11, "Aging," (M. M. B. Kay, J. Galpin, and T. Makinodan, eds.), pp. 33–78, Raven Press, New York.

Kay, M. M. B., 1981a, The IgG autoantibody binding determinant appearing on senescent cells resides on a 62,000 MW peptide, *Acta Biol. Med. Ger.* **40:**385.

Kay, M. M. B., 1981b, Isolation of the phagocytosis inducing IgG-binding antigen on senescent somatic cells, *Nature* (*London*) **289:**491.

Kay, M. M. B., 1981c, Immunodeficiency in old age, *in* "Immunodeficiency Disorders" (R. K. Chandra, ed.), Churchill Livingstone, Edinburgh, in press.

Kay, M. M. B., and Makinodan, T., 1978, Physiologic and pathologic autoimmune manifestations as influenced by immunologic aging, *in* "Clinical Immunochemistry: Chemical and Cellular Bases and Applications in Disease" (S. Natelson, A. J. Pesce, and A. A. Dietz, eds.), pp. 192–199, Am. Assoc. Clin. Chem., Washington, D.C.

Kluskens, L., and Kohler, H., 1974, Regulation of immune response by autogenous antibody against receptor, *Proc. Natl. Acad. Sci. USA* **71:**5083.

Kochwa, S., and Rosenfield, R., 1964, Immunochemical studies of the Rh system. I. Isolation and characterization of antibodies, *J. Immunol.* **92:**682.

Levine, P., 1978, Self-nonself concept for cancer and diseases previously known as "autoimmune" diseases, *Proc. Natl. Acad. Sci. USA* **75:**5697.

Lutz, H. U., and Kay, M. M. B., 1981, An age specific cell antigen is present on senescent human red blood cell membranes, *Mech. Ageing Dev.* **15:**65.

Lutz, H. U., von Daniken, A., Semenza, G., and Bachi, T. H., 1979, Glycophorin-enriched vesicles obtained by a selective extraction of human erythrocyte membranes with a non-ionic detergent, *Biochim. Biophys. Acta* **552:**262.

Makinodan, T., and Albright, J. F., 1962, Cellular variation during the immune response: One possible model of cellular differentiation, *J. Cell. Comp. Physiol.* **60**(Suppl. 1)**:**129.

Marchesi, V. T., and Andrews, E. P., 1971, Glycoproteins: Isolation from cell membranes with lithium diiodosalicylate, *Science* **174:**1247.

Martin, W. J., and Martin, S. E., 1975, Thymus reactive IgM autoantibodies in normal mouse sera, *Nature* (*London*) **254:**716.

Mihăescu, S., Lenkei, R., and Ghitie, V., 1981, Radioimmunoassay of anti-albumin autoantibodies in human sera, *J. Immunol. Methods* **42:**187.

Murphy, J. R., 1973, Influence of temperature and method of centrifugation on the separation of erythrocytes, *J. Lab. Clin. Med.* **82:**334.

Najjar, V. A., 1974, The physiological role of gamma-globulin, *in* "Advances in Enzymology," Vol. 41 (A. Meister, ed.), pp. 129–178, Wiley, New York.

Roder, J. C., Bell, D. A., and Singhal, S. K., 1978, Regulation of the autoimmune plaque-forming cell response to single-stranded DNA (sDNA) *in vitro, J. Immunol.* **121:**38.

Rodkey, L. S., 1974, Studies of idiotypic antibodies. Production and characterization of autoantiidiotypic antisera, *J. Exp. Med.* **139:**712.

Saunders, J. W., 1966, Death in embryonic systems, *Science* **154:**604.

Saunders, J. W., Jr., and Fallon, J. F., 1966, Cell death in morphogenesis, *in* "Major Problems in Developmental Biology" (M. Locke, ed.), pp. 289–314, Academic Press, New York.

Schooley, J. C., 1961, Autoradiographic observations of plasma cell formation, *J. Immunol.* **86:**331.

Schrater, F. A., Goidl, E. A., Thorbecke, G. J., and Siskind, G. W., 1979, Production of auto-anti-idiotypic antibody during the normal immune response to TNP-Ficoll. I. Occurrence in AKR/J and BALB/c mice of hapten-augmentable, anti-TNP plaque-forming cells and their accelerated appearance in recipients of immune spleen cells, *J. Exp. Med.* **150:**138.

Scribner, D. J., Weiner, H. L., and Moorhead, J. W., 1978, Anti-immunoglobulin stimulation of murine lymphocytes. V. Age-related decline in Fc receptor-mediated immunoregulation, *J. Immunol.* **121:**377.

Shearer, G. M., 1974, Cell-mediated cytotoxicity to trinitrophenyl-modified syngeneic lymphocytes, *Eur. J. Immunol.* **4:**527.

Silverberg, M., Furthmayr, H., and Marchesi, V. T., 1976, The effect of carboxymethylating a single methionine residue on the subunit interactions of glycophorin A, *Biochemistry* **15:**1448.

Steele, E. J., and Cunningham, A. J., 1978, High proportion of Ig-producing cells making autoantibody in normal mice, *Nature* (*London*) **274:**483.

Swartzendruber, D. C., 1964, Phagocytized plasma cells in mouse spleen observed by light and electron microscopy, *Blood* **24:**432.

Syllm-Rapoport, I., Jacobasch, G., Prehn, S., and Rapoport, S., 1969, On a regulatory system of the adenine level in the plasma connected with red cell maturation and its effect on the adenine nucleosides of the circulating erythrocyte. Lack of relation between ATP-level and life span of the erythrocyte, *Blood* **33:**617.

Tannert, Ch., 1978, Untersuchungen zum altern roter blutzellen, Ph.D. Dissertation, Humbolt University, Berlin, German Democratic Republic.

Tata, J. R., 1966, Requirement for RNA and protein synthesis for induced regression of the tadpole tail in organ culture, *Dev. Biol.* **13:**77.

Tsoukas, C., Carson, D., Fong, S., Pasquali, J.-L., and Vaughan, J., 1980, Cellular requirements for pokeweed mitogen-induced autoantibody production in rheumatoid arthritis, *J. Immunol.* **125:**1125.

Tung, K. S. K., 1975, Human sperm antigens and antisperm antibodies. I. Studies on vasectomy patients, *Clin. Exp. Immunol.* **20:**93.

Tyler, J. M., Hargreaves, W. R., and Branton, D., 1979, Purification of two spectrin-binding proteins: Biochemical and electron microscopic evidence for site-specific reassociation between spectrin and bands 2.1 and 4.1, *Proc. Natl. Acad. Sci. USA* **76:**5192.

Vos, G. H., 1977, Anti-Fab' antibodies in human sera. I. A study of their distribution in health and disease, *Vox Sang.* **33:**16.

Weber, R., 1962, Induced metamorphosis in isolated tails of *Xenopus* larvae, *Experientia* **18:**84.

Wegner, G., Tannert, Ch., Maretzki, D., Schössler, W., and Strauss, D., 1980, IgG binding to glucose depleted and preserved erythrocytes, *in* "Abstracts, IXth International Symposium on Structure and Function of Erythroid Cells, Berlin, German Democratic Republic," p. 57.

Zwilling, E., 1964, Controlled degeneration during development, *in* "Cellular Injury" (A. V. S. de Reuck and M. P. Cameron, eds.), pp. 352–362, Churchill, London.

Chapter 6

Alien Histocompatibility Antigens

Ruta M. Radvany

Department of Surgery
Northwestern University Medical School
Evanston, Illinois

I. INTRODUCTION

The existence of foreign antigens on tumor cells has attracted the attention of investigators in multiple disciplines. In the early days of immunology, Paul Erlich (Himmelweit, 1957) speculated that tumor cells could have antigens which are foreign to the immunological system of the host and that such antigens could play a role in the treatment of cancer by immunological means. During the years that followed, foreign antigens have been demonstrated on a variety of tumor cells. These antigens either can be unique to a tumor or can be variants of antigens present on normal cells. Some tumor antigens are differentiation antigens—antigens present on normal fetal cells, but not expressed on normal adult cells. Still other tumors show a loss rather than a gain of a particular antigenic specificity. The antigen lost or gained by a particular tumor can be an organ-specific antigen or alloantigen—blood group or histocompatibility antigen (HA) which is foreign to the tumor-bearing animal. For the purpose of this review on alien (HA), the discussion will be confined to these antigens only. Other tumor antigens will be discussed only in the context of alien HA.

Alien HA are additional HA which appear on the membrane surface of a cell. These antigens have been found on neoplastic cells and may play a role in the immunology and genetics of oncological diseases. The existence of alien HA has not yet been fully accepted, due to contradictory or incomplete observations. Early findings must be confirmed, and experiments must be performed to clarify and explain contradictory findings, before these antigens can be accepted with certainty. At the same time, it seems unlikely that all of the reported experiments on alien HA are studies of laboratory artifacts. It seems more likely that alien HA, while present on neoplastic cells, probably are not expressed on all types of neoplastic cells nor during all phases of a cell cycle.

II. HISTOCOMPATIBILITY ANTIGENS

Regular HA, as found on cell membranes, are glycoprotein molecules coded for by genes in the major histocompatibility complex (MHC) (see Fig. 1). The more commonly known HA are referred to as Class I and Class II HA. Class I HA consist of a heavy chain (mol. wt. 45,000) which carries the HA specificity, and a smaller polypeptide chain (mol. wt. 12,000) that has been identified as beta-2 microglobulin and is not coded for by the MHC. Class I HA have been found on all nucleated cells studied and are referred to as the H-2K, H-2D, and H-2L antigens in the mouse and HLA-A -B, and -C locus antigens in man. Class II HA consist of two glycoprotein chains (mol. wt. 33,000 and 28,000). The distribution of Class II HA is more restricted—they are found on subpopulations of lymphocytes, endothelial cells, sperm, and monocytes. Class II HA are referred to as the Ia or HLA-D,DR antigens in the mouse and man, respectively.

HA belong to a polymorphic system—there are many alleles (variants

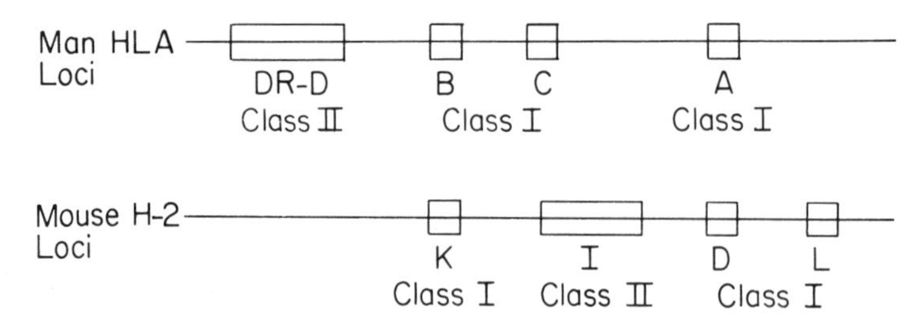

Fig. 1. The MHC of man and mouse on chromosomes 6 and 17 respectively. Class 1 antigens are expressed on all nucleated cells; class II antigens are expressed on a restricted population of cells.

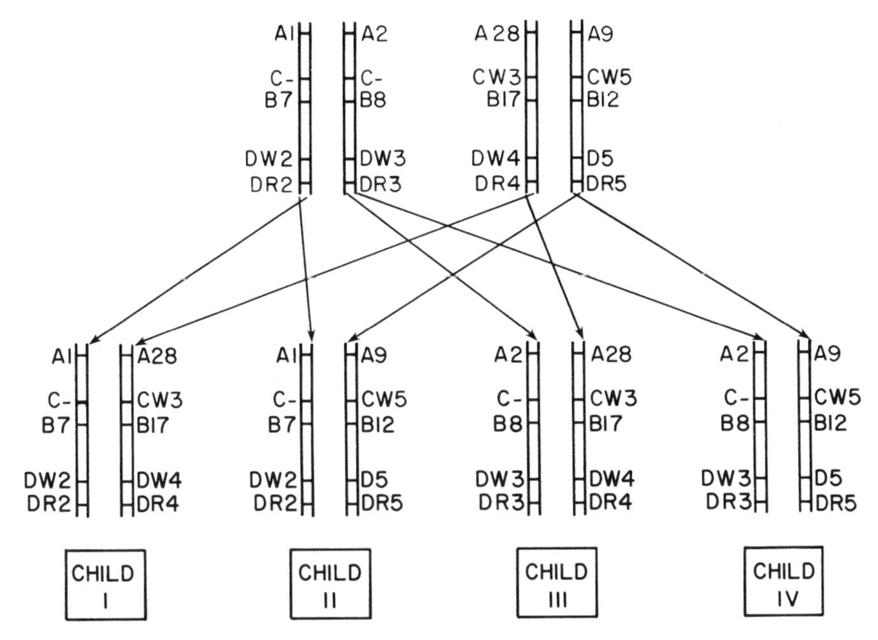

Fig. 2. Inheritance of HLA antigens in man. HLA antigens are inherited *en bloc* with the sixth chromosome transmitted to the offspring. Each offspring inherits one maternal and one paternal chromosome. There are four possible combinations.

of a gene) at many loci which are inherited following Mendelian rules of inheritance (see Fig. 2). Alleles of the MHC are codominant, which means that all of the alleles present in the genome are expressed on the cell surface and there are no "silent" or suppressed alleles. The rules have been confirmed over and over by evaluating family pedigrees where half of the antigens present on healthy paternal and maternal lymphocytes are transmitted to the offspring with the chromosome carrying genes of the MHC. If the existence of alien HA is confirmed, an important question will have to be answered: from where does the cell obtain the genetic information which codes for alien HA? Does the neoplastic cell acquire new genetic information, or does the chromosome carry alleles which are not expressed on the surface of a normal cell?

Several theories have been proposed to explain the sudden appearance of alien specificities on abnormal cells. It has been suggested (Amos, 1971) that several separate genes are present at each histocompatibility locus and that in a normal cell only one gene per locus is functional. It has also been proposed (Bodmer, 1973) that the chromosome carries groups of genes (pseudoalleles) which code for families of HA, and an operator gene, governed by a genetic control mechanism, determines which "alleles" are expressed on the cell surface. Under altered conditions, such as exist in the neoplastic virus-infected or cultured cell, the operator gene mechanism could break down. As a result, other (inappropriate) HA would be expressed. Alternate mechanisms have also been suggested: a virus could incorporate its DNA in the chromosome and introduce new HA (Nowinski and Klein, 1975), or a mutation could occur (Lengerova, 1977). All of the above suggestions are real possibilities, and, if confirmed by experimental evidence, would lead to the re-evaluation of present day concepts about the inheritance of HA.

III. ALIEN HISTOCOMPATIBILITY ANTIGENS

A. First Evidence

The best evidence for the presence of alien HA comes from studies of mouse tumors, and the first studies involved tumor transplantation experiments. Martin *et al.*(1973), using a chemically induced lung tumor, and Invernizzi and Parmiani (1975), using a chemically induced fibrosarcoma (ST2), were among the first to observe tumor regression following allosensitization (see Fig. 3). When mice of strain A were immunized with tissues of another mouse strain B differing in HA, and then transplanted with a sarcoma of strain A, the growth of the tumor was considerably slowed down or the tumor was rejected. Nonimmunized strain A mice succumbed to the tumor. The protective ability of the immunized mice resided in lymphocytes and could be transferred with lymphocytes to another mouse of the same strain. Serological experiments showed that tumor cells of strain A carried antigens characteristic of strain B in addition to the appropriate strain A antigens. These findings suggested that the new specificities were encoded by the mouse MHC, but the studies were not performed in congenic strains, and it was not possible to determine the effect of nonhistocompatibility (non-H-2) loci.

Experiments were performed to determine if alien HA were identical or similar to the tumor-specific antigens (TSTA) commonly identified on tumor cells. Contradictory results were obtained: some experiments showed that alien HA and TSTA are identical (Invernizzi and Parmiani, 1975), but later experiments did not confirm the first observation (Parmiani *et al.*, 1978). The question of the relationship between alien HA and TSTA could not be settled from the above experiments.

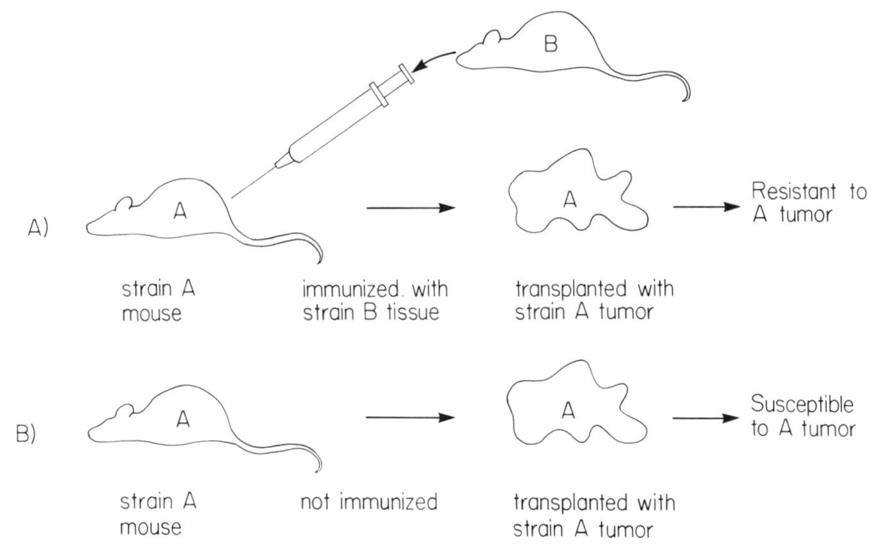

Fig. 3. (A) Strain A tumor does not grow in a strain A mouse which has been immunized with strain B tissue. (B) Strain A tumor grows in a strain A mouse not immunized with strain B tissue and eventually kills the host.

In later studies other investigators, using mouse lymphoma (SL2) (Garrido *et al.*, 1976b; Matossian-Rogers *et al.*, 1977), fibrosarcoma (C-1) (Meschini *et al.*, 1977), and lung tumor (Martin *et al.*, 1976), also reported that tumors were able to induce specific tumor immunity and histocompatibilitylike inappropriate (alien) antigens could be found on tumor cells (Garrido *et al.*, 1976b; Matossian-Rogers *et al.*, 1977). Further search for alien antigens showed that these antigens were not found on all tumors (Pliskin and Prehn, 1978). Reagents used to detect alien antigens caused problems, and Klein (1975), using immunofluorescence assays, found that his antisera detected proteins of murine leukemia virus on the surfaces of the tumor cells tested (Nowinski and Klein, 1975).

Early transplantation experiments showed that alloimmunization provided specific immunity to some tumors. Subsequent experiments provided preliminary evidence for the presence of alien specificities on at least some types of tumors. It became evident that alien antigens must be studied in congenic mice strains using well-identified, monospecific sera. Finally, a relationship between alien HA and TSTA was suggested, but not confirmed. In spite of the fact that these early findings were a bit ambiguous, alien antigens aroused the curiosity of tumor immunologists and opened a new field— the study of alien antigens.

B. Criteria for Defining Alien Antigens

Several criteria had to be met in order to analyze early observations and to prove the existence of alien antigens:

1. Identification of alien antigens on surfaces of tumor cells and tumor lines using monospecific and thoroughly analyzed reagents as well as a sensitive assay system.
2. Induction of specific immunological responses (cellular and/or humoral) *in vivo* by tumor cells carrying alien antigens.
3. Demonstration of a specific *in vitro* reactivity of the *in vivo*-induced immunological response (humoral or cellular) which is directed to the alien antigen.
4. Isolation and biochemical identification of molecules carrying alien specificities.
5. Demonstration of specific *in vivo* (stimulation of humoral or cellular responses, inhibition of tumor growth) or *in vitro* (inhibition of cellular and serological reactions) biological activity of isolated molecules with alien specificity.

It was not possible to meet all of the above criteria with all tumors carrying alien specificities, but interesting findings were described by those who attempted to fully analyze these strange specificities.

C. Later Studies of Alien Antigens

1. In the Mouse

a. Immunological Studies. Immunological experiments attempted to demonstrate alien histocompatibility antigens by

1. *In vitro* tests
   a. Serology—using known reagents to detect alien HA on tumor cells and to precipitate solubilized alien HA;
   b. Absorption studies—ability of tumor cells to remove antibodies to alien HA from known reagents;
   c. Cellular immunology—to demonstrate that specific cytotoxic lymphocytes kill cells bearing alien HA, and to show that corresponding soluble alien HA block the reaction

2. *In vivo* tests: ability of tumor cells and soluble alien HA to induce humoral and/or cellular responses to the corresponding alien HA.

Standard serological techniques were not always sensitive enough in detecting alien HA and Garrido *et al.* (1977), using a specially devised "post-labeling assay," investigated their initial observations (Garrido *et al.*, 1976b) in detail. They confirmed the presence of alien Class I and Class II HA on one type of tumor (SL2) cells, but even with a more sensitive method, failed to find all appropriate HA on another tumor (Lymphosarcoma). Well-identified sera containing antibodies to the alien HA were absorbed with normal and tumor cells and then back-tested with normal and tumor cells. The results confirmed the presence of alien HA on SL2 and the absence of proper HA on lymphosarcoma.

Cytotoxic cells of strain A were produced against strain B tissues. These cytotoxic cells were able to kill tumor cells of strain A which indicated that the tumor carried alien HA. Blocking experiments showed specific blocking of anti B cytotoxic activity by SL2 tumor cells. The authors (Garrido *et al.*, 1976a) concluded that alien HA on SL2 tumor cells can be detected with specific, cytotoxic lymphocytes as well as with specific antibodies. Later, Matossian-Rogers *et al.* (1980) using a more sensitive assay, observed similar blocking by lymphosarcoma tumor cells, previously thought not to carry alien HA.

Alien histocompatibility antigens were also demonstrated in another tumor (mouse leukemia K36) and with another serological technique—the immunoprecipitation technique (Festenstein *et al.*, 1980; Schmidt *et al.*, 1979; Schmidt *et al.*, 1980). Well-identified monospecific mouse antisera were used to immunoprecipitate [^{3}H]arginine-labeled, isolated HA (K36 glycoprotein). Antisera precipitated a peak of molecular weight of 45,000 (heavy chain of Class I HA) of strain A as well as strain B glycoprotein (alien HA). These solubilized antigens induced cytotoxic lymphocyte production against the alien HA and possibly the production of a specific antibody against the alien specificity (Testorelli *et al.*, 1980; Matossian-Rogers *et al.*, 1980). This series of experiments demonstrated the presence of alien HA in a tumor and the ability of isolated HA molecules to induce cellular and humoral responses.

Strong humoral and cellular responses against alien HA were demonstrated by still another set of experiments—by transplantation and immunization with tumor cells of a methylcolanthracene (MCA)-induced mouse tumor (Schirrmacher *et al.*, 1980) and reticulum cell sarcoma (Roman and Bonavida, 1979; Bonavida *et al.*, 1980). But immunization with fibrosarcoma produced only humoral response—a specific antibody which could precipitate the alien HA specificity *in vitro* (Meshini and

Parmiani, 1978; Carbone *et al.*, 1978). Cytotoxic cells directed against the alien HA of fibrosarcoma were not produced (Parmiani *et al.*, 1978).

In summary, immunological studies demonstrated alien HA on tumor cell surfaces, the ability of alien HA to react with specific antisera *in vitro*, and the ability of alien HA (on tumor cells and in a solubilized state) to induce specific cellular and humoral responses *in vivo*. However, not all of the tumors studied yielded positive responses to all of the immunological tests, and, with the existing techniques, alien HA were not demonstrated on many of the tumors. In addition to alien HA, loss of some proper HA was demonstrated on some tumors.

b. Biochemical Studies. Only biochemical purification and characterization of alien HA can provide the proof that these antigens, which appear on abnormal cells, are a separate entity. The methods used to solubilize and purify alien HA are those used to study ordinary or proper HA. Histocompatibility antigens are hydrophobic glycoproteins which can be isolated after detergent solubilization or limited proteolytic digestion of cell membranes. Whereas the molecular weight of detergent-solubilized Class I molecules is 45,000, smaller molecules of 37,000 are obtained by proteolytic digestion. Solubilized HA are purified by immunoprecipitation or on gel filtration or affinity columns and analyzed by polyacrylamide gel electrophoresis (PAGE).

The aim of biochemical studies was to separate serologically active alien HA and to compare their biochemical characteristics and serological activity to that of regular host HA. To achieve this aim, Rogers *et al.* (1979) purified deoxycholate (DOC)-solubilized fibrosarcoma (C-1) membranes by gel filtration and applied the resulting HA activity-containing fractions to affinity columns prepared from *Lens culinaris* lectin. Normal and alien HA were eluted from the column with α-methylmannoside. The eluates were further purified by radioimmunoprecipitation using highly specific antisera. Resulting antigen–antibody complexes were precipitated with protein A bound to Sepharose, eluted, and analyzed by PAGE. The results showed that molecules carrying the alien specificity B of the tumor were identical to normal HA molecules with B specificity. The alien antigens, however, had a lower specific activity, were present in lower amounts on the cell surfaces, and were less stable than normal HA. Alien antigens were also sensitive to papain in contrast to normal HA which are routinely separated from cell membranes by papain. Roman and Bonavida (1979) reported a very similar finding. Using sodium dodecyl sulfate (SDS)–PAGE analysis, these investigators showed that solubilized membranes of SJL tumor contained 45,000-dalton molecules which carried the alien specificity. Thus, it appears that, except for their sensitivity to papain and lower specific activity, alien HA are very similar to ordinary HA specificities.

Other investigators reported admixture of viral antigens in their preparations. When mouse leukemia (K36) cell membranes were solubilized with a detergent, and HA immunoprecipitated with specific reagents, SDS–PAGE gels showed three peaks with molecular weights of 70,000, 45,000, and 12,000. The first two peaks contained regular and alien HA as well as viral antigens while the 12,000-dalton peak was thought to be a beta-2 microglobulin (Schmidt and Festenstein, 1980). These results indicate that 45,000-dalton molecules carrying viral antigens could be isolated with anti-HA antisera containing antibodies to viral antigens and possibly falsely identified as alien specificities.

c. Relationship between TSTA and Alien HA. Gross (1943) was one of the first investigators to demonstrate the presence of TSTA on cell surfaces of chemically induced tumors. These sutdies were confirmed and extended by others (Foley, 1953; Prehn and Main, 1957), and further studies showed that as a rule the TSTA of chemically induced tumors are unique for each tumor line (Old and Boyse, 1964; Baldwin, 1973), while virally induced tumors have TSTA that are common to all tumors induced by the same virus (Sjögren, 1965; Klein, 1966). It has been suggested that alien HA and TSTA are identical molecules. In order to investigate this suggestion, alien HA and TSTA must be separated by biochemical methods. Law *et al.* (1980) solubilized C-1 fibrosarcoma membranes with detergent, applied solubilized fractions to an Ultragel AcA34 gel-filtration column, and collected fractions based on an optical density profile at O.D. 280 and HA activity determined in the radioimmunoprecipitation assay. The investigators showed that when mice were immunized with the fraction containing HA activity and then challenged with tumor, the HA activity-containing fractions did not inhibit tumor growth *in vivo*. They then subjected Nonidet P-40 (NP40)-solubilized, filtrated membrane fractions to *L. culinaris* lectin chromatography and showed that material containing TSTA did not bind to the column and showed greater than 50% tumor inhibitory activity. On the contrary, material containing alien HA activity bound to the column, and, after elution with α-methyl-mannoside, failed to inhibit tumor growth. Law and colleagues conclude that TSTA and HA are separate molecules and that TSTA, but not alien HA, are able to inhibit tumor growth. In contrast, Fujiwara *et al.* (1978), utilizing mammary carcinoma of mice, were unable to separate TSTA and alien HA activities. Further studies are needed, but it is possible that the relationship between TSTA and alien HA varies in each particular tumor.

In summary, studies on alien HA in the mouse have shown that well-defined antisera can identify alien HA specificities on tumor cells, and alien HA on tumor cells are able to induce specific cellular and humoral responses *in vivo* and *in vitro*. Alien HA can be separated by biochemical methods and separated molecules are also able to induce specific cellular and humoral responses. Biochemically separated alien HA are very similar to ordinary HA except for their sensitivity to papain and lower specific activity. Immunity to alien antigens is not always associated with slower tumor growth or with tumor rejection. Finally, whether alien HA and TSTA are identical or not still remains an open question.

2. *In Man*

a. Blood Group Antigens on Tumor Cells. Red cell antigens, not coded for by genes within the MHC of man, have also been referred to as HA because of their role in transplantation of solid organ grafts. Loss as well as gain of blood group antigens on tumor cells has been reported. Oh-Uti (1949) analyzed pooled human gastric cancer tissue and found a loss of group A and B antigens in human gastric cancer. Similarly, Kay and Wallace (1961) and Nairn (1962) were not able to detect A and B antigens on tumor cells of human bladder and adenocarcinoma cells, respectively. Deletion of H antigen in metastic cervical cancer was demonstrated by Davidsohn *et al.* (1969). Parallel to these studies Weiler (1956) demonstrated a lack of kidney-specific antigens in kidney cancer thereby showing that a tumor may lack any type of antigen or antigens.

Lack of blood group antigens, however, was not a universal observation, and Tellem *et al.* (1962) showed that ABH antigens are present on mammary tumor cells. "Extra" blood group antigens, foreign to the host bearing the tumor, were also observed. Levine *et al.* (1951) reported a new blood group antigen (Tj^a) on gastric tumor cells but not on normal cells of the same patient. Since this antigen did not show a Mendelian inheritance pattern in the patient's family, Levine speculated that it could have arisen as a result of a mutation. Hakamori and Andrews (1970) found Le^a antigen on tumors of a host with Le^b type. These authors explained these observations on the basis of earlier findings (Hakomori and Jeanloz, 1964) that blocked synthesis of Le^b glycolipid in tumor cells resulted in the accumulation of a precursor glycosphingolipid with Le^a antigenic activity. Hakomori's group also found low amounts of Le^b-active glycolipid in individuals with Le^a type but were not able to offer an easy explanation for this finding. Hakomori *et al.* (1967) described a glycolipid with blood group A activity in tumors of blood group O individuals. An A-like antigen on tumors of blood group B- and O-bearing hosts was also described by H. Häkkinen (1970), and it was thought to be a Forsman-like antigen. A carcinoembryonic antigen with blood group A-like properties was described by Gold *et al.* (1972), but immunochemical studies showed that the hapten structure of this antigen was different. Most likely all of these antigenic specificities associated with glycolipids arise as a result of incomplete synthesis of many carbohydrate chains because of malignant transformation (Hakomori and Murakami, 1968).

b. Histocompability Antigens on Tumor and Cultured Cells. It is more difficult to perform well-controlled studies on alien HA of man than of mice. Therefore studies on alien HA of man are limited to the *in vitro* detection of foreign specificities on tumor cells, cell lines, and normal fibroblasts. While earlier studies describe "extra" reactions of antisera with cultured fibroblasts and tumor lines, later studies show a more organized approach—in addition to tumor cells, HA are determined on peripheral lymphocytes of the tumor donor. The results are compared and analyzed for additional reactions. Investigators have acknowledged for years "extra" serological reactions when neoplastic and cultured cells have been tested against known reagents. These reactions were often interpreted as background reactions due to fragile cultured cells, or polyspecific reagents (pregnancy sera) containing antibodies to MHC and non-MHC antigenic determinants present on fetal cells.

Those who first observed "extra" reactions, recognized that, in order to define a new antigen on a cell surface by serology, the first requirement is the purity of the reagent: sera must be properly analyzed, tested with the appropriate number and composition of panel cells, and rendered monospecific by proper absorptions. Using such well-identified (but not monospecific) reagents, Bernoco *et al.* (1969) and Dick *et al.* (1972) were among the first to report "extra" reactions with cultured cells, and Sasportes *et al.* (1971) reported the appearance and disappearance of HLA antigens on human fibroblasts *in vitro*.

Cann *et al.* (1974), using cultured amniotic fluid cells, observed weak expression of some HLA antigens as well as "extra" HLA antigens (Table I). These "extra" antigens were of either maternal or paternal origin: fibroblasts of one fetus expressed two paternal HLA-A locus antigens (instead of one as expected) and fibroblasts of the second fetus, two maternal HLA-B locus antigens. The former could be explained by cross-reactive antibodies to both antigens and were interpreted as such. The additional maternal antigens found on the second set of cells were thought to be due to a possible contamination with maternal cells. This suspicion could not be tested by chromosome studies because fetal cells had the 46XX (female) karyotype. Scattered additional reactions were observed with other reagents and interpreted as being due to corssreacting antibodies in the serum. Another amniotic fluid-cultured cell line (Radvany, not published), which showed three HLA-A locus and four HLA-B locus antigens, was tested for its ability to absorb reagents containing antibodies

Table I
Expression of HLA Antigens on Amniotic Fluid Cells

Mother's lymphocytes	HLA antigens Father's lymphocytes	Fetal amniotic fluid cells
1. HLA-A1, A-; B8, BW35	HLA-A10, A11; BW35, BW16	HLA-A1, A10, A11; B8, (BW35) [a]
2. HLA-A3, A-; B7, B12	HLA-A2, A11; BW35, BW21	HLA-A3, A11; B7, BW35, (B12)

Note: Case 1. One maternal HLA-A locus antigen (A1) but two paternal A locus antigens (A10, A11) are expressed on fetal cells. Crossreactive antibodies in reagents could possibly be the cause. Maternal HLA-B locus antigen (BW35) is weakly expressed.
Case 2. Two maternal B locus antigens (B7 and B12) and one paternal B locus antigen (BW35) are detected on fetal cells. Contamination with maternal cells in suspected.
[a] Weakly expressed antigen in parentheses.

against these antigens. The results showed (Table II) that this cell line was able to absorb antibodies from reagents detecting the two HLA-A locus and two HLA-B locus antigens present on parental lymphocytes but failed to absorb antibodies from reagents directed to the "extra" specificities. Pellegrino *et al.* (1976) showed that skin fibroblasts infected with SB40 virus also showed "extra" HLA antigens; but, perhaps the most interesting earlier studies were performed by Pious and Sonderland (1974), who showed that cultured lymphoid cells could be induced to display new antigens by mutagens, whereas spontaneously occurring variants of their lines did not express "extra" antigens.

McAlack (1980) reported her attempts to determine HLA antigens on lymphocytes, fibroblasts, and cell lines of patients with neuroblastoma. Extra specificities and missing specificities were found on tumor cell lines and McAlack suggests that since she was not able to remove the "extra" specificity from the reagent with normal cells the "extra" specificity could be an alien HA.

Pollack *et al.* (1980a) reported a well-controlled study which contradicts the finding of McAlack. Using human tumor cell lines, fibroblasts, and peripheral blood lymphocytes of the donor, these authors also found by direct HLA typing "extra" antigens on fibroblasts and tumor lines, but these specificities were removed by absorption with normal lymphocytes. At the present time, there is no convincing evidence that the "extra" specificities found on tumor cell lines are indeed alien HA.

As methods detecting Class II (HLA-DR) antigens became available, it became clear that many human sera which showed "extra" specificities in reality contained HLA-DR antibodies which caused these reactions. This finding led to the discovery of Class II antigens on some tumor cells but not on their normal counter parts. These Class II antigens were not alien HA. These antigens were genetically appropriate and alien only in the sense that they were expressed on the abnormal cell, but not on the normal cell of the same type. For example, Winchester *et al.* (1978) and Wilson *et al.* (1979) reported that cultured human melanoma cells express Class II antigens, whereas normal melanocytes fail to express these antigens (Rowden *et al.*, 1977; Klareskog *et al.*, 1977). The above findings were confirmed by Pollack *et al.* (1980b) in their studies with bladder cancer: appropriate Class II antigens were found on neoplastic, but not on normal cells of the bladder. The above findings on Class II antigens are of interest because they demonstrate two instances where HA coded by the genome becomes derepressed in the malignant cell. To date, alien HA, as defined in mouse experiments, have not been convincingly described in man.

c. Biochemical Studies. Biochemical studies have been performed on Class II antigens which appear on melanoma tumor cells but are genetically appropriate molecules. Wilson *et al.* (1980) immunoprecipitated radiolabeled HA molecules from detergent-solubilized long-term melanoma tissue culture cell membranes and determined molecular weights

Table II
HLA Antigens on Cultured Amniotic Fluid Cells Detected by Microcytotoxicity and Absorption

Microcytotoxicity

Mother: HLA-A2, A26; B27, BW35;
Father: HLA-A1, A2, B17, B27:
Cultured fetal amniotic fluid cells: HLA-A1, A26, A28; B15, BW39, B17, BW35;

Absorption[a]

	Antibody titers in sera								
	Not absorbed				Absorbed with amniotic fluid cells				
Antibodies in reagent	1:1	1:2	1:4	1:8	1:1	1:2	1:4	1:8	HLA antigens on test cells
Anti-A1	+	+	+	−	±	−	−	−	A1
Anti-A26	+	+	+	±	−	−	−	−	A26
Anti-A28	+	+	+	−	+	+	+	−	A28
Anti-B15	+	+	−	−	+	+	−	−	B15
Anti-BW39	+	−	−	−	+	−	−	−	BW39
Anti-B17	+	+	+	−	±	−	−	−	B17
Anti-BW35	+	+	+	+	+	±	−	−	BW35
Anti-B27[b]	+	+	+	−	+	+	+	−	B27

[a] Three A locus and four B locus antigens are detected on cultured fetal cells by absorption, but only two antigens for each locus are found by absorption.
[b] Control.

by SDS–PAGE. These investigators found that both chains of Class II molecules on melanoma cells had the same molecular weights as Class II molecules on normal cells. Human melanoma cells expressed generally lower levels of Class II antigens than normal lymphoid cells. Authors speculate that this reduced expression of Class II antigens may be the result of reduced synthesis of these antigens or incomplete insertion in the cell membrane. As in the mouse studies, HA appearing on malignant cells are very similar to ordinary HA molecules.

D. Changes in Histocompatibility Antigens

In vitro environment alters surface antigens. Partial loss of HA accompanied by an increase in TSTA has been described in mouse systems (Cikes *et al.*, 1973; Jacobs and Huseby, 1967; Jones and Feldman, 1975; Kon *et al.*, 1976) and in human systems (Pious and Sonderland, 1974). Bonavida *et al.* (1980) reported the failure of mouse histocompatibility anti-sera to kill tumor cells (RCS) maintained *in vitro* as compared to good killing of freshly explanted cells. Carbone *et al.* (1978) found that alien and original HA are reduced after *in vitro* culture of mouse tumor cells, but Pellegrino *et al.* (1976) did not find a reduction of HA on human fibroblast lines. The time in culture as well as the type of cell and the culture medium used are all important variables in tissue culture and play a role in the expression of HA.

Present evidence indicates that all tumors do not express alien HA. Parmiani *et al.* (1975) found alien antigens on three of nine MCA-induced sarcomas on one strain (BALB/c) and no alien HA on seven sarcomas produced in two different strains of mice (C57BL65 and C3H/HEDP). Waksal (1980) found alien HA on 1% of the tumors he studied, but he only examined 40 tumors. All human melanoma tumors did not express Class II antigens either (Pollack *et al.*, 1980b). It is possible that the methods used are not sensitive enough to detect small amounts of HA, but it is also possible that presently available reagents can not detect some of the alien HA. *In vitro* artifacts could render the detection of alien HA difficult, and variation of HA with cell cycle could produce inconsistencies and contradictory resullts.

E. Major Objections against the Evidence for Alien HA

Some of the major objections raised against the evidence in favor of alien HA are:

1. Contamination of Cell Lines with Another Line

Contamination with another line is not a rare occurrence and has been reported in tumor lines of both mouse (Nelson-Rees and Flandermeyer, 1977) and man (Pollack *et al.*, 1980a). To assure purity of cell lines, Pollack *et al.* (1980b) recommended that HA of the tumor donor should always be determined from peripheral lymphocytes or fibroblasts. If any discrepancies are observed between the tumor HA and peripheral lymphocyte HA, the tumor should be tested for contamination by determining isoenzymes. Contamination with another cell line may also cover up any true alien specificities. In an attempt to grow fibroblasts from placenta tissues (Radvany, not published) it was observed that failure to detect paternal antigens on placenta fibroblasts was often due to heavy contamination or complete overgrowth of fetal cells by maternal fibroblasts. Similarly, normal fibroblasts present in the tumor tissue could overgrow tumor cells thus obscuring the detection of alien HA (or lack of antigens) on tumor cells. The HLA type of the culture would be like that of the tumor cell donor, and any additional specificities present on tumor cells could escape detection. Contamination with another line is something that should be kept in mind as a possible source of false results. However, it seems unlikely that all studies describing alien HA are studies of contaminated cell lines. Investigators are becoming increasingly more aware of this problem, and in some cases clear evidence has been presented which assures that contamination by another cell line did not occur.

2. Multispecific and Poorly Defined Reagents (Sera)

Sera containing antibodies to viral or fetal antigens could cause confusion in the identification of alien HA. Sera containing low-titer antibodies to other than the main specificity would be "operationally" monospecific only with some techniques, and the use of a more sensitive technique would bring out additional specificities. Again, most investigators are very much aware of these problems and, especially in later studies, use carefully selected and identified sera. Most human reagents contain additional antibodies to Class I and Class II antigens as well as antibodies to free virus. Some of the earlier findings of "extra" reactions have already been identified as reactions caused by antibodies to genetically appropriate Class II antigens. It is possible that well-identified and monospecific reagents for human HA will only become available with the identification of more monoclonal antibodies directed to human HA.

F. Explanations Offered for the Appearance of Alien HA

1. Alien Antigens Are Hidden Specificities Which Are Uncovered on Tumor Lines

In man, "extra" serological reactions could, in some cases, be explained by uncovered crossreactive specificities, because "extra" reactions often are observed with sera detecting specificities crossreacting with the expressed, gentically appropriate specificity. Relatively few human tumors have been studied, and it is difficult to predict how common these crossreacting specificities are on human tumor cells. In the mouse, alien HA are not a standard occurrence even with the same type of tumor. If there would be hidden specificities, it seems that they should have to be uncovered and detected with greater frequency.

2. Mutant Genes Are Coding for a Specificity Which Crossreacts with a Normal HA

This possibility has been recognized in mouse (Nabholtz *et al.*, 1975; Chiang and Klein, 1978) and man (Pious and Sonderland, 1974). The problem with this theory is that mutation could produce an alteration of the existing HA in some cases and appearance of known HA, foreign to the host, in other cases. Altered HA have not been reported. It is possible that, a slight alteration could pass unnoticed—a possible explanation of why alien HA are detected on a few tumors only.

3. Viral Products May Associate with MHC Products and Produce a New Determinant

This theory has been suggested by Fineberg *et al.* (1978) and it states that a viral product in association with an HA may produce the alien specificity which crossreacts with an existing specificity. Simpson *et al.* (1978) have shown that histocompatibility-restricted responses are highly specific and do not crossreact—a contradiction which opposes Fineberg's theory.

4. The Gene Derepression Theory

This theory implies that genes of MHC are normally inherited as a family of structural genes which are controlled by a regulatory gene. The regulatory gene determines the expression of particular alleles according to the haplotype of the strain (Amos, 1971; Festenstein, 1978). A virus or another agent could interfere with the repression system and normally silent genes would be expressed, or the production of a normally expressed HA could be suppressed.

Derepression of genes coding for embryonal enzymes or antigens has been documented (Baldwin *et al.*, 1974; Knox, 1974; Springer and Desai, 1977). Derepression of alloantigens on neoplastic cells absent from normal cells has been shown by the TL, ML, and MMA systems (Old and Stockert, 1977; Chang *et al.*, 1972) and HLA-DR (Pollack *et al.*, 1980a; Winchester *et al.*, 1978). To explain the appearance of alien antigens, one would have to speculate the presence of unexpected antigens not characteristic to the host. Several recent findings show the appearance of foreign antigens: (1) a foreign IgG allotype in a mouse (Bosma and Bosma, 1974); (2) expression of the inappropriate alloantigens in rabbits after *Micrococcus lysodeikticus* immunization (Strossberg *et al.*, 1974); (3) expression of the wrong allotype on human cells transplanted to a hamster (Pothier *et al.*, 1974); (4) expression of new HLA antigens following SV40 infection of tumor lines (Pellegrino *et al.*, 1976), and others.

The evidence presented does not permit firm conclusions as to the existence of alien HA. A great deal of systematic research will be needed before alien HA will be fully understood in terms of their structure and dynamics. However, the study of alien HA is a fast-moving field, important findings have already been made, and further data are forthcoming.

IV. CONCLUSION

In conclusion, studies on tumor cells show both a loss of HA as well as a gain of "extra" HA. Alien HA are the "extra" HA which appear on malignant cells. Studies on red cell antigens (carbohydrate antigens) of man show that tumor cells may either lack or show decreased amounts of antigens appropriate to host cells, or an extra red cell antigen, foreign to the host, may also appear on malignant cells. Extensive biochemical studies of red cell antigens on normal and tumor cells suggest that alien red cell antigens are a result of incomplete synthesis of carbohydrate chains because of malignant transformation. Studies on red cell antigens have concentrated on the synthesis, biochemistry, and *in vitro* activity of these antigens, but sufficient data are not available on the immunogenicity of alien red cell antigens.

Alien HA (H-2 antigens) in the mouse have been demonstrated on cells of some tumors. Although biochemistry and synthesis of alien H-2 antigens has not been studied in such detail as that of human red cell antigens, evidence based on isolated fractions indicates that alien HA of the mouse are similar but not identical to HA which are native to host cells. Immunologically mouse alien HA can induce either antibody formation, cellular cytotoxicity, or both. The differences in the type of immunological responses produced may well depend on the chemistry, the size, and the nature of the antigen when presented to the host (on cell surface or solubilized).

Studies on alien HA (HLA) in man have been performed by typing malignant cells, tumor cell lines, and fibroblasts. Extra reactions have been found, but these studies do not provide sufficient data for the existence of alien antigens on cultured or malignant cells.

Criticism of studies on alien HA (H-2 and HLA) include the possibility that cell lines are contaminated with another cell line carrying different antigens and that reagents used in testing are not monospecific, but contain antibodies to other specificities including those of viruses. Biochemical studies have suggested that molecules carrying viral antigens could possibly be falsely identified as alien HA by multispecific reagents.

Several theories have been proposed to explain the presence of alien HA. Alien HA could be "hidden specificities" which become uncovered on malignant cells or crossreactive antigens or as a result of interactions between viral products and HA. More likely explanations are that normally "silent" (repressed) genes coding for HA become derepressed in malignant cells or that an incomplete synthesis of the HA protein molecule renders it a new specificity which is foreign to the host.

The evidence presented does not permit firm conclusions as to the existence of alien HA. Clearly, more biochemical and immunological studies are needed before alien HA will be fully understood.

ACKNOWLEDGMENTS. The author would like to thank Dr. Roger Melvold for critical evaluation of the manuscript and for helpful suggestions.

V. REFERENCES

Amos, B., 1971, Genetic control of the human HL-A histocompatibility system: Alternatives to the two sublocus hypothesis, *Transplant. Proc.* **3**:71.

Baldwin, R. W., 1973, Immunological aspects of chemical carcinogenesis, *Adv. Cancer Res.* **18**:1.

Baldwin, R. W., Embleton, M. J., Price, M. R., and Vose, B. M., 1974, Embryonic antigen expression on experimental rat tumors, *Transplant. Rev.* **20**:77.

Bernoco, D., Glade, P., Broder, S., Miggiano, V., Hirschhorn, K., and Ceppelini, R., 1969, Stability of HL-A and appearance of other antigens (LIVA) on the surface of lymphoblasts grown in vitro, *Folia Haematol.* **54:**795.

Bodmer, W. F., 1973, A new genetic model for allelism at histocompatibility and other complex loci: Polymorphism for control of gene expression, *Transplant. Proc.* **5:**1471.

Bonavida, B., Roman, J. M., and Hutchinson, I. V., 1980, Inappropriate Alloantigen-like specificities detected on reticulum cell sarcomas of SJL/J mice: Characterization and biologic role, *Transplant. Proc.* **12**(1):59.

Bosma, H., and Bosma, G., 1974, Congenic mouse strains. The expression of hidden immunoglobulin allotype in a congenic partner strain of BALB/C mice, *J. Exp. Med.* **139:**512.

Cann, H., Radvany, R. M., and Payne, R. O., 1974, HLA phenotype of fetus detected on cultured amniotic fluid cells, *Pediatr. Res.* **8:**388/114.

Carbone, G., Invernizzi, G., Meschini, A., and Parmiani, A., 1978, In vitro and in vivo expression of original and foreign H-2 antigens and of the tumor associated transplantation antigens of a murine fibrosarcoma, *Int. J. Cancer* **21:**85.

Chang, S., Nowinski, R. C., Nishioka, K., and Irie, R. F., 1972, Immunological studies on mouse mammary tumors. VI. Further characterization of a mammary tumor antigen and its distribution in lymphatic cells of allogenic mice, *Int. J. Cancer* **9:**409.

Chiang, C. K., and Klein, J., 1978, Immunogenetic analysis of H-2 mutations. VII. H-2 associated recognition of minor histocompatibility antigens in H-2K^{b} mutants, *Immunogenetics* **6:**333.

Cikes, M., Fribert, S., Jr., and Klein, G., 1973, Progressive loss of H-2 antigens with concomitant increase of cell surface antigen(s) determined by Moloney leukemia virus in cultured murine lymphomas, *J. Natl. Cancer Inst.* **50:**347.

Davidsohn, I., Kovarik, S., and Ni, L. Y., 1969, Isoantigens A, B and H in benign and malignant lesions of the cervix. *Arch. Pathol.* **87:**306.

Dick, H., Steel, C. M., and Crichton, W., 1972, HL-A typing of cultured peripheral lympho blastoid cells, *Tissue Antigens* **2:**85.

Festenstein, H., 1978, Distinctive mechanisms in naturally occurring genetic resistance to leukemogenesis, *in* "Natural Resistance Systems Against Foreign Cells, Tumors and Microbes" (G. Cudkowicz, M. Landy, and G. Shearer, eds.), pp. 223–231, Academic Press, New York/London.

Festenstein, H., Schmidt, W., Testorelli, C., Marelli, O., and Simpson, S., 1980, Biologic effects of the altered MHS profile on the K36 tumor, a spontaneous leukemia of AKR, *Transplant. Proc.* **12**(1):25.

Finberg, R., Burakoff, S. J., Cantor, H., and Benacerroff, B., 1978, Biological significance of alloreactivity: T cells stimulated by syngeneic cells specifically lyse allogeneic target cells, *Proc. Natl. Acad. Sci. USA* **75:**5145.

Foley, E. J., 1953, Antigenic properties of methylcholanthrene induced tumors in mice of the strain of origin, *Cancer Res.* **13:**835.

Fujiwara, H., Aoki, H., Tsuchida, T., and Hamaoka, T., 1978, Immunologic characterization of tumor associated transplantation antigens on MM102 mammary tumor eliciting preferentially helper T cell antivity, *J. Immunol.* **121:**1591.

Garrido, F., Festenstein, H., and Schirrmacher, V., 1976a, Further evidence for derepression of H-2 and Ia like specificities of foreign haplotypes in mouse tumor cell lines, *Nature* (*London*) **261:**704.

Garrido, F., Schirrmacher, V., and Festenstein, H., 1976b, H-2 line specificities of foreign haplotype appearing on a mouse sarcoma after vaccinia virus infection, *Nature* (*London*) **259:**228.

Garrido, F., Schirrmacher, V., and Festenstein, H., 1977, Studies on H-2 specificities on mouse tumor cells by a new microradio assay, *J. Immunogen.* **4:**15.

Garrido, F., Schmidt, W., and Festenstein, H., 1978, Immunogenetic studies on Meth-A-Vaccinia tumor cells *in vivo* and *in vitro*, *J. Immunogen.* **4:**115.

Gold, J. M., Freedman, S. O., and Gold, P., 1972, Detecting human anti-CEA antibodies, *Nature (London)* **239:**70.

Gross, L., 1943, Intradermal immunization of C3H mice against a sarcoma that originated in an animal of the same line, *Cancer Res.* **3:**326.

Häkkinen, I., 1970, A-like blood group antigen in gastric cancer cells of patients in blood groups O or B, *J. Natl. Cancer Inst.* **44:**1183.

Hakomori, S., and Andrews, H. D., 1970, Sphinoglycolipids with Le^b activity, and the co-presence of Le^a, Le^b glycolipids, *Biochim. Biophys. Acta* **202:**225.

Hakomori, S., and Jeanloz, R. W., 1964, Isolation of a glycolipid containing fucose, galactose, glucose and glucosamine, *J. Biol. Chem.* **239:**PC3606.

Hakomori, S., and Murakami, W. T., 1968, Glycolipids of hamster fibroblasts and derived malignant transformed cell lines, *Proc. Natl. Acad. Sci. USA* **59:**254.

Hakomori, S., Koscielak, J., Bloch, H., and Jeanloz, R. W., 1967, Immunologic relationship between blood group substances and a fucose containing glycolipid of human adenocarcinoma, *J. Immunol.* **98:**31.

Hendricksen, O., Robinson, E. A., and Apella, E., 1978, Structural characterization of H-2 antigens purified from mouse liver, *Proc. Natl. Acad. Sci. USA* **75:**3322.

Himmelweit, F. (ed.), 1957, "The Collected Papers of Paul Erlich" Pergamon, New York.

Invernizzi, G., and Parmiani, G., 1975, Tumor associated transplantation antigens of chemically induced sarcoma cross reacting with allogeneic histocompatibility antigens, *Nature (London)* **254:**713.

Jacobs, B. B., and Huseby, R. A., 1967, Growth of tumors in allogeneic hosts following organ culture explantation, *Transplant. Proc.* **5:**410.

Jones, J. M., and Feldman, J. D., 1975, Alloantigen expression of rat Moloney sarcoma, *J. Natl. Cancer Inst.* **55:**995.

Kay, H. E. H., and Wallace, B. M., 1961, A and A antigens of tumors arising from urinary epithelium, *J. Natl. Cancer Inst.* **26:**1349.

Klareskog, L., Malmnäs-Tjernlund, U., Forsum, U., and Peterson, P., 1977, Epidermal Langerhans cells express Ia antigens, *Nature (London)* **268:**248.

Klein, G., 1966, Tumor antigens, *Annu. Rev. Microbiol.* **20:**223.

Klein, P., 1975, Anamalous reactions of mouse alloantisera with cultured tumor cells. I. Demonstration of widespread occurrence using reference typing sera, *J. Immunol.* **115:**1254.

Knox, W. E., 1974, The graded enzymic immaturity of transplanted neoplasms, *Cancer Res.* **34:**2102.

Kon, N. D., Forbes, J. T., and Klein, P. A., 1976, Ability of delayed-type hypersensitivity reactions to distinguish tumor associated antigens and histocompatibility antigens in soluble extracts from murine fibrosarcomas, *Int. J. Cancer* **17:**613.

Law, L. W., DuBois, G. C., Rogers, M. J., Appella, E., Pierotti, M. A., and Parmiani, G., 1980, Tumor rejection activity of antigens isolated from the membranes of a Methylcholanthrene-induced sarcoma, C-1 bearing alien H-2 antigens, *Transplant. Proc.* **12**(1):46.

Lengerova, E., 1977, "Membrane Antigens," pp. 58–75, Fischer, Jena.

Levine, P., Bobbitt, O. B., Waller, R. K., and Kuhmichel, A., 1951, Isoimmunization by a new blood factor in tumor cells, *Proc. Soc Exp. Biol Med.* **77:**403.

Martin, W. J., Esber, E., Cotton, W. G., and Rice, J. M., 1973, Depression of alloantigens in malignancy: Evidence for tumor susceptibility alloantigens and for possible self-reactivity of lymphoid cells active in the microcytoxicity assay, *Br. J. Cancer* **28**(Suppl. 1):48.

Martin, W. J., Gipson, T. G., Martin, S. E., and Rice, J. M., 1976, Depressed alloantigen on transplacentally induced lung tumor coded for by H-2 linked gene, *Science* **194:**532.

Matossian-Rogers, A., Garrido, F., and Festenstein, H., 1977, Emergence of foreign H-2 like cytotoxicity and transplantation targets on Vaccinia and Moloney virus infected Meth. A tumor cells, *Scand. J. Immunol.* **6:**541.

Matossian-Rogers, A., di Giorgi, L., and Festenstein, H., 1980, H-2^d like specificities on Gardner (H-2^k) tumor detected in a restricted anti tumor reaction, *J. Immunogen.* **7**(1):99.

McAlack, R. F., 1980, Normal HLA phenotypes and neo-HLA-like antigens on cultured human neuroblastoma, *Transplant. Proc.* **12**(1):107.

Meschini, A., and Parmiani, G., 1978, Anti H-2 alloantibodies elicited by syngeneic immunizations with a chemically induced fibrosarcoma, *Immunogenetics* **6:**117.

Meschini, A., Invernizzi, G., and Parmiani, G., 1977, Expression of alien H-2 specificities on chemically induced BALB/C fibrosarcoma, *Int. J. Cancer* **20:**271.

Nabholtz, M., Young, H., Meo, T., Miggiano, U., Rijnbeck, A., and Shreffler, D. C., 1975, Genetic analysis of an H-2 mutant B6-C-H-2^{ba} using cell mediated lymphoysis: T and B cell dictionaries for histocompatibility determinants are different. *Immunogenetics* **1:**457.

Nairn, R. C., 1962, Loss of gastro-intestinal-specific antigen in neoplasia, *Br. Med. J.* **1:**1791.

Nelson-Rees, W. A., and Flandermeyer, R. R., 1977, Inter- and intraspecies contamination of human breast tumor cell lines HBC and BRCa5 and other cell cultures, *Science* **195:**1343.

Nowinski, R. C., and Klein, P. A., 1975, Anomalous reactions of mouse alloantisera with cultured tumor cells. II. Cytotoxicity is caused by antibodies to leukemia viruses, *J. Immunol.* **115:**1261.

Oh-Uti, K., 1949, Polysaccharides and a glycidamin in the tissue of gastric cancer, *Tohoku J. Exp. Med.* **51:**297.

Old, L. J., and Boyse, E. A., 1964, Immunology and experimental tumors. *Annu. Rev. Med.* **15:**167.

Old, I. J., and Stockert, E., 1977, Immunolgenetics of cell surface antigens of mouse leukemia. *Annu. Rev. Genet.* **11:**127.

Parmiani, G., Meschini, A., and Invernizzi, G., 1975, Alien histocompatibility determinants on the cell surface of sarcomas induced by methyl cholantrene. I. In vivo studies, *Int. J. Cancer* **16:**756.

Parmiani, G., Meschini, A., Invernizzi, G., and Carbone, G., 1978, Tumor associated transplantation antigen distinct from H-2^k like antigens on a BALB/C (H-2^d) fibrosarcoma. *J. Natl. Cancer Inst.* **61:**1229.

Pelligrino, M. A., Ferrone, S., Brautbar, C., and Hayflick, L., 1976, Changes in HL-A antigen profiles on SV40-transformed human fibroblasts. *Exp. Cell Res.* **97:**340.

Pious, D., and Soderland, C., 1974, Expression of new antigens by HLA variants of cultured lymphoid cells. *J. Immunol.* **113:**1399.

Pliskin, M. E., and Prehn, R. T., 1978, Are tumor-associated transplantation antigens of chemically induced sarcomas related to alien histocompatibility antigens? *Transplantation* **26:**19.

Pollack, M. S., Heagney, S., and Fogh, J., 1980a, HLA typing of cultured human tumor cell lines: The detection of genetically appropriate HLA-A, B, C and DR alloantigens, *Transplant. Proc.* **12**(1):134.

Pollack, M. S., Livingston, P. O., Fogh, J., Carey, T. E., Dupont, B., and Oettgen, H. F., 1980b, Genetically appropriate expression of HLA and DR (IA) alloantigens on human melanoma cell lines, *Tissue Antigens* **15**(3):249.

Pothier, L., Borel, H., and Adams, R., 1974, Expression of IJG allotypes in human lymphoid tumor lines serially transplantable in the neonatal Syrian hamster, *J. Immunol.* **113:**1984.
Prehn, R. T., and Main, D., 1957, Immunity to methylcholanthrene, *J. Natl. Cancer Inst.* **18:**768.
Rogers, M. J., Apella, E., Pierotti, M. A., Invernizzi, G., and Parmiani, G., 1979, Biochemical characterization of alien H-2 antigens expressed on a methylcholanthrene induced tumor, *Proc. Natl. Acad. Sci. USA* **76:**1415.
Roman, J. M., and Bonavida, B., 1979, Expression of inappropriate H-2 antigens on SJL reticulum cell tumors, *Transplant. Proc.* **11:**1365.
Rowden, G., Lewis, M. G., and Sullivan, A. K., 1977, Ia antigen expression on human epidermal Langerhans cells, *Nature (London)* **268:**247.
Sasportes, M., Dehay, C., and Fellous, M., 1971, Variations of the expression of HLA antigens on human diploid fibroblasts in vitro, *Nature (London)* **233:**332.
Schirrmacher, V., Garrido, F., Hubsch, D., Garcia-Olivares, E., and Koszinowski, U., 1980, Foreign H-2 like molecules on a murine tumor (MCG4): Target antigens for alloreactive cytolytic T lymphocytes (CTL) and restricting elements for virus-specific CTL, *Transplant. Proc.* **12**(1):32.
Schmidt, W., and Festenstein, H., 1980, Serological and immunochemical studies of allospecificities on K36, a syngeneic tumor of AKR, *J. Immunogen.* **7:**7.
Schmidt, W., Atfield, G., and Festenstein, H., 1979, Loss of H-2K^k gene products, *Immunogenetics* **8:**311.
Schmidt, W., Festenstein, H., and Atfield, G., 1980, Serologic and immunochemical studies of the cell membrane alloantigens of K36 and AKR spontaneous leukemia, *Transplant. Proc.* **12**(1):29.
Simpson, E., Mobraaten, L., Chandler, P., Hetherington, C., Hurme, M., Brunner, C., and Bailey, D., 1978, Cross-reactive cytotoxic responses, *J. Exp. Med.* **148:**1478.
Sjögren, H. O., 1965, Transplantation methods as a tool for detection of tumor specific antigens, *Prog. Exp. Tumor Res.* **6:**289.
Springer, G. F., and Desai, P. R., 1977, Cross-reacting carcinoma associated antigens with blood group and precursor specificities, *Transplant. Proc.* **9:**1105.
Strossberg, D., Hamers, C., Van der Loo, W., and Hamers, R. J., 1974, A rabbit with the allotypic phenotype: ala2a3 b465b6, *Immunology* **113:**1313.
Tellem, M., Plotkin, H. R., and Merance, D. R., 1962, Studies of blood group antigens in benign and malignant human breast tissue, *Cancer Res.* **23:**1528.
Testorelli, C., Marelli, O., Schmidt, W., and Festenstein, H., 1980, Changes in H-2 antigen expression on a murine spontaneous leukemia (K36) detected by cell-mediated cytotoxicity assay, *J. Immunogen.* **7:**19.
Waskal, 1980, (Discussion) *Transplant. Proc.* **12**(1):36.
Weiler, E., 1956, Antigen differences between normal hamster kidney and stilboestrol induced kidney carcinoma: Histological demonstration by means of fluorescing antibodies, *Brit. J. Canc.* **10:**533.
Wilson, B. S., Indiveri, F., Pellegrino, M. A., Ferrone, S., 1979, DR (Ia like) Antigens on human melanoma cells serological detection and immunochemical characterization. *J. Exp. Med.* **149:**658.
Wilson, B. S., Indiveri, F., Molinaro, G. A., Wuaranta, V., and Ferrone, S., 1980, Characterization of DR antigens on cultured melanoma cells by using monoclonal antibodies, *Transplant Proc.* **12**(1):125–129.
Winchester, R. J., Wang, C. Y., Gibofsky, A., Kunkel, H., Lloyd, K., Old, L., 1978, Expression of Ia like antigens on cultured human malignant melanoma cell lines. *Proc. Natl. Acad. Sci. USA* **75:**6235.

Chapter 7

Blood Group Antigens in Tumor Cell Membranes

John S. Coon and
Ronald S. Weinstein

Department of Pathology
Rush Medical College
and Rush Presbyterian–St. Luke's Medical Center
Chicago, Illinois

I. INTRODUCTION

This review is concerned with the expression by human tumor cells of the so-called blood group antigens, a group of carbohydrate antigen structures originally detected on erythrocytes at the turn of the century (Race and Sanger, 1975). The term *blood group antigens* is somewhat of a misnomer since these antigens are also normal plasma membrane components in the endothelia and epithelia of many human organs (Weinstein *et al.*, 1981b). Further, they have no known special functional significance in the hematopoietic system. As compared with the bulk of cell surface glycoprotein and glycolipids of unknown function (called antigens for lack of a better word), the blood group antigens are reasonably well characterized with respect to their biochemistry, genetics, and the reagents which specifically bind to them (Watkins, 1978; Hakomori, 1981). The effects of malignant transformation on their expression in epithelia has been investigated by researchers with an interest in oncofetal membrane markers of neoplasia.

Two major categories of alterations in blood group antigens in carcinoma cells are the deletion of normally expressed blood group antigenic reactivity at the cell surface and the emergence of additional antigens, not present on normal epitehlial cells (Davidsohn, 1979; Springer *et al.*, 1977). Many of the additional antigens are probably cryptantigens, masked by terminal sugar residues in normal tissue. The biosynthesis of cell surface carbohydrate antigens involves the sequential addition of sugar residues to precursor structures by specific glycosyltransferases (Watkins, 1978). Loss of normal blood group antigens due to the incomplete synthesis of oligosaccharide chains by neoplastic cells can result in the expression of new antigens, representing the incompletely synthesized oligosaccharide chains (Hakamori, 1975b). Alternatively, their expression may result from the action of specific glycosidases or other unmasking enzymes (Limas and Lange, 1980).

Various biochemical and immunohistochemical methods have been used to study the expression of blood group antigens on tumor cells. Each method has certain advantages and its own set of limitations. Biochemical analyses of fresh tissues permit quantitation of blood group substances and precise delineation of their molecular structure. Further, biochemical studies of glycosyltransferase activities have provided insights into the pathogenesis of blood group antigen abnormalities in some settings (Hakomori, 1975b). An important disadvantage of many biochemical methods is the requirement for relatively large amounts of fresh tumor tissue. This restricts the number and types of human carcinomas available for analysis, since only relatively large tumors are likely to yield sufficient tissue for this type of study. Another drawback is that biochemical methods are essentially averaging techniques, producing data reflecting the mean level of blood group antigen expression within a tumor, including the contributions of blood vessels and entrapped hematopoietic elements. Heterogeneity of antigen expression within a tumor cell population, a common occurrence (Hart and Fidler, 1981) cannot be detected. Another drawback is that unless quantitative microscopy studies are performed on representative samples of the tumor, which is tedious, requires the participation of a pathologist, and is beyond the capabilities of most biochemistry laboratories, the partial volume of the tumor mass representing tumor cells per se is unknown.

Advantages of studying blood group antigens in tissue sections by immunohistological methods include (1) the small quantity of tissue required; (2) the large number of tumor specimens that can be processed in a batch mode; (3) the capability to relate the distribution of the antigens to the histopathology and cytopathology; and (4) the extreme sensitivity of the methods. Large numbers of carcinomas from virtually any site can be readily analyzed by surgical pathology laboratories since paraffin-embedded tissue obtained for diagnostic purposes can be used. This permits the characterization of antigen expression in preneoplastic lesions and the correlation of expression with the subsequent clinical course of the malignant dyscrasia. The ease and convenience of immunohistologic methods have contributed significantly to their usefulness in subclassifying tumors with respect to type and prognosis on the basis of marker analyses (Weinstein *et al.*, 1981b). Principal drawbacks of immunohistologic methods are (1) a critical dependence upon the availability of appropriate, specific staining reagents; (2) the semiquantitative nature of the methods; and (3) an inability to identify the carrier of the antigenic determinants (e.g., glycoproteins vs. glycolipids). Furthermore, tissue processing for histopathology partially degrades and solubilizes blood group antigen-active substances (Limas and Lange, 1980) although efforts are being made to circumvent these pitfalls.

In summary, both the biochemical and immunohistologic approaches have advantages in the study of blood group antigen expression by malignant cells. Because of space limitations, this review will focus on immunohistologic studies. The relationship of tissue blood group antigen expression to the biological behavior of tumors will be examined, as well as the potential usefulness of antigen modifications as a measure of the aggressive potential of human cancers.

II. STRUCTURE AND GENETICS

A. ABH Antigens

Several recent reviews summarize the extensive literature on the genetics and biochemistry of the ABH antigens on human erythrocytes (Hakamori, 1981; Watkins, 1978). Far less is known about the biochemistry of these cell surface antigens on normal or neoplastic epithelial cells. As an introduction to the topic, relevant data on the biochemistry and genetics of these antigens on erythrocytes will be briefly summarized and related to the relatively small body of information on blood group antigens on neoplastic epithelial cells.

The structure of the A, B, and H blood group determinants is shown in Fig. 1. The A and B determinants differ with respect to having either *N*-acetylgalactosamine or galactose added to the H structure, the common precursor of both antigens A and B (Watkins, 1978). This structure is constant, regardless of whether the carrier molecule is a glycolipid or glycoprotein. However, the structure of the proximal segments of the oligosaccharide chains, apart from the terminal nonreducing blood group-active portions, are highly variable and related to the specific glycoproteins or glycolipids bearing the determinants (Watkins, 1978; Hakomori, 1981; Rauvala and Finne, 1979).

At least three genetic loci influence the expression of the A, B, and

ANTIGEN	STRUCTURE
A	α-GalNAc(1→3)β-Gal(1→3 or 4)GlcNAc ↑1,2 α-Fuc
B	α-Gal(1→3)β-Gal(1→3 or 4)GlcNAc ↑1,2 α-Fuc
H	β-Gal(1→3 or 4)GlcNAc ↑1,2 α-Fuc

Fig. 1. Structures of the A, B, and H antigenic determinants; GalNac, *N*-acetylgalactosamine; Gal, galactose; GlcNAc, *N*-acetylglucosamine; Fuc, fucose.

H antigens (Watkins, 1978). The *ABO* locus contains two functional alleles, *A* and *B*, and a nonfunctional allele, *O*. The *A* and *B* genes code for specific glycosyltransferases that specifically add *N*-acetylgalactosamine, for group A, or galactose, for group B, to the specific precursor, the H antigen structure. Since the *O* allele is nonfunctional, patients of blood group O have only the H antigen. The H antigen is formed by a glycosyltransferase specified by a gene *H* at a different genetic locus *Hh*. The *h* allele is silent; therefore, patients who are *hh* do not form the H antigen structure, and thus neither A- nor B-active substances, regardless of what genes are present at the *ABO* locus. The *hh* genotype is rare. A third independent genetic locus which may be important in the expression of the ABH antigens on epithelial cells is the *Sese* locus which determines the individual's *ABO* secretor status. The *Hh* locus in certain tissues with secretory functions, but not hematopoietic tissue, is under the control of the *Sese* locus. Patients who are homozygous *sese* are nonsecretors, i.e., their secretions such as saliva, milk, and intestinal juice do not contain ABH-active glycoproteins, whereas those who are secretors (i.e., *SeSe* or *Sese*) do have ABH-active substances in their secretions. The expression of the ABH antigens by certain epithelial cells with prominent secretory activity is strongly influenced by the patient's secretor status (Szulman, 1960, 1962). Thus, some types of gastrointestinal epithelial cells have ABH antigens easily detectable in intracellular secretions if the patient is an ABO secretor. The influence of secretor status on the expression of cell surface epithelial ABH antigens is uncertain. A number of investigators have reported quantitative differences in antigen expression between secretors and nonsecretors (Limas and Lange, 1980; Vedtofte *et al.*, 1981; Dabelsteen, 1972) but this has not been a universal finding (Coon and Weinstein, 1981a).

Several other factors have a bearing on the expression of the ABH antigens by normal epithelial cells. Persons of blood group A are subtyped into A_1 and A_2 on the basis of structural differences in the antigen (Hakomori, 1981). The expression of antigen A is somewhat weaker in A_2 than in A_1 individuals, at least on erythrocytes (Watkins, 1978). Specific antisera and lectins which recognize the A determinant on erythrocytes in A_1 individuals will not do so in A_2 individuals. The influence of subgroup status has not been systematically investigated in neoplastic epithelial cells. It is noteworthy that individuals with blood groups A, B, and AB also express the H antigen (Race and Sanger, 1975), since H is the precursor to A and B and glycosylation is incomplete. The quantity of residual H antigen expression varies with *ABO* genotype (Race and Sanger, 1975). For example, A_2 individuals have more H than A_1 individuals on their red blood cells (Race and Sanger, 1975).

The nature of the carrier molecules of the ABH antigens on erythrocytes is controversial, and the identity of the specific carrier molecules on epithelial cells is unknown (Hakomori, 1981). However, the glycoproteins and glycolipids carrying the ABH antigens on epithelial cells are probably different from those on erythrocytes (Hakomori, 1981). With respect to this point, membrane-associated blood group antigen substances of epithelial cells must be distinguished from secretions of epithelial cells, which are glycoproteins and have been extensively characterized (Watkins, 1978). In human erythrocytes, some investigators believe that the majority of the ABH-active substances are glycolipids (Dejter-Juszynski *et al.*, 1978; Watkins, 1978; Koscielak, 1977; Hakomori, 1981), whereas others argue that glycoproteins carry the majority of the ABH determinants and glycolipids account for a small fraction of the total (Karhi and Gahmberg, 1980; Schenkel-Brunner, 1980; Wilczynska *et al.*, 1980). A variety of ABH-carrying glycolipids have been described with oligosaccharide components of various size (Dejter-Juszynski *et al.*, 1978; Watkins, 1978; Koscielak, 1977; Hakomori, 1981). Those with very large oligosaccharide components, termed *macroglycolipids* or *polyglycosylceramids*, are highly water soluble and may have been misidentified as glycoproteins in some studies (Dejter-Juszynski *et al.*, 1978). Those who favor glycoproteins as the major ABH antigen species propose that most of the ABH determinants are on the major anion transport protein of erythrocytes, called band 3 (Fig. 2), with the oligosaccharide component N-linked to asparagine residues (Viitala *et al.*, 1981; Hakomori, 1981; Karhi and Gahmberg, 1980; Fukuda *et al.*, 1979a). ABH determinants

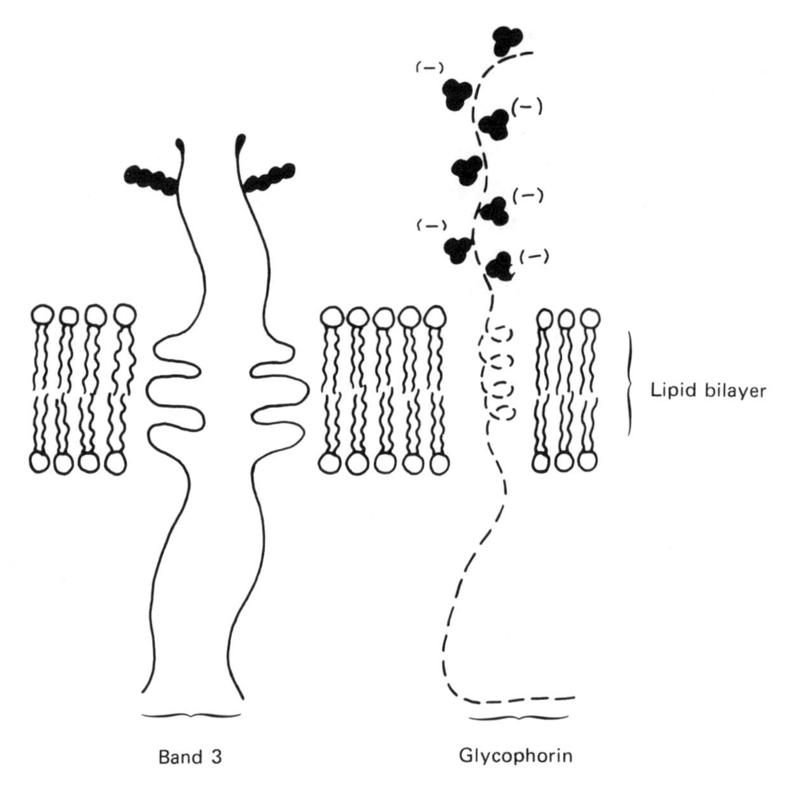

Fig. 2. Diagram of band 3 protein and glycophorin in the erythrocyte membrane. Both proteins carry blood group antigen determinants: band 3 carries A, B and H, and glycophorin carries M, N, T and Tn. Negative charges in diagram refer to sialyl residues.

have also been reported in oligosaccharides O-linked to hydroxyamino acids in glycophorin (Fig. 2) (Hakomori, 1981; Rauvala and Finne, 1979). The N-linked oligosaccharide chains bearing the ABH determinants on erythrocytes display increasingly complex branching during fetal development (Hakomori, 1981; Fukuda *et al.*, 1979a). Whether comparable differences exist in the oligosaccharide core of the ABH antigens between benign and malignant epithelial cells is unknown.

The expression of tissue ABH antigens in normal human organs is summarized in Table I. Normal epithelium in many organs expresses the ABH antigens, whereas vascular endothelium is universally positive for the antigens, as determined by immunologic methods. Tissue lacking the ABH antigens include connective tissue, smooth and striated muscle, and central nervous system parenchyma. Lymphocytes apparently absorb small amounts of ABH antigens from the plasma (Oriol *et al.*, 1981).

Expression by some epithelial cells appears to be a marker for specific stages of fetal developmental (Szulman, 1960, 1962, 1964, 1965). Thus, in the fetus the epithelium lining the entire colon expresses the ABH antigens, whereas in adults the ABH antigens are expressed exclusively in the proximal colon (Abdelfattah-Gad and Denk, 1980; Cooper and Haesler, 1978; Denk *et al.*, 1974).

Table I
Presence of Isoantigens A, B, and H in Normal Adult Human Tissues

Location	Positive	Negative	Reference
Adrenal		Cortex and medulla	Davidsohn (1979)
Blood vessels	Endothelial cells		Davidsohn (1979)
Breast	Glands, ducts		Lill *et al.* (1979)
Bronchus	Pseudostratified columnar epthelia	Serous glands	Davidsohn (1979)
	Mucous glands		Davidsohn (1979)
Central nervous system		Nerve cells and glial cells	Davidsohn (1979)
Colon	Proximal colon	Distal colon	Cooper *et al.* (1980)
Connective tissue		Collagen, bone, and cartilage	Davidsohn (1979)
Duodenum	Epithelia		Lill *et al.* (1979)
	Brunners glands		Lill *et al.* (1979)
Endometrium	Epithelia (proliferative and secretory)		Lill *et al.* (1979)
Esophagus	Squamous epithelia	Basal cell layer	Davidsohn (1979)
	Mucous glands		Davidsohn (1979)
Exocervix	Squamous epithelia		Lill *et al.* (1976)
Fallopian tube	Columnar epithelia		Davidsohn (1979)
Gallbladder	Columnar epithelia		Davidsohn (1979)
Kidney	Glomeruli	Convoluted tubules	Davidsohn (1979)
	Collecting tubules		Davidsohn (1979)
Larynx	Squamous epithelia		Davidsohn (1979)
	Mucous glands	Serous glands	Davidsohn (1979)
Liver	Kupffer cells	Hepatocytes	Holborow *et al.* (1960)
Lung	Mucous glands	Serous glands	Davidsohn and Ni (1969)
Muscles		Smooth and striated	Davidsohn (1979)
Pancreas	Exocrine glands	Islets of Langerhans	Davidsohn (1979)
Pituitary	Colloid, pars intermedia		Lill *et al.* (1979)
Placenta	Trophoblast[a]	Trophoblast Cytotrophoblast	Loke and Ballard (1973) Stejskal *et al.* (1973)
Prostate	Glands		Lill *et al.* (1979)
Skin	Horny layer of squamous epithelium	Cells of basal and Malpighian layers	Davidsohn (1979)

continued

Table I
(*Continued*)

Location	Positive	Negative	Reference
	Sweat glands	Sebaceous glands	Davidsohn (1979)
Small intestine	Columnar epithelia		Davidsohn (1979)
	Brunner's glands		Davidsohn (1979)
	Paneth cells		Davidsohn (1979)
Stomach	Columnar epithelia		Davidsohn (1979)
Testis		Tubules	Lill *et al.* (1979)
Tongue	Squamous epithelia		Davidsohn (1979)
	Mucous glands	Basal layer	Kovarik *et al.* (1968)
Trachea	Pseudostratified columnar epithelia		Davidsohn (1979)
	Mucous glands	Serous glands	Davidsohn (1979)
Urinary system	Urinry bladder, ureter		Coon and Weinstein (1981a)
Vagina	Squamous epithelia	Basal layer	Davidsohn (1979)
	Mucus glands		Kovarik *et al.* (1968)

[a] Detected by immunoferritin electron microscopy.

B. MN Blood Group Antigen-Related Structures

The M, N, T (Thomsen–Friedenreich), and Tn antigens may be considered as a group because they are structurally related blood group antigens. The structures and relationships of these antigens are diagrammed in Fig. 3. Expression of the M and N antigens requires linkage of appropriate oligosaccharide side chains to a specific sequence of N-terminal amino acids, as shown in Fig. 3 (Rolih, 1980; Anstee, 1981). The difference between M and N is determined by amino acid differences in the sequence specified by the M and N genes. A specific N-terminal peptide sequence is required to create the conformation of oligosaccharide chains required for the M and N antigenic determinants. In erythrocytes, M and N specificities are carried at the N terminus of glycophorin (Rolih, 1980; Anstee, 1981). As is the case for ABH antigens, the M and N antigens are represented in tissues other than erythrocytes. Springer *et al.* (1977) detected

→

Fig. 3. Diagram of structure and interrelationships of M, N, T, and Tn antigens. NANA, sialic acid; Gal, galactose; GalNAC, *N*-acetylgalactosamine; Ser, serine; Thr, threonine. Diagrams of M and N antigens represent N terminus of glycophorin. For further structural details, see text.

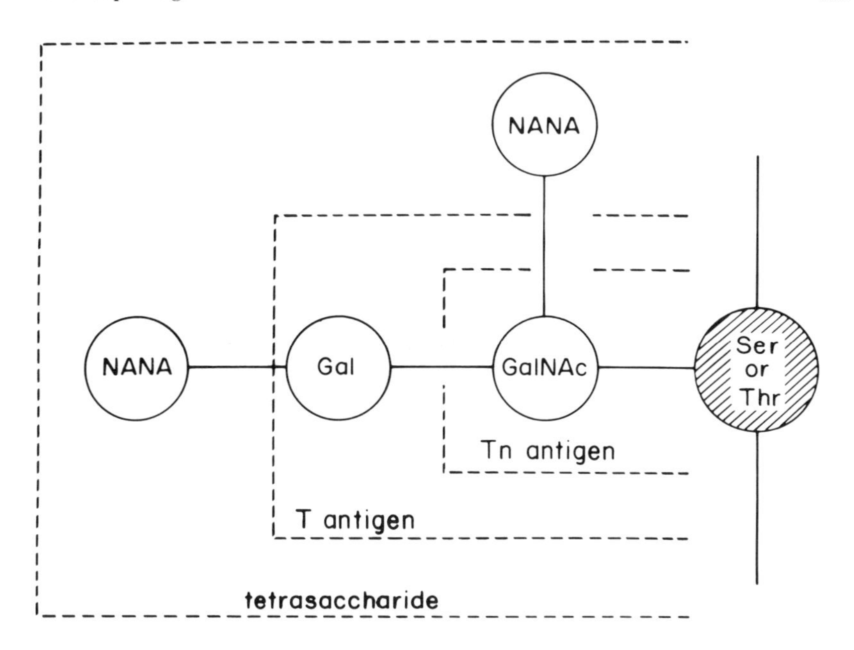
NANA
NANA
Gal
GalNAc
Ser
or
Thr
Tn antigen
T antigen
tetrasaccharide

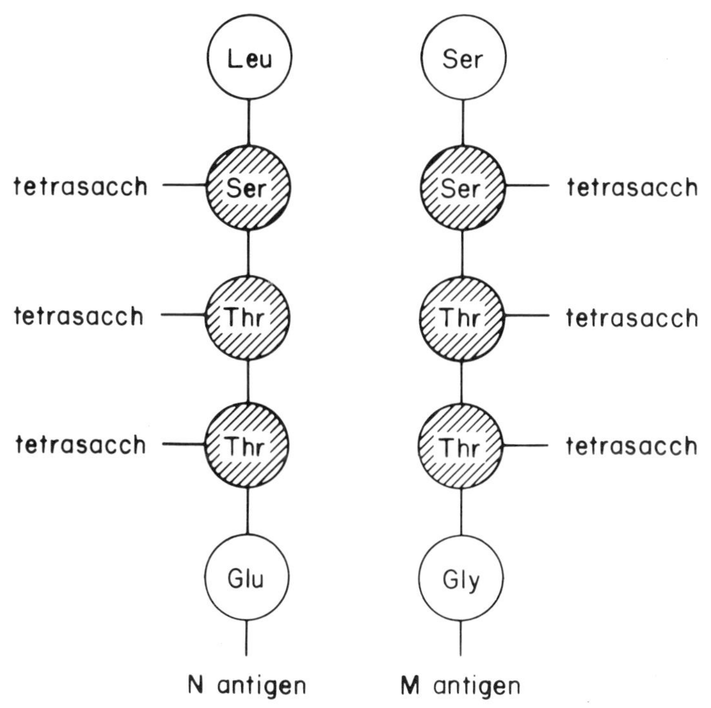
Leu
Ser
tetrasacch
Ser
Ser
tetrasacch
tetrasacch
Thr
Thr
tetrasacch
tetrasacch
Thr
Thr
tetrasacch
Glu
Gly
N antigen
M antigen

them in normal breast and gastrointestinal tissue by hemagglutination inhibition. They are not present in normal human urinary bladder epithelium (Coon *et al.*, 1982). Differences in M and N antigen expression in normal and neoplastic epithelium have not been systematically examined.

The T antigen was originally described as a rare erythrocyte antigen in patients whose erythrocytes were agglutinated by normal human serum irrespective of the serum donor's blood group (Thomsen, 1927; Friedenreich, 1930). The immunodominant structure has been established by inhibition studies to be terminal dissacharide galactose β(1-3) *N*-acetylgalactosamine (Rolih, 1980). The lectin, peanut (*Arachis hypogaea*) agglutinin (PNA), has a specificity very similar to anti-T antiserum and is used routinely in blood banking for detecting the T antigen (Bird, 1977; Lis and Sharon, 1977). The T antigen was originally demonstrated to be a precursor structure for the M and N blood group antigens by showing that it could be derived from the M and N antigens on normal erythrocytes by removal of their terminal sialic acid residues with neuraminidase. It is now known that the T antigen structure is also a core structure in oligosaccharides which have other sugar residues substituted for sialic acid (Beyer *et al.*, 1979; Rauvala and Finne, 1979). Also, many T structures masked by sialic acid are present on glyçoproteins lacking the peptide structure required for M and N antigen expression (Springer *et al.*, 1979; Rolih, 1980; Anstee, 1981). On erythrocytes each molecule of glycophorin carries many masked T structures but only one M or N structure, since a specific sequence of N-terminal amino acids is required for M or N expression (Anstee, 1981). The T antigen is also a component of membrane glycolipids (Rauvala and Finne, 1979; Springer *et al.*, 1979). Thus, the chemistry of the T antigen is complex. In some normal tissues, it may be expressed overtly at the terminus of an oligosaccharide chain where it will combine with specific anti-T ligands, but, more commonly, it is in a cryptic form masked by sialic acid or other sugar residues.

The Tn antigen, like the T antigen, was initially described on rare, pathological erythrocytes. It is defined by its ability to specifically combine with certain human antisera and with Tn-specific lectins such as *Salvia sclerea* agglutinin (Bird, 1977). The Tn antigenic determinant is thought to include terminal *N*-acetylgalactosamine α-linked to a hydroxyamino acid (Springer *et al.*, 1979; Anstee, 1981) (Fig. 3). However, the precise structure of the antigenic determinant, including the role of peptides is unclear. The Tn antigen may thus be considered to be a precursor of the T antigen and the M and N antigens. As with the T antigen, however, Tn antigen-active structures can occur in molecules without the required amino acid composition for expression of antigens M or N (Springer *et al.*, 1979; Anstee, 1981). Springer *et al.*, (1977) have reported that the Tn antigen is present in many carcinomas but not in normal tissue as determined by hemagglutination inhibition. Immunohistologic studies on the Tn antigen have not been reported.

C. The I and i Antigens

Both the I and i antigens are families of carbohydrate antigens of related structure which are defined by their ability to combine with certain human antisera that are cold agglutinins (Watanabe *et al.*, 1979; Childs *et al.*, 1980). The Ii antigens are generally considered to be precursor structures to the ABH antigens and have been identified on band 3 protein in erythrocytes (Fukuda *et al.*, 1979a). The precise structure of certain of these antigens which combine with monoclonal anti-I and anti-i antisera has been defined (Childs *et al.*, 1980). Considerable differences exist in the exact structures recognized by different antisera. I antigens differ from i antigens by having a complexly branched, rather than linear oligosaccharide chain (Fukuda *et al.*, 1979a). I is found on adult erythrocytes, but i is not (Childs *et al.*, 1980). Increased amounts of both I and i have been found in various human carcinomas compared with corresponding normal epithelium by immunofluorescence or quantitative precipitin assays using glycoprotein extracts of human tumors (Feizi *et al.*, 1975; Picard *et al.*, 1978). The relationship of Ii antigenic expression to tumor biological behavior has not been explored.

III. EXPRESSION OF BLOOD GROUP ANTIGENS ON TUMORS

A. ABH Antigens

Many types of benign and malignant human tumors have been surveyed for expression of ABH antigens (Table II). Carcinomas arising from epithelia in which the ABH antigens are normally expressed often show complete or partial loss of antigen expression (Davidsohn, 1979). The relationship of ABH antigen expression to the degree of tumor differentiation as expressed in the histologic grade is variable (Limas and Lange, 1980; Davidsohn, 1979). In tissues in which the ABH antigens are expressed in fetal development but not in adult life, carcinomas derived from such tissue may express the ABH antigens as oncofetal antigens, as has been convincingly demonstrated for adenocarcinomas of the left

Table II
ABH Antigens in Human Tumors

Organ	Tumor histology	Reference
Blood vessels	Angioma, angiosarcoma	Feigl *et al.* (1976)
Breast	Adenocarcinoma	Gupta and Schuster (1973)
Colon	Adenocarcinoma	Davidsohn *et al.* (1966); Stellner *et al.* (1973); Denk *et al.* (1974); Kim and Isaacs (1975); Kim *et al.* (1974); Abdelfattah-Gad and Denk (1980)
Fallopian tube	Adenocarcinoma	England and Davidsohn (1973)
Larynx	Squamous carcinoma	Dabelsteen *et al.* (1974); Lin *et al.* (1977)
Lung	Squamous carcinoma, adenocarcinoma	Davidsohn and Ni (1969)
Oral epithelium	Squamous carcinoma	Prendergast *et al.* (1968); Dabelsteen and Pindborg (1973); Liu *et al.* (1974); Dabelsteen *et al.* (1975)
Ovary	Cystadenoma, cystadenocarcinoma	Davidsohn *et al.* (1968)
Pancreas	Adenocarcinoma	Davidsohn *et al.* (1971, 1972)
Prostate	Adenocarcinoma	Gupta *et al.* (1973)
Skin	Epidermal appendage tumors	England *et al.* (1979)
Stomach	Adenocarcinoma	Cowan (1962); Eklund *et al.* (1963); Sheahan *et al.* (1971); Denk *et al.* (1974); Picard *et al.* (1978)
Urinary bladder	Transitional cell carcinoma	Decenzo *et al.* (1975); Kato (1977); Alroy *et al.* (1978); Bergman and Javadpour (1978); Lange *et al.* (1978); Limas and Lange (1980); Limas *et al.* (1979; 1982); Weinstein *et al.* (1979, 1981a,b); Johnson and Lamm (1980); Richie *et al.* (1980); Newmann *et al.* (1980); Coon and Weinstein (1981b); Stein *et al.* (1981)
Uterine cervix	Squamous carcinoma	Davidson *et al.* (1969, 1973); Bonfiglio and Feinberg (1976); Lill *et al.* (1976)

colon (Abdelfattah-Gad and Denk, 1980; Cooper and Haesler, 1978; Denk *et al.*, 1974). Inappropriate ABH expression has also been reported in carcinomas. There are several reports of the expression of antigen A in carcinomas arising in patients of blood group B or O, particularly in gastrointestinal tract tumors (Cooper *et al.*, 1980; Picard *et al.*, 1978). Since these studies employed immunohistological methods, the possibility that the positive staining reaction for group A antigen was caused by a substance which cross-reacted with antigen A, such as the Tn antigen (Anstee, 1981) or Forssman antigen (Hakomori *et al.*, 1977), must be ruled out.

Alterations in the blood group antigens may be an early event in tumorigenesis. ABH deletion has been described in carcinoma *in situ* in larynx (Lin *et al.*, 1977) oral mucosa (Dabelsteen *et al.*, 1975), and urinary bladder (Weinstein *et al.*, 1979). The ABH antigens are deleted not only in areas of frank carcinoma *in situ* but also in surrounding histologically normal epithelium. In this setting, ABH antigen deletion may be indicative of the presence of a malignant diathesis in normal-appearing epithelium prior to the morphologic expression of the neoplastic lesion (Weinstein *et al.*, 1979).

B. ABH Expression in Human Urinary Bladder Carcinomas

Study of ABH antigen expression by transitional cell carcinomas of the urinary bladder has been of special interest ever since Decenzo *et al.* (1975), in a retrospective study, reported a relationship between ABH antigen expression and the biological behavior of human bladder carcinomas (Figs. 4 and 5). Subsequently, many laboratories have confirmed

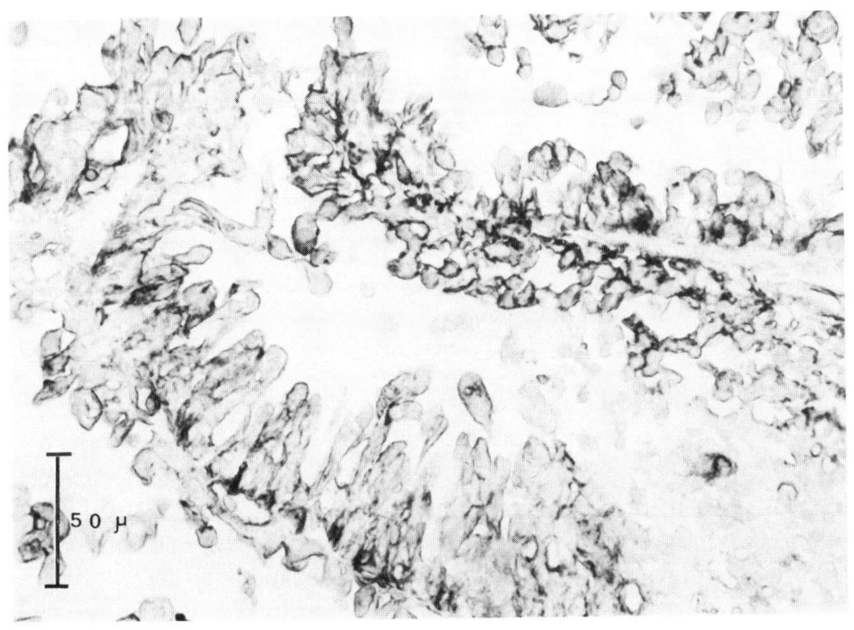

Fig. 4. Immunohistologic demonstration of the H antigen in low-grade human bladder cancer from a patient of blood group O. Immunoperoxidase stain for the H antigen. Tumor cell membranes are strongly positive. Nuclear fast red counterstain.

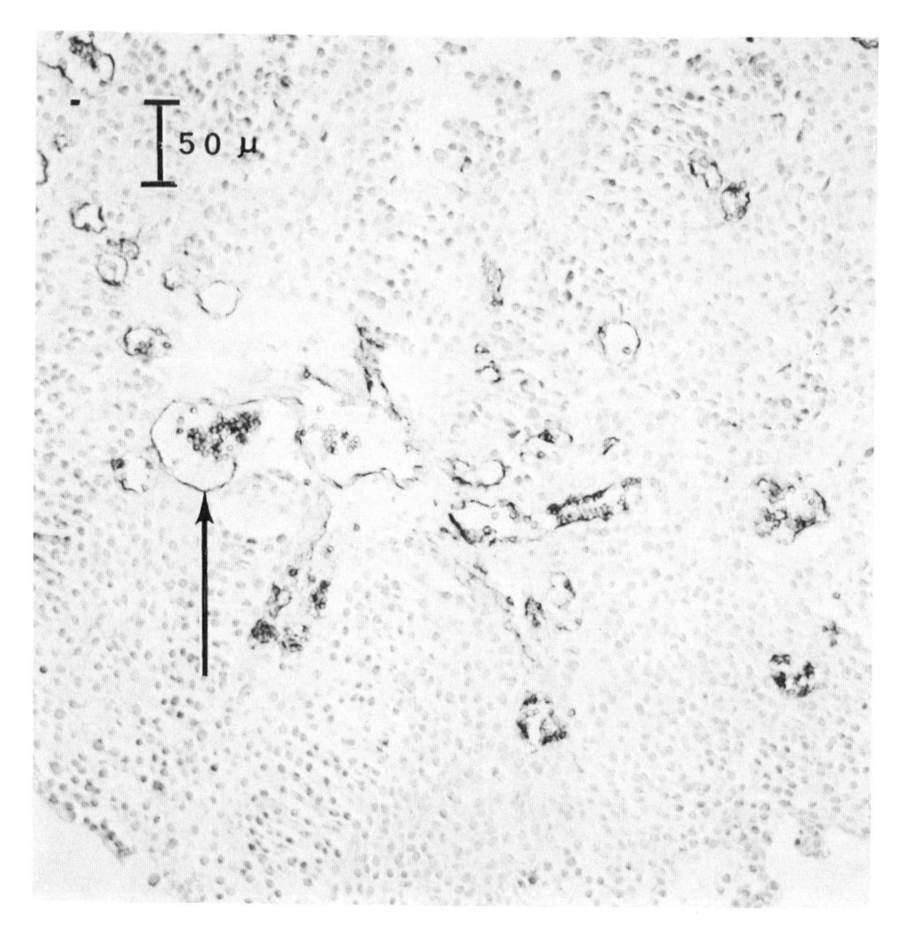

Fig. 5. Photomicrograph of low-grade human bladder carcinoma from a patient of blood group O. Immunoperoxidase stain for the H antigen. Tumor is negative but vascular endothelium (arrow) and intravascular erythrocytes are positive. Nuclear fast red counterstain.

the seminal observations that patients with early transitional cell carcinomas which express the ABH antigens infrequently develop invasion of the bladder wall, whereas patients whose tumors do not express these antigens often become invasive (Weinstein *et al.*, 1981b) (Fig. 6). This is of considerable interest to cancer cell biologists and is highly relevant in the clinical setting. The need to identify prognostically significant tumor markers for transitional cell carcinomas of the urinary bladder has been urgent because of the variability of the clinical course of cancer patients who present initially with noninvasive bladder carcinoma and because of the morbidity and mortality associated with cystectomy (Weinstein, 1979; Whitmore, 1977; Friedell *et al.*, 1976). Many patients with small superficial papillary transitional cell carcinomas have a protracted clinical course with appearance of recurrent tumors over a period of many years. Others have no recurrences, whereas still others with histologically similar tumors rapidly develop invasive tumors that are life-threatening. Current therapy for urinary bladder carcinoma is based on the assessment of tumor grade and stage by the pathologist and the urologist, the coexistence of carcinoma *in situ* elsewhere in the bladder, and the tolerance of the patient for frequent follow-up examinations (Farrow, 1979). It is currently believed that the determination of ABH antigen expression in urinary bladder carcinomas may be helpful in discriminating between those patients with superficial bladder carcinomas who are at low risk

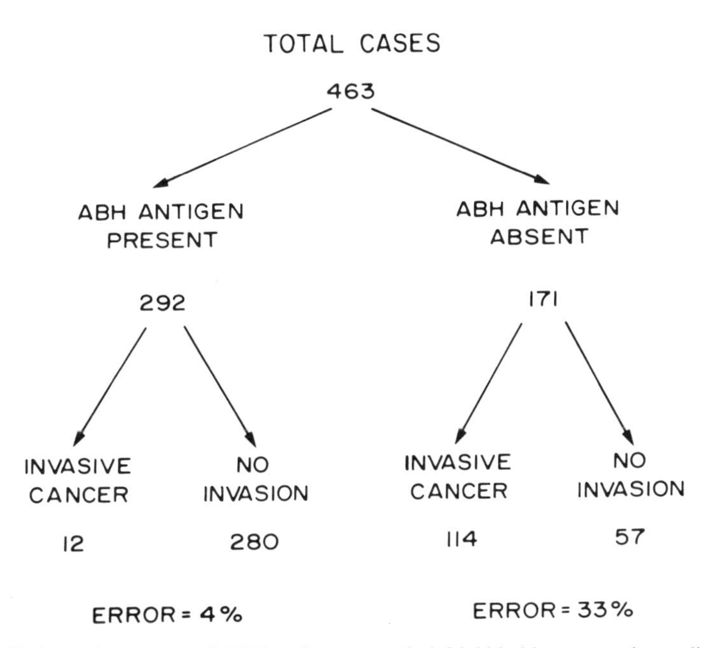

Fig. 6. Estimated accuracy of ABH antigen status in initial bladder tumors in predicting the subsequent aggressiveness of recurrences. ABH antigen expression was highly predictive of noninvasion, whereas ABH deletion was somewhat predictive of subsequent invasion. Data compiled from Decenzo *et al.* (1975), Lange *et al.* (1978), Emmott *et al.* (1979), Richie *et al.* (1980), Johnson and Lamm (1980), Newman *et al.* (1980).

for subsequent tumor invasion from those at relatively high risk for the development of invasion and for whom early aggressive therapy might be justified, although appropriately controlled prospective studies are still needed to firmly establish the clinical value of such determinations.

C. The T Antigen in Human Carcinomas

Springer and his co-workers developed the concept that the T antigen is expressed on many human carcinoma cells but is masked by sialic acid on homologous normal epithelial cells (Springer and Desai, 1977; Springer

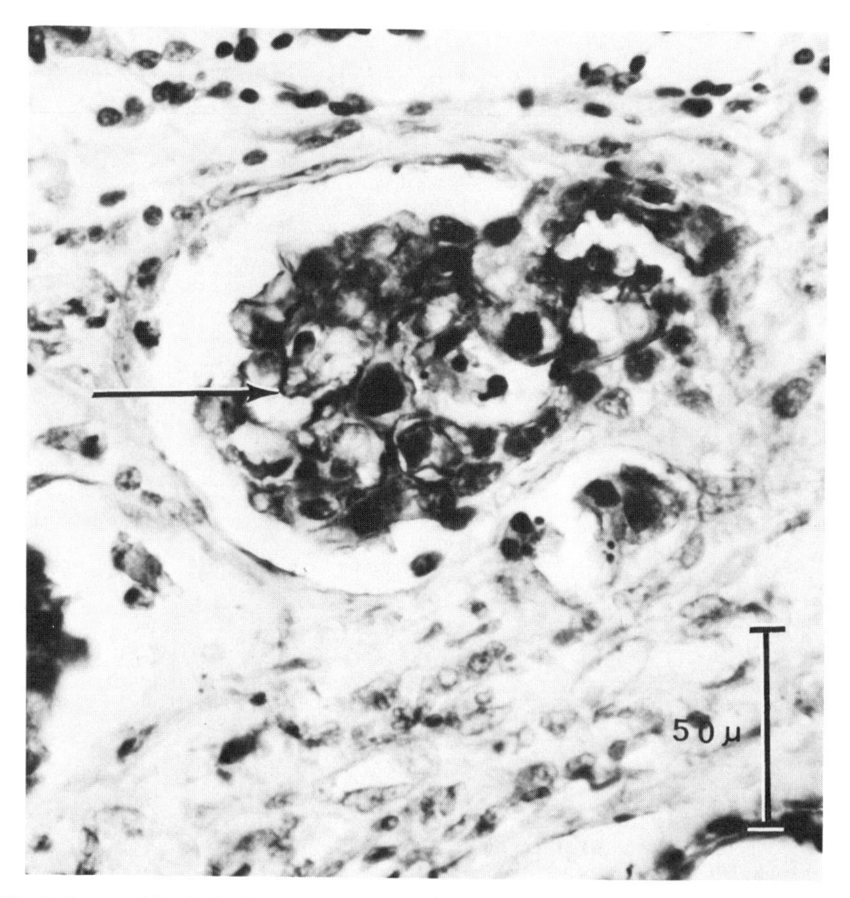

Fig. 7. Immunohistologic demonstration of the T (Thomsen–Friedenreich) antigen in human carcinoma of the breast. Lectin–immunoperoxidase stain for the T antigen. The cell membranes of most of the tumor cells are positive (arrow). Hematoxylin counterstain.

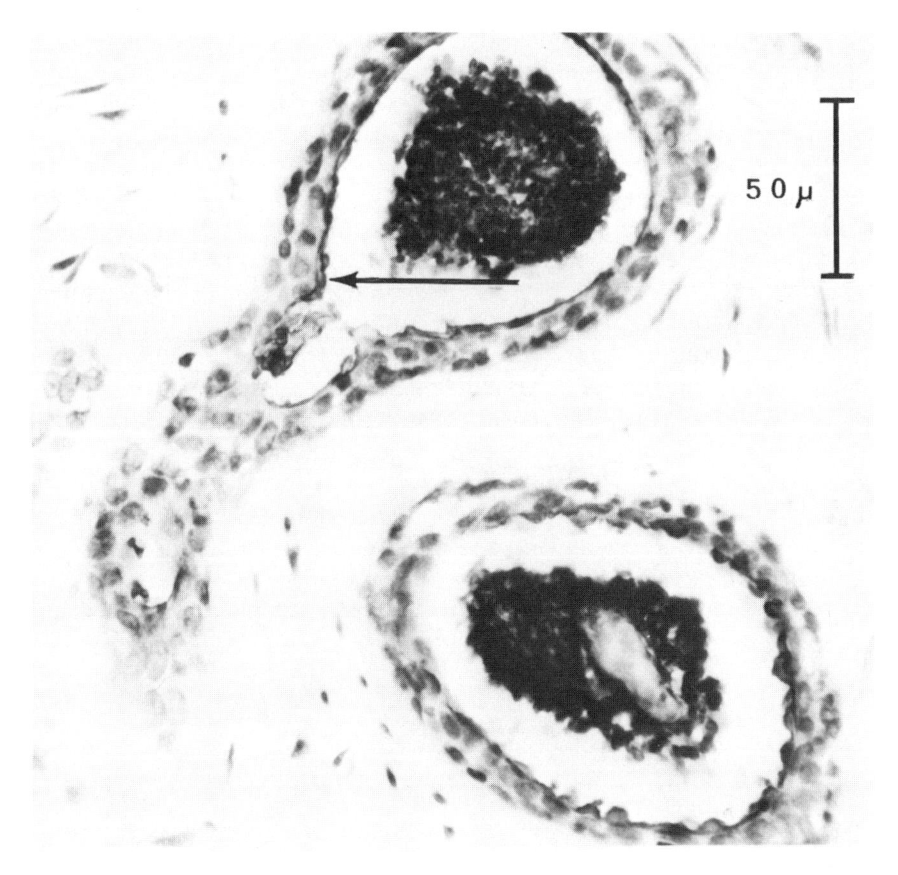

Fig. 8. Immunohistologic demonstration of the T antigen in normal human breast tissue. Lectin–immunoperoxidase stain for the T antigen. Only the lumenal membrane (arrow) and intralumenal secretions are positive. Hematoxylin counterstain.

et al., 1977; Springer *et al.*, 1979). He employed hemagglutination inhibition of T-active erythrocytes by anti-T antiserum or lectin with extracts of human carcinomas to detect the presence of T antigen-active substances in the carcinomas (Springer *et al.*, 1977). Springer found that nearly all breast and gastrointestinal carcinomas express the T antigen. Although this was not a normal component of the tissues of origin, the T antigen could be uncovered in normal tissues by neuraminidase treatment. Serological studies showed that normal individuals have antibody to the T antigen, whereas in cancer patients, anti-T antibody titers were lowered (Springer and Desai, 1977; Springer *et al.*, 1977). Anti-T antibody levels rose after surgical removal of the primary tumor (Springer and Desai, 1977; Springer *et al.*, 1977).

Recently developed immunohistochemical methods make it possible to localize the T antigen at the cellular level and to study antigen expression in minute biopsy specimens (Coon *et al.*, 1982). The T-specific lectin, PNA, is visualized in tissue sections with an anti-lectin immunoperoxidase method (Coon *et al.*, 1982). To date, T antigen expression has been examined in benign and malignant tumors in human breast (Howard and Taylor, 1979; Howard and Batsakis, 1980; Howard *et al.*, 1981) and in urinary bladder carcinomas (Coon *et al.*, 1982). Most breast carcinomas show strong cytoplasmic and cell membrane staining for the T antigen (Fig. 7), whereas benign breast tissue shows staining only along the lumenal membrane of ducts or lobules (Fig. 8). However, undifferentiated breast carcinomas are often completely negative for the T antigen. Neuraminidase pretreatment increases the intensity of staining in both benign and malignant breast tissue and introduces widespread membrane and cytoplasmic staining into benign breast tissue (Howard *et al.*, 1981). Newman *et al.* (1979) reported that undifferentiated breast carcinoma may be negative for the T antigen with and without neuraminidase pretreatment, a finding we have confirmed by immunoperoxidase methods.

Human urinary bladder has proven to be a suitable organ in which to

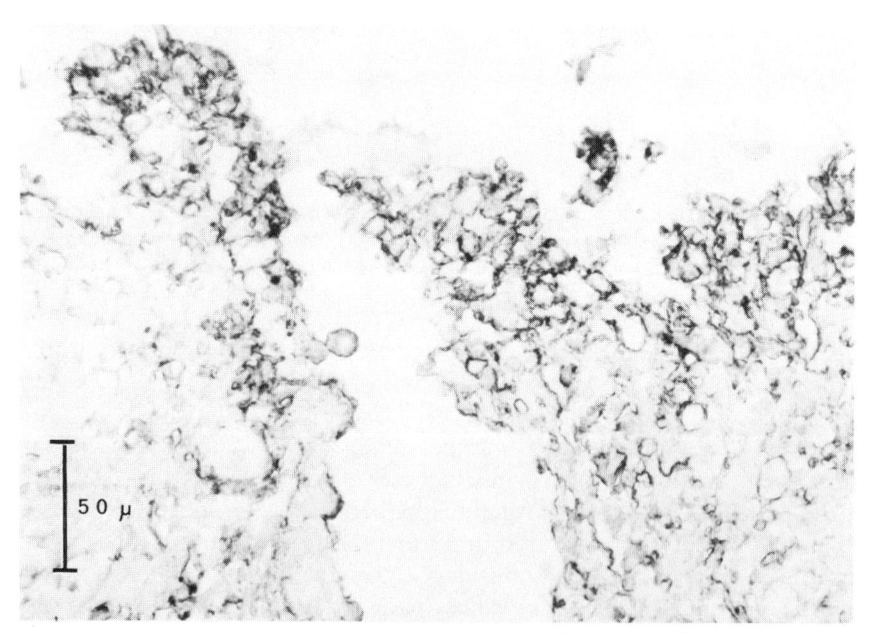

Fig. 9. T antigen in high-grade carcinoma of the urinary bladder. Lectin–immunoperoxidase stain for the T antigen. Tumor cell membranes are strongly positive. Nuclear fast red counterstain.

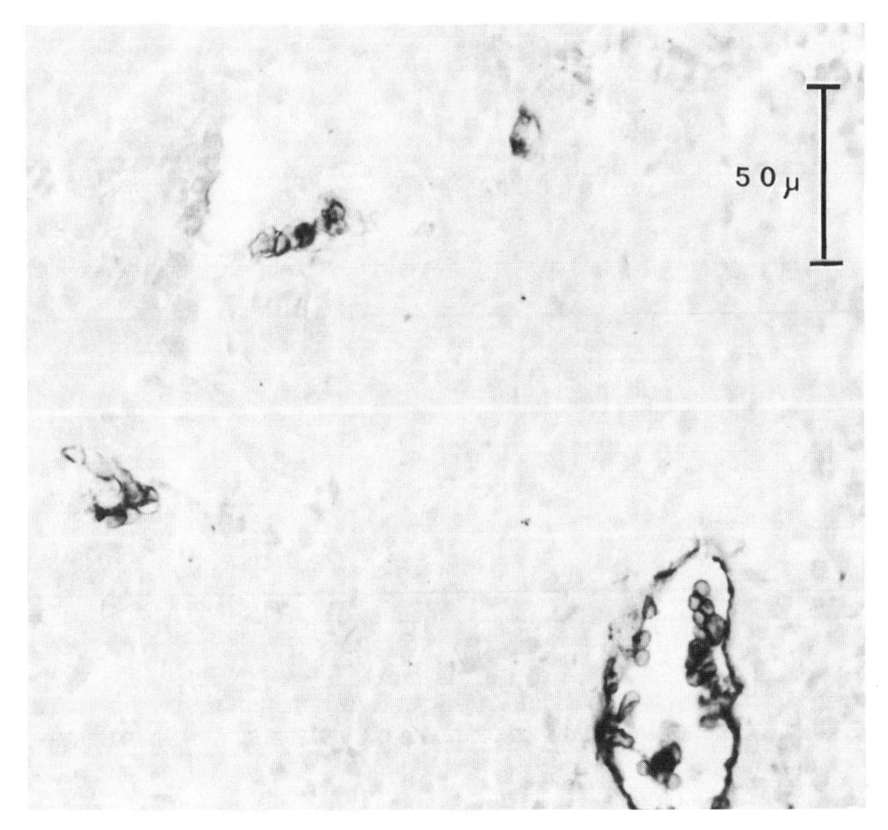

Fig. 10. High-grade human bladder carcinoma. Section treated with neuraminidase before lectin-immunoperoxidase stain for the T antigen. Tumor is negative (cryptic T negative), but vascular endothelium and erythrocytes are positive. Nuclear fast red counterstain.

examine the relationship of T antigen expression to the biological behavior of carcinomas (Coon *et al.*, 1982). Urinary bladder carcinomas have been subclassified into three categories on the basis of their expression of the T antigen: (1) cryptic T antigen positive, as in normal urothelium; (2) T antigen positive (Fig. 9); (3) T antigen negative with and without neuraminidase pretreatment (cryptic T antigen negative) (Fig. 10). Several relationships have been established between T antigen status and the morphology and clinical behavior of the human urinary bladder tumors. The correlation between T antigen status and histologic grade (i.e., degree of cytological anaplasia) is highly significant. The majority of low-grade tumors are cryptic T antigen positive, whereas high-grade tumors are generally T antigen positive or cryptic T antigen negative, reflecting a parallelism between morphological differentiation at the light microscopic

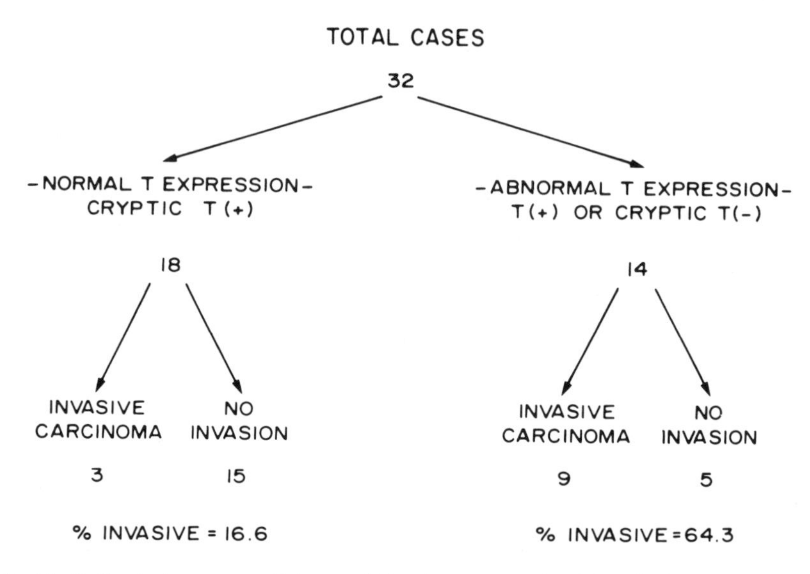

Fig. 11. Estimated accuracy of abnormal T antigen expression in predicting tumor invasion in low-grade bladder cancer patients with ABH-negative initial tumors. Abnormal T antigen status heralded a relatively poor prognosis. Adapted from Coon *et al.* (1982).

level and biochemical differentiation of the cell membranes. Abnormalities of biochemical differentiation reflected in T antigen expression are a significant predictor of biological behavior of some "look-alike" tumors, which are indistinguishable histologically. Relatively low-grade tumors which are cryptic T antigen positive have a significantly lower incidence of invasion and recurrence than low-grade T antigen-positive or cryptic T antigen-negative tumors (Fig. 11).

D. Heterogeneity of Blood Group Antigen Expression in Tumors

Tumor expression of blood group antigens may reflect phenotypic diversity within tumors. Although some experimental animal tumors have been shown to be polyclonal in origin (Reddy and Fialkow, 1979), the majority of tumors are thought to be monoclonal (Fialkow, 1972, 1976). Such monoclonal tumors nevertheless contain subpopulations of cells which vary widely with respect to many parameters including the ability to invade and metastasize (Poste and Fidler, 1980; Hart and Fidler, 1981). Identifying antigenic or biochemical properties of subpopulations which correlate with biologic behavior has proven difficult (Hart and Fidler, 1981).

Patients with transitional cell carcinoma of the urinary bladder commonly have multiple recurrent tumors at noncontiguous sites in the bladder (Weinstein, 1979; Whitmore, 1977). Although these tumors were previously regarded as multifocal in origin, Falor and Ward (1976) have presented cytogenetic evidence that multifocal bladder tumor may indeed be monoclonal in origin in some instances (Weinstein, 1979). Immunohistologic studies have shown bladder tumors are highly variable with respect to expression of the blood group antigens (Weinstein *et al.*, 1981b; Coon and Weinstein, 1981b; Coon *et al.*, 1982). Some tumors have histologically identical areas that are strongly positive and completely negative for ABH antigen expression. In other tumors, antigen is expressed only in the basal or superficial layers of cells in papillary structures. In still other tumors, isolated clusters of cells express the antigens (Fig. 12). The extent to which this heterogeneity reflects the establishment of phenotypically distinct subpopulations of cells within a tumor or is a measure of antigen expression paralleling cellular biochemical differentiation remains to be determined.

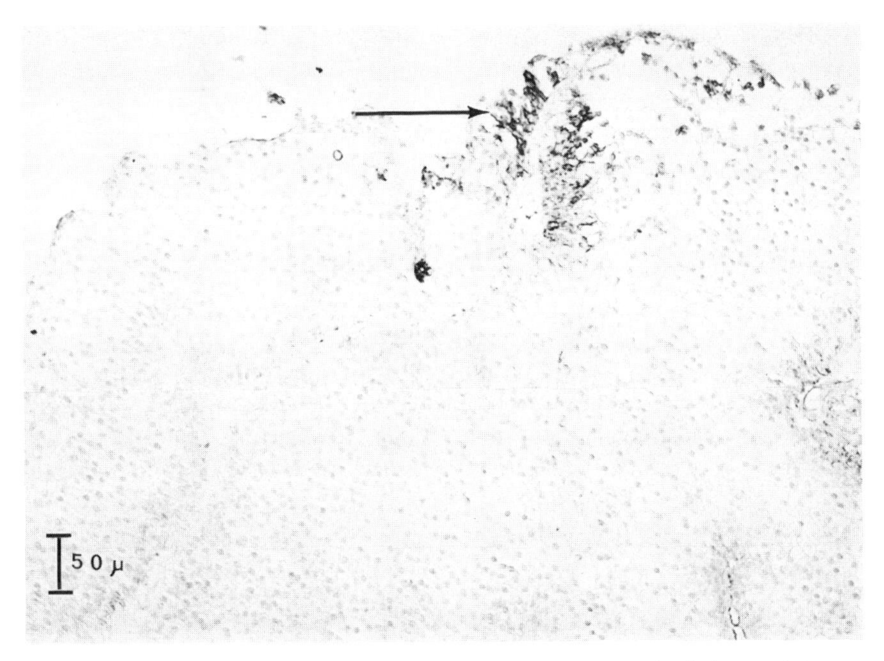

Fig. 12. Immunohistologic demonstration of focal H antigen expression in low grade human bladder cancer from patient of blood group O. Immunoperoxidase stain for the H antigen. One small focus of tumor cells (arrow) shows definite staining. Nuclear fast red counterstain.

The capacity of a tumor to invade and metastasize may be determined by the properties of a subpopulation constituting a minority of the malignant cells in a tumor (Poste and Fidler, 1980; Hart and Fidler, 1981). Since biologic behavior in human bladder cancer appears related to ABH antigen expression, the possibility of a direct link between the malignant phenotype and ABH antigen expression in subpopulations of human bladder carcinoma cells is worth investigating.

E. Mechanism of Blood Group Antigen Alterations in Tumors

Several theories have been advanced to account for the abnormalities of blood group antigen expression in tumors. Some theories attempt to explain both the disappearance of normal antigens (e.g., ABH) and appearance of additional antigens (e.g., T) by the same mechanism. Theories backed by experimental evidence are the following:

1. Tumors have abnormalities in specific glycosyltransferases, blocking the incorporation of sugar residues at or near the nonreducing terminus of oligosaccharide chains (Stellnor *et al.*, 1973; Kim *et al.*, 1974).
2. Tumors are unable to synthesize certain classes of glycoconjugates, e.g., glycoproteins, because of inability to form the necessary precursor substances (Krag, 1979; Reitman *et al.*, 1980).
3. Tumors or tumor stroma have highly active glycosidases or proteases acting at the tumor cell surface causing abnormal degradation of cell surface glycoconjugates (Fukuda *et al.*, 1979b; Rolih, 1980; Pauli *et al.*, 1980).
4. Tumors have abnormal terminal glycosylation of oligosaccharide chains, masking normal antigenic determinants (Limas and Lange, 1980).

There are many reports of increased amounts of blood group antigen precursor substances in tumors, providing indirect support for theories suggesting defective glycosyltransferase activity or excessive glycosidase activity. An increase in H-active substances in bladder carcinomas, relative to corresponding normal epithelium, in patients of blood group A and B has been reported, using a semiquantitative immunohistologic approach (Limas and Lange, 1980). Increased amounts of both I- and i-active substances, immediate precursors of H-active substances, have been detected in various human carcinomas compared with corresponding normal epithelium by immunofluorescence or quantitative precipitin assays, using tissue glycoprotein extracts (Feizi *et al.*, 1975; Picard *et al.*, 1978).

Hakomori (1975a) and others have quantitatively analyzed tumors for multiple oligosaccharide structures representing several arrest points in the synthesis of oligosaccharide chains which may carry the ABH antigens. Thus increased amounts of the glycolipid, β-GlcNAc(1-3)β-Gal(1-4)β-Glc-ceramide, was found in colon carcinoma compared to normal colonic mucosa, whereas several other precursor structures were not increased (Watanabe and Hakomori, 1976). Glycoproteins extracted from human colon carcinoma had more H precursor substance than glycoproteins from normal colonic mucosa as judged by reactivity with *Ricinus communis* agglutinin I (Kim *et al.*, 1974).

The simplification of cell surface oligosaccharide chains in carcinomas relative to corresponding normal tissues does not appear to be confined to molecules carrying the blood group antigens but may be a general phenomenon in neoplastic transformation. A general simplication of olgiosaccharides on glycolipids has been found in a number of experimental animal tumors in tissue culture (reviewed by Hakomori, 1975b, Critchly, 1979). Critchly (1979) reported a correlation between simplification of oligosaccharide chains on cell surface glycolipids and presence of malignant morphology using cells transformed with a temperature-sensitive oncogenic virus. Glycoconjugates of tumors of high and low malignancy have been compared to attempt to relate cell surface carbohydrates to malignant behavior. Skipski *et al.* (1980) reported that a highly metastatic human tumor had less total sialic acid than a morphologically comparable nonmetastatic tumor. Morre *et al.* (1978) noted a progressive simplication of gangliosides with increasing malignancy in a series of carcinogen-induced premalignant and malignant hepatic tumors in the rat. Simplification of cell surface oligosaccharides is not an invariant in malignant tumors (reviewed by Critchly, 1979). For example, Siddiqui *et al.* (1978) found no general simplification of oligosaccharide chains in tumors in which the ABH antigens had been deleted. Narasimhan and Murray (1979) found little simplification of oligosaccharides on glycolipids of human lung cancers.

Alterations of the glycosyltransferase activities responsible for the formation of the blood group antigens have been reported in several tumor systems. Stellnor *et al.* (1973) and Kim *et al.* (1974) found reduced activity of the enzyme responsible for converting the H structure to A or B in human colon cancer compared to normal mucosa. Decreased amounts of other glycosyltransferases have been found in several tumors in experimental animals (reviewed by Hakomori, 1975b; Critchly, 1979). Measurement of glycosyltransferase activity is technically difficult, necessitating considerable caution in the interpretation of data (Hakomori, 1975b).

Some evidence exists that tumors cells may have altered synthesis of cell surface oligosaccharides for reasons other than impaired glycosyltransferase activity. Krag (1979) reported a hamster cell line with a defect in the synthesis of the oligosaccharide–lipid intermediate required to synthesize the oligosaccharide component of glycoproteins (Schwarz and Datema, 1980). The cells were markedly deficient in glycosylation of cell surface proteins. Reitman *et al.* (1980) described two mouse lymphoma cell lines unable to incorporate fucose into oligosaccharide chains despite intact fucosyltransferase activity. One line was unable to convert GDP-mannose to GDP-fucose. Theoretically, defects such as these could be responsible for the failure of tumor cells to form the ABH antigens as well.

In contrast to the evidence for defective synthesis of oligosaccharide chains in tumors, evidence that alterations in surface oligosaccharides are due to excessive degradation is less convincing. Loss of ABH antigen expression from erythrocytes has been observed following treatment with *endo*-β-galactosidase from *Escherichia freundii* (Fukuda *et al.*, 1979b). Similarly, the T antigen can be unmasked on normal erythrocytes with neuraminidase (Rolih, 1980). Proof is lacking that expression of blood group antigens by tumors is linked to glycosidase activity. Some tumor cells are known to secrete powerful proteases (Pauli *et al.*, 1980; Liotta *et al.*, 1980) which might influence blood group antigen expression. In the erythrocyte, proteolytic treatment with chymotrypsin cleaves band 3 protein, but the outer surface fragment remains associated with the membrane because of the insertion of a second hydrophopic domain into the lipid bilayer (Weinstein *et al.*, 1978). Therefore, in this case, cleavage might not affect ABH expression. M and N antigen activity on glycophorin, in contrast, is highly sensitive to proteolysis (Springer *et al.*, 1977). Protease activity might cause a selective loss of blood group antigens in tumors but this remains to be demonstrated experimentally.

The concept that normal blood group antigens may be masked by extra terminal sugar residues, such as sialic acid, has been advanced by several investigators, Limas and Lange (1980) were unable to unmask ABH antigen activity in ABH-negative transitional cell carcinomas of the urinary bladder, using neuraminidase to cleave off terminal sialic acid residues. We have confirmed their findings using a sensitive immunoperoxidase technique (unpublished observation). However, increased cell surface sialic acid content has been correlated with malignant behavior in various murine tumors including variants of the B-16 melanoma. Several sublines of this tumor have been derived which vary considerably in their ability to metastasize to different organs in the mouse (Poste and Fidler, 1980; Hart and Fidler, 1981). Dobrossy *et al.* (1981) studied sublines with different ability to metastasize to the lung and found a correlation between the degree of lung implantation and the amount of tumor cell sialic acid accessible to neuraminidase cleavage and tumor cell surface sialyltransferase activity. In related studies, Yogeeswaran and Salk (1981) studied several virus or carcinogen induced and spontaneous murine tumors of different metastatic ability. They found that the ability of tumor cells to metastasize from subcutaneous sites was correlated with the total and surface-exposed sialic acid, and especially with the degree of sialylation of cell surface galactosyl and *N*-acetylgalactosaminyl residues.

F. Relationships of Multiple Alterations in Blood Group Antigen Expression in Malignant Cells to Each Other and to Other Tumor Markers

Because of our interest in predictive markers for human transitional cell carcinoma of the urinary bladder, interrelationships of multiple marker systems are being studied in this laboratory. Hyperploidy and marker chromosomes have previously been identified as predictors of a poor prognosis in human bladder carcinoma (Summers *et al.*, 1981; Weinstein *et al.*, 1981a). We recently examined four tumor marker systems—ABH antigen deletion, T antigen status, ploidy, and presence of marker chromosomes—in a series of patient biopsies to assess interrelationships between these markers (Summers *et al.*, manuscript in preparation). Each biopsy was from a patient with early noninvasive transitional cell carcinoma of the urinary bladder. Clinical follow-up was for 3 or more years after this initial biopsy. Each marker system correlated to various extents with tumor behavior. In general, the following parameters heralded a relatively favorable prognosis: (1) presence of ABH antigen; (2) cryptic T antigen; (3) diploid, pseudodiploid, or hypoploid chromosome modal number; and (4) absence of a marker chromosome. Although subclassifications based on single parameters were statistically significant, there were striking differences in the marker profiles for individual patients (Table III). In general, patients whose tumors has more than one abnormal marker had a much higher incidence of subsequent invasion than patients whose tumors had a single abnormal marker. ABH antigen expression was statistically unrelated to T antigen expression (Coon *et al.*, 1982). Patients whose tumors were blood group antigen negative and T antigen positive or cryptic T antigen negative also had a significantly worse prognosis than patients whose tumors had only one abnormal blood group antigen marker (Coon *et al.*, 1982).

Table III
Relation of Four Markers to Each Other and to Clinical Course in Patients with Bladder Cancer

	ABH Ag present		TAg status		
	Yes	No	CrT+	T+	CrT−
Ploidy					
46	8(0)[a]	4(0)	10(0)	2(0)	0
<46	5(0)	3(0)	5(0)	3(0)	0
>46	5(2)	14(11)	5(0)	11(10)	3(3)
Marker chromosome					
No	9(0)	5(0)	12(0)	2(0)	0
Yes	9(2)	16(11)	8(0)	14(10)	3(3)
ABH Ag present?					
Yes			13(0)	4(1)	1(1)
No			7(0)	12(9)	2(2)

[a] Number in parentheses refers to patients who have developed deep invasion.

Since the mechanisms of expression of each of these markers in tumors is not known, the biological significance of a particular constellation of these markers can only be approached empirically at the present time. The T antigen structure has been described within some oligosaccharide chains that have ABH antigen activity (Rauvala and Finne, 1979; Rolih, 1980), raising the possibility of a direct relationship between the disappearance of the ABH antigens and the appearance of T antigen. However, the lack of correlation between ABH and T antigen status in our studies argues against this relationship. There are no proven relationships between specific cytogenetic abnormalities and blood group antigen alterations in human bladder carcinomas. Theoretical considerations aside, our data do indicate that for detection of the subpopulation of bladder cancer patients who are truly at high risk for the development of subsequent invasion, a combination of mulitple blood group antigen markers, or blood group antigen abnormalities and cytogenic markers may be useful.

IV. CONCLUSION

The study of blood group antigen expression in human malignant tumors is still largely at a descriptive stage. Although the evidence linking blood group antigen expression with the biologic potential of human urinary bladder tumors for invasion is impressive, nothing is known about the mechanism of such a relationship. Whether a particular combination of cell surface carbohydrate antigens might unequivocally identify a cell as truly malignant with the capacity to invade stroma and to metastasize remains to be proven. Numerous investigators have suggested that cell surface oligosaccharides are important in controlling a cell's interaction with its environment and with other cells (Cooper *et al.*, 1977; Critchly, 1979; Weir, 1980; Stanley and Sudo, 1981). Study of blood group antigen expression by human tumors in tissue culture and organ culture systems in which antigen expression can be directly related to quantifiable parameters of tumor cell behavior may further elucidate these issues.

V. REFERENCES

Abdelfattah-Gad, M., and Denk, H., 1980, Epithelial blood group antigens in human carcinoma of the distal colon: Further studies on their pathologic significance, *J. Natl. Cancer Inst.* **64:**1025.

Alroy, J., Teramura, K., Miller, A. W., III, Pauli, B. U., Gottesman, J. E., Flanagan, M., Davidsohn, I., and Weinstein, R. S., 1978, Isoantigens A, B and H in urinary bladder carcinomas following radiotherapy, *Cancer* **41:**1739.

Anstee, D., 1981, The blood group MNSs-active sialoglycoproteins, *Semin. Hematol.* **18:**13.

Bergman, S., and Javadpour, N., 1978, The cell surface antigen A, B or O (H) as an indicator of malignant potential in stage A bladder carcinoma, *J. Urol.* **119:**49.

Beyer, T., Rearick, J., Paulson, J., Prieels, J., Sadler, J., and Hill, R., 1979, Biosynthesis of mammalian glycoproteins. Glycosylation pathways in the synthesis of the nonreducing terminal sequences, *J. Biol. Chem.* **254:**12531.

Bird, G., 1977, Blood banking, *in* "CRC Handbook Series in Clinical Laboratory Science" (T. Greenwalt, ed.), pp. 459–470, CRC Press, Boca Raton, Florida.

Bonfiglio, T. A., and Feinberg, M. R., 1976, Isoantigen loss in cervical neoplasia. Demonstration by immunofluorescence and immunoperoxidase techniques, *Arch. Pathol. Lab. Med.* **100:**307.

Childs, R., Kapadia, A., and Feizi, T., 1980, Expression of blood group I and i active carbohydrate sequences on cultured human and animal cell lines assessed by radioimmunoassays with monoclonal cold agglutinins, *Eur. J. Immunol.* **10:**379.

Coon, J., and Weinstein, R. S., 1981a, Variability in the expression of the O(H) antigen in human transitional epithelium, *J. Urol.* **125:**301.

Coon, J., and Weinstein, R. S., 1981b, Detection of ABH tissue isoantigens by immunoperoxidase methods in normal and neoplastic urothelium. Comparison with the red cell adherence method, *Am. J. Clin. Pathol.* **76:**163.

Coon, J., Weinstein, R. S., and Summers, J., 1982, Blood group precursor T antigen expression in human urinary bladder carcinoma, *Am. J. Clin. Pathol.* **77:**692.

Cooper, A., Morgello, S., Miller, D., and Brown, M., 1977, Role of surface glycoproteins in tumor growth, *in* "Cancer Invasion and Metastasis, Biologic Mechanisms and Therapy" (S. B. Day, ed.), pp. 49–64, Raven Press, New York.

Cooper, H., and Haesler, W., 1978, Blood group substances as tumor antigens in the distal colon, *Am. J. Clin. Pathol.* **69:**594.

Cooper, H., Cox, J., and Pathefsky, A. S., 1980, Immunohistologic study of blood group substances in polyps of the distal colon. Expression of a fetal antigen, *Am. J. Clin. Pathol.* **73**:345.

Cowan, W. K., 1962, Blood group antigens on human gastrointestinal carcinoma cells, *Br. J. Cancer* **16**:535.

Critchley, D., 1979, Glycolipids as membrane receptors important in growth regulation, *in* "Surface of Normal and Malignant Cells" (R. Hynes, ed.), pp. 64–101, Wiley, New York.

Dabelsteen, E., 1972, Quantitative determination of blood group substance A of oral epithelial cells by immunofluroescence and immunoperoxidase methods, *Acta Pàthol. Microbiol. Scand. Sect. A* **80**:847.

Dabelsteen, E., and Pindborg, J. J., 1973, Loss of epithelial blood group substance A in oral carcinomas, *Acta Pathol. Microbiol. Scand. Sect. A* **81**:435.

Dabelsteen, E., Mygind, N., and Henriksen, B., 1974, Blood group substance A in carcinoma of the larynx, *Acta Otolaryngol.* **77**:360.

Dabelsteen, E., Roed-Petersen, B., and Pindborg, J. J., 1975, Loss of blood group epithelial antigens A and B in oral premalignant lesions, *Acta Pathol. Microbiol. Scand. Sect. A* **83**:292.

Davidsohn, I., 1979, The loss of blood group antigens A, B and H from cancer cells, *in* "Immunodiagnosis of Cancer" (R. B. Heberman and K. R. McIntire, eds.), pp. 644–669, Dekker, New York.

Davidsohn, I., and Ni, L. Y., 1969, Loss of isoantigens A, B and H in carcinoma of the lung, *Am. J. Pathol.* **57**:307.

Davidsohn, I., Kovarik, S., and Lee, C. L., 1966, A, B and O substances in gastrointestinal carcinoma, *Arch. Pathol.* **81**:381.

Davidsohn, I., Kovarik, S., and Stejskal, R., 1968, Ovarian cancer: Immunological aspects. Influence on prognosis and treatment, *in* "U.I.C.C. Monograph Series," Vol. II. "Ovarian Cancer," (A. C. Jungueira and F. Gentil, eds.) pp. 105–121, Springer-Verlag, Berlin.

Davidsohn, I., Kovarik, S., and Ni, L. Y., 1969, Isoantigens A, B and H in benign and malignant lesions of the cervix, *Arch. Pathol.* **87**:306.

Davidsohn, I., Ni, L. Y., and Stejskal, R., 1971, Tissue isoantigens A, B and H in carcinoma of the pancreas, *Cancer Res.* **31**:1244.

Davisohn, I., Stejskal, R., and Lill, P., 1972, Immunopathologic diagnosis and prognosis in carcinoma of the pancreas, *in* "Proceedings, VII Congress of Anatomic and Clinical Pathology," pp. 18–23, Excerpta Medica, Amsterdam.

Davidsohn, I., Norris, H. J., Stejskal, R., and Lill, P., 1973, Metastatic squamous cell carcinoma of the cervix. The role of immunology in its pathogenesis, *Arch. Pathol.* **95**:132.

Decenzo, J. M., Howard, P., and Irish, C. E., 1975, Antigenic deletion and prognosis of patients with stage A transitional cell bladder carcinoma, *J. Urol.* **114**:874.

Dejter-Juszynski, M., Harpaz, N., Flowers, A., and Sharon, N., 1978, Blood-group ABH-specific macroglycolipids of human erythrocytes: Isolation in high yield from a crude membrane glycoprotein fraction, *Eur. J. Biochem.* **83**:363.

Denk, H., Tappeiner, G., Davidovits, A., Eckerstorfer, R., and Holzner, J., 1974, Carcinoembryonic antigen and blood group substances in carcinomas of the stomach and colon, *J. Natl. Cancer Inst.* **53**:933.

Dobrossy, L., Pavelic, Z., and Bernock, R., 1981, A correlation between cell surface sialyl transferase, sialic acid, and glycosidase activities, and the implantability of B16 murine melanoma, *Cancer Res.* **41**:2262.

Eklund, A. E., Guilbring, B., and Lagerlof, B., 1963, Blood group specific substances in human gastric carcinoma: A study using fluorescent antibody technique, *Arch. Pathol. Microbiol. Scand.* **59:**447.

Emmott, R. C., Javadpour, N., Bergman, S., and Soares, T., 1979, Correlation of the cell surface antigens with stage and grade in cancer of the bladder, *J. Urol.* **121:**37.

England, D., and Davidsohn, I., 1973, Isoantigens A, B and H in carcinoma of the fallopian tube, *Arch. Pathol.* **96:**350.

England, D., Solie, B., and Winkelmann, R., 1979, Isoantigens A, B, H in normal skin and tumors of the epidermal appendages, *Arch. Pathol. Lab. Med.* **103:**586.

Falor, W., and Ward, R., 1976, Fifty-three month persistance of ring chromosome in non-invasive bladder carcinoma, *Acta. Cytol.* **20:**272.

Farrow, G. M., 1979, Pathologist's role in bladder cancer, *Semin. Oncol.* **6:**198.

Feigl, W., Denk, H., Davidouits, A., and Holzner, J. H., 1976, Blood group isoantigens in human benign and malignant vascular tumors, *Virchows Arch. A* **370:**323.

Feizi, T., Ruberville, C., and Westwood, J., 1975, Blood-group precursors and cancer-related antigens, *Lancet* **2:**391.

Fialkow, P., 1972, Use of genetic markers to study cellular origin and development of tumors in human females, *Adv. Cancer Res.* **15:**191.

Fialkow, P., 1976, Clonal origin of human tumors, *Biochim. Biophys. Acta* **458:**283.

Freidell, G. H., Bell, J. R., Burney, S. W., Soto, E. A., and Tiltman, A. J., 1976, Histopathology and classification of urinary bladder carcinoma, *Urol. Clin. N. Am.* **3:**53.

Friedenreich, V., 1930, "The Thomsen Hemagglutination Phenomenon," Levin & Munksgaard, Copenhagen.

Fukuda, M., Fukuda, M., and Hakomori, S., 1979a, Developmental change and genetic defect in the carbohydrate structure of band 3 glycoprotein of human erythrocyte membrane, *J. Biol. Chem.* **254:**3700.

Fukuda, M., Fukuda, M., and Hakomori, S., 1979b, Cell surface modification by endo-β-galactosidase. Changes of blood group activities and release of oligosaccharides from glycoproteins and glycosphingolipids of human erythrocytes, *J. Biol. Chem.* **254:**5458.

Gupta, R. K., and Schuster, R., 1973, Isoantigens A, B and H in benign and malignant lesion of breast, *Am. J. Pathol.* **72:**253.

Gupta, R. K., Schuster, R., and Christian, W., 1973, Loss of isoantigens A, B and H in prostate, *Am. J. Pathol.* **70:**439.

Hakomori, S. I., 1975a, Fucolipids and blood group glycolipids in normal and tumor tissue, *Prog. Biochem. Pharmacol.* **10:**167.

Hakomori, S., 1975b, Structures and organization of cell surface glycolipids. Dependency on cell growth and malignant transformation, *Biochim. Biophys. Acta* **417:**55.

Hakomori, S., 1981, Blood group ABH and Ii antigens of human erythrocytes: Chemistry, polymorphism, and their developmental change, *Semin. Hematol.* **18:**39.

Hakomori, S., Wang, M., and Young, W., 1977, Isoantigenic expression of Forssman glycolipid in human gastric and colon mucosa: Its possible identity with "A-like Antigen" in human cancer, *Proc. Natl. Acad. Sci. USA* **74:**3023.

Hart, I., and Fidler, I., 1981, The implications of tumor heterogeneity for studies on the biology and therapy of cancer metastasis, *Biochim. Biophys. Acta* **651:**37.

Holborow, E. J., Brown, P. C., Glynn, L. E., Hawes, M. D., Gresham, G. A., O'Brien, T. F., and Coombs, R. R., 1960, The distribution of the blood group A antigen in human tissue, *Br. J. Exp. Pathol.* **41:**430.

Howard, D., and Batsakis, J., 1980, Cytostructural localization of a tumor-associated antigen, *Science* **210:**201.

Howard, D., and Taylor, C., 1979, A method for distinguishing benign from malignant breast lesions utilizing antibody present in normal human sera, *Cancer* **43:**2279.

Howard, D., Ferguson, P., and Batsakis, J., 1981, Carcinoma-associated cytostructural antigenic alterations: Detection by lectin binding, *Cancer* **47:**2872.

Johnson, J. D., and Lamm, D. L., 1980, Prediction of bladder tumor invasion with the mixed cell agglutination test, *J. Urol.* **123:**25.

Karhi, K., and Gahmberg, C., 1980, Identification of blood group A-active glycoproteins in the human erythrocyte membrane, *Biochim. Biophys. Acta* **622:**344.

Kato, T., 1977, Detection of A, B and H (O) antigens in normal and neoplastic epithelium of the urinary bladder by the specific red cell adherence test (SRCA), *Tohoku J. Exp. Med.* **121:**239.

Kim, Y., Isaacs, R., and Perdomo, J., 1974, Alterations of membrane glycoproteins in human colonic adenocarcinoma, *Proc. Natl. Acad. Sci. USA* **71:**4869.

Kim, Y. S., and Isaacs, R., 1975, Glycoprotein metabolism in inflammatory and neoplastic disease of human colon, *Cancer Res.* **35:**2092.

Koscielak, J., 1977, Chemistry and biosynthesis of erythrocyte membrane glycolipids with A, B, H and I blood group activities, *in* "Human Blood Groups, 5th International Convocation on Immunology, Buffalo, N. Y., 1976," pp. 143–149, Karger, Basel.

Kovarik, S., Davidsohn, I., and Stejskal, R., 1968, ABO antigen in cancer. Detection with the mixed cell agglutination reaction, *Arch. Pathol.* **86:**12.

Krag, S., 1979, A concanavallin A-resistant Chinese hamster ovary cell line is deficient in the synthesis of [^{3}H] glucosyl oligosaccharidelipid, *J. Biol. Chem.* **254:**9167.

Lange, P. H., Limas, C., and Fraley, E. E., 1978, Tissue blood group antigens and prognosis in low stage transitional cell carcinoma of the bladder, *J. Urol.* **119:**52.

Lill, P. H., Norris, H. J., Rubenstone, A. I., Chango-Lo, M., and Davidsohn, I., 1976, Isoantigens ABH in cervical intraepithelial neoplasia, *Am. J. Clin. Pathol.* **66:**767.

Lill, P. H., Stejskal, R., and Milsna, J., 1979, Distribution of H antigen in persons of blood groups A, B and AB, *Vox Sang.* **36:**159.

Limas, C., and Lange, P., 1980, Altered reactivity for A, B, H antigen in transitional cell carcinomas of the urinary bladder. A study of the mechanisms involved, *Cancer* **46:**1366.

Limas, C., Lange, P., Fraley, E., and Vessella, R. L., 1979, A, B, H antigens in transitional cell tumors of the urinary bladder. Correlation with the clinical course, *Cancer* **44:**2099.

Limas, C., Coon, J., Lange, P., and Weinstein, R. S., 1982, ABH antigens in urinary bladder carcinomas: Detection and clinical applications, *in* "Bladder Cancer" (W. Bonney and O. Prout, eds.), AUA Monograph, Vol. I, pp. 69–79, Williams & Wilkins, New York.

Lin, F., Liu, P. I., and McGregor, D. H., 1977, Isoantigens A, B and H in morphologically normal mucosa and in carcinoma of the larynx, *Am. J. Clin. Pathol.* **68:**372.

Liotta, L., Tryggvason, K., Garbisa, S., Hart, I., Foltz, C., and Shafie, S., 1980, Metastatic potential correlates with enzymatic degradation of basement membrane collagen, *Nature* (*London*) **284:**67.

Lis, H., and Sharon, N., 1977, Lectins: Their chemistry and application to immunology, *in* "The Antigens" (M. Sela, ed.), Chap. 7, Academic Press, New York.

Liu, P. I., McGregor, D. H., Liu, J. G., Dunlap, D. L., Jinks, W. I., Lin, F., Przybylski, C., and Miller, L. A., 1974, Carcinoma of the oral cavity evaluated by specific red cell adherence test, *Oral Surg.* **38:**56.

Loke, Y. W., and Ballard, A. C., 1973, Blood group A antigens on human trophoblast cells, *Nature* (*London*) **245:**329.

Morre, D., Kloppel, T., Merritt, W., and Keenan, T., 1978, Glycolipids as indicators of tumorigenesis, *J. Supramol. Struct.* **9:**157.

Narasimhan, R., and Murray, R., 1979, Neutral glycosphingolipids and gangliosides of human lung and lung tumors, *Biochem. J.* **179:**199.

Newman, R., Klein, P., and Rudland, P., 1979, Binding of peanut lectin to breast epithelium, human carcinomas, and a cultured rat mammary stem cell: Use of the lectin as a marker of mammary differentiation, *J. Natl. Cancer Inst.* **63:**1339.

Newman, A., Carlton, E., and Johnson, S., 1980, Cell surface A, B or O(H) blood group antigens as an indicator of malignant potential in Stage A bladder carcinoma, *J. Urol.* **124:**27.

Oriol, R., LePendu, J., Sparkes, R., Sparkes, M., Crist, M., Galo, R., Terasaki, P., and Bernoro, M., 1981, Insights into the expression of ABH and Lewis antigens through human bone marrow transplantation, *Am. J. Hum. Genet.* **33:**551.

Pauli, B., Anderson, S., Memoli, V., and Kuettner, K., 1980, Development of an *in vitro* and *in vivo* epithelial tumor model for study of invasion, *Cancer Res.* **40:**4571.

Picard, J., Edward, D., and Feizi, T., 1978, Changes in the expression of the blood group A, B. H, Le^a and Le^b antigens and the blood group precursor associated I (MA) antigen in glycoprotein-rich extracts of gastric carcinomas, *J. Clin. Lab. Immunol.* **1:**119.

Poste, G., and Fidler, I., 1980, The pathogenesis of cancer metastasis, *Nature* (*London*) **283:**139.

Prendergast, R. C., Tot, P. D., and Gargiulo, A. W., 1968, Reactivity of blood group substances of neoplastic oral epithelium, *J. Dent. Res.* **47:**306.

Race, R., and Sanger, R., 1975, "Blood Groups in Man," Chap. 2, Blackwell, Oxford.

Rauvala, H., and Finne, J., 1979, Structural similarity of the terminal carbohydrate sequences of glycoproteins and glycolipids, *FEBS Lett.* **97:**1.

Reddy, A., and Fialkow, P., 1979, Multicellular origin of fibrosarcomas in mice induced by the chemical carcinogen 3-methylcholanthrene, *J. Exp. Med.* **150:**878.

Reitman, M., Trowbridge, I., and Kornfeld, S., 1980, Mouse lymphoma cell lines resistant to pea lectin are defective in fucose metabolism, *J. Biol. Chem.* **255:**9900.

Richie, J. P., Blute, R. B., Jr., and Waisman, J., 1980, Immunologic indicators of prognosis in bladder cancer: The importance of cell surface antigens, *J. Urol.* **123:**22.

Rolih, S., 1980, Erythrocyte antigens of the MN system and related structures, *in* "A Seminar on Antigens on Blood Cells and Body Fluids" (C. Bell, ed.), Chap. 8, American Association of Blood Banks, Washington, D. C.

Schenkel-Brunner, H., 1980, Blood group-ABH antigens of human erythrocytes. Quantitative studies on the distribution of H antigenic sites among different classes of membrane components, *Eur. J. Biochem.* **104:**529.

Schwarz, R., and Datema, R., 1980, Inhibitors of protein glycosylation, *Trends Biochem. Sci.* **5:**65.

Sheahan, D. G., Horowitz, S. A., and Zamcheck, N., 1971, Deletion of epithelial ABH isoantigens in primary gastric neoplasms and in metastatic cancer, *Dig. Dis.* **16:**961.

Siddiqui, B., Whitehead, J., and Kim, Y., 1978, Glycosphingolipids in human colonic adenocarcinoma, *J. Biol. Chem.* **253:**2168.

Skipski, V., Gitterman, C., Prendergast, J., Betit-Yen, K., Lee, G., Luell, S., and Stock, C., 1980, Possible relationship between glycosphingolipids and the formation of metastasis in certain human experimental tumors, *J. Natl. Cancer Inst.* **65:**249.

Springer, G., and Desai, P., 1977, Cross reacting carcinoma-associated antigens with blood group and precursor specificities, *Transplant. Proc.* **9:**1105.

Springer, G., Desai, P., Yang, H., and Murthy, M., 1977, Carcinoma-associated blood group MN precursor antigens against which all humans possess antibodies, *Clin. Immunol. Immunopathol* **7:**426.

Springer, C., Desai, P., Murthy, M., Yang, H., and Scanlon, E., 1979, Precursors of the blood group NM antigens as human carcinoma associated antigens, *Transfusion* **19:**233.

Stanley, P., and Sudo, T., 1981, Microheterogeneity among carbohydrate structure at the cell surface may be important in recognition phenomena, *Cell* **23:**763.

Stein, B., Reyes, T., Peterson, R., McNellis, D., and Kendall, A., 1981, Specific red cell adherence: Immunologic evaluation of random mucosal biopsies in carcinoma of the bladder, *J. Urol.* **126:**37.

Stejskal, R., Lill, P. H., and Davidsohn, I., 1973, A, B, and H isoantigens in the human fetus, *Dev. Biol.* **34:**274.

Stellner, K., Hakomori, S., and Warner, G. A., 1973, Enzymic conversion of "H_1-glycolipid" to A- or B-glycolipid and deficiency of these enzyme activities in adenocarcinoma, *Biochem. Biophys. Res. Commun.* **55:**439.

Summers, J., Falor, W., and Ward, R., 1981, A 10-year analysis of chromosomes in noninvasive papillary carcinoma of the bladder, *J. Urol.* **125:**177.

Szulman, A. E., 1960, The histological distribution of the blood group substances A and B in man, *J. Exp. Med.* **111:**785.

Szulman, A. E., 1962, The histological distribution of blood group substances in man as disclosed by immunofluorescence. II. The H antigen and its relation to A and B antigens, *J. Exp. Med.* **115:**977.

Szulman, A. E., 1964, The hisotlogical distribution of the blood group substances in man as disclosed by immunofluorescence. III. The A, B and H antigens in embryos and fetuses from 18 nm in length, *J. Exp. Med.* **119:**503.

Szulman, A. E., 1965, The ABH antigens in human tissues and secretions during embryonal development, *J. Histochem. Cytochem.* **13:**752.

Thomsen, O., 1927, Ein vermehrungs Fahiges Agens als Veranderer des isoagglutinatorischen Verhalten der roton Blufkorpershen, eine bisher unbekannte Quelle der Fehlbestimmung, *Z. Immunitaetsforsch.* **52:**85.

Vedtofte, P., Hansen, H., and Dabelsteen, E., 1981, Distribution of blood group antigen H in human buccal epithelium of secretors and non-secretors, *Scand. J. Dent. Res.* **89:**188.

Viitala, J., Karhi, K., Gahmberg, C., Finne, J., Jarnefelt, J., Myllyla, J., and Kraus, T., 1981, Blood group A and B determinants are located in different polyglycosyl peptides isolated from human erythrocytes of blood-group AB, *Eur. J. Biochem.* **113:**259.

Watanabe, K., and Hakomori, S. I., 1976, Status of blood group carbohydrate chains in ontogenesis and in oncogenesis, *J. Exp. Med.* **144:**644.

Watanabe, K., Hakomori, S., Childs, R., and Feizi, T., 1979, Characterization of a blood group I-active ganglioside. Structural requirements for I and i specificities, *J. Biol. Chem.* **254:**3221.

Watkins, W., 1978, Genetics and biochemistry of some human blood groups, *Proc. R. Soc. London.* **202:**31.

Weinstein, R. S., 1979, Origin and dissemination of human urinary bladder cancer, *Semin. Oncol.* **6:**149.

Weinstein, R. S., Khodadad, J., and Steck, T., 1978, Fine structure of the band 3 protein in human red cell membranes: Freeze-fracture studies, *J. Supramol. Struct.* **8:**325.

Weinstein, R. S., Alroy, J., Farrow, G. M., Miller, A. W., III, and Davidsohn, I., 1979, Blood group isoantigen deletion in carcinoma *in situ* of the urinary bladder, *Cancer* **43:**661.

Weinstein, R. S., Coon, J., Summers, J., Ward, R., and Falor, W. H., 1981a Prognostic value of ABH (O) antigen deletion and marker chromosomes in human urinary bladder carcinoma (Abstract), *Lab Invest.* **44:**74A.

Weinstein, R. S., Coon, J., Alroy, J., and Davidsohn, I., 1981b, Tissue-associated blood group antigens in human tumors, *in* "Diagnostic Immunohistochemistry" (R. D. DeLellis, ed.), pp. 239–261, Masson, New York.

Weir, D., 1980, Surface carbohydrates and lectins in cellular recognition, *Immunol. Today* **1**:45.

Whitmore, W., 1977, Special article: Summary of all phases of bladder carcinoma, *J. Urol.* **119**:77.

Wilczynska, Z., Miller-Podraza, H., and Koscielak, J., 1980, The contribution of different glycoconjugates to the total ABH blood group activity of human erythrocytes, *FEBS Lett.* **112**:277.

Yogeeswaran, G., and Salk, P., 1981, Metastatic potential is positively correlated with cell surface sialylation of cultured murine tumor cell lines, *Science* **212**:1514.

Chapter 8

Cell-Surface Macromolecular and Morphological Changes Related to Allotransplantability in the TA3 Tumor

John F. Codington and David M. Frim

Laboratory for Carbohydrate Research
Massachusetts General Hospital
and Departments of Biological Chemistry and Medicine
Harvard Medical School
Boston, Massachusetts

I. INTRODUCTION

The phenomena of allo- and xenotransplantability in experimental tumors may be related to properties of metastatic cancers in humans, particularly the capacity of individual metastatic cells to survive in a potentially hostile immunological environment. It was not surprising, therefore, that a number of laboratories began investigations of allotransplantability in the TA3 tumor, as soon as such investigations became feasible (Hauschka *et al.*, 1971; Friberg, 1972a,b; Grohsman and Nowotny, 1972; Lippman *et al.*, 1973; Sanford *et al.*, 1973; Nowotny *et al.*, 1974; Codington *et al.*, 1973).

It was fortuitous that studies in the laboratory of Dr. T. S. Hauschka left a legacy of TA3 ascites cells of both strain-specific and strain-non-specific characteristics.

The TA3 mammary carcinoma arose spontaneously in a female A/HeHa mouse (Hauschka *et al.*, 1971). The strain-specific subline, TA3-St, which was developed in Dr. Hauschka's laboratory (Klein, 1951), was preserved over many years at the Karolinska Institutet in Stockholm. A second strain-specific subline, TA3-Ha, which was converted to the ascites form during this same period, remained at Roswell Park Memorial Institute, either under active investigation or in the frozen tumor cell bank. As will be described later (Section II.A), this subline, TA3-Ha, later lost its strain specificity and became capable of growth, not only in foreign mouse strains but also in some foreign species. Accounts of the origin and early development of the TA3 ascites sublines have been published (Hauschka *et al.*, 1971; Friberg, 1972a). Studies of the characteristics of these two sublines have ranged over many scientific disciplines. During recent years, however, studies of additional allotransplantable TA3 sublines have added greater understanding, as well as greater complexity, to the problem.

The TA3 story may be important, not only because these studies have presented a physicochemical basis for the biological differences exhibited by strain-specific and strain-nonspecific sublines but also because of the extensive investigations, physical, chemical, and biological, which have been performed on this one particular tumor system. Although the complexities of malignancy in TA3 tumor cells, like the complexities of other biological processes in living cells, are only partly understood, what is already known suggests a deep involvement of macromolecular structures at the cell surface in the response of the cell to immune factors of the host. Studies of the TA3 tumor subline have shown that changes in allotransplantability are accompanied by demonstrable changes in glycoprotein structures at the cell surface (Codington *et al.*, 1973, 1975a, 1979a; Codington, 1978, 1980; Cooper *et al.*, 1979; Van den Eijnden *et al.*, 1979). These changes appear to be directly related to the control of the immune response. In this article, we have attempted to review the immunochemical and physicochemical data, as well as morphological studies performed with the electron microscope, which appear to be related to the differences in the biological characteristics of strain-specific and strain-nonspecific TA3 sublines. Particular attention has been given to the properties of the large endogenous membranebound glycoprotein, epiglycanin, which has been found in high concentration at the surfaces of these allotransplantable TA3 sublines. Earlier reviews of some of this work have been published (Jeanloz and Codington, 1973, 1974, 1976; Codington and Jeanloz, 1974; Codington *et al.*, 1974; Codington, 1975, 1978, 1980).

II. ORIGINS OF THE SUBLINES OF THE TA3 TUMOR

A. The TA3-St and TA3-Ha Sublines

The occurrence of the TA3 tumor and the origins of the TA3 sublines are illustrated in Fig. 1. The historical background of this tumor has been described previously (Hauschka *et al.*, 1971; Friberg, 1972a). It seems noteworthy that the tumor arose spontaneously in a female A/HeHa

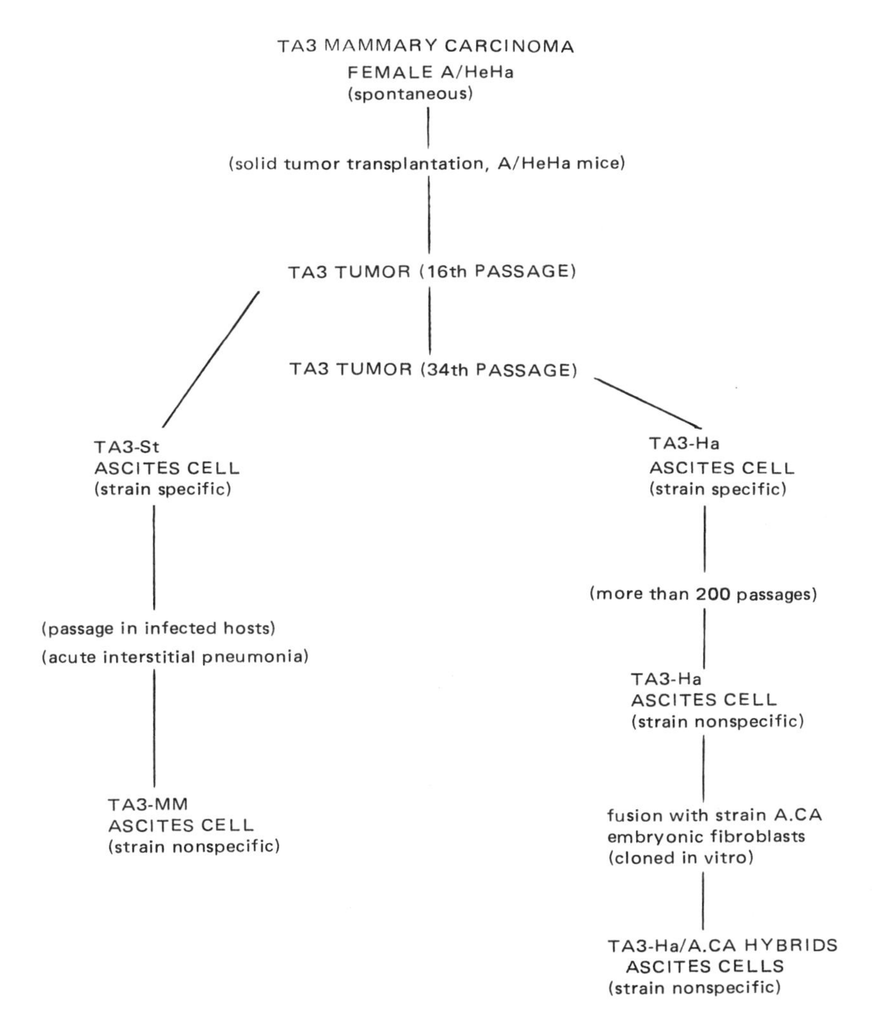

Fig. 1. Outline of the relationship between strain-specific and strain-nonspecific sublines of the TA3 tumor. Connecting lines illustrate direct descent of each cell.

mouse. Since the mouse mammary tumor virus is indigenous to this mouse (Ringold *et al.*, 1975), one may speculate that this virus may have been involved in the etiology of the tumor. The loss of strain specificity in the TA3-Ha ascites tumor occurred spontaneously, i.e., in the absence of any known selection pressure. This event, which occurred *in vivo* after more than 200 passages in ascites form, was apparently unnoticed at the time, and no account of the circumstances surrounding this change to allotransplantability is available.

B. The TA3-MM Sublines

The development of the TA3-MM ascites tumor was fortuitous. Routine passage of the TA3-St ascites cell was made in A/HeHa mice affected with pneumonia-associated microorganisms (Cooper *et al.*, 1979; Codington, 1980). The presence of the infection was not recognized at the time, as the mice had only recently arrived from the supplier. As the disease progressed to epidemic proportions it was characterized as acute interstitial pneumonia (Cooper *et al.*, 1979). The majority of stock mice infected with the disease later died. Two microorganisms were identified in the sick mice, *Pasteurella pneumotropica* and *P. multocida*. Although no evidence was available to implicate the viruses in the disease, their possible involvement in both the disease and the later tumor cell metamorphosis cannot be excluded. It was concluded that a qualitative change in the TA3-St cells occurred after only one passage in ascites form in the infected mice (Cooper *et al.*, 1979).

Receptors for the lectin from *Vicia graminea* seeds (Uhlenbruck and Dahr, 1971), a lectin with specificity for a glycopeptide structure present in the glycoprotein epiglycanin of the TA3-Ha ascites cell (Springer *et al.*, 1972; Cooper *et al.*, 1974; Codington *et al.*, 1975a), were observed in high concentration at the cell surface (Codington *et al.*, 1979a; Cooper *et al.*, 1979). Such receptors are normally not present at the TA3-St cell surface or present in very low concentrations. TA3-MM cells were placed in spinner culture, where a high concentration of *V. graninea* receptors persisted for more than 2 months (Cooper *et al.*, 1979). Two *in vivo* variant lines, TA3-MM/1 and TA3-MM/2, were established from the cells grown in spinner culture. A detailed account of the origin of the two TA3-MM sublines has been published (Codington *et al.*, 1979a; Cooper *et al.*, 1979). Both TA3-MM/1 and TA3-MM/2 cells were found to be allotransplantable.

C. The TA3-Ha/A.CA Hybrids

Allotransplantable TA3-Ha ascites cells were fused *in vitro* with normal embryonic fibroblasts of the A.CA mouse in the presence of inactivated Sendai virus (Wiener *et al.*, 1971). Cells were cloned *in vitro*, and nine lines were selected on the basis of colony morphology (Klein *et al.*, 1972). Although these lines were originally incapable of growth in either parental line, strain A or strain A.CA (Klein *et al.*, 1973), after a process of selection, growth in strain A mice in ascites form was possible. During this process, these lines lost H-2^f histocompatibility antigens, characteristic of the A.CA strain but retained the H-2^a antigens of the strain A mouse (Friberg *et al.*, 1973). During early passages in the strain A mouse, the hybrid cells were capable of consistent growth only in this strain, but during later passages they were found to be allotransplantable (Codington *et al.*, 1978a, 1979c). Six hybrid lines were chosen for investigation in our laboratory, TA3-Ha/A.CA/3B, /4, /6, /7, /10, and /11.

III. BIOLOGICAL CHARACTERISTICS

A. Transplantability of TA3 Ascites Cells

It has been shown that tumor cells, like normal cells, contain genetically determined histocompatibility antigens capable of initiating immune rejection reactions by the host (Gorer, 1937; Friberg, 1972a). Tumor cells injected into a foreign host are thus susceptible to the immune response of the host. Unlike most tumor cells, the TA3-Ha ascites cell has the capacity to grow in ascites form in allogeneic mice. Transplantability data obtained from experiments performed with ascites cells in our laboratory and results obtained in collaboration with others (Sanford *et al.*, 1973; Codington *et al.*, 1978a; Cooper *et al.*, 1979) are presented in Table I. Similar data have been obtained by others (Friberg, 1972a; Hauschka *et al.*, 1971).

It has been reported also that the TA3-Ha ascites cell will grow progressively in foreign species, the rat (Hauschka *et al.*, 1971; Friberg 1972a) and the hamster (Friberg, 1972a). By contrast, the TA3-St ascites cell, which was derived from the same original carcinoma, is unable to grow in either foreign strains or species. Table I also includes data on the transplantability of the TA3-MM ascites cell, a cell which resulted from an *in vivo* change in the TA3-St ascites cell, as described in Section II.B. The TA3-MM cell, which possesses characteristics distinct from those of the TA3-Ha cell, nevertheless, shares several important properties with this cell, among them the capacity to grow allogeneically. A further comparison of the characteristics of the TA3-MM and TA3-Ha cells will be described later. Also included in Table I are the transplantation data for six hybrid lines resulting from the fusion of TA3-Ha ascites cells and normal embryonic fibroblasts of the A.CA mouse, as described in Section II.C. These lines share the property of allotransplantability with the parental TA3-Ha cell. Further investigations of the characteristics of the hybrid lines, which have enriched our understanding of strain nonspecificity in the TA3 tumor system, will also be described later.

Table I
Progressive Growth (Mice Dead/Mice Injected) of TA3 Sublines in Syngeneic and Allogeneic Mouse Strains

Cell line	No. cells injected	Strain A	C57/B1	C3H	CBA	DBA/2	A.SW	A.CA	A.BY	BALB
TA3-St	10^4	41/42	0/8	0/22	0/12					0/6
	10^5	8/8	0/14	0/10	0/11	1/9	0/4	0/4		
TA3-Ha	10^4	30/30		23/32						6/6
	10^5	11/11	5/15	13/20	4/4	12/12	4/4	4/4		
TA3-MM/1	10^4	4/4		4/6	2/4					6/6
	10^5									1/6
TA3-MM/2	10^4	6/6	1/8		7/8					
	10^5	7/7	2/5		9/10	7/7				
TA3-Ha/A.CA/3B	10^6	1/1	2/2	2/2	1/2	2/2	2/2	2/2	2/2	
4	10^6	1/1	2/2	0/1	1/2	2/2	2/2	2/2	2/2	
6	10^6	1/1	2/2	2/2	1/2	1/2	1/2	2/2	1/2	
7	10^6	4/4	2/2	1/2	1/2	1/1	2/2	1/2	2/2	
10	10^6		0/2	2/2	0/2	1/2	2/2	1/2	2/2	
11	10^6	1/1	0/2	0/2	1/2	0/2	0/1	1/2	0/2	

B. Protection against TA3 Cell Growth in Allogeneic Mice

Despite reduced immunogenicity and increased immunoresistance, as compared to the TA3-St cell, allogeneic mice could be protected against growth of the TA3-Ha cell. Presensitization to cell growth was accomplished by grafting skin of strain A mice onto C3H mice (Sanford *et al.*, 1973) and ICR mice (Grohsman and Nowotny, 1972). Successful allografts of strain A liver were also performed. Protection against TA3-Ha ascites cell growth was also accomplished by injection of either lyophilized or formaldehyde-treated TA3-Ha ascites cells (Nowotny *et al.*, 1974). It was similarly found that simultaneous injection of viable TA3-St and TA3-Ha ascites cells into allogeneic mice protected the mice against the growth of TA3-Ha ascites cells (Sanford *et al.*, 1973). The same degree of protection could be accomplished by preinjection of TA3-St cells, alone (Lippman *et al.*, 1973).

C. Chromosome Analysis

Allotransplantability in TA3 ascites tumors does not appear to be related to chromosome number. Karyotypes of the TA3-Ha and TA3-St cells have been compared by several laboratories (Friberg, 1972a; Hauschka *et al.*, 1971; Cooper *et al.*, 1979). The TA3-Ha ascites cell is near diploid with a mode number of chromosomes of 41 or 42. The TA3-MM, which is similar to the TA3-Ha cell in transplantation characteristics, is tetraploid, with a mode chromosome number of 82, twice that of the TA3-Ha cell. The TA3-St ascites cell, from which the TA3-MM cell line was derived, was found to possess an intermediate number of chromosomes, 69, three more than previously reported (Friberg, 1972a).

D. Growth Rates and Mortality in Syngeneic Mice

Although the rate of growth of TA3-Ha ascites cells in syngeneic strain A mice was found to be greater than that of TA3-St cells during the first 2 days of growth (Friberg, 1972a), there was little difference in growth rate thereafter between the two sublines. Consistent with previous results (Friberg, 1972a), we have found that in routine passages, the number of TA3-St cells at a ten-fold greater inoculum (1×10^6 cells) after 7 days (the usual day of harvest) was approximately equal to the number of TA3-Ha cells that resulted from an inoculum of 1×10^5 cells. The number of TA3-MM cells harvested 7 days following an injection of 1×10^5 cells, however, was similar to that obtained from TA3-Ha cells.

More interesting perhaps is a comparison of the mortality of A/HeJ mice with time after introperitaneal injections of TA3-Ha, TA3-MM, or TA3-St ascites cells. With an inoculum of 10^4 cells, half of the TA3-Ha- or TA3-MM-injected mice were dead by Day 10, but it was not until about Day 20 that an equal proportion of the TA3-St-injected mice had died of the tumor (Cooper *et al.*, 1979). It is thus clear that the capacity of the allotransplantable lines to kill the syngeneic host was not strictly proportional to the growth rate. These results in syngeneic mice may be related to those described earlier (Purdom *et al.*, 1958). It was reported that a subline of a tumor with higher net negative charge caused a more rapid death of the hosts—and also gave rise to more metastases per animal. The TA3-Ha ascites cell possesses about 2.5-fold and the TA3-mm/2 cell about 4.5-fold more sialic acid than the TA3-St cell (See Section VI.B) (Codington *et al.*, 1979b), and it has been shown that the negative charge on most tumor cells is due mainly to sialic acid residues (Cook and Stoddart, 1973). However interesting this correlation may appear, it makes no distinction between sialic acid per se and the macromolecules containing sialic acid residues. As described later (Section VI.B), the absorption of anti-H-2^a antibody by TA3 ascites sublines was not affected by neuraminidase treatment (Sanford and Codington, 1971), which removed the sialic acid but not epiglycanin.

IV. ESCAPE MECHANISMS IN THE TA3 ASCITES TUMORS

A. Measurement of Escape Mechanisms

In studies of the capacity of tumor cells to escape the hosts' immune defenses several parameters have been considered. The most significant of these, the capacity to absorb antibody to the major histocompatibility antigen (H-2^a), the capacity to grow progressively in foreign mouse strains until the death of the host (allotransplantability), and the concentration of cell-surface epiglycanin, are plotted as bar graphs in Fig. 2 for most of the TA3 sublines. The absorptive capacities, plotted in the reverse order of magnitude, are compared in Fig. 2A with the allotransplantabilities and in Fig. 2B with the concentrations of epiglycanin. Other cell characteristics, such as karyotypes, or the concentrations or compositions of cell surface sialic acid could not usually be related directly to the escape capacity. As will be described in Section V, however, cell morphology, as related particularly to the topographical features of the cells, appeared to be directly related to allotransplantability.

B. Effect of Cell Disruption

In early studies (Sanford *et al.*, 1973), we confirmed the previous results (Friberg, 1972b) that the TA3-St cell consistently absorbed manyfold more H-2^a antibody than did the TA3-Ha cell. Later results with A.CA anti-H-2^a antiserum (Codington *et al.*, 1978a) are shown in Fig. 2B as described (Section IV.D). Absorptions of antisera from two ends of the H-2 spectrum, anti-d and anti-k, were performed. These results were consistent with the observation (Gorer, 1948) that immunoresistant lines generally absorb less antibody to isoantigens than immunosensitive lines. This was termed *antigen simplification*. It was observed (Sanford *et al.*, 1973) that lyophilized TA3-St cells actually absorbed slightly less antibody from each antiserum than did intact cells. Surprisingly, lyophilized TA3-Ha cells absorbed far more of the antibodies from either antiserum than did the intact cells. Indeed, the amount of antibodies absorbed by disrupted TA3-Ha cells was found to be approximately equal to that absorbed by intact TA3-St cells (Sanford *et al.*, 1973). Other investigators reported similar findings (Friberg and Lilliehöök, 1973). In addition to disruption by lyophilization, these workers found that sonication of the TA3-Ha cells also increased manyfold the capacities of these cells to absorb anti-H-2^a antibody but had little or no effect upon the absorptive capacity of the strain-specific TA3-St cell.

These results suggested that both the strain-specific TA3-St cell and the strain-nonspecific TA3-Ha cell actually possessed approximately equal concentrations of histocompatibility antigens (H-2^a) on their surfaces, but in the case of the TA3-Ha cell (but not the TA3-St cell), these antigens were not available for binding to the antibodies. We suggested, as a result of these and other experiments, to be described later (Section IV.C), that histocompatibility antigens on the intact TA3-Ha cell surface were covered or masked, and that disruption of the cells resulted in unmasking the antigens, thus enabling the cells to bind the antibodies.

C. Cell-Surface Glycoproteins: Antigen Masking

The finding of a large-molecular-weight glycoprotein in high concentration at the cell surface of the strain-nonspecific subline TA3-Ha, but not at the surface of the strain-specific TA3-St cell, occurred during the same period in which we found a marked increase in H-2^a antibody absorption after disruption of the TA3-Ha cell structure (Section IV.B). It was suggested at that time (Codington *et al.*, 1973; Sanford *et al.*, 1973) that this glycoprotein, later called epiglycanin, served to mask histocom-

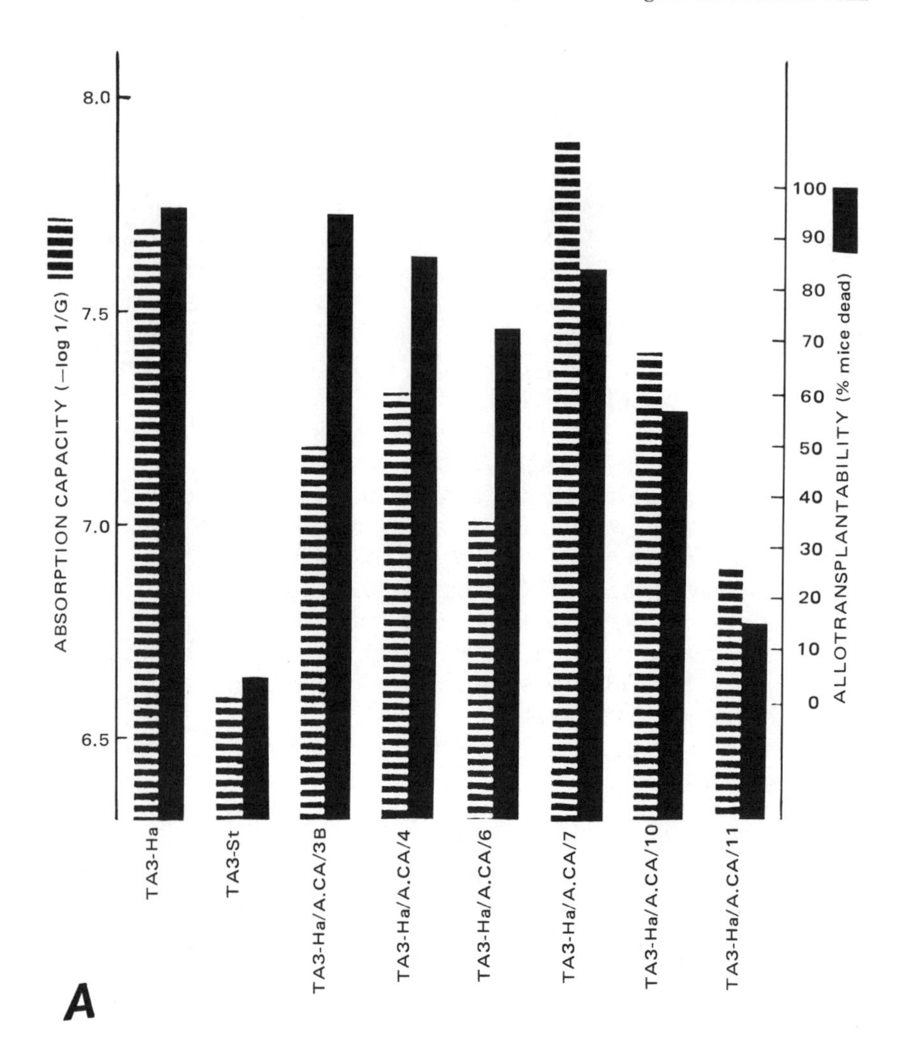

Fig. 2. (A) Comparison of anti-H-2^a antibody absorptive capacities (log $1/G$) and allotransplantability. Allotransplantability was measured as the number of mice dead from progressively growing tumors, divided by the number of mice of seven foreign strains inoculated with 10^6 cells for six TA3-Ha/A.CA hybrid and 10^5 cells for the TA3-Ha and TA3-St cell lines (solid bars); for comparison with allotransplantability, absorptive capacities were plotted in reverse order of magnitude, with values decreasing as distance on the y axis increases (striped bars). (B) Comparison of the anti-H-2^a antibody absorptive capacities (log $1/G$) and the amount of epiglycaninlike glycoprotein detected at the cell surface of the TA3-Ha, TA3-St, and six TA3-Ha/A.CA hybrid cell lines. Glycoprotein amounts are expressed in milligrams per 10^9 cells tested (solid bars); absorptive capacities are expressed as in A (striped bars).

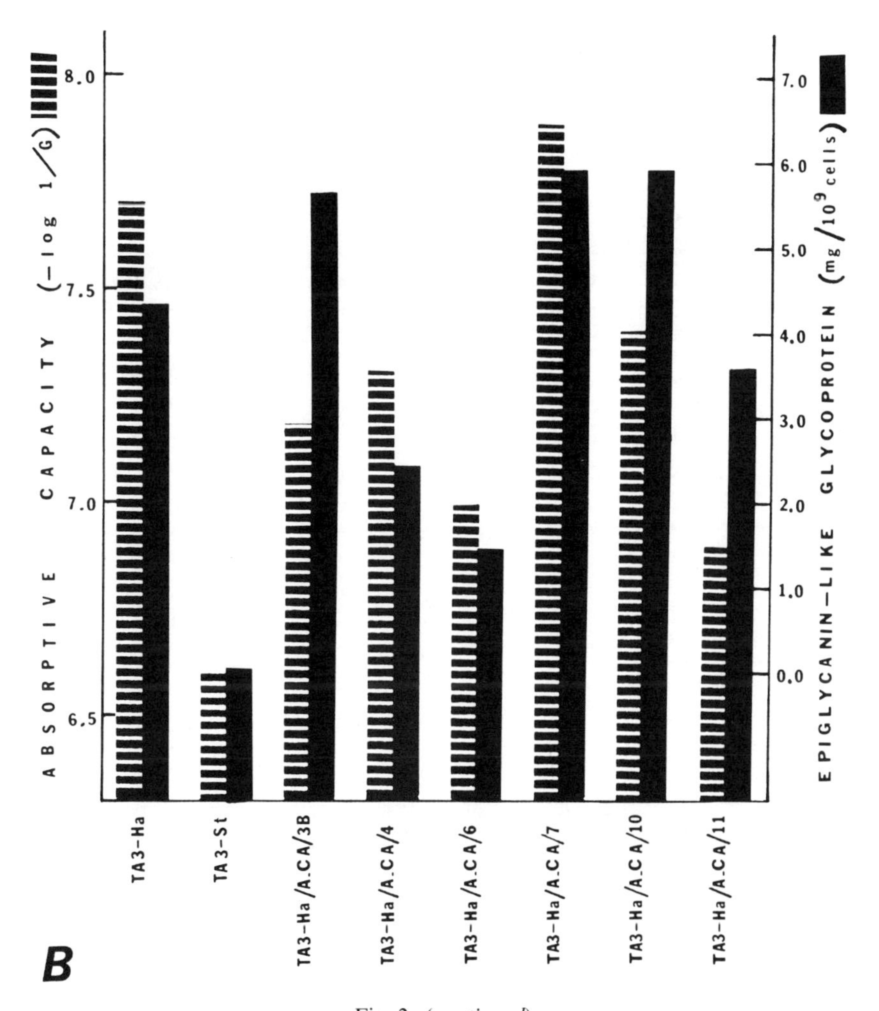

Fig. 2. (*continued*)

patibility antigens at the surface of the TA3-Ha ascites cell and physically protected antigenic sites from the binding of antibody. The masking role hypothesis further suggested that protection of antigens from immune lymphocytes *in vivo* could explain, at least in part, allotransplantability and xenotransplantability in the TA3-Ha ascites cell.

Epiglycanin was obtained as large glycopeptide fragments after fractionation of material removed by either L-1-tosyl-2-phenylethylchloromethyl ketone (TPCK)–trypsin (Codington *et al.*, 1972) or papain from viable TA3-Ha cells. Epiglycanin was later isolated in similar fashion from viable TA3-MM cells (Codington *et al.*, 1979a). Incubations were performed in the cold (4°C) and at low concentrations of enzymes (18 μg/ml trypsin; 93 μg/ml papain). Fractionation of material removed from the TA3-Ha ascites cells and from two TA3-MM variants on a column of Bio-Gel P-100 gave epiglycanin material in the void volume, Fraction A (Codington *et al.*, 1972, 1973, 1979a), as illustrated in Fig. 3. The carbohydrate and amino acid compositions of Fraction A isolated from cells of the three sublines are given in Table II. The compositions of this fraction from the TA3-Ha and the TA3-MM/1 were almost identical, but the proportion of carbohydrate residues in the fraction from the TA3-MM/2 subline was distinctly different. The major characteristics of the material are the same from the three lines, however. The high proportion of *N*-acetylgalactosamine (GalNAc), serine, and threonine suggested the presence of *O*-

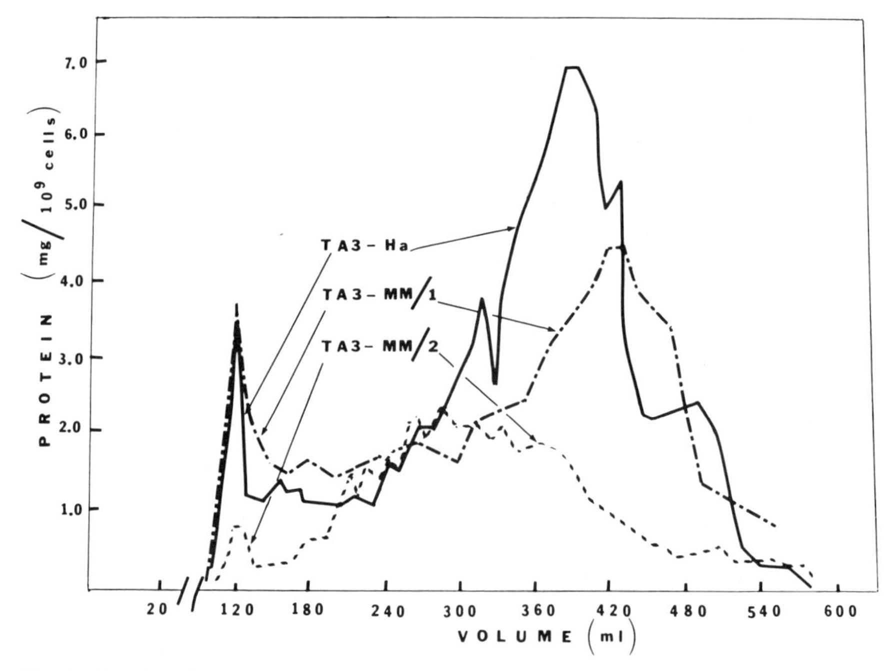

Fig. 3. Fractionation on a column of Bio-Gel P-100 (2.5 × 93 cm) of glycoproteins removed by TPCK–trypsin from viable TA3-Ha, TA3-MM/1, and TA3-MM/2 ascites cells. The eluent was 0.05 M pyridine acetate (pH 5.3), and fractions of 6.0 ml were collected at 4°C.

Table II
Compositions of Fractions A Removed from Viable TA3 Ascites Cells by Proteolysis

Components	Compositions of Fractions A removed from viable ascites cells		
	TA3-Ha	TA3-MM/1	TA3-MM/2
Carbohydrate components	Moles relative to *N*-acetylgalactosamine		
N-Acetylgalactosamine	1.0	1.0	1.0
Galactose	1.3	1.3	1.45
N-Acetylglucosamine	0.2	0.2	0.45
N-Acetylneuraminic acid	0.6	0.6	0.7
Mannose	0.01	0.03	0.01
Amino acid components	Residues per 1000 residents		
Alanine	118	127	121
Valine	22	21	26
Glycine	79	81	109
Isoleucine	9	6	5
Leucine	38	37	40
Proline	112	120	84
Threonine	313	352	302
Serine	261	220	228
Phenylalanine	3	Trace	Trace
Aspartic acid	19	16	32
Glutamic acid	26	19	54

Note. Determinations by gas–liquid chromatography.

glycosyl-linked carbohydrate chains (Section VII.H). The small proportion of mannose mitigated against the presence of more than three *N*-glycosyl-linked (asparagine-linked) chains (see Section VII.J). The yield of epiglycanin fragments from either the TA3-Ha or TA3-MM/1 cell was 2–3 mg/10^9 cells. This represents approximately 1% of the TA3-Ha dry weight. The elution profile of material obtained in a similar fashion from the TA3-St cell also gave a peak at the void volume. Whereas Fraction A from the strain-nonspecific cells (Table II) was 75–85% carbohydrate, Fraction A from the TA3-St cell was only about 3% carbohydrate, and this was of a totally different composition. Indeed, as will be seen later, by use of lectin and antibody binding experiments (Sections VII.C and VII.D), the TA3-St ascites cell can contain no more than 0.5% as much epiglycanin as the TA3-Ha cell. In other experiments, such as hemagglutination inhibition studies (Codington *et al.*, 1975a, 1979a; Cooper *et al.*, 1974), no epiglycanin like material was detected in material isolated from the TA3-St cell.

As described elsewhere (Sections IV.F, V.D, and VII.A), the glycopeptides of epiglycanin cleaved by proteolysis give an average molecular weight of about 200,000 and appear to be composed of two major components of molecular weights: A, 450,000, and C, 100,000. The native epiglycanin, i.e., that present at the cell surface, is considered to be of approximately the same size as that shed from the molecules into the ascites fluid, with molecular weight of 500,000 and length of 450–500 nm (Section VII.A). Epiglycanin solubilized by detergents or liberated by freeze–thawing appears to have similar characteristics, insofar as size is concerned.

In addition to consideration of the size of the epiglycanin molecule (approximately 500 nm), the masking hypothesis is also based upon the high concentration of epiglycanin at the cell surface. With approximately 4×10^6 molecules per cell, it is estimated that these molecules may be no more than 20 nm apart (Codington, 1975).

D. Anti-H-2^a Absorption and Epiglycanin Concentration in TA3-Ha/A.CA Hybrid Cells

The six TA3-Ha/A.CA hybrid cells lines which we investigated (Codington *et al.*, 1978a, 1979c) were transplantable in inbred mouse strains other than strain A, although this capacity to grow in foreign strains differed considerably among the hybrid lines, as recorded in Table I. Each of the hybrid lines was found to contain on its surface high concentrations of material of properties similar to those of epiglycanin. In Fig. 2B the concentration of this material in each of the hybrid lines (per 10^9 cells) is compared with the restriction on the capacity to absorb anti-H-2^a antibody, i.e., the absorptive capacity plotted in reverse order of magnitude; in like manner the absorptive capacity is plotted against the allotransplantability in Fig. 2A. From this figure, it is suggested that an inverse relationship probably exists between allotransplantability and the capacity to absorb antibody and that a direct relationship may exist between allotransplantability and epiglycanin concentration (Codington *et al.*, 1978a).

E. Possible Blocking Effect of Epiglycanin

It has been demonstrated that TA3 ascites cells bearing epiglycanin shed that glycoprotein into the ascites fluid of their hosts (Cooper *et al.*, 1974, 1979; Codington *et al.*, 1979a). The concentration of epiglycanin in solution has been determined by the inhibition of the agglutination of neuraminidase-treated human erythrocytes (blood group N-specific) by the lectin from *V. graminea* seeds. As will be discussed later (Section VII.D), this lectin binds specifically to a glycopeptide structure present in epiglycanin (Springer *et al.*, 1972; Codington *et al.*, 1975a). After 7 days of growth in the ascites form in A/HeJ mice inoculated with 10^5 TA3-Ha cells, the concentration of epiglycanin in 2.0–2.5 ml of ascites fluid varied from 44 to 510 μg/ml, with the median value near 150 μg/ml. The average TA3-Ha cell count was about 2×10^8 cells per mouse. No inhibitory activity (<5 ng/ml epiglycanin) was detected in the ascites fluid of A/HeJ mice inoculated with TA3-St cells (inoculum of 10^6 cells) under similar conditions (Cooper *et al.*, 1974).

The concentration of epiglycanin that appeared in the serum of tumor-bearing mice was also determined by the inhibition of hemagglutination by *V. graminea* lectin. Detectable concentrations of epiglycanin were usually found 3 to 5 days after injection of tumor cells. By the use of a continuous flow automated hemagglutination inhibition apparatus (Cooper *et al.*, 1974), concentrations of epiglycanin as low as 5 ng/ml could be accurately determined. Materials with inhibitory activity in both the ascites fluid and serum of TA3-Ha tumor-bearing mice were found to possess the same apparent molecular weight, as determined by gel-filtration chromatography (Cooper *et al.*, 1974). On a Sepharose 4B column, epiglycaninlike material shed into the body fluids was eluted well ahead of material (Fraction A, see Fig. 3) removed from viable cells by proteolysis.

Although no evidence to support a biological role for *in vivo* circulating epiglycanin in allogeneic mice has been obtained, we have suggested the possibility that serum epiglycanin may form blocking factors (Hellström and Hellström, 1977) in allogeneic mice. Circulating blocking factors are believed to consist of membrane-derived antigens (glycoproteins) alone or antibodies to these antigens or antigen–antibody complexes (Hellström and Hellström 1974, 1977). Such factors in the sera of allogeneic mice could conceivably bind to immune lymphocytes, thus removing cytotoxic activity. In support of this possibility is the recent observation that the injection of intact epiglycanin-bearing ascites cells into rabbits produced an antibody to epiglycanin in high titer (see Section VII.B). It seems probable that a similar type of antibody may be formed in allogeneic mice.

Whereas a possible role for epiglycanin at the cell surface in masking surface histocompatibility antigens is supported by the results of several different experimental approaches, no direct experimental data support a serological blocking role for this glycoprotein. The observation that epiglycanin is shed from the cell surface into the body fluids and that the molecule, when attached to intact TA3-Ha cells, is immunogenic suggests that this possibility should be further investigated.

To this end, in separate experiments (Miller *et al.*, 1978; Miller and Cooper, 1978), TA3-Ha cells were found to release epiglycanin material into *in vitro* growth medium; the kinetics of this release was studied (Miller and Cooper, 1978). The released material was found to be cell-surface epiglycanin, as determined by labeling studies.

A related mechanism, also based upon observations of the TA3-Ha ascites cell, was suggested by Nowotny *et al.* (1974). It was observed (Grohsman and Nowotny, 1972) that a particulate fraction in the ascites fluid of TA3-Ha cells growing in ascites form in mice of foreign strains stimulates the growth of this tumor in normal mice. They (Nowotny *et al.*, 1974) suggested that this action was due to decay products from dying cells, which served as primary targets for host immune factors. The effective defenses of the host were thus spent against the cell products and not against the living cells.

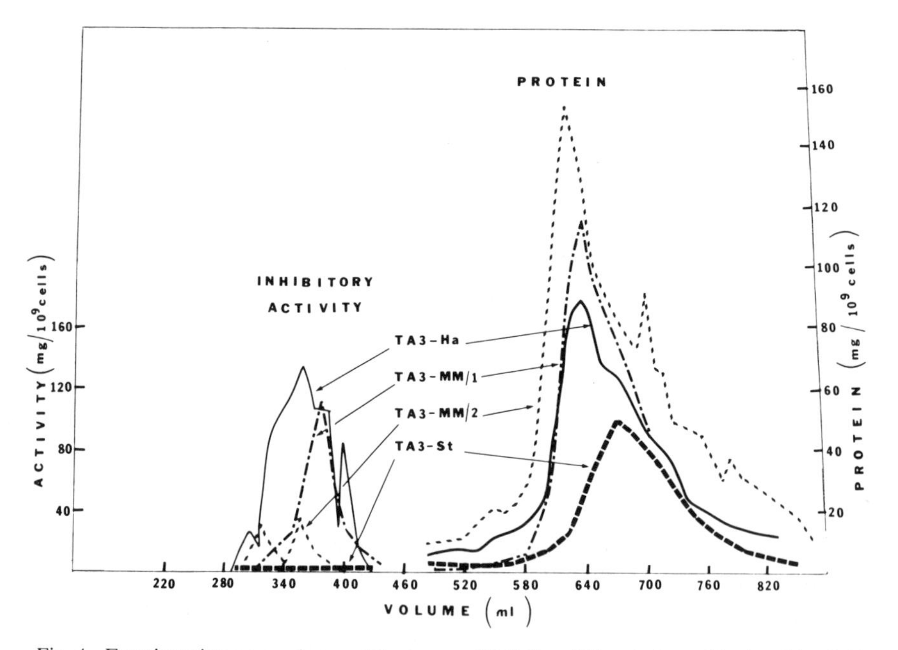

Fig. 4. Fractionation on a column of Sepharose 4B (2.5 × 170 cm) of perchloric acid-soluble glycoproteins from the ascitic fluid of A/HeJ mice bearing TA3-St, TA3-MM/1, and TA3-MM/2 ascites cells. The eluent was 0.05 M pyridine acetate (pH 5.3), and fractions of 6.0 ml were collected at 4°C.

Table III
Compositions of Fractions II in Sepharose 4B Column[a] from Ascitic Fluid of A/HeJ Mice Bearing TA3 Ascites Cells[b]

Components	Compositions of fractions II from:		
	TA3-Ha	TA3-MM/1	TA3-MM/2
Carbohydrate components	Moles relative to N-acetylgalactosamine		
N-Acetylgalactosamine	1.0	1.0	1.0
Galactose	1.4	1.4	1.6
N-Acetylglucosamine	0.6	0.3	0.2
N-Acetylneuraminic acid	0.4	0.6	0.4
Mannose	ND[c]	ND	ND
Amino acid components	Residues per 1000 residues		
Alanine	131	131	117
Valine	26	12	25
Glycine	81	92	106
Isoleucine	1	11	5
Leucine	33	49	39
Proline	93	107	81
Threonine	307	306	293
Serine	252	223	221
Phenylalanine	1	4	Trace
Aspartic acid	24	31	31
Glutamic acid	32	36	52
Methionine	ND	ND	9
Lysine	12	ND	ND
Histidine	1	ND	7
Arginine	1	ND	15

[a] See Fig. 4.
[b] Gas–liquid chromatography was used for carbohydrate components and an amino acid analyzer was used for amino acid components.
[c] Not determined.

F. Shed Epiglycanin Molecules

After the separation of tumor cells, the ascites fluid was clarified by centrifugation at 30,000*g*, and stored at −20°C. Pooled batches of fluid were briefly treated at 0°C with 0.3 M perchoric acid to precipitate contaminating proteins and glycoproteins. The perchlorate-free solution was concentrated and fractionated by gel-filtration chromotography (Codington *et al.*, 1979a). The elution profiles, based upon the protein content and the capacity to inhibit the agglutination of neuraminidase-treated human erythrocytes by the *V. graminea* lectin (a measure of epiglycanin concentration), are plotted in Fig. 4 for the TA3-St, TA3-Ha, TA3-MM/

1, and TA3-MM/2 ascites fluids. Similar profiles were obtained by fractionation of the ascites fluids from mice bearing several TA3-Ha/A.CA hybrid cells (Codington *et al.*, 1978a). All inhibitory activity was found in the high-molecular-weight region of the effluent. This region also contained a high proportion of bound sialic acid, as illustrated for experiments involving TA3-Ha/A.CA hybrid cell material (Codington *et al.*, 1978a). Yields of epiglycanin (Fraction II, Fig. 4) were 200–300 $\mu g/10^9$ cells, approximately 10% as high as could be obtained by removal of epiglycanin fragments from the cell surface by TPCK–trypsin (Codington *et al.*, 1978a, 1979a), as illustrated in Fig. 3. The carbohydrate and amino acid compositions of Fraction II from the TA3-Ha ascites cell are given in Table III. These values are similar to those presented in Table II for a purified fraction (Fraction A) from material cleaved from viable cells by proteolytic enzymes. The compositions of a similar fraction from the ascites fluids of mice bearing TA3-MM/1 ascites cells were almost identical to those of TA3-Ha material, but Fraction II from the TA3-MM/2 cell differed slightly in carbohydrate composition from the TA3-Ha and TA3-MM/1 materials (Codington *et al.*, 1979a).

Epiglycanin isolated by gel filtration after removal from viable cells grown in spinner culture by incubation with TPCK–trypsin was similar in amino acid composition to that of epiglycanin obtained from cells grown *in vivo* (Table II) (Codington *et al.*, 1979a). The carbohydrate composition was similar except for a smaller proportion of sialic acid.

V. MORPHOLOGY OF THE TA3 ASCITES SUBLINES

A. Scanning Electron Microscopy

The topography of the strain-specific TA3-St and the strain-nonspecific TA3-Ha sublines are markedly different. Scanning electron micrographs show the TA3-St cell surface to be irregular with folds from which filopodia project (Miller *et al.*, 1977), as illustrated in Fig. 5A. The TA3-Ha cell, by contrast, appears to possess a much more regular topography with long microvilli distributed at regular intervals over the surface (Miller *et al.*, 1977), as shown in Fig. 5B. The intervillous space appears to be smooth. The microvilli in the TA3-Ha ascites cell appear to extend outward a distance equal to the cell diameter. Although this feature was not readily seen in fixed cells, it was observed by dark-field microscopy of the cells suspended in aqueous media (Miller and Codington, unpublished observations).

B. Transmission Electron Microscopy

Transmission electron micrographs of TA3-St and TA3-Ha ascites cells, fixed routinely with glutaraldehyde and osmium tetroxide, are presented in Fig. 6. The internal features of the two cells appear to be similar (Miller *et al.*, 1977). Each contains easily identifiable virus particles. The characteristics of the cell surfaces, however, show the same general differences as seen by scanning electron microscopy: the folded and curved microextensions of the TA3-St cell being clearly different from the more rigid and lengthy microvilli (most of them cut at a short distance from the plasma membrane) of the TA3-Ha cell (Miller *et al.*, 1977).

C. Viruses Present in TA3 Ascites Cells

No differences could be detected in the appearance of virus particules, as viewed by transmission electron microscopy, in TA3-St, TA3-Ha (Miller *et al.*, 1977), or TA3-MM/2 (Miller and Codington, unpublished observations) cells.

Two different types of particles were observed. In a radioimmunoassay (Sheffield *et al.*, 1977), performed by Dr. Sheffield of the Institute for Cancer Research, the concentrations of both mouse mammary tumor virus and Rauscher mouse leukemia virus were determined in the TA3-Ha, TA3-MM/2, and TA3-St ascites cells (Cooper *et al.*, 1979). Each cell line possessed a significant amount of protein from each virus, but the TA3-Ha gave higher values than the other two cells. A serological screening procedure for 11 viruses, performed by Drs. M. J. Collins, Jr., and J. C. Parker of Microbiological Associates, indicated the presence of the antibody to the pneumonia virus of mice in the sera from mice bearing each TA3 cell (Cooper *et al.*, 1979).

D. High-Magnification Transmission Electron Microscopy

Under high magnification, i.e., 50,000–100,000×, cell-surface features in the two sublines appeared to be dramatically different (Miller *et al.*, 1977). Whereas the surface of the TA3-St cell appeared to be smooth and free of filamentous material, the surface of the TA3-Ha cell was covered with long filaments (Fig. 7). This was especially evident in the neighborhood of high concentrations of microvillous fragments, which

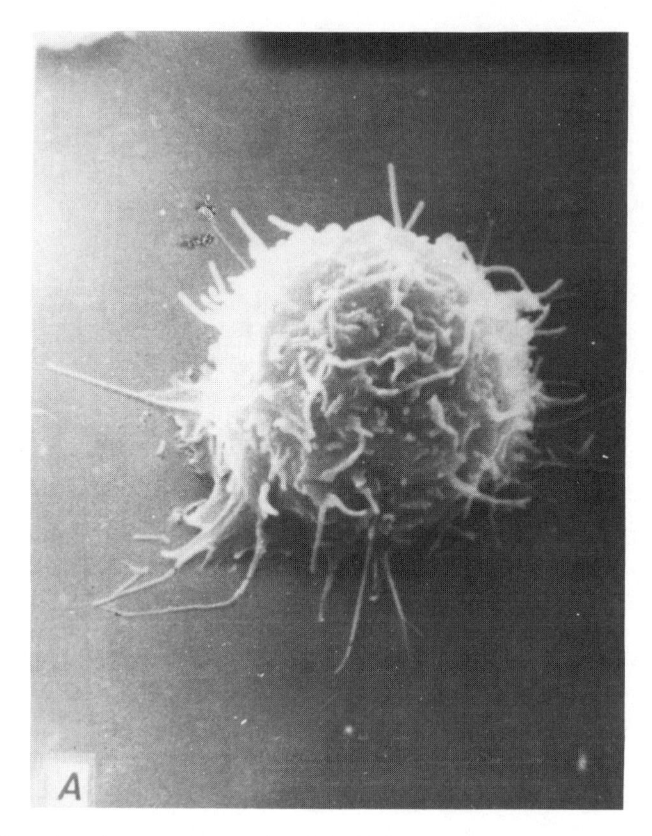

Fig. 5. Scanning electron micrographs of a TA3-St cell (A), as well as a TA3-Ha cell (B), both attached by slender filopodia to a glass coverslip. Long, regular microvilli are char-

appeared to stabilize these structures. Similar types of filaments were observed at the surface of the TA3-MM/2 cell (Miller *et al.*, 1982). Although the identity of this filamentous material cannot be made with absolute certainty, its characteristics suggest that it probably consists of epiglycanin aggregates. Evidence to support this contention was obtained by the use of polycationic ferritin, followed by fixing for transmission electron microscopy (Miller *et al.*, 1977) Polycationic ferritin was abserved to adhere in much greater amounts to the surface of the TA3-Ha than to the surface of the TA3-St ascites cell and to extend a distance greater that 200 nm beyond the plasma membrane, a result consistent with adherence to the filamentous material. After treatment of the TA3-Ha cell with neuraminidase (to remove approximately 80% of the cell-

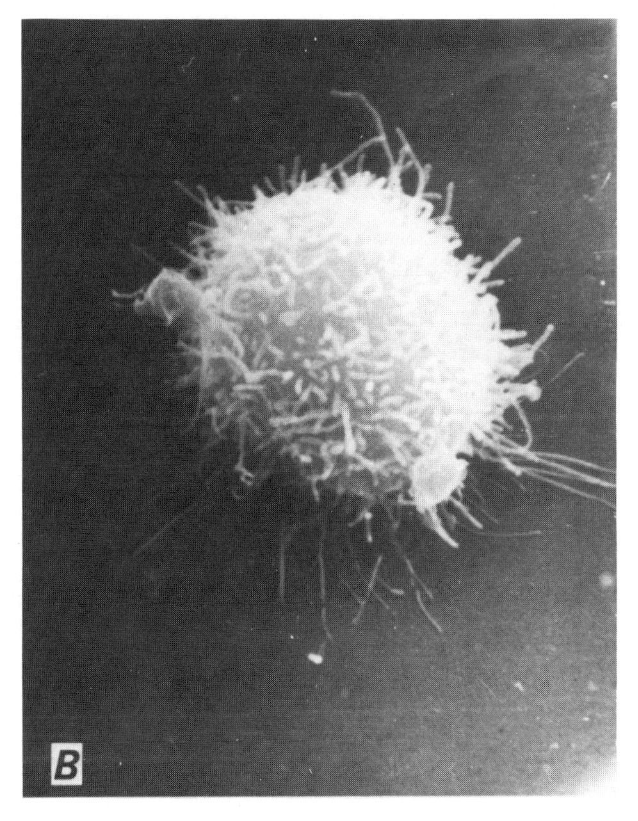

acteristic of the surface of the Ha cell, while the filopodia of the St cell tend to fall back on the surface of the cell. ×5000.

surface sialic acid), only a small amount of the polycationic ferritin adhered to the TA3-Ha cell, and none of this was observed beyond the plasma membrane. This suggests that the thin filaments visualized by electron microscopy are composed of epiglycanin, since this glycoprotein contains about 60% of the total cell-surface sialic acid. The long TA3-Ha filaments cannot be visualized without prior fixing with glutaraldehyde, suggesting that they are of protein composition. Their length from the outer membrane (200–400 nm) is approximately equal to that of epiglycanin molecules (400–500 nm) (Slayter and Codington, 1981; Codington *et al.*, 1979a), as described later (Section VII.A). Electron micrographs of the two cell lines support the conclusions derived from chemical and immunochemical data (described later in detail), namely, that the allo-

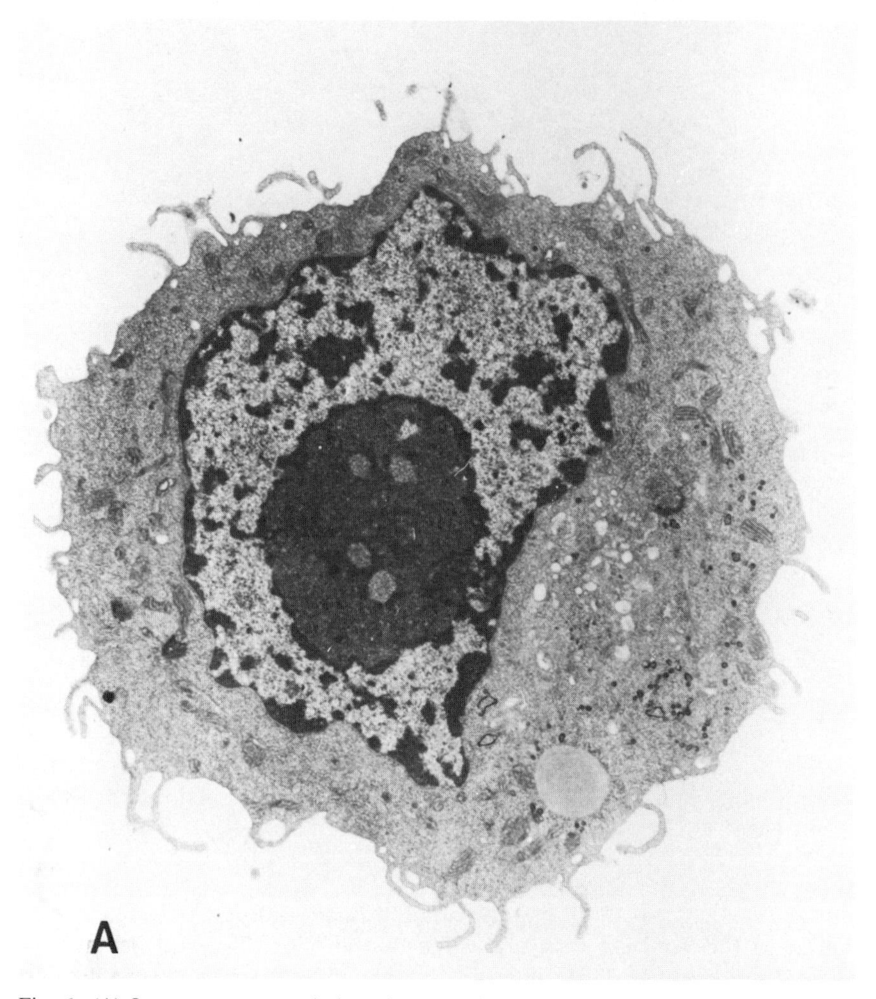

Fig. 6. (A) Low-power transmission electron micrograph of a section of a TA3-St cell showing the characteristic irregular surface contour and large eccentric nucleus with prominent nucleolus. The cytoplasm contains elongated mitochondria, abundant free ribosomes, and, to the right of the nucleus, a Golgi zone. The numerous cytoplasmic processes projecting from the cell tend to be bent and folded. Glutaraldehyde–osmium tetroxide fixation. ×9400. (B) Low-power transmission electron micrograph of a section of a TA3-Ha cell. The internal features of this cell are similar to those of the TA3-St cell, i.e., eccentric nucleus, Golgi zone, and abundant free ribosomes. The processes projecting from the cell surface are generally straighter and more uniform in diameter than those found on the TA3-St cell. Glutaraldehyde–osmium tetroxide fixation. ×12,200.

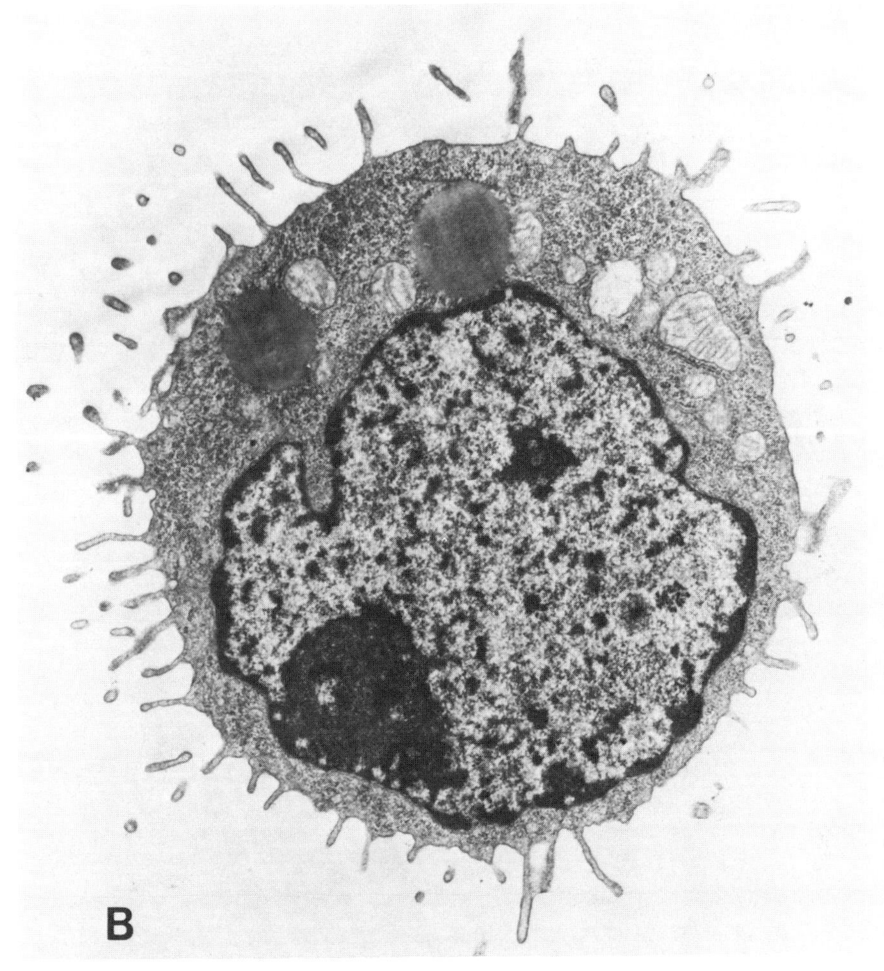

Fig. 6. (*continued*)

transplantable TA3-Ha subline possesses large membrane-bound glycoprotein molecules in high concentration at its surface, whereas the strain-specific TA3-St cell does not possess surface molecules of this type.

VI. CELL-SURFACE SIALIC ACID

A. Masking by Sialic Acid

Investigation of the role played by sialic acid in malignancy and especially in allotransplantability has been pursued throughout the course of our work with the TA3 tumor. Our interest began with the report

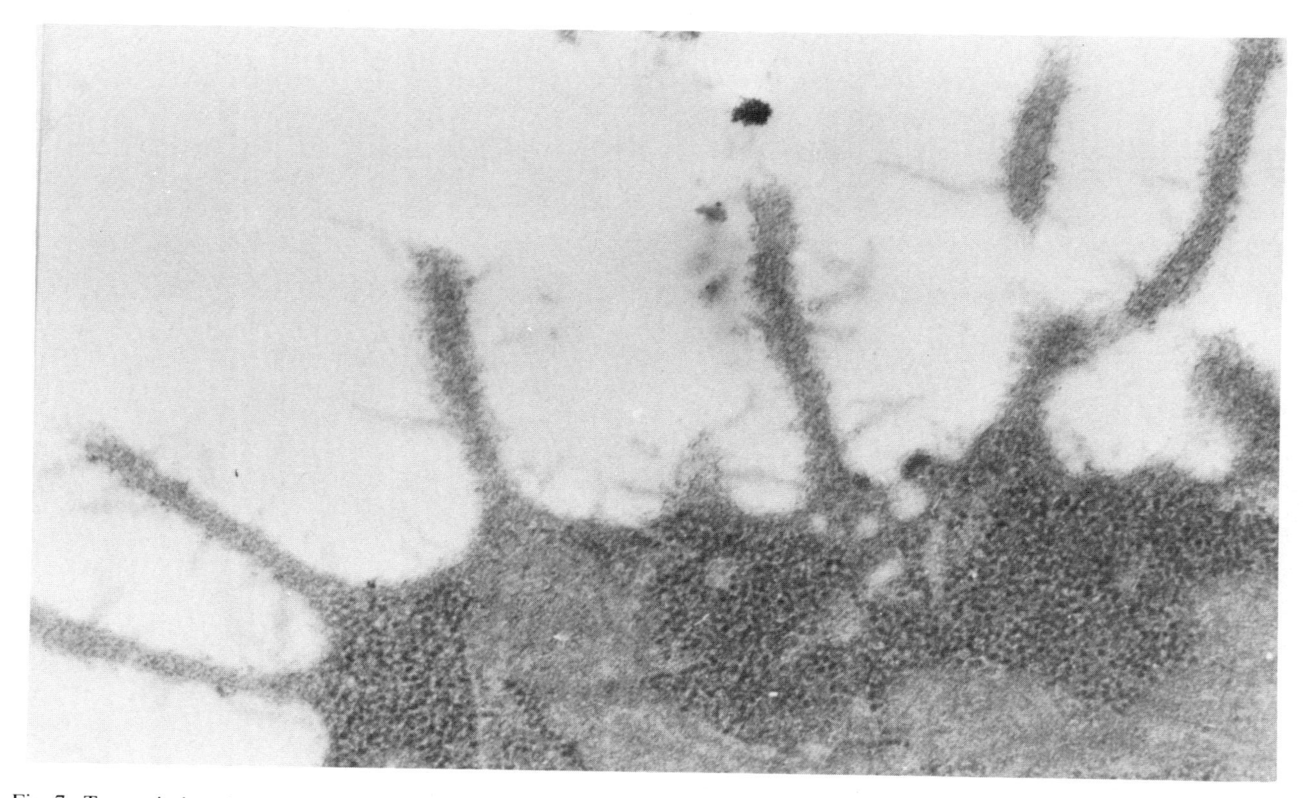

Fig. 7. Transmission electron micrograph of the TA3-Ha cell surface. Extramembranous structures 200–500 nm in length are observed on the cell surface. Glutaraldehyde–osmium tetroxide fixation. $\times 43,000$.

Table IV
Toxicity of Mouse and Guinea Pig Serum under Various Conditions for Neuraminidase-Treated and Untreated TA3-Ha Ascites Cells

	Percentage of dead cells[a]	
	Untreated cells	Neuraminidase-treated cells
Undiluted C3H serum + guinea pig complement[b]	10	73–98
Guinea pig complement	10	10
Undiluted A/He mouse serum + guinea pig complement	10	6–24
Undiluted serum from irradiated thymectomized C3H mice + guinea pig complement	10	10
Undiluted guinea pig serum	10	81
Undiluted guinea pieg serum (heated at 56°C for 30 min)	10	10
Undiluted guinea pig serum (absorbed with 4 × 10^8/ml neuraminidase treated TA3-Ha cells)	10	10

[a] Cells (4–5 × 10^4/ml) were incubated with serum and/or complement at 37°C for 30 min. The percentage dead cells was determined by trypan blue uptake.
[b] Guinea pig serum used as a source of complement had been absorbed with 4 × 10^8/ml neuraminidase-treated TA3-Ha cells.

(Sanford, 1967) that treatment of TA3-Ha ascites cells with *Vibrio cholerae* neuraminidase under mild conditions reduced their transplantability in the allogeneic C3H mouse. It was later demonstrated that this effect, which was observed only with small inocula of tumor cells, was not due to masking of surface histocompatibility (H-2) antigens by sialic acid, since equal amounts of monospecific serum directed against subcomponents 3, 4, 11, and 28 were absorbed by untreated and neuraminidase-treated TA3-Ha cells (Sanford and Codington, 1971). The effect appeared rather to be due to the exposure of new galactose-terminated antigenic determinants on the cell surface. We were able to demonstrate the presence of a factor in the serum of guinea pigs, as well as C3H mice, which, in the presence of complement, was cytotoxic to neuraminidase-treated, but not untreated, TA3-Ha cells. The results of these studies are presented in Table IV. This effect is not limited to cells with epiglycaninlike cell-surface glycoproteins. Furthermore, antibodies cytotoxic to neuraminidase-treated cells appear to exist in the serum of many mammalian species (Hughes *et al.*, 1973; Rosenberg and Schwarz, 1974; Rosenberg and Rogentine 1972). Although the early work (Sanford and Codington, 1971; Simmons and Rios, 1973) suggested that the phenomenon was dependent upon the immune response, as well as upon the presence of a cytotoxic factor, perhaps a naturally occurring antibody, other mechanisms for the effect have also been suggested (Jeanloz and Codington, 1976; Ferguson *et al.*, 1978). [Several reviews of this work have appeared (Ray, 1977; Sedlacek *et al.*, 1977; Jeanloz and Codington, 1976; Ferguson *et al.*, 1978).]

It should be emphasized that this transplantation effect is dependent upon the removal of only small fractions of the total cell surface sialic acid and is observed with only small (10^3–10^4 cells) inocula of tumor cells. This phenomenon appears to be unrelated to the masking effect of large cell-surface glycoproteins (epiglycanin), as described in Section IV.C. Allotransplantability in those experiments (Hauschka *et al.*, 1971; Sanford and Codington, 1971; Sanford *et al.*, 1973) was not affected by the removal of sialic acid.

B. Total Cell-Surface Sialic Acid

Values for the total amount of cell-surface sialic acid that can be cleaved from viable cells at 37°C by *V. cholerae* neuraminidase vary widely for TA3 ascites cells, as shown in Table V. It seems probable that these values represent approximately 75–80% of the total sialic acid of the cell. This was found to be the case for the TA3-Ha cell (Codington *et al.*, 1970).

Although all allotransplantable lines examined contained more surface sialic acid than the nonallotransplantable TA3-St cell, no direct cor-

Table V
Amount and Composition of Sialic Acid Removed from Mouse Ascites Cells by *V. cholerae* Neuraminidase

Cell line	Proportion NeuNGl (%)	Total sialic acid (μg/10^9 cells)
Mammary carcinomas		
TA3-St	20	270
TA3-Ha	7	620
TA3-MM/1	12	850
TA3-MM/2	12	1200
Hybrid cells		
TA3-Ha/A.CA/3B	13	700
TA3-Ha/A.CA/4	19	870
TA3-Ha/A.CA/6	17	1180
TA3-Ha/A.CA/7	38	910
TA3-Ha/A.CA/10	17	470
TA3-Ha/A.CA/11	6	850

relation appears to exist between these values and the degree of allotransplantability (Codington *et al.*, 1979b). In a series of polyoma-induced renal carcinoma variants, and in mouse melanoma lines, however, a correlation between metastatic potential and the total concentration of cell-surface sialic acid was found (Yogeeswaran and Salk, 1978; Yogeeswaran *et al.*, 1978). Nevertheless, other factors seem to be of greater importance in allotransplantability, as suggested above for the TA3-Ha ascites cell, since neuraminidase treatment had no effect upon the H-2^a antibody absorption (Sanford and Codington, 1971). Therefore, as sialic acid-containing glycoproteins do appear to play an important role in determining transplantation capabilities, sialic acid concentration per se may often be indicative of this capability.

C. Composition of Sialic Acid

No evidence has been found in any of the TA3 sublines for the existence of O-acetylated sialic acid residues (Codington *et al.*, 1979b). All of the bound sialic acid appears to exist as both *N*-acetylneuraminic acid and *N*-glycolylneuraminic acid. Table V gives the composition of the sialic acid removed by neuraminidase from viable cells of all the TA3 sublines examined. Values range from as low as 6% *N*-glycolylneuraminic acid for the TA3-Ha/A.CA/11 hybrid cell and 7% *N*-glycolylneuraminic acid for the TA3-Ha cell to as high as 38% *N*-glycolylneuraminic acid for TA3-Ha/A.CA/7. The composition of sialic acid varied from fraction to fraction of glycopeptide isolated from the same cell line (Codington *et al.*, 1979b). The fractions designated A to F in Fig. 8A were analyzed for their sialic acid compositions. The results are shown graphically in Fig. 8B.

Although it appeared that the proportion of *N*-glycolylneuraminic acid decreased with the increase in apparent molecular weight of the fractions (i.e., Fraction A has a lower proportion of this type of sialic acid than Fraction F), a more fundamental relationship was found to exist. It was noted that the proportion of mannose (indicative of the concentration of asparagine-linked carbohydrate chains) decreased with the apparent size of the fractions (Fig. 9A), and the proportion of *N*-acetylgalactosamine (indicative of the proportion of *O*-glycosyl-linked chains) increased with the apparent sizes of the fractions (Fig. 9B). Thus, Fraction A has the lowest proportion of mannose, the highest proportion of *N*-acetylgalactosamine, and the lowest proportion of *N*-glycolylneuraminic acid. As will be described in detail later, epiglycanin from the TA3-Ha cell, which represents the material in Fraction A, contains per molecule of 500,000 molecular weight, more than 500 *O*-glycosyl-linked carbohydrate

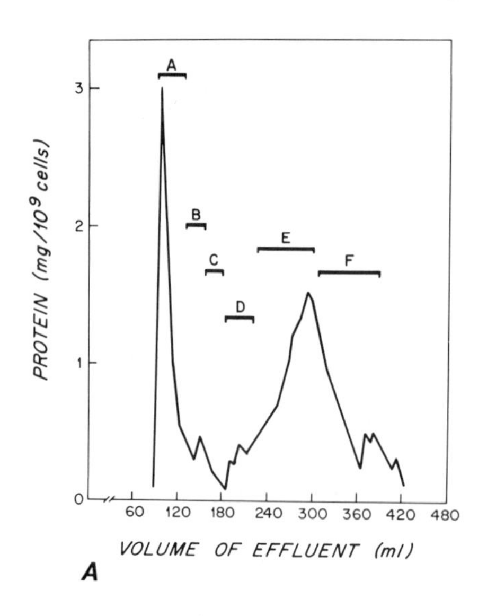

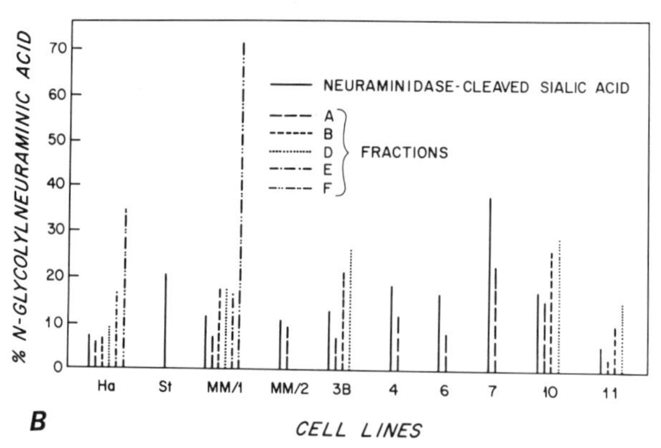

Fig. 8. (A) Elution profile (protein by Lowry method) of TA3-Ha glycopeptides, removed by TPCK–trypsin from 10^9 viable cells, and fractionated on a column of Bio-Gel P-100 (2.5 × 90 cm); the eluent was 0.05 M pyridine acetate (pH 5.3). (B) Proportion of *N*-glycolylneuraminic acid (%) in total cell-surface sialic acid (——) and in glycopeptide fractions of Bio-Gel P-100 columns (A) of trypsin-cleaved material. Fraction letters are those from A; cell lines tested were the TA3-Ha, TA3-St, and TA3-MM/1 and -MM/2, as well as the six TA3-Ha/A.CA hybrid strains, TA3-Ha/A.CA/3B, /4, /6, /7, /10, /11.

chains and probably only one to three *N*-glycosyl-linked chains. Although no biological role for *N*-glycolylneuraminic acid, distinct from that of *N*-acetylneuraminic acid, has been described, qualitative differences in the lectin binding activity have been reported (Bhavanandan and Katlic, 1979). These workers reported that terminal residues of *N*-acetylneuraminic acid in sialic acid-containing glycoproteins were capable of binding wheat germ agglutinin, whereas terminal *N*-glycolylneuraminic acid residues were not capable of binding this lectin. It seems plausible that in mammalian cells *in vivo* the additional hydroxyl group in *N*-glycolylneuraminic acid at the cell surface could possibly serve either to form hydrogen bonds with carbonyl groups present in neighboring peptide linkages, or it could give additional hydrophilic character to key sialic acid residues, which might serve to repel hydrophobic, and attract hydrophilic,

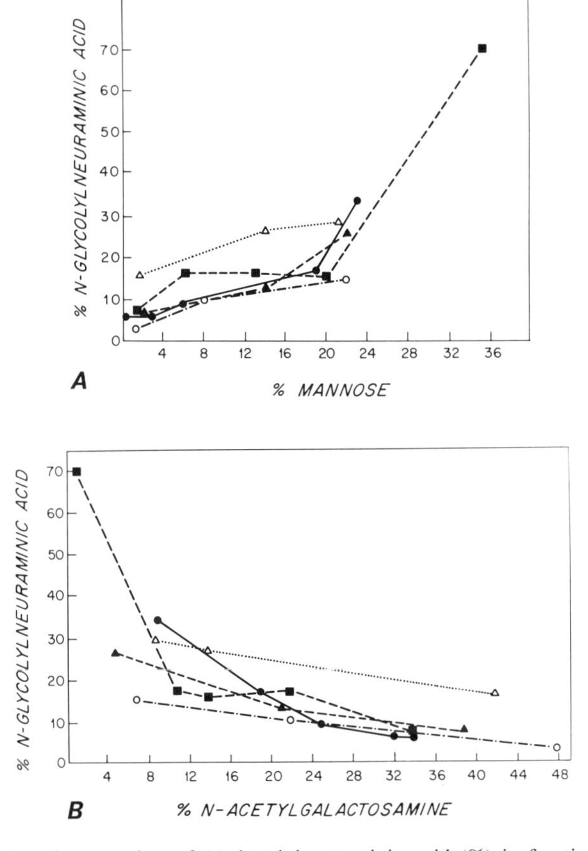

Fig. 9. (A) Plot of proportion of *N*-glycolylneuraminic acid (%) in fractions cleaved by trypsin from viable cells and isolated from a column of Bio-Gel P-100 (see Fig. 8) vs. the proportion of mannose in the carbohydrate moiety, as determined by gas–liquid chromatography. (B) Plot of the proportion of *N*-glycolylneuraminic acid (%) in proteolytically derived fractions (Fig. 8) vs. the proportion of GalNAc in the carbohydrate moiety, as determined by gas–liquid chromatography. Cell lines: TA3-Ha (●---●); TA3-MM/1 (■---■); TA3-Ha/A.CA/3B (▲---▲); TA3-Ha/A.Ca/10 (△···△); TA3-Ha/A.CA/11 (○-·-○).

groups. In either event, the result might be a change in the conformation of a cell-surface antigen or a modification of the environment in the neighborhood of the antigen, resulting in a modification of the activity of the antigenic determinant.

VII. STRUCTURE OF EPIGLYCANIN

A. Physical Properties

Several studies (Slayter and Codington, 1973, 1981; Codington *et al.*, 1979a) have reported the dimensions, conformation, and molecular weights of several epiglycanin fragments. In early studies (Slayter and Codington, 1973), fractions cleaved by proteolysis were examined by electron microscopy (shadow casting technique). The effluent from the fractionation of Fraction A (Fig. 3) on a column of Sepharose 4B (Slayter and Codington, 1973) was divided into three fractions: A, B, and C. Electron micrographs of the three fractions are shown in Fig. 10. Epiglycanin particles in each fraction appeared by electron microscopy to be

Table VI
Physical Dimensions and Molecular Weights of Various Epiglycanin Fractions

Cell line	Fraction	Width (nm)	Length (nm)[d] No. Ave.	Wt. Ave.	Peak	Molecular weight: Based on Lw[a]	Based on sedimentation quilibrium
TA3-Ha	Shed, ascites fluid, Fraction II (Fig. 4)	2.5	384	440	440	490,000	480,000
TA3-MM/2	Shed, ascites fluid, Fraction II (Fig. 4)	2.5	383	450	440	500,000	510,000
TA3-Ha	Protease Fraction A	2.5	320	390	350 310 255	300,000 to 500,000 150,000	450,000 (170,000)[b]
TA3-Ha	Protease Fraction B	2.5	170	220		150,000 to 200,000	ND[c]
TA3-Ha	Protease Fraction C	2.5	94	120	50 80 120	100,000	100,000

[a] Lw, Weight average length.
[b] A minor component, not considered an accurate value.
[c] Not determined.
[d] For a discussion of number and weight average lengths and molecular weights, see Tanford (1961).

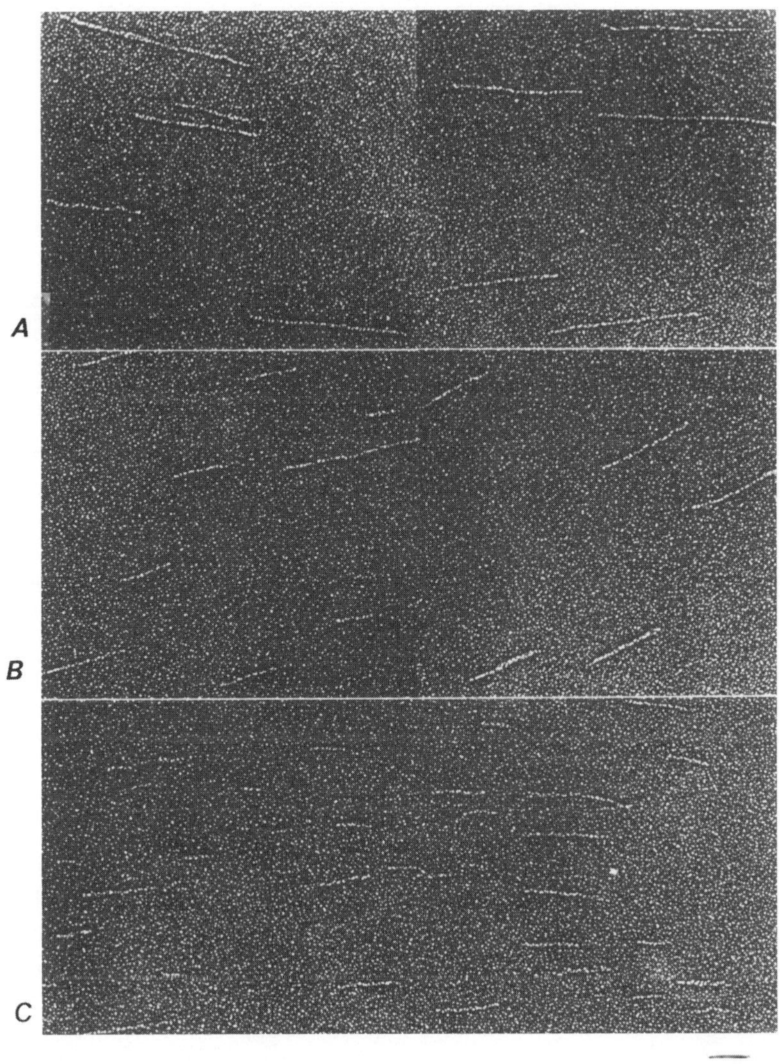

Fig. 10. Electron micrographs of platinum replicas of glycoprotein fractions derived proteolytically from the surface of TA3-Ha cells and fractionated at 4°C on a column of Sepharose 4B (2.6 × 87 cm; eluent: pyridine acetate, 0.05 M, pH 5.3). (A) Fraction A eluted in ml 140–190; Fraction B in ml 190–260; and Fraction C in ml 260–330. Note that all species appear as straight rods. Magnification in all three plates, ×100,000.

highly asymmetric rods similar in appearance to those isolated by fractionation of material shed from TA3-Ha or TA3-MM/2 cells while growing in A/HeJ mice in the ascites form (Codington *et al.*, 1979a) (see Fig. 4). The dimensions and molecular weights of each epiglycanin sample are given in Table VI. All molecules appear to have similar shapes, with identical widths of approximately 2.5 nm.

Based upon the assumption that epiglycanin consists of a single extended polypeptide chain of about 0.36 nm per amino acid residue (Pauling *et al.*, 1951), that it is about 23% protein, and that it has an average amino acid residue weight of 93, the calculated molecular weights (Table VI) were similar to those found by sedimentation equilibrium (Table VI) (Slayter and Codington, 1973; Codington *et al.*, 1979a). This was considered a reasonably sound basis for assigning an extended conformation to epiglycanin, particularly since the width of only 2.5 nm is not sufficient to account for any helical structures and is consistent with a single peptide backbone to which are attached multiple short carbohydrate chains.

No evidence has been found to suggest aggregation in any of the fractions described in Table VI, and the molecular weights listed are considered fairly accurate for the monomeric forms of the molecules. A molecular weight of 500,000 is considered fairly accurate for material shed from the cells into the ascites fluid. It should be stated, however, that epiglycanin solubilized from membranes by detergents appears to form aggregates in solution.

B. Detection Methods

Two specific and highly sensitive methods, one involving an antibody and the other a lectin, have been developed for the detection of epiglycanin, The specificity of each depends upon the presence of the same disaccharide structure, β-D-Gal(1 → 3)GalNAc, which is attached to serine or threonine in the polypeptide backbone, and each also appears to depend upon the presence of a particular sequence of amino acids, probably at the point of the glycopeptide linkage. It is not yet known, however, whether or not each protein (lectin or antibody) binds to the same glycopeptide linkage in epiglycanin. Indeed, this would appear to be improbable, since, as described later, there are over 500 carbohydrate chains of this type in epiglycanin of molecular weight 500,000.

C. Antibody to Epiglycanin

An antibody with specificity for epiglycanin and epiglycanin glycopeptides was induced in New Zealand white rabbits by the repeated in-

Table VII
Concentrations (ng/50 μl) of Epiglycanin Fractions and Other Carbohydrate-Containing Substances Required to Inhibit the Reaction: ^{125}I-Epiglycanin + Anti-epiglycanin Antiserum

Inhibitor	Concentration for 50% inhibition (ng/50 μl)
Epiglycanin (Fraction II, TA3-Ha shed into ascites fluid)	2
Epiglycanin (Fraction C, protease cleaved, Sepharose 4B fraction)	0.3
Neuraminidase treated	
Epiglycanin (Fraction C, protease cleaved)	0.2
Bovine lubricating glycoprotein	500
Glycophorin	>5000[a]
Fetuin	>5000[a]
Antifreeze glycoprotein (Antarctic fish)	>5000
α_1-Acid glycoprotein	>5000
Epiglycanin disaccharide	>500
Epiglycanin disaccharide (reduced)	>5000
Lactosamine	>5000
Methyl 2-acteamido-2-deoxy-α-D-galactose	>5000

[a] Partial (but less than 50%) inhibition observed with 5000 ng of these substances.

jection of TA3-Ha ascites cells (Codington *et al.*, 1981). By the use of ^{125}I-labeled epiglycanin and the anti-epiglycanin antiserum in a radioimmunoassay, it was possible to detect picogram quantities of epiglycanin. The radioimmunoassay could be used either for the determination of antibody activity or for the quantitation of the antigenic determinant. After incubation of the inhibitor (epiglycaninlike material) with the diluted antiserum, the mixture was incubated with radioactive epiglycanin. A second antigen–antibody system was used to complete the precipitation. The inhibition of the binding of ^{125}I-labeled epiglycanin to the antibody, as compared to the inhibition of standard samples of epiglycanin, was used as a measure of the antigenic determinant. Table VII gives the inhibitory activities of epiglycanin fractions and other glycoproteins and oligosaccharides in the radioimmunoassay. Except for the epiglycanin, the only substances with detectable inhibitory activity were those containing the same disaccharide as found in epiglycanin (Codington *et al.*, 1981), and the most active of these, asialobovine lubricating glycoprotein (Garg *et al.*, 1980), was less than 1% as active as epiglycanin. The involvement of the carbohydrate moiety in the antigenic determinant was shown by the complete loss of activity after periodate oxidation. Incubation with *endo*-α-*N*-acetylgalactosaminidase (Bhavanandan *et al.*, 1976), an enzyme found to remove only the disaccharide (about 75% complete from epiglycanin) (Bhavanandan and Codington, in preparation), destroyed 99% of the activity. Evidence that the peptide moiety was involved in the binding activity was shown by the observation that the disaccharide itself, cleaved from epiglycanin by this same enzyme and isolated by gel-filtration chromatography, had no detectable activity (Codington *et al.*, 1981), a result also found for the reduced disaccharide removed by chemical means (Van den Eijnden *et al.*, 1979). Interestingly, no activity was found for the Antarctic fish glycoprotein (Osuga and Feeney, 1978), but this glycoprotein consists entirely of the disaccharide attached to a repeating tripeptide unit.

The TA3-Ha and TA3-St ascites cells were tested for their capacity to absorb anti-epiglycanin antibody. Weak absorption was found for the TA3-St cell, but this absorption was less than 0.5% that of the TA3-Ha cell.

The radioimmunoassay has been used to measure the amount of epiglycanin shed from TA3-Ha cells into the ascites fluid of A/HeJ mice bearing the tumor. After growth of TA3-Ha cells for 7 days, values have ranged from 100 to 500 μg/ml, as determined by this method. These values are in agreement with those determined earlier (Cooper *et al.*, 1974) by inhibition of hemagglutination by *V. graminea* lectin (Section IV.E).

D. *Vicia graminea* Lectin

The presence of glycopeptide structures in epiglycanin that are specific for the lectin from *V. graminea* seeds was demonstrated by hemagglutination inhibition for fragments cleaved by proteolytic enzymes from the TA3-Ha ascites cell (Springer *et al.*, 1972; Codington *et al.*, 1972, 1975a, 1979a), the TA3-MM cells (Codington *et al.*, 1979a), and the TA3-Ha/A.CA hybrid cells (Codington *et al.*, 1978a). Strong inhibitory activity was also determined in epiglycanin shed into the ascites fluid of A/HeJ mice bearing either TA3-Ha or TA3-MM ascites cells (Codington *et al.*, 1979a), as described in Section IV.E. These data suggested the probable presence in epiglycanin from each of these allotransplantable ascites lines of the disaccharide, β-D-Gal(1 → 3)GalNAc, attached by an *O*-glycosyl linkage to either serine or threonine, as proposed previously (Uhlenbruck and Dahr, 1971), and of a polypeptide of a specific sequence (Lisowska and Wasniowska, 1978). The quantitation of these determinations was made possible by the development of an automated procedure (Cooper *et al.*, 1974) with which concentrations of epiglycanin as low as 5 ng/ml could be measured.

Table VIII
Inhibitory Activity to Hemagglutination by *V. graminea* Lectin Removed from 10^9 TA3-Ha, TA3-MM/1, or TA3-MM/2 Cells by 18 μg/ml TPCK–Trypsin in Five 20-min Incubations at 18°C

No. of Incubations	Inhibitory activity released (μg epiglycanin/10^9 cells)					
	Untreated			Neuraminidase treated		
	TA3-Ha	TA3-MM/1	TA3-MM/2	TA3-Ha	TA3-MM/1	TA3-MM/2
1	1,700	1,100	464	2,940	6,770	780
2	350	650	130	680	4,010	372
3	240	150	64	350	1,270	302
4	60	60	10	120	640	208
5	60	20	10	120	380	126
Total	2,410	2,030	680	4,210	13,070	1,790

The inhibitory activities were expressed as microgram-equivalents of the protease-cleaved epiglycanin fraction, Fraction C (Slayter and Codington, 1973), as shown in Fig. 10. Epiglycanin equivalents cleaved by proteolysis from TA3-Ha, TA3-MM/1, and TA3-MM/2 cells are presented in Table VIII. Values are given for neuraminidase-treated and untreated glycopeptides. Removal of sialic acid significantly increased the activity for each trypsinate but especially for the TA3-MM/1 material, where the activity was increased sixfold. These results suggested that although the disaccharide is present in all fractions of epiglycanin examined, a significant proportion of these structures may be masked by sialic acid residues. As described earlier (Section VI.A), this conclusion was substantiated by chemical experiments.

Proteolysis also removed inhibitory activity from each of the TA3-Ha/A.CA hybrid lines, but the quantity removed varied substantially from cell line to cell line (among the different cell lines) (Codington *et al.*, 1978a). These values and those arrived at by two other types of experiments were incorporated into the values used in Fig. 2B for epiglycanin concentration in the hybrid cells.

In similar procedures performed with the strain-specific TA3-St ascites cell, no inhibitory activity could be detected (Codington *et al.*, 1973, 1975a, 1978a) in the trypsinates. In experiments performed with both the TA3-Ha and the TA3-St ascites cells (Codington *et al.*, 1975a), the inhibitory activities of fractions from Bio-Gel P-4 and P-100 columns were tested. Although the material of the highest apparent molecular weights from the TA3-Ha cell possessed high activities, no activity was detected in any of the column fractions from the TA3-St-derived material.

Table IX
Specific Inhibitory Activities (μg/ml) to Hemagglutination by *V. graminea* Lectin

Source of sample	Column fraction	Neuraminidase treatment	Inhibitory activity of epiglycaninlike material from: TA3-Ha	TA3-MM/1	TA3-MM/2
Proteolysis	Fraction C[a] Sepharose 4B	No	1.0		
	Fraction A	No	0.88	0.53	0.13
	Bio-Gel P-100	Yes	1.5	1.5	0.50
Ascites fluid	Fraction I	No	0.18		0.03
	Sepharose 4B	Yes	0.90		0.85
	Fraction II	No	1.3	0.25	0.66
	Sepharose 4B	Yes	2.0	1.3	1.1

[a] This sample, derived by proteolysis and eluted from columns of Bio-Gel P-100 and Sepharose 4B, was used as a standard (see Fig. 10).

As expected, considerable variation was observed in the specific activities of epiglycanin isolated from different sources and indeed from the same fraction isolated from different batches of cells. Whether these variations are due to small changes in the cells or to the altered metabolic processes of the host animals has not been determined. Table IX presents the specific inhibitory activities of the various fractions of epiglycanin. In all experiments, the same purified sample of epiglycanin (Fraction C, molecular weight 100,000) was used. In all samples, activity was enhanced by the removal of sialic acid, but the amount of increase varied with each sample.

Table X gives the amounts of *V. graminea* lectin (in μg-equiv of epiglycanin) adsorbed by neuraminidase-treated and untreated TA3-Ha

Table X
Adsorption of the Lectin from *Vicia graminea* Seeds by Intact Cells

Cell type	Adsorption by *V. graminea* lectin (mg-equiv. epiglycanin/10^9 cells): Nontreated cells	Neuraminidase-treated cells
TA3-Ha ascites cells	5.0–9.0	15–21
TA3-St Ascites cells	0.02–0.05	0.2–0.3
Normal strain A spleen leukocytes	0.002	
Normal strain A erythrocytes	0.002	
Human erythrocytes of MN specificity	0.02	

and TA3-St cells (Codington *et al.*, 1975a), as well as values for some normal human and mouse cells. Since previous experiments had demonstrated that in the TA3-Ha ascites cell only the epiglycanin-containing fractions possessed inhibitory activity to agglutination by this lectin, it seems probable that all specific adsorption was due to epiglycanin. By this procedure, however, about twice the amount of epiglycanin (averaging 7 mg/10^9 cells) as could be removed by proteolysis under the most favorable conditions (3 mg) was detected. Some of this may be nonspecific, but since the amount of lectin adsorbed by the other cells tested was so little, by comparison, it seems probable that the amount of nonspecific adsorption was not significant. Adsorption by the TA3-St ascites cell was detectable, however, and represented approximately 0.5% as much epiglycaninlike material as was detected on the TA3-Ha cell. Since no detectable inhibitory activity could be removed from this cell by proteolysis, however, it is conceivable that this adsorption was not specific. Interestingly, as described in Section VII.C, the TA3-St ascites cell also adsorbed slightly less than 0.5% as much antiepiglycanin antibody as did the TA3-Ha cell. This similarity between the relative amounts of adsorption of the two proteins specific for the epiglycanin molecule suggests that they may share the same binding sites(s).

E. Adsorption of ^{125}I-Labeled Lectins by Intact Cells

The TA3-Ha ascites cell was found to bind 1.5-fold more ^{125}I-labeled *Ricinus communis* agglutinin than did the TA3-St cell (Matsumoto *et al.*, 1980), which suggested the presence of terminal galactose residues linked β-1,4 to *N*-acetylhexosamine residues (Codington *et al.*, 1975b). This lectin was shown to bind to epiglycanin (Sections VII.F and VII.G), which may explain its greater adsorption by the TA3-Ha cell than by the TA3-St cell. Indeed, by electron microscopy each epiglycanin molecule was found to bind an average of 16 lectin molecules (Slayter and Codington, 1981).

The binding of fourfold more eel serum agglutinin, a lectin specific for H blood group activity (which requires a terminal fucose residue) by the TA3-Ha (14×10^6 binding sites) than by the TA3-St cell was surprising, as there was no prior chemical evidence for the presence of fucose in epiglycanin. Consistent with these results, however, was the demonstration (Matsumoto *et al.*, 1980) that the agglutination of the TA3-St cells by eel serum agglutinin could be inhibited by epiglycanin (Table XI).

Both TA3-Ha and TA3-St ascites cells were shown to bind equal numbers of concanavalin A molecules (1×10^6) (Matsumoto *et al.*, 1980),

Table XI
Inhibition by Neuraminidase-Treated or Untreated Epiglycanin (μg/ml) of the Agglutination of Erythrocytes or Tumor Cells by Lectins

Lectin	Cell	Neuraminidase treatment	Epiglycanin (μg/ml) required to inhibit agglutination			
			Fraction A (P-100)[a]	Fraction A[b]	Fraction B[b]	Fraction C[b]
Bauhinia purpuria	RBCs[c]	No		28	18	73
		Yes		28	13	73
Iberis amara	RBCs[c]	No			4	
		Yes			4	
Arachis hypogaea	RBCs[d]	No		2	3	3
		Yes		2	1	5
Wistaria floribunda	RBCs[c]	No		55	56	76
		Yes		28	18	36
Glycine max	RBCs[c]	No		>440	320	290
		Yes		>440	160	290
Ricinus communis	RBCs[c]	No		>440	320	>580
		Yes		220	80	>580
	TA3-Ha[e]	No	5			
		Yes	5			
	TA3-St[e]	No	110			
		Yes	4			

Phaseolus limensis	RBC[c]	No			>125	
		Yes			>125	
Phaseolus vulgaris	RBCs[c]	No		>440	>640	>580
		Yes		>440	>640	>580
Concanavalin A	RBCs[f]	No	6			
	TA3-Ha[e]	No				
	TA3-St[e]	No	100			
Eel serum	TA3-Ha[e]	No	>200			
		Yes	>200			
	TA3-St[e]	No	200			
		Yes	150			
Wheat germ	TA3-Ha[e]	No	15			
		Yes	50			
	TA3-St[e]	No	50			
		Yes	50			
Potato	TA3-Ha[e]	No	150			
		Yes	150			
	TA3-St[c]	No	35			
		Yes	35			

[a] Epiglycanin removed from TA3-Ha cells and fractionated as described by Codington *et al.* (1979a).
[b] Fractions A, B, and C obtained from fractionation on Sepharose 4B as described by Slayter and Codington (1973) (see Fig. 10).
[c] Human red blood cells prepared as described by Matsumoto and Osawa (1970).
[d] Human red blood cells treated with neuraminidase.
[e] Concentration 2×10^6 cells/ml.
[f] Rabbit erythrocytes at 3×10^7 cells/ml.

a result consistent with that previously reported (Friberg *et al.*, 1974). The presence of concanavalin A binding sites in epiglycanin was demonstrated by the specific binding of epiglycanin molecules to concanavalin A-Sepharose affinity columns, and by electron microscopy (Slayter and Codington, 1981).

F. Agglutination Inhibition

The inhibition by epiglycanin fractions of the agglutination of human erythrocytes has been reported (Codington *et al.*, 1975b), and the inhibition of the agglutination of TA3-Ha and TA3-St cells by lectins has been described (Matsumoto *et al.*, 1980). Some of these results are compiled in Table XI. They suggest the presence of several different oligosaccharide structures in epiglycanin. The first four lectins in the table have been reported to bind to short carbohydrate chains attached to the peptide moiety by *O*-glycosyl linkages, similar, or related to that required for binding to the *V. graminea* lectin.

By these experiments, epiglycanin was shown to possess binding sites not demonstrable by chemical means, i.e., a binding site for eel serum agglutinin. Inhibition of agglutination by *R. communis* agglutinin was consistent with the presence of structures suggested by cell adsorption studies. Terminal lactosamine structures, probably the binding sites for *R. communis* agglutinin, were confirmed by chemical experiments (Van den Eijnden *et al.*, 1979; Codington *et al.*, 1975b, 1978b) and electron microscopy (Slayter and Codington, 1981). Due to the probability that only small proportions of the epiglycanin molecule constitute the binding sites for either concanavalin A or eel serum agglutinin, however, the use of chemical methods has not been feasible. The demonstration of a binding site in epiglycanin for eel serum agglutinin was consistent with the results of lectin binding by intact cells (Section VII.E). Since fucose, a carbohydrate residue involved in this binding site (Watkins, 1972), was not detected in any of the *O*-glycosyl-linked chain types isolated from epiglycanin (Section VII.G), it is suggested that this structure is probably present on an asparagine-linked carbohydrate chain. Evidence for the presence of a concanavalin A binding site in epiglycanin (Table XI) has been supported by other experiments, notably those involving the incorporation of (radiolabeled) mannose into epiglycanin (Frim, Codington, and Jeanloz, in preparation), and those involving the visualization of epiglycanin–concanavalin A complexes by electron microscopy (Slayter and Codington, 1981).

G. Active Sites on Epiglycanin Molecules

In consideration of the long rodlike character of the isolated epiglycanin molecules (Fig. 10) and the large number of lectins known to bind specifically to epiglycanin (Table XI), this glycoprotein appeared to represent an ideal model system for determination of the loci of different types of carbohydrate chains along an extended polypeptide chain (Slayter and Codington, 1981). Electron micrographs of complexes of epiglycanin and two proteins, *R. communis* toxin and concanavalin A, are given in Fig. 11. *R. communis* appears to bind usually at irregular intervals along the polypeptide chain (Fig. 11A) suggesting considerable heterogeneity in the locations of these sites. Although there was considerable variation in the number of toxin molecules bound per epiglycanin molecule, the mean value was 16. The maximum number observed for any complex was 23, a value in agreement with the calculated value of 20–25 chains terminating in a β-D-Gal(1 → 4)-β-D-GlcNAc sequence (Slayter and Codington, 1981).

Complexes with concanavalin A (Fig. 11B) appeared to contain only one lectin molecule, and this was always situated at one end of the epiglycanin molecule, which in view of the resolution of the method, probably means that it was attached to 1 of the first 20 amino acid residues. Only about half of the epiglycanin molecules appeared to bind the lectin. It is not known whether this was due to the absence of active sites (a possibility suggested by the inability of all epiglycanin molecules to bind to an affinity column), failure of the reaction going to completion, or the loss of attached molecules during preparation for electron microscopy. In any case, it appears that there is no heterogeneity in the location of the concanavalin A binding site when it occurs but that it may not be present in all molecules.

H. *O*-Glycosyl Chains

Upon analysis, approximately 29% of the weight of epiglycanin was found in *N*-acetylgalactosamine (Codington *et al.*, 1979a), which represents about 700 residues per molecule of molecular weight 500,000. The other probable components of the *O*-glycosyl bond, serine and threonine, represented 57–64% of the total number of amino acids, or about 780 residues per molecule (Codington *et al.*, 1975b, 1979a). It was not surprising, therefore, that treatment of protease-cleaved epiglycanin (Fraction A, Fig. 3) with dilute sodium hydroxide (0.1 M) and sodium boro-

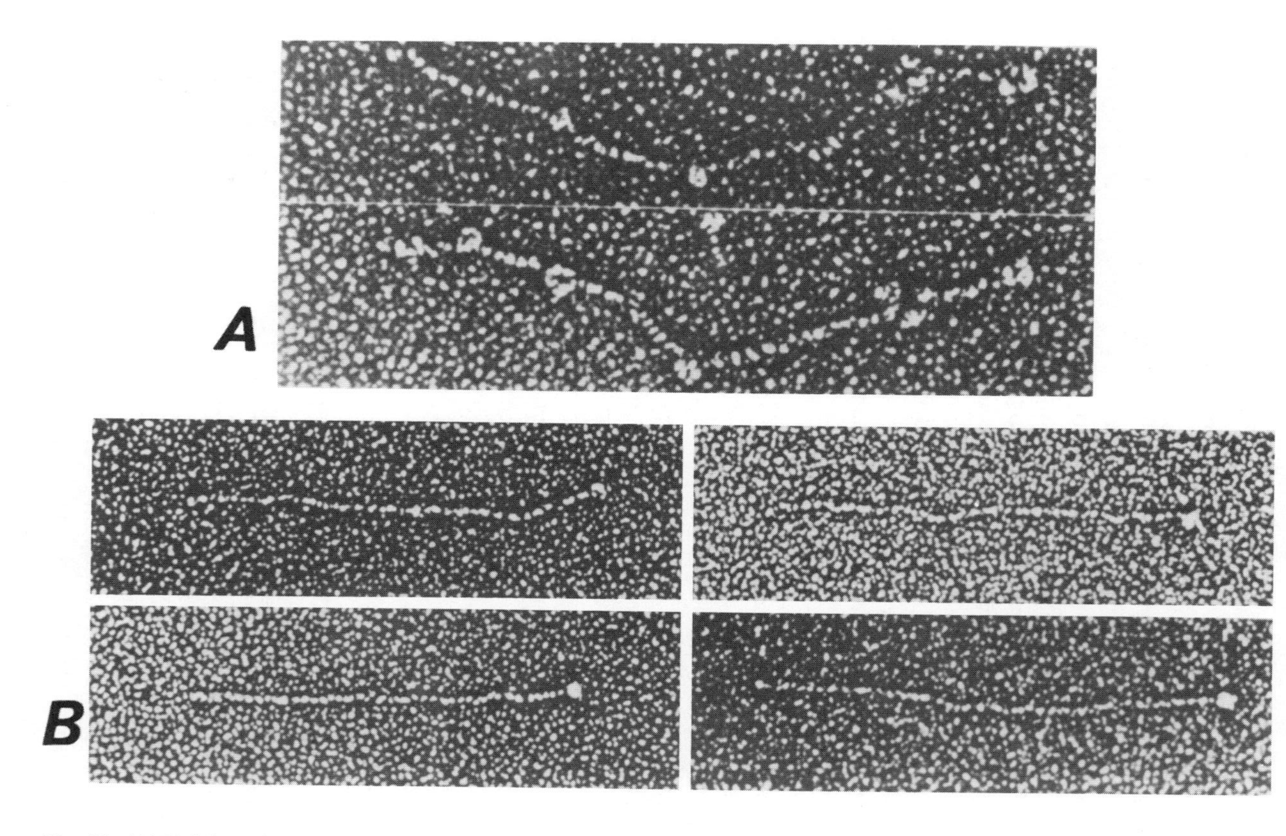

Fig. 11. (A) Epiglycanin molecules complexed with *Ricinus communis* toxin and rotary shadow cast with platinum. $\times$130,000. (B) Epiglycanin molecules complexed with concanavalin A. $\times$90,000.

```
                                                                       RELATIVE MOLES
                                                                            (%)
                                                 α-D-GalNAc-Ser (Thr)       19
                                  β-D-Gal-(1→3)-α-D-GalNAc-Ser (Thr)        60
                               (2 Gal, 1 GlcNAc)-α-D-GalNAc-Ser (Thr)        5
                  α-NeuNAc-(2→3)-β-D-Gal-(1→3)-α-D-GalNAc-Ser (Thr)        13
β-D-Gal-(1→4)-β-D-GlcAc-(1→3)-β-D-Gal-(1→3)-α-D-GalNAc-Ser (Thr)            3
     ↑                                          ↑
   (2→3)                                      (2→6)
     |                                          |
α-NeuNAc                                   α-NeuNAc
```

Fig. 12. Proposed structures for *O*-glycosyl-linked carbohydrate chains in epiglycanin.

hydride (0.3 M) for 5 days at 24°C resulted in the loss of 73% of the serin and threonine residues (570 residues) (Codington *et al.*, 1975b), with the appearance of 2-aminobutyric acid and an increased proportion of alanine (Codington *et al.*, 1975b).

With a higher concentration of sodium borohydride (1.0 M) at 37°C for 72 hr, 75–85% of the *O*-glycosyl chains were released (Van den Eijnden *et al.*, 1979). Fractionation by ion-exchange and gel-filtration chromatography separated the mixture into five reduced oligosaccharides. The chemical composition of each chain was determined by gas–liquid chromatography; and the fine structures of four of the chains (Fig. 12) were elucidated by a combination of procedures, including use of specific glycosyl hydrolases, chemical reactions, which included periodate oxidations, mass spectroscopy of the permethylated derivatives (Van den Eijnden *et al.*, 1979), and nuclear magnetic resonance (Codington *et al.*, 1979c). The details of the work performed in the elucidation of these structures are described in several papers (Codington *et al.*, 1975b, 1978b, 1979c) Van den Eijnden *et al.*, 1979).

J. *N*-Glycosyl Chains

Mannose constitutes 0.3–0.4% of epiglycanin (Codington, 1975; Codington *et al.*, 1975b) suggesting that in a molecule of molecular weight 500,000 there may be 9–12 mannose residues. Depending upon the proportion of mannose present (i.e., whether they are of the complex or high mannose type), this would represent from one to four *N*-glycosyl (asparagine)-linked chains. Positive evidence for the existence of at least one asparagine-linked chain was obtained by agglutination inhibition studies (Table XI). Epiglycanin glycopeptides exhibited inhibition of the agglutination by concanavalin A of both rabbit erythrocytes and TA3-St cells (Matsumoto *et al.*, 1980). These experiments could not be performed with the TA3-Ha ascites cells, since they could not be agglutinated by concanavalin A (Matsumoto *et al.*, 1980; Friberg, 1972a). The possibility that both the presence of mannose and the binding to concanavalin A was due to a contaminating agent and not to epiglycanin itself was disproved in three ways: (1) the binding of epiglycanin to concanavalin A-Sepharose columns and the specific elution with methyl α-D-mannopyranoside (Slayter and Codington, 1981); (2) the demonstration that material containing a [2-^{3}H]mannose label could be bound to the anti-epiglycanin antibody (Frim, Codington, Herscovics, and Jeanloz, in preparation); and (3) the visualization by electron microscopy of concanavalin A–epiglycanin complexes (Slayter and Codington, 1981).

Approximately 50% of epiglycanin (Fraction II, Fig. 4) shed from TA3-Ha ascites cells (Codington *et al.*, 1979a) was bound to a concanavalin A affinity column and specifically eluted with methyl α-D-mannoside (Slayter and Codington, 1981). The amount of mannose in the bound and unbound fractions was approximately the same, suggesting that there was heterogeneity in the structures of these chains. It has been shown (Narasimhan *et al.*, 1979) that even small changes in the structure can affect binding affinity to concanavalin A.

The presence of a high-mannose type carbohydrate chain in epiglycanin was established by reaction with endo-*N*-acetylglucosaminidase H (Frim, Codington, Herscovics, and Jeanloz, in preparation). Purified glycopeptide fractions obtained after extensive reaction of (2-[^{3}H]-mannose)-epiglycanin with pronase were incubated with endo-*N*-acetylglucosaminidase H (Tarentino and Maley, 1974). Fractionation of the product on a column of Bio-Gel P-2 produced a peak with an apparent molecular weight lower than the glycopeptides. Fractionation of the new peak by high-pressure liquid chromatography gave three peaks which possessed the same retention times as standard oligosaccharides of 7, 8, and 9 mannose residues.

VIII. BIOSYNTHETIC STUDIES RELATED TO EPIGLYCANIN

The rates of incorporation of radiolabeled *N*-acetylglucosamine into cell-surface glycoproteins of neuraminidase-treated or untreated TA3-Ha ascites cells in culture were found to be similar (Hughes *et al.*, 1972). After removal of proteins from the surfaces of intact cells with trypsin, the trichloroacetic acid-soluble fractions (probably containing epiglycanin) were studied. It was calculated that half of the maximum amount of label ($t_{1/2}$) was incorporated in 2–3 hr. Miller *et al.* (1978) found a more rapid rate of epiglycanin synthesis ($t_{1/2} = 0.75$ hr) in TA3-Ha cells grown in culture. A study of the incorporation of radiolabeled mannose into epiglycanin (Frim, Codington, Herscovics, and Jeanloz, in preparation) seemed to point to complete turnover of epiglycanin at the cell surface in 5–10 hr. Miller and Cooper (1978) reported the rate of shedding of labeled epiglycanin into the culture medium.

Investigation of sialyltransferase activity (in TA3-Ha ascites cells in culture) for several exogenous substrates (Miller, 1975; Miller and Jeanloz, in preparation) showed that the highest activity could be attributed to a lactosamine terminal structure in asparagine-linked chains of several glycoproteins after removal of sialic acid. Although approximately 60% of the sialic acid of the TA3-Ha cell surface is bound to epiglycanin (Codington *et al.*, 1975b), asialoepiglycanin was found to be a poor substrate in these experiments. Although no lactosamine structures in asparagine-linked chains are known to exist in epiglycanin, more than 20 such structures per molecule are present in *O*-glycosyl-linked chains (Section VII.G; Fig. 12).

An investigation of the involvement of lipid-linked intermediates in the synthesis of *N*-glycosyl-linked oligosaccharides in the TA3-Ha ascites cell (Frim, 1981) employed [2-^{3}H]mannose as a precursor. Isolated intermediates comigrated on silica gel plates with standards of dolichol–phospate–mannose and dolichol–pyrophosphate–$GlcNAc_2$–Man_9Glc_3, suggesting the presence of these intermediates in the synthesis of N-linked carbohydrate chains in the TA3-Ha cell. Although the presence of at least one asparagine-linked chain in epiglycanin has been demonstrated (Section VII.J), no evidence has been presented to implicate lipid-linked intermediates in epiglycanin biosynthesis.

IX. CONCLUSIONS

The most probable mechanism for the TA3 ascites tumors' escape from the immune defenses of allogeneic hosts appears to be antigen masking by membrane-bound epiglycanin molecules. Although other suggested mechanisms, namely, deactivation by blocking of immune lymphocytes due to epiglycanin–anti-epiglycanin antibody complexes (Cooper *et al.*, 1974) and activated lymphocyte depletion by attack on cell membrane decay products (Nowotny *et al.*, 1974), appear to be plausible and may possibly be involved, there is no reported direct evidence to support them. On the other hand, the presence of a high concentration of epiglycanin at the surface of the TA3-Ha ascites cell is supported by physical, chemical, and immunochemical data, and a masking role for this material is suggested by the manyfold increase in the amount of anti-H-2^a antibody absorbed after disruption of the cells by lyophilization (Sanford *et al.*, 1973) or sonication (Friberg and Lillihöök, 1973).

It has been suggested (Hagmar and Ryd, 1977) that the failure of TA3-Ha ascites cells to grow progressively as subcutaneous tumors in mice of foreign strains argued against the validity of the TA3 tumor system as a model for human cancer studies. Cooper *et al.* (1977) attributed this inability to grow as a solid tumor in allogeneic hosts to the absence of epiglycanin at the cell surface. These workers (Cooper *et al.*, 1977) found that by repeated subcutaneous passage of the same TA3-Ha tumor in syngeneic hosts the concentration of cell-surface epiglycanin increased progressively with each passage. The resulting subcutaneous tumor, in solid form, was transplantable as a solid tumor in allogeneic hosts. The importance of this finding in relation to the stepwise, long-term development of immune resistance in most human cancers needs no further explanation.

To our knowledge, no evidence to support antigen masking in human cancer cells has been reported. Nevertheless, the occurrence of cell-surface glycoproteins with *O*-glycosyl-linked carbohydrate chains similar to those found in epiglycanin has been described for human mammary carcinoma (Springer *et al.*, 1979; Bray *et al.*, 1981). The occurrence of antigen masking has been suggested, however, for other experimental tumors. Evidence to support masking of cell-surface antigens in an IgA-synthesizing plasmacytoma cell of the BALB/c mouse was based upon decreased antibody absorption (Ohno *et al.*, 1975). The use of proteases caused a marked increase in the absorption of antibody. Many analogies to the TA3 tumor system have been found in the 13762 rat mammary carcinoma ascites cell (Carraway *et al.*, 1978; Buck *et al.*, 1975a). Cells (MAT-C1) that are transplantable across species histocompatibility barriers possessed high concentrations of large endogenous sialic acid-containing glycoprotein molecules, similar in many respects to epiglycanin (Sherblom *et al.*, 1980). Cells (MAT-B1) that are nonxenotransplantable also contained a large glycoprotein, but this material differed in carbohydrate composition and contained much less sialic acid than the glycoprotein from the MAT-C1 line.

Attempts to discover a physicochemical basis for a biological phenomenon, such as allotransplantability, may face a situation in which no simple solution will suffice. In our experience, the change from strain specificity to strain nonspecificity has been accompanied by more than one physicochemical transformation. True, epiglycanin has appeared dramatically, and the unique character of its physicochemical structure, as well as its unusually high concentration, leave little doubt that it must play a major role at the cell surface. The topographies of the allotransplantable cells are also different from those of the nonallotransplantable cells (Miller *et al.*, 1977). Furthermore, the composition of sialic acid (Codington *et al.*, 1979b), as well as its concentration (Codington *et al.*, 1979b) is altered. The chemical compositions of other isolated glycopeptide fractions from the cell surface appear to be qualitatively different (Codington *et al.*, 1973). The karyotype of each cell line appears to be distinct (Friberg, 1972a; Cooper *et al.*, 1979). The significance to the cancer problem of these various changes is by no means obvious. Can it be that a tumor cell, such as the TA3-Ha or TA3-MM cell, in order to survive the synthesis of a new glycoprotein (epiglycanin) in concentration at least tenfold greater than any of its previously existing glycoproteins, must compensate by developing a new and different topography and perhaps the discontinuation of the synthesis of some surface components? Perhaps the initiation or acceleration of the synthesis of glycoproteins (or perhaps glycolipids) previously absent or present in low concentration is an example of a response of this type.

Many functions have been attributed to glycoproteins bound to the plasma membranes of tumor cells. One of these, masking of cell-surface antigens, has received relatively little attention. Is this because the occurrences of cell-surface glycoproteins, such as epiglycanin or the equally large glycoprotein at the surface of the MAT-C1 rat mammary carcinoma ascites cell, are rare? We would like to suggest that antigen masking may indeed be a common occurrence but only in those tumors which have developed a propensity to defend themselves against host immune factors, such as tumors developed after long growth periods or metastatic cancers. A glycoprotein with the physicochemical properties of epiglycanin may perhaps occur only in the mouse, however, and analogous cell-surface macromolecules which perform similar functions in other species may exhibit different physicochemical properties.

ACKNOWLEDGMENTS. The authors would like to thank all of those in our laboratory who have contributed their technical services during portions, some large and some small, of this investigation: Jane Caldwell, Debra M. Darby, Philippe Douyon, Ann M. Heos, Marianne Jahnke, Lorna A. Lampert, Keyes B. Linsley, Mary D. Maxfield, Mary McDonough, William Rea, Cyla Silber, Linda Sullivan, and John Wells. We are grateful to Drs. E. D. Hay, S. C. Miller, and H. S. Slayter for the use of electron micrographs. We thank Dr. Roger W. Jeanloz for his years of collaboration and for his encouragement and guidance. We gratefully acknowledge the financial support to this laboratory from the National Cancer Institute, U. S. Public Health Service, which made this work possible: CA08418 (to R. W. J. and to J. F. C.) and CA18600 (to J. F. C.).

X. REFERENCES

Bhavanandan, V. P., and Katlic, A. W., 1979, The interaction of wheat germ agglutinin with sialoglycoproteins, *J. Biol. Chem.* **254:**4000.

Bhavanandan, V. P., Umemoto, J., and Davidson, E. A., 1976, Characterization of an endo-α-N-acetylgalactosaminidase from *Diplococcus pneumoniae, Biochem. Biophys. Res. Commun.* **70:**738.

Bray, J., Lemieux, R. U., and McPherson, T. A., 1981, Use of a synthetic hapten in the demonstration of the Thomsen-Friedenreich (T) antigen on neuraminidase-treated human red blood cells and lymphocytes, *J. Immunol.* **126:**1966.

Buck, R. L., Sherblom, A. P., and Carraway, K. L., 1979, Sialoglycoprotein differences between xenotransplantable and nonxenotransplantable ascites sublines of the 13762 rat mammary adenocarcinoma, *Arch. Biochem. Biophys.* **198:**12.

Carraway, K. L., Huggins, J. W., Sherblom, A. P., Chestnut, R. W., Buck, R. L., Howard, S. P., Ownby, C. L., and Carraway, C. A. C., 1978, Membrane glycoproteins of rat mammary gland and its metastasizing and nonmetastasizing tumors, *in* "Glycoproteins and Glycolipids in Disease Processes" (E. G. Walborg, Jr., ed.), pp. 432–445, Amer. Chem. Soc., Washington, D. C.

Codington, J. F., 1975, Masking of cell-surface antigens on cancer cells, *in* "Cellular Membranes and Tumor Cell Behavior," 28th Annual Symposium, M. D. Anderson Hospital and Tumor Institute, pp. 399–419, Williams & Wilkins, Baltimore.

Codington, J. F., 1978, Masking of cell-surface antigens by ectoglycoproteins, *in* "Glycoproteins and Glycolipids Disease Processes" (E. F. Walborg, Jr., ed.), pp. 277–294, Amer. Chem. Soc., Washington, D. C.

Codington, J. F., 1981, The masking of cancer cell surface antigens. *Handb. Cancer Immunol.* **8:**171.

Codington, J. F., and Jeanloz, R. W., 1974, The chemistry of glycoproteins at the surfaces of tumor cells, *in* "Connective Tissues, Biochemistry and Pathophysiology" (R. Fricke, ed.), pp. 56–61, Springer-Verlag, Heidelberg.

Codington, J. F., Sanford, B. H., and Jeanloz, R. W., 1970, Glycoprotein coat of TA3 cell. I. Removal of carbohydrate and protein material from viable cells, *J. Natl. Cancer Inst.* **45:**637.

Codington, J. F., Sanford, B. H., and Jeanloz, R. W., 1972, Glycoprotein coat of the TA3 cell. Isolation and partial characterization of a sialic acid-containing glycoprotein fraction, *Biochemistry* **11:**2559.

Codington, J. F., Sanford, B. H., and Jeanloz, R. W., 1973, Cell-surface glycoproteins of two sublines of the TA3 tumor. *J. Natl. Cancer Inst.* **51:**585.

Codington, J. F., Tuttle, B., and Jeanloz, R. W., 1974, Methods for the removal, isolation, and characterization of glycoprotein fragments from the TA3 tumor cell surface, *in* "Methodologie de la Structure et du Metabolisme de Glycoconjugues," Colloques Internationaux du CNRS, No. 221, pp. 793–801, Paris.

Codington, J. F., Cooper, A. G., Brown, M. C., and Jeanloz, R. W., 1975a, Evidence that the major cell surface glycoprotein of the TA3-Ha carcinoma contains the *Vicia graminea* receptor sites. *Biochemistry* **14:**855.

Codington, J. F., Linsley, K. B., Jeanloz, R. W., Irimura, T., and Osawa, T., 1975b, Immunochemical and chemical investigations of the structure of glycoprotein fragments obtained from epiglycanin, a glycoprotein at the surface of the TA3-Ha cancer cell. *Carbohydr. Res.* **40:**171.

Codington, J. F., Klein, G., Cooper, A. G., Lee, N., Brown, M. C., and Jeanloz, R. W., 1978a, Further studies on the relationship between large glycoprotein molecules and allotransplantability in the TA3 tumor: Studies on segregating TA3-Ha/Hybrids, *J. Natl. Cancer Inst.* **60:**811.

Codington, J. F., Van den Eijnden, D. H., and Jeanloz, R. W., 1978b, Structural studies on the major glycoprotein of the TA3-Ha ascites tumor cell, *in* "Cell Surface Carbohydrate Chemistry" (R. E. Harmon, ed.), pp. 49–66, Academic Press, New York.

Codington, J. F., Cooper, A. G., Miller, D. K., Slayter, H. S., Silber, C., and Jeanloz, R. W., 1979a, Isolation and partial characterization of an epiglycanin-like glycoprotein from a new nonstrain-specific TA3 subline. *J. Natl. Cancer Inst.* **63:**153.

Codington, J. F., Klein, G., Silber, C., Linsley, K. B., and Jeanloz, R. W., 1979b, Variations in the sialic acid compositions in glycoproteins of mouse ascites tumor cell surfaces. *Biochemistry* **18:**2145.

Codington, J. F., Klein, G., Cooper, A. G., Lee, N., Brown, M. C., and Jeanloz, R. W., 1979c, Relationship between allotransplantability and cell-surface glycoproteins in TA3 ascites mammary carcinoma cells, *in* "Glycoconjugate Research" (J. D. Gregory and R. W. Jeanloz, eds.), pp. 517–519, Academic Press, New York.

Codington, J. F., Yamazaki, T., Van den Eijnden, D. H., Evans, N. A., and Jeanloz, R. W., 1979, Unequivocal evidence for a β-D-configuration of the galactose residue in the disaccharide chain of epiglycanin, the major glycoprotein of the TA3-Ha tumor cell. *FEBS Lett.* **99:**70–72.

Codington, J. F., Bhavanandan, V. P., Silber, C., Jeanloz, R. W., and Bloch, K. J., 1981, The characteristics of an antibody to epiglycanin, a glycoprotein of the TA3-Ha mouse tumor cell, *in* "Glycoconjugates" (T. Yamakawa, T. Osawa, and S. Handa, eds.), pp. 310–311, Japan Scientific Societies Press, Tokyo.

Cook, G. M. W., and Stoddart, R. W., 1973, "Surface Carbohydrates of the Eukaryotic Cell," Academic Press, New York.

Cooper, A., Morgello, S., Miller, D., and Brown, M., 1977, Role of surface glycoproteins in tumor growth, *in* "Cancer Invasion and Metastasis: Biologic Mechanisms and Therapy" (S. B. Day *et al.*, eds.) pp. 49–64, Raven Press, New York.

Cooper, A. G., Codington, J. F., and Brown, M. C., 1974, *In vivo* release of glycoprotein I from the Ha subline of TA3 murine tumor into ascites fluid and serum, *Proc. Natl. Acad. Sci. USA* **71:**1224.

Cooper, A. G., Codington, J. F., Miller, D. K., and Brown, M. C., 1979, Loss of strain specificity of the TA3-St subline: Evidence for the role of epiglycanin in mouse allogeneic tumor growth. *J. Natl. Cancer Inst.* **63:**163.

Ferguson, R. M., Schmidtke, J. R., and Simmors, R. L., 1978, Immunotherapy of experimental animals, *in* "Mechanisms of Tumor Immunity" (I. Green, S. Cohen, and R. T. McCluskey, eds.), pp. 193–214, Wiley, New York.

Friberg, S., Jr., 1972a, Comparison of an immunoresistant and an immunosusceptible ascites subline from murine tumor TA3. I. Transplantability, morphology, and some physicochemical characteristics. *J. Natl. Cancer Inst.* **48:**1463.

Friberg, S., Jr., 1972b, Comparison of an immunoresistant and an immunosusceptible ascites subline from murine tumor TA3. Immunosensitivity and antibody-binding capacity *in vitro*, and immunogenicity in allogeneic mice. *J. Natl. Cancer Inst.* **48:**1477.

Friberg, S., Jr., and Lilliehöök, B., 1973, Evidence for nonexposed H-2 antigens in immunoresistant murine tumor. *Nature New Biol.* **241:**112.

Friberg, S., Jr., Klein, G., and Wiener, F., 1973, Hybrid cells derived from fusion of TA3-Ha ascites carcinoma with normal fibroblasts. II. Characterization of isoantigenic variant sublines, *J. Natl. Cancer Inst.* **50:**1269.

Friberg, S., Jr., Molnar, J., and Pardoe, G. I., 1974, Tumor cell-surface organization: Differences between two TA3 sublines, *J. Natl. Cancer Inst.* **52:**85.

Frim, D. M., 1981, Characterization and Biosynthesis of a Large Cell Surface Glycoprotein, Thesis, Harvard University.

Garg, H. G., Swann, D. A., and Glasgow, L. R., 1980, The structure of the O-glycosidically linked chains of LPG-I glycoprotein present in articular lubricating fraction of bovine synovial fluids, *Carbohydr. Res.* **78:**79.

Gorer, P. A., 1937, The genetic and antigenetic basis of tumor transplantation, *J. Pathol. Bacteriol.* **44:**691.

Gorer, P. A., 1948, The significance of studies with transplanted tumors, *Br. J. Cancer* **2:**103.

Grohsman, J., and Nowotny, A., 1972, The immune recognition of TA3 tumors, its facilitation by endotoxin, and abrogation by ascites fluid, *J. Immunol.* **109:**1090.

Hagmar, G., and Ryd, W., 1977, Site dependency of TA3-Ha allotransplantability. *Transplantation* **23:**93.

Hauschka, T. S., Weiss, L., Holdridge, B. A., Cudney, T. L., Zumpft, M., and Planinsek, J. A., 1971, Karotypic and surface features of murine TA3 carcinoma cells during immunoselection in mice and rats. *J. Natl. Cancer Inst.* **47:**343.

Hellström, K. E., and Hellström, I., 1974, Lymphocyte-mediated cytotoxicity and blocking serum activity to tumor antigens, *Adv. Immunol.* **18:**209.

Hellström, K. E., and Hellström, I., 1977, Immunologic enhancement of tumor growth, *in* "Mechanisms of Tumor Immunity" (I. Green, S. Cohen, and R. T. McCluskey, eds.), pp. 147–174, Wiley, New York.

Hughes, R. C., Sanford, B., and Jeanloz, R. W., 1972, Regeneration of the surface glycoproteins of a transplantable mouse tumor cell after treatment with neuraminidase, *Proc. Natl. Acad. Sci. USA* **69:**942.

Hughes, R. C., Palmer, P. D., and Sanford, B. H., 1973, Factors involved in the cytotoxicity of normal guinea pig serum for cells of murine tumor TA3 sublines treated with neuraminidase. *J. Immunol.* **111:**1071.

Jeanloz, R. W., and Codington, J. F., 1973, Les glycoprotéines de surface des cellules cancéreuses TA3, *in* "Exposés Annuels de Biochimie Médicale" (P. Louisot and J. Polonovski, eds.), pp. 49–58, Masson, Paris.

Jeanloz, R. W., and Codington, J. F., 1974, Glycoproteins at the cell surface of sublines of the TA3 tumor, *in* "Biology and Chemistry of Eucaryotic Cell Surfaces" (Y. C. Lee and E. E. Smith, eds.), pp. 241–257, Academic Press, New York.

Jeanloz, R. W., and Codington, J. F., 1976, The biological role of sialic acid at the surface of the cell, *in* "Biological Roles of Sialic Acid" (A. Rosenberg and C. L. Schengrund, eds.), pp. 201–238, Plenum Press, New York.

Klein, G., 1951, Comparative studies of mouse tumors with respect to their capacity for growth as "ascites tumors" and their average nucleic acid content per cell, *Exp. Cell Res.* **2:**518.

Klein, G., Friberg, S., Jr., and Harris, H., 1972, Two kinds of antigen expression in tumor cells revealed by cell fusion, *J. Exp. Med.* **135:**839.

Klein, G., Friberg, S., Jr., and Wiener, F., 1973, Hybrid cells derived from fusion of TA3-Ha ascites carcinoma with normal fibroblasts. I. Malignancy, karyotype, and formation of isoantigenic variants. *J. Natl. Cancer Inst.* **50:**1259.

Lippmann, M. M., Venditti, J. M., Kline, I., and Elam, D. L., 1973, Immunity to a TA3 tumor subline that grows in allogeneic hosts elicited by strain-specific TA3 tumor cells, *Cancer Res.* **33**:679.

Lisowska, E., and Wasniowska, K., 1978, Immunochemical characterization of cyanogen bromide degradation products of M and N blood-group glycopeptides, *Eur. J. Biochem.* **88**:247.

Matsumoto, I., Codington, J. F., Jahnke, M. R., Jeanloz, R. W., and Osawa, T., 1980, Comparative lectin binding and agglutination properties of the strain-specific TA3-Ha murine mammary carcinoma ascites sublines. Further studies of receptors in epiglycanin. *Carbohydr. Res.* **80**:179.

Miller, D. K., 1975, Sialyltransferase Activity in the TA3 Murine Mammary Adenocarcinoma, Thesis, Harvard University.

Miller, D. K., and Cooper, A. G., 1978, Kinetics of release of glycosamine-labeled glycoproteins from the TA3-Ha murine adenocarcinoma cell, *J. Biol. Chem.* **253**:8798.

Miller, D. K., Cooper, A. G., and Brown, M. C., 1978, Cellular origin of glucosamine-labeled glycoproteins released from the TA3-Ha tumor cell, *J. Biol. Chem.* **253**:8804.

Miller, S. C., Hay, E. D., and Codington, J. F., 1977, Ultrastructural and histochemical differences in cell surface properties of strain-specific and nonstrain-specific TA3 adenocarcinoma cells, *J. Cell Biol.* **72**:511.

Miller, S. C., Codington, J. F., and Klein, G., 1982, Further studies on the relationship between allotransplantability and the presence of the cell-surface glycoprotein epiglycanin in the TA3-MM mouse mammary carcinoma ascites cell, *JNCI* **68**:403.

Narasimhan, S., Wilson, J. R., Martin, E., and Schachter, H., 1979, A structural basis for four distinct elution profiles on concanavalin A-Sepharose affinity chromatography of glycopeptides, *Can. J. Biochem.* **57**:83.

Nowotny, A., Grohsman, J., Abdelnoor, A., Rote, N., Yang, C., and Waltersdorff, R., 1974, Escape of TA3 tumors from allogeneic immune rejection: Theory and experiments, *Eur. J. Immunol.* **4**:73.

Ohno, S., Natsu-ume, S., and Migita, S., 1975, Alteration of cell-surface antigenicity of the mouse plasmacytoma. I. Immunologic characterization of surface antigens masked during successive transplantations, *J. Natl. Cancer Inst.* **55**:569.

Osuga, D. R., and Feeney, R. E., 1978, Antifreeze glycoproteins from antarctic fish, *J. Biol. Chem.* **253**:5338.

Pauling, L., Corey, R. B., and Branson, H. R., 1951, The structure of proteins: Two hydrogen-bonded helical configurations of the polypeptide chain, *Proc. Natl. Acad. Sci. USA* **37**:205.

Purdom, L., Ambrose, E. J., and Klein, G., 1958, A correlation between electrical surface charge and some biological characteristics during the stepwise progression of a mouse sarcoma, *Nature (London)* **181**:1586.

Ray, P. K., 1977, Bacterial neuraminidase and altered immunological behavior of treated mammalian cells. *Adv. Appl. Microbiol.* **21**:227.

Ringold, G., Lasfargues, E. Y., Bishop, J. M., and Varmus, H. E., 1975, Production of mouse mammary tumor virus by cultured cells in the absence and presence of hormones: Assays by molecular hybridization, *Virology* **65**:135.

Rosenberg, S. A., and Rogentine, G. N., Jr., 1972, Natural human antibodies to hidden membrane components. *Nature New Biol.* **239**:203.

Rosenberg, S. A., and Schwarz, S., 1974, Murine auto-antibodies to a cryptic membrane antigen: Possible explanation for neuraminidase-induced increase in cell immunogenicity. *J. Natl. Cancer Inst.* **52**:1151.

Sanford, B. H., 1967, An alteration in tumor histocompatibility induced by neuraminidase, *Transplantation* **5**:1273.

Sanford, B. H., and Codington, J. F., 1971, Further studies on the effect of neuraminidase on tumor cell transplantability. *Tissue Antigens* **1:**153.

Sanford, B. H., Codington, J. F., Jeanloz, R. W., and Palmer, P. D., 1973, Transplantability and antigenicity of two sublines of the TA3 tumor, *J. Immunol.* **110:**1233.

Sedlacek, H. H., Seiler, F. R., and Schwick, H. G., 1977, Neuraminidase and tumor immunotherapy. *Klin. Wochenschr.* **55:**199.

Sheffield, J. B., Daly, T., Dion, A. S., and Tarashi, N., 1977, Procedures for radioimmunoassay of the mouse mammary tumor virus, *Cancer Res.* **37:**1480.

Sherblom, A. P., Huggins, J. W., Chesnut, R. W., Buck, R. L., Ownby, C.L., Dermer, G. B., and Carraway, K. L., 1980, Cell surface properties of ascites sublines of the 13762 rat mammary adenocarcinoma, *Expt. Cell Res.* **126:**417.

Simmons, R. L., and Rios, A., 1973, Differential effect of neuraminidase on the immunogenicity of viral associated and private antigens of mammary carcinomas, *J. Immunol.* **111:**1820.

Slayter, H. S., and Codington, J. F., 1973, Size and configuration of glycoprotein fragments cleaved from tumor cells by proteolysis, *J. Biol. Chem.* **248:**3405.

Slayter, H. S., and Codington, J. F., 1981, Distribution of receptor sites on large glycoprotein molecules by electror microscopy: Application to epiglycanin. *Biochem. J.* **193:**203.

Springer, G. F., Codington, J. F. and Jeanloz, R. W., 1972, Surface glycoprotein from a mouse tumor cell as specific inhibitor of antihuman blood-group N agglutinin, *J. Natl. Cancer Inst.* **49:**1469.

Springer, G. F., Desai, P. R., Murthy, M. S., Yang, H. J., and Scanlon, E. F., 1979, Precursors of the blood group MN antigens as human carcinoma-associated antigens, *Transfusion* **19:**233.

Tarentino, A.L., and Maley, F., 1974, Purification and properties of an endo-β- N- acetyl glucosaminidase from *Streptomyces griseus, J. Biol. Chem,* **249:**811.

Uhlenbruck, G., and Dahr, W., 1971, Studies on lectins with a broad agglutination spectrum, *Vox Sang.* **21:**338.

Van den Eijnden, D. H., Evans, N. A., Codington, J. F., Reinhold, V., Silber, C., and Jeanloz, R. W., 1979, Chemical structure of epiglycanin, the major glycoprotein of the TA3-Ha ascites cell. The carbohydrate chains, *J. Biol. Chem.* **254:**12,153.

Watkins, W. M., 1972, Blood-group specific substances, *in* "Glycoproteins, Their Composition, Structure and Function" (A. Gottschalk, ed.), pp. 830–891, Elsevier, New York.

Weiner, G., Klein, G., and Harris, H., 1971, The analysis of malignancy by cell fusion. III. Hybrids between diploid fibroblasts and other tumor cells, *J. Cell Sci.* **8:**681.

Yogeeswaran, G., and Salk, P., 1978, Cell surface sialyl components of metastatic variant PW20 tumor cell lines, *Fed. Proc., Fed. Am. Soc. Exp. Biol.* **37:**1299.

Yogeeswaran, G., Stein, B. S., and Sebastian, H., 1978, Altered cell surface organization of ganglioside and sialoglycoproteins of mouse metastatic melanoma variant lines selected *in vivo* for enhanced lung implantation, *Cancer Res.* **38:**1336.

Chapter 9

Simian Virus 40-Coded Antigens and the Detection of a 55K-Dalton Cellular Protein in Early Embryo Cells

Peter T. Mora and K. Chandrasekaran

National Cancer Institute
National Institutes of Health
Bethesda, Maryland

I. INTRODUCTION

The complete nucleotide sequence of the simian virus 40 (SV40) genome, consisting of 5226 base pairs (Fiers *et al.*, 1978; Reddy *et al.*, 1978), as well as the details of the transcription and the messenger RNA processing (for a review see Ziff, 1980) has been determined. In SV40-transformed cells, only the "early" half of the circular SV40 genome is transcribed (Fig. 1), producing two messenger RNA species, which synthesize the two early proteins: the ~94-kilodalton large T antigen and the ~17-kilodalton small t antigen. The messenger RNAs (mRNAs) for the SV40 T and t antigens have identical 5′ and 3′ termini but are different in their splicing (Berk and Sharp, 1978). The RNA encoding t is colinear with the DNA sequence starting with an AUG at coordinate 66 in the genetic map near the *BglI* site (which is also at the origin for the SV40 DNA synthesis in the lytic or productive infection), until reaching a terminator codon at coordinate 55, after which a downstream splice follows which is within an untranslated region up to coordinate 17 (see the legend of Fig. 1 for explanations). The messenger RNA for T is common with the t messenger RNA at the 5′ site, but a large splice starting at coordinate 59 removes the terminator at 55, after which the RNA transcription switches to a second frame which is free of terminator codons TAG, TGA, and TAA. This allows T translation to the 3′ end at its coordinate

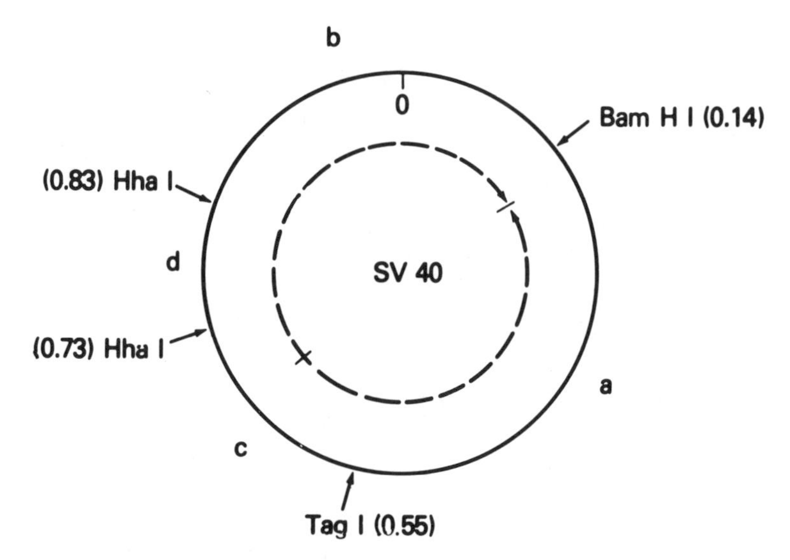

Fig. 1. A simplified map of the SV40 genome and of its transcription and the restriction endonuclease fragments studied. The zero coordinate of the genetic map is at the *Eco*RI site, then by convention the map proceeds clockwise 0–100 (or 0–1). The origin of DNA replication in lytic infection and also of the RNA transcription (in both lytic infection and in the non-producing infections when cells are transformed by SV40) is at a site at approximately coordinate 66 indicated on the inner circle. From there both the DNA replication and the transcription (the dot in the inner circle) proceeds first counterclockwise to coordinate 17; and this area is designated as the early half of the SV40 genome. The DNA fragments we obtained and studied by restriction endonucleases are indicated by the letters *a–d*; their presence or absence in one transformed cell and in several derivative T antigen-negative tumor cell lines and clones are discussed in Section II.A. The early half of the SV40 DNA has only one of the nucleotide triplet readout frame open for transcription at any given region, and this defines a unique amino acid sequence for the early antigens: T and t (for details, see Introduction). This fact, and comparative peptide studies between the SV40-encoded large T antigen and of a 55K protein we have observed to precipitate with the SV40 tumor serum (see text), led to the conclusion that the 55K protein is of cellular origin (Section III).

17, where it is terminated by the sequence TAA. Thus the t and T proteins share amino-terminal sequences (Paucha *et al.*, 1978) but are different in size and in the carboxy termini. The third nucleotide frame of the SV40 DNA (just as the first frame) is rich in terminating codons from 59 to 17. Thus, the splicing and processing of the RNA is versatile, but only the synthesized t and T antigens can be coded by the SV40 as substantial size proteins. These t and T antigens are called the "early" proteins, and are the only SV40-coded proteins which were observed in the SV40-transformed nonproductive cells. Normally at least one whole SV40 genome is inserted in the cellular genome of SV40-transformed cell (below we will discuss exceptions which are rare revertants). The late or second half of the SV40 is transcribed and translated only in the virus productive or lytic infections, which occur in the natural host cells (monkey) and also frequently in the cells of a related species (human). The late proteins are virion structural proteins.

The transformation or nonproductive infection with SV40 virus or SV40 DNA is a laboratory artifact of tissue-cultured cells from murine species such as mouse, rat, and hamster, (Mora, 1982). It is generally detected by differential cell growth in tissue culture (Martin, 1981) and by the synthesis of the above-mentioned t and T antigens. The T antigen is concentrated in the nucleus: one of the simplest techniques is to stain the nucleus with fluorescent-labeled antibody to the SV40 T antigen. Sera of hamsters bearing SV40-induced tumors (SV40 tumor or T serum) contain such antibodies.

Most of the laboratory infection by SV40 virus, or transfection by SV40 DNA, is generally carried out on established murine fibroblast cell lines or derivative clones. Such established murine cell lines or clones both before and after the transformation by SV40 are repeatedly passaged in tissue culture. Murine cells, when established in culture and cultivated, often become tumorigenic "spontaneously," i.e., without SV40 virus infection or without transfection with SV40 DNA. Some conditions, such as crowded cultivation of cells, O_2 tension, and exposure to light, may increase the frequency of the spontaneous transformation to tumorigenicity, but as the term *spontaneous* indicates, the cause and mechanism of such transformation is not known. The spontaneous transformation to tumorigenicity may well be the consequence of somatic mutation, the frequency of which can vary from cell to cell. We have observed that a certain AL/N strain mouse cell line is especially apt to become tumorigenic even after relatively short cultivation involving only ~40 tissue culture transfers (Mora *et al.*, 1969). The high tumorigenicity thus attained becomes then a heritable property, which was retained in cells cloned by single-cell technique (Mora *et al.*, 1977a; Winterbourne and Mora, 1978). Other AL/N mouse cell clones can remain relatively nontumorigenic even after prolonged cultivation (Winterbourne and Mora, 1978). Cells from other strains of mice (cf. Balb/c) also can become tumorigenic during prolonged cultivation.

First we shall review our work on the balance between spontaneously arising cellular tumorigenicity and of the apparently opposing effects of the SV40 T antigen—expressed on the cell surface—after the spontaneously transformed highly tumorigenic mouse cells were further transformed by SV40. We shall also discuss properties of essentially normal nontumorigenic mouse cells when infected and transformed by SV40. Then we shall review how our work led to the discovery of an ~55K-dalton SV40-induced cellular protein which is not encoded by the SV40 genome, and how we found that this protein is also expressed (constitutively) in embryo cells not infected with SV40.

II. THE TWO MAIN BIOLOGICAL EFFECTS OF THE PRODUCTS OF THE EARLY GENE OF SV40

As already mentioned, murine cells, such as of mouse, rat, and hamster, are commonly used to study the effects of transformation by SV40 on the cell growth properties in tissue culture. Such cell growth properties generally are considered as following a hierarchy of "transformed" phenotypes: the ability to grow in low serum, to reach high saturation density on plastic, to overgrow monolayers and to form foci, to grow in suspension (without "anchorage") in viscous medium, and finally, as generally assumed, to form tumors, These are thought to be the more "stringent" criteria of transformation, meaning that the latter abilities are expressed in the more "fully" transformed cells and then can be detected only in a subset of cells exhibiting the former abilities. The SV40 early gene expression (the T antigen synthesis), when correlated with a change in a biologic phenotype, for example, with the ability of cells to form dense foci on plastic, is thought to be coupled biochemically with an increase in cellular DNA synthesis, part of a "mitogenic" effect (Martin, 1981).

It is seldom recognized by those who are oriented toward molecular biology studies that the SV40 virus is tumorigenic only in the newborn hamster and the mastomys (the latter is an African rodent intermediate in size between the rat and the mouse). Neither the SV40 virus nor the cells transformed in culture by SV40 virus or its DNA are as a rule tumorigenic in the mouse (Black and Rowe, 1963), the rat (Alstein *et al.*, 1967), the immunocompetent adult hamser (Lewis and Cook, 1980), or, as far as it is known, any other species including the natural host of the virus: the simian monkey or other primates, such as man (cf. Mora, 1982). In the mouse, tumors have been obtained from SV40-transformed non-producer cells, but only if a considerable amount of time and a large number (>50) of tissue culture transfers (tct) followed the SV40 transformation. Even so a large number of cells (>10^6) had to be injected (Kit *et al.*, 1969; Wesslen, 1970; Tevethia and McMillan, 1974). Most of these primary tumors were temporal and were eventually rejected by the mice

(Kit *et al.*, 1969). In immunologically compromised mice, such as by antilymphocyte serum, neonatal thymectomy, or irradiation, SV40-positive primary tumors may be produced more readily than in the immunologically competent mature mice, further implying immunological rejection of the SV40-transformed cells (for current studies cf. Tevethia *et al.*, 1980; Mora, 1982).

Considering the above, two groups of related questions can be posed in regard to the tumorigenic transformation of mouse cells: (1) Is the spontaneous transformation (somatic mutation?) of a few cells the primary cause of the tumorigenic phenotype and not the SV40 early gene expression, or could it be that the two are somehow connected and may reinforce each other? (2) If the two main cellular effects of the SV40 early gene(s) (the mitogenic or growth promoting and the immunologic changes leading to recognition and rejection) indeed oppose each other, which is on balance dominant? This question should be investigated by using cells which do not undergo spontaneous transformation. We studied some aspects of these questions in two types of cell families: one starting from a highly tumorigenic spontaneously transformed AL/N mouse clonal cell, the other from a very low essentially nontumorigenic clonal cell. Both clones started from the same established mass cell line.

A. SV40 Early Gene Expression in Highly Tumorigenic Spontaneously Transformed Mouse Cells

We observed that an AL/N strain mouse fibroblast mass cell line when transformed by SV40 (denoted as SV AL/N line) had very low tumorigenicity (the median tumorigenic dose TD_{50} was $>10^8$ cells in the immunologically competent syngeneic mouse). This low tumorigenicity prevailed after prolonged cultivation up to ~300 tissue culture transfers (Smith *et al.*, 1970). We were able to show by inference, using an indirect immunologic technique, that on the surface of the SV AL/N cells there were SV40-specific antigens, because the SV AL/N cells were killed in the presence of complement by serum from SV40 tumor-bearing mice as measured by the release of ^{51}Cr. The reaction was inhibited by incubation of the antisera with any SV40-transformed cells and with their subcellular fractions (Chang *et al.*, 1977b). We surmised (Smith *et al.*, 1970)and later proved (Chang *et al.*, 1979b; Luborsky *et al.*, 1978) that such inhibitory action was due to virus-specific transplantation antigens (SV40 TSTA), which are capable of preventing tumor formation by SV40-transformed cells. During this phase of our work the serum-mediated cytotoxic assay was developed and used as a rapid and sensitive method to detect and purify the solubilized SV40 antigens from SV40-transformed cells (Chang *et al.*, 1977a,b; Luborsky *et al.*, 1976, 1978; Pancake and Mora, 1974, 1976).

The AL/N mass cell line without the SV40 infection after relatively short cultivation (~40 tct) under routine conditions became highly tumorigenic: less than 10^2 cells/mouse ($TD_{50} < 10^2$) caused rapidly (<9 weeks) lethal fibrosarcomas. This cell line was then denoted T AL/N and was our standard "spontaneously" transformed mass cell line. The T AL/N cell or its subcellular fractions did not inhibit the SV40-specific serum-mediated cytotoxic assay and did not inhibit the take of SV40-transformed tumor cells (Smith *et al.*, 1970; Mora *et al.*, 1977a).

A clone of the T AL/N mass cell, now denoted 104 C, retained the high tumorigenicity (Mora *et al.*, 1977a). This clone 104 C was infected by SV40 and was immediately recloned: two independently transformed SV40 T antigen-positive subclones and three T antigen-negative subclones were obtained and characterized (Mora *et al.*, 1977a). Table I shows that the tumorigenicity of the 104 C was reduced about 100-fold in the SV40-transformed daughter clones but remained high in their T antigen-negative sisters, showing that the SV40 transformation and the expression of the T antigen is associated with lowered cellular tumorigenicity in these spontaneously transformed mouse fibroblasts. We have shown that the SV40-specific antigens indeed function as a virus-specific transplantation antigen (Mora *et al.*, 1977a,b; Anderson *et al.*, 1977b) by (1) immunization with the SV40 T antigen-positive cells followed by *in vivo* challenge in the AL/N syngeneic system with an SV40 T antigen-positive AL/N tumor cell line, or (2) by immunization with subcellular fractions of SV40-transformed cells from any strain of mice and challenging the Balb/c strain mouse with *its* syngeneic SV40 T antigen-positive tumor-forming Balb/c cells [an mKSA TU5 (Kit *et al.*, 1969) ascites]. The appropriate negative controls were included in these experiments. This antigenic activity can be abbreviated to a tumor-specific transplantation antigen induced by SV40 (SV40 TSTA) in a loose use of the word. Keep in mind, however, that the original tumorgenicity of these cells was not caused by SV40.

Incidentally, it is instructive to note from Table I that the highly tumorigenic spontaneously transformed 104 C clone, and what for all practical matters can be considered subclones 109, 110, and 111 CSC, had very low or no detectable growth in suspension: the anchorage independent growth is not a necessary or even a very general correlate of high tumorigenicity in spontaneous transformation of mouse fibroblasts (Mora *et al.*, 1977a). In the family of cells presented in Table I the expression of the SV40 antigens correlated with the anchorage-independent growth (for contrasting data, however, see Section II.B). *Thus, there is*

Table I
Characterization of Spontaneously and SV40-Transformed Mouse Cells[a]

Cell line	SV40 antigens			Tumorigenicity		Tissue culture growth property	
	T	S[b]	TSTA[c]	TD_{50}[d] (cells per mouse)	Time to death[e] (weeks)	Saturation density[f] (cells $cm^{-2} \times 10^5$)	Colony formation in viscous Methocell suspension[e] (%)
T AL/N mass line	–	–	–	$<10^2$	9	5.5	2
104C	–	–	–	10^2	14	3.0	0.007
SV40-infected subclone							
106CSC	+	+	+	10^4	13	1.0[g]	20
107CSC	+	+	+	10^4	14	0.6[g]	10
109CSC	–	–	–	10^2	17	2.4	0.0001
110CSC	–	–	–	$<10^1$	21	3.5	0.6
111CSC	–	–	–	10^2	16	3.1	0.03
SV AL/N mass line	+	+	+	$>10^8$	—	5.3	10

[a] Most of the data are collated from earlier publications (Mora *et al.*, 1969, 1977a; Smith *et al.*, 1970; McFarland *et al.*, 1975).
[b] Surface antigen determined by a serum-mediated cytolytic assay, measured by ^{51}Cr release.
[c] *In vivo* immunization and rejection tests for virus (tumor)-specific transplantation antigens.
[d] Tumorigenicity was tested by intramuscular (i.m.) injection of cells, at least in three dilutions each differing ten-fold, into groups of 6 to 8-week-old AL/N mice, ten mice in each group. TD_{50} is the cell dose which caused in 50% of the animals large ($\geq$1-cm diameter) and eventually lethal, tumors. It was calculated as follows: $TD = 10^{d+x}$, where d is the log number of cells which caused tumor in just below 50% of the mice, and $x = (a - 50)/(a - b)$, where a is the percentage of mice with tumors within that group of ten mice which had nearest to but above 50% tumors, and b is the percentage nearest to but below 50%. The TD_{50} values were reproducible within $\pm$ 0.5 log at 10^2 cell dose and $\pm$ 0.2 log at 10^4 cell dose.
[e] Time when substantially all the animals, injected i.m. with a cell dose rounded up to the nearest integer of the exponential of TD_{50}, die off due to progressive tumor growth. There were no regressing tumors.
[f] Determined in the presence of 10% fetal calf serum.
[g] Cells after reaching confluence tend to detach from the substratum.

no simple or general correlation between the tissue culture growth properties of the cells, the transformation by SV40, and the tumorigenicity.

After injecting 10^4 or 10^5 T antigen-positive 106 CSC cells/mouse, tumor lines were re-established in cell culture. All the cells in such derived tumor mass cell lines were found to be SV40 T antigen negative by nuclear immunofluorescence (Mora *et al.*, 1977a). On the average these tumor lines were very highly tumorigenic when injecting various doses of cells ($TD_{50} \leq 10^2$; Table II). The sublcones of a tumor line 124 CSCT were also all T antigen negative but had some fluctation in the TD_{50} values (Table II). Also note that again the highly tumorigenic cells did not grow at all—or grew only very poorly—in suspension in viscous medium. During these experiments, we noted a fortunate occurrence, which proved to us an important point, namely that the 106 CSC was a single-cell clone. Fortuitously the 106 CSC clone acquired marker chromosomes absent in the 104 C parent clone before the SV40 transformation, and all of the T antigen-negative derivative tumor lines of 106 CSC and their subclones possessed the marker chromosomes (Mora *et al.*, 1977a). Then by banding a sufficient number of the various cells for chromosomal analysis, we concluded that the T antigen-negative tumor lines must have come from revertants of the 106 CSC cells and not from possible contaminating 104

Table II
Properties of Tumor Cell Lines from 106 CSC and of Clones of Tumor Line 124CSCT

Tumor cell line	Tumorigenicity (TD_{50})	Colony formation in viscous Methocell suspension (%)
113 CSCT	NT[a]	10
124 CSCT	10^2	0.18
127 CSCT	$<10^2$	0
128 CSCT	$<10^2$	0.01
Clones of 124 CSCT		
129 CSCTC	10^2	0.01
130 CSCTC	10^2	0.15
131 CSCTC	10^3	0
132 CSCTC	$>10^4$	0
133 CSCTC	10^3	0.12
134 CSCTC	$>10^4$	0

Note. All tumor cells were SV40 T and surface antigen negative. All the clones of 124 were tested only for SV40 T antigen; they were all negative. The tumor line 124 CSCT was cloned by a single-cell technique. Tumorigenicity and colony formation in viscous Methocell were determined as in Table I. Some of these data were presented earlier (Mora *et al.*, 1977a).

[a] Not tested.

C cells. The above findings then could be explained if the expected immunologic pressures operate very efficiently in the immunologically competent syngeneic mouse against those cells which possess SV40 antigens, and only those rare revertant cells form the tumors which do not express the SV40 antigens (Mora *et al.*, 1977a). Such tumor cell lines then should show again the high tumorigenicity of the parent spontaneously transformed cell 104 C.

Further confirmation and some elucidation of the reversion mechanism came from studies on the nucleic acid level. When highly labeled whole SV40 DNA or DNA pieces (obtained by the restriction endonucleases as indicated in Fig. 1) were used in re-association kinetics with the cellular DNA of the 106 CSC, the T antigen-negative derivative tumor lines, and the derivative tumor clones, it was found that the 106 CSC had a full copy equivalent of SV40 DNA, while the tumor lines 113 and 124 CSCT had only about 0.5 copy number equivalent (Küster *et al.*, 1977). Furthermore, 106 CSC had 1.1 copy number equivalent of the fragment *a*—which encompasses the major portion of the early half of the SV40 genome—and this fragment *a* was barely detectable if at all (0.1–0.2 copy number) in all the tumor lines and derivative tumor clones. Thus, the tumor cell lines and clones were indeed depleted or devoid in the region of the early SV40 DNA, apparently by some recombination event between the cellular DNA and the inserted SV40 DNA. Only these cells were allowed to grow as tumors, which as a consequence, could not express SV40 T, surface antigen, and TSTA, as previously detected (Mora *et al.*, 1977a). That the tumor cell lines and clones were true revertants of the 106 CSC was proved by the finding of substantial amounts (0.5–3.6 copy number) of fragments *b*, *c*, and *d* corresponding to the later half of the SV40 DNA. These results were consistent with other experiments devised to give information on transcription. In the latter experiments, hydridization was between the minus (transcribed) strand of the *in vivo* ^{32}P-labeled SV40 DNA and its *Hind*II and III fragments to the total cellular RNA, and the assay was by hydroxyapatite chromatography. An essentially full complement of the early SV40 RNA was synthesized in 106 C cells, but there was no detectable SV40 RNA in the revertant cells (Küster *et al.*, 1977).

Thus, *in vivo* selection pressures indeed operate dominantly against the SV40-transformed cells which express the T antigen, when relatively low numbers of cells (10^4–10^5) are injected into the immunologically competent AL/N mouse. This allows then only those rare cells to grow as tumors which are early SV40 DNA minus and T antigen negative revertants. That the tumor cell lines recover the original high tumorigenicity (TD_{50} = 10^2) of the spontaneously transformed parent clone 104 C suggests that the spontaneously arising tumorigenicity is a nonsegregating heritable property. Incidentally, the early SV40 DNA minus but late SV40 plus revertant tumor cell lines are susceptible to re-infection with SV40, after which the tumorigenicity again decreases ($TD_{50} < 10^6$!) concomitant with the re-expression of the T antigen. Again by *in vivo* transplantation, the whole cycle can be repeated, and T antigen negative revertant tumor lines can be again obtained, again with the very high tumorigenicity ($TD_{50} = 10^2$) of the original spontaneously transformed great-grandparent 104 C cell. Thus, the cellular tumorigenicity which was acquired in a spontaneous transformation event is indeed retained through several generation and selection procedures as a highly heritable nonsegregating property (McFarland and Mora, in preparation).

Apparently the presence and expression of the early SV40 DNA mitigates against the tumorigenicity in this family of AL/N mouse cells (Mora *et al.*, 1977b). Our observation agrees with the findings of others which show in other strains of mice (Wesslen, 1970) that mouse cells give tumors only after prolonged (>50 tct) cultivation following the SV40 infection, and even then only at high cell doses (Kit *et al.*, 1969; Tevethia and McMillan, 1974). After SV40-positive tumors are obtained once in the syngeneic mouse, such tumors become more transplantable, and when established in culture the cellular tumorigenicity becomes higher, i.e., the TD_{50} becomes lower (McFarland *et al.*, 1975).

All of our results are consistent with the simplest explanation: by transplantation we apply selection pressure *for* cellular tumorigenicity and *against* the cells which are recognized and rejected because of the SV40 antigen(s). We conclude then from these studies that SV40 early gene is not involved in the tumorigenic transformation in the mouse cells which have a propensity for "spontaneous" transformation for tumorigenicity in tissue culture. On the contrary, the SV40 transformation causes these cells to be rejected by the immunologically competent syngeneic mouse.

Incidentally, our studies on the subcellular localization, isolation, and characterization of the various solubilized SV40 antigens led us into a line of collaborative research, in which it was demonstrated that the SV40 T antigen molecule (which is mainly in the nucleus), when purified to homogeneity on polyacrylamide gel electrophoresis, is the most potent SV40-specific TSTA in the mouse (Anderson *et al.*, 1977b; Chang *et al.*, 1979a). Thus, the T antigen, or a part of it, must be accessible on the (mouse) cell surface and also recognized as TSTA by whatever immunologic mechanism operates in rejection (Tevethia *et al.*, 1980; Chang *et al.*, 1982). Below we will summarize current studies on the detection of the SV40 antigens on the cell plasma membrane.

B. Effect of SV40 on Nontumorigenic Mouse Cells

When another clone (210 C) was isolated from the same AL/N mouse embryo established cell line from which the above-described highly tumorigenic clone 104 C was obtained, it was found to be very low in its tumorigenic property ($TD_{50} = 10^{6.4}$). This very low cellular tumorigenicity of 210 C remained constant after prolonged tissue culturing and also essentially was the same in the reclones. We shall then consider that this cell has very low or—if 10^6 cells can be used as a cutoff point—essentially a nontumorigenic heritable phenotype. After SV40 infection and immediate recloning of the 210 C cells, we obtained two independent T antigen-positive SV40-transformed subclones, 214 CSC and 215 CSC, and one T antigen-negative subclone, 213 CSC. This latter can be considered as a reclone of the parent 210 C cell (Winterbourne and Mora, 1978). Table III presents the summary of the growth properties of these clones in culture and *in vivo*, the latter both in the immunologically competent syngeneic mouse and an immunologically impaired (T cell deficient) nude mouse. The two SV40 T Ag-positive clones 214 and 215 CSC do not differ appreciably from the parent 210 C in tumorigenicity in the syngeneic mouse: the tumorigenicity remained very low ($TD_{50} \simeq 10^6$). Apparently in those mouse cells which are very low in tumorigenicity, no further possible *decrease* in tumorigenicity may be noticed (in contrast to the spontaneously transformed, highly tumorigenic cells, see above). Note, however, that neither is there an increase in the tumorigenicity which could have been more easily noticed. The latter would be expected if

Table III
Characteristics of the Cell Lines Based on the Nontumorigenic Parent Cell[a,b]

Cell line	SV40 T antigen	Saturation density (10^5 cells/cm^2)	Tumorigenicity (TD_{50}) Syngeneic	Nude	Methocell plating efficiency (%)
210 C	−	1.9 ± 0.6(4)[c]	$10^{6.4}$	$10^{5.5}$	1–2
213 CSC	−	0.7 ± 0.2(2)	$10^{5.5}$	$10^{5.4}$	1–2
214 CSC	+	8.0	10^6	10^7	—
215 CSC	+	3.5 ± 0.4(3)	$10^{6.4}$	$10^{5.5}$	10–20
219 CT	−	4.8	$<10^2$	$10^{3.5}$	~2.0

[a] Saturation densities, Methocell colony formation (plating efficiency), and tumorigenicity are determined as in Table I.
[b] Some of these data have been presented (Winterbourne and Mora, 1978); the observations on 219 CT cells and the TD_{50} measurements on nude mice are unpublished (McFarland and Mora, in preparation).
[c] Mean ± S.D. Number of separate experiments in parentheses.

SV40 regularly endows a dominant tumorigenic potential on such non-tumorigenic cells in the syngeneic recipients, which obviously is not the case. Observations in other strains of mice, also using cell hybrids, confirmed and extended this finding (Gee and Harris, 1979).

To avoid complications from the immunologic recognition and rejection of the SV40 antigens, experiments for tumorigenicity were also done in the immunologically incompetent nude mice. The two T antigen-positive subclones gave disparate results the 215 CSC had equal tumorigenicity ($TD_{50} = 10^{5.5}$) to that of the parent 210 C or to that of the T antigen-negative sister 213 CSC, but the second T antigen-positive 214 CSC clone had *lower* tumorigenicity ($TD_{50} = 10^{7}$). Note that in this set of cells there was *no* higher tumorigenicity observed which would be generally expected as SV40 transformation is thought to increase cellular tumorigenicity (cf. Martin 1981). This could have been easy to detect considering the cell doses employed. It appears that the SV40-transformed cells may be recognized and rejected in the nude mice also, apparently by other than T-cell-mediated mechanisms, such as by natural killer cells. Similar data have been obtained now in numerous related SV40-transformed mouse cells (McFarland and Mora, in preparation). Note that in all of the derivative cells from 210 C, there was again no noticeable general correlation among T antigen, the cell growth properties in cluture, and cellular tumorigenicity in both syngeneic and nude mice. In fact the 214 CSC cell which grew best *in vitro*, including in Methocell, was not more tumorigenic in the nude mouse than 210 C; thus, the high anchorage-independent growth capacity does not necessarily correlate with cellular tumorigenicity. Furthermore, the highly tumorigenic cells 219 CT (see also the numerous examples in Section II.A) do not grow in viscous medium. Obviously, the *in vivo* tumor growth is controlled by further factors, which may not operate in any *in vitro* growth test, including the anchorage-independent growth.

We tried to repeat in the experiments reported in Section II.A on tumorigenic cells for the selection of the T antigen-negative revertants, employing the method of *in vivo* passage of T antigen-positive clones in the syngeneic immunologically competent mouse. We injected ten mice, each with the necessarily high (10^{7}) number of the T antigen-positive 215 CSC cells. We obtained four tumors: three gave T antigen-positive tumor lines (221, 222, and 224 CSCT) and only one gave a T antigen-negative (revertant) tumor cell line (223 CSCT). At this high cellular dose of injection not all of the SV40 T antigen-positive cells were rejected, possibly because of the immunological competence of all the mice was not necessarily equal and high. These cells then grew out as the T antigen-positive tumor lines (McFarland and Mora, unpublished observation). Such a phenomenon was also observed in other mouse strains more regularly (such as the Balb/c mice from which the T antigen-positive tumorigenic mouse cells were obtained earlier by other investigators (Kit *et al.*, 1969; Wesslen, 1970). The T antigen-negative 223 CSCT may be a revertant of the 215 CSC cells, but we had no chromosomal markers to go by and have not carried out any nucleic acid hybridization or Southern blot experiments to detect possible residual SV40 DNA in the tumor cell lines, as we did in the 100 family series (Section II.A). However, the parent T antigen-positive 215 CSC cell was obtained by a single-cell technique (Winterbourne and Mora, 1978), so it is probable that in obtaining 223 CSCT, selection pressures again led to an (at least early) SV40 DNA minus and T antigen minus revertant cell.

This time, instead of attempting to detect SV40 nucleic acid fragments, we decided to check the various cells for various sizes of T antigens. After labeling with [^{35}S-]methionine in culture and solubilization of the cells, we looked for any detectable protein by an immunoprecipitation procedure: using the SV40 tumor antiserum in the presence of the *Staphylococcus aureus* protein A, and analyzing the solubilized immunoprecipitates by gel electrophoresis and autoradiography. Under these conditions, the SV40 T antigen-positive cells normally show the large T antigen at about 94K daltons and the small t antigen at about 17K daltons, as these are the SV40-coded early proteins which specifically precipitate with the SV40 tumor antiserum (Prives *et al.*, 1977). We scrutinized very carefully the T antigen-positive tumor lines 221, 222, and 224 CSCT for intermediary-size proteins, just in case some of these tumor lines contain partially abbreviated T antigens which could be the result of a "partial reversion" in which the DNA rearrangement may not excise out the whole early region of the SV40 DNA. In the above-described radioimmunoassay all of the T antigen-positive tumor lines (and their clones) gave the strong 94K T bands and the usually weaker 17K t bands, but also a ~55K-dalton-size intermediate band (Chang *et al.*, 1979b). However, when we included for control a T antigen-positive *nontumor* cell, such as the parent SV40-transformed 215 CSC clone, to our surprise this cell also gave the extra 55–56K band. Naturally, we then rapidly tested all the accessible SV40-transformed cells, and we confirmed that in all (~25)of the T antigen-positive murine cells tested there was a ~55K cross-reacting protein, which precipitated with the SV40 tumor antiserum (Chang *et al.*, 1979b). This was our original observation, which we communicated to an International Fogarty Conference on Early Functions of Tumorigenic DNA Viruses in February, 1978. We analyzed the peptides of the 55K protein in the summer of 1978, and we found that this 55K protein is an SV40-induced but cellular DNA-encoded protein (Chang, *et al.*, 1979b). We will discuss our current work on this protein in Sections III and IV.

C. Selection of Spontaneously Transformed Tumorigenic Variant (Mutant) Cells

Returning to the 210 C parent cell, we found a simple way to select for highly tumorgenic cells: after injecting 10^7 cells/mouse and re-establishing the tumor in culture, a simple passage through the syngeneic mouse will result in tumor cell lines, with a very high tumorigenicity ($TD_{50} = 10^2$). An example of the 219 CT cells is given in Table III. The cells are of course T antigen negative, as they never saw SV40 virus. Apparently a small subpopulation of the 210 C clone contains highly tumorigenic variants (mutants?), which are then selected out by growth *in vivo* for high tumorigenicity (McFarland and Mora, in preparation). Similar experiments have been repeated and found to produce without exceptions highly tumorigenic cell lines (Mora and Waters, unpublished). These tumor lines were cloned and are now ready for further biological and biochemical characterization to detect possible common phenotypic biochemical markers of the tumorigenicity. We also plan to determine if the tumorigenicity is transferable by DNA transfection to the nontumorigenic parent 210 C clone [for a possible method, see Shih *et al.* (1981)].

One challenging biochemical observation on such a tumor line (the 219 CT) pertains to the cell surface glycosaminoglycans, which were systematically investigated and compared with the glycosaminoglycans obtained from other cells originating from 210 C as included in Table III. What we found is a very specific biochemical change: decrease in the 6-O-sulfated glucosamine residues in the cell surface heparan sulfates. Furthermore, we found that this is common to the two SV40-transformed brother clones (215 CSC, 214 CSC) and to the tumorigenic 219 CT cells but not to the SV40-negative 213 CSC brother clone (Winterbourne and Mora, 1981). Dr. Winterbourne is continuing the biochemical work and attempting to define the anabolic or catabolic enzymatic changes which lead to this alteration in heparan sulfates in the cell surface membranes. He speculates that the altered heparan sulfates which are present on the cell surface somehow feed back and interfere with certain enzymes in the cellular DNA synthesis. Indeed heparan sulfate is a potent inhibitor of DNA synthesis *in vitro*, and the high inhibitory activity to DNA polymerase depends in part on specific structural features, such as the amount and type of the sulfate groups (Winterbourne and Salisbury, 1981). This then may be behind the common change we observed in both the spontaneously transformed and in the SV40-transformed cells.

III. A 55K-DALTON CELLULAR PROTEIN IS INDUCED IN SV40-TRANSFORMED CELLS

As described above, in all T antigen-positive SV40-transformed mouse cells there is a 55K-dalton protein which precipitates with the SV40 tumor antiserum. Figure 2 shows examples of this on cells described in previous sections. All of the T antigen-negative cells, such as the parent 210 C, its reclone 213 CSC, and the T antigen minus tumor cell line 223 CSCT, show no difference in the fluorograms of the ^{35}S-labeled proteins, which precipitate with the SV40 tumor antiserums (T) or with the normal (N) pre-immune hamster serum. However, the ~55K protein, together

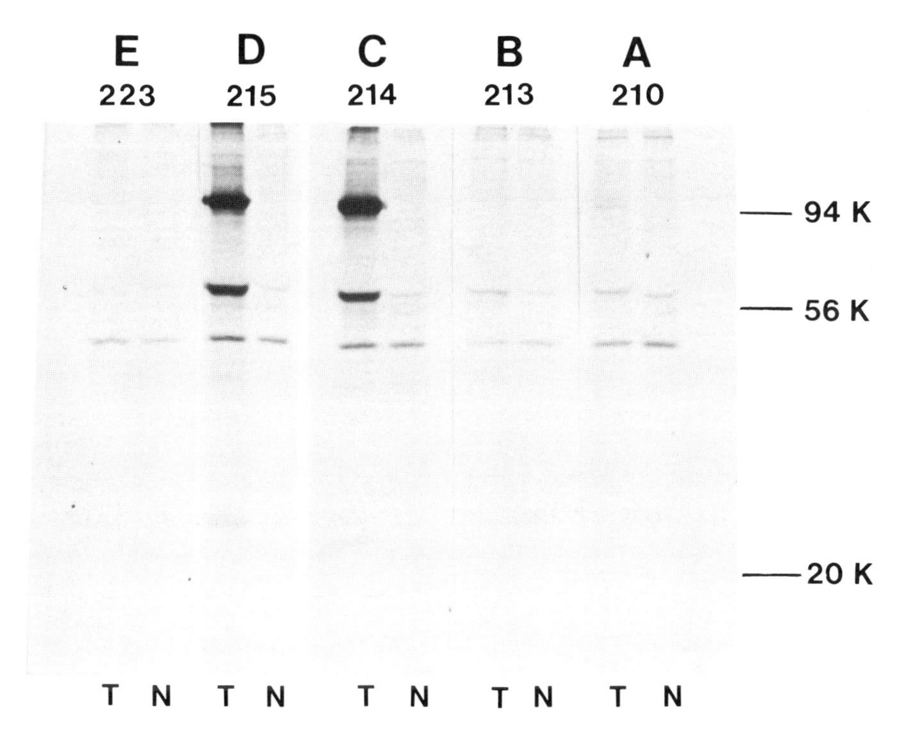

Fig. 2. Detection of the ~55K-dalton protein in SV40-transformed mouse cells. All cells were grown in monolayers and labeled in methionine-free Eagle's minimum essential medium, 2% dialyzed fetal calf serum, and 50 μCi L-[^{35}S]methione (800–1000 Ci/mmole)/ml for 3 hr at 37°C. Extraction, immunoprecipitation, sodium dodecyl sulfate)–polyacrylamide gel electrophoresis, and fluorography was as described (Chang *et al.*, 1979b). Precipitation was either with normal hamster serum (N) or hamster SV40 tumor serum (T). The cells are described in the text. Only 214 CSC and 215 CSC possessed SV40 T antigen (as detected by nuclear immunofluorescence), and only these cells show the specific precipitation of the SV40-coded T and t antigens and also the cellular ~55K protein.

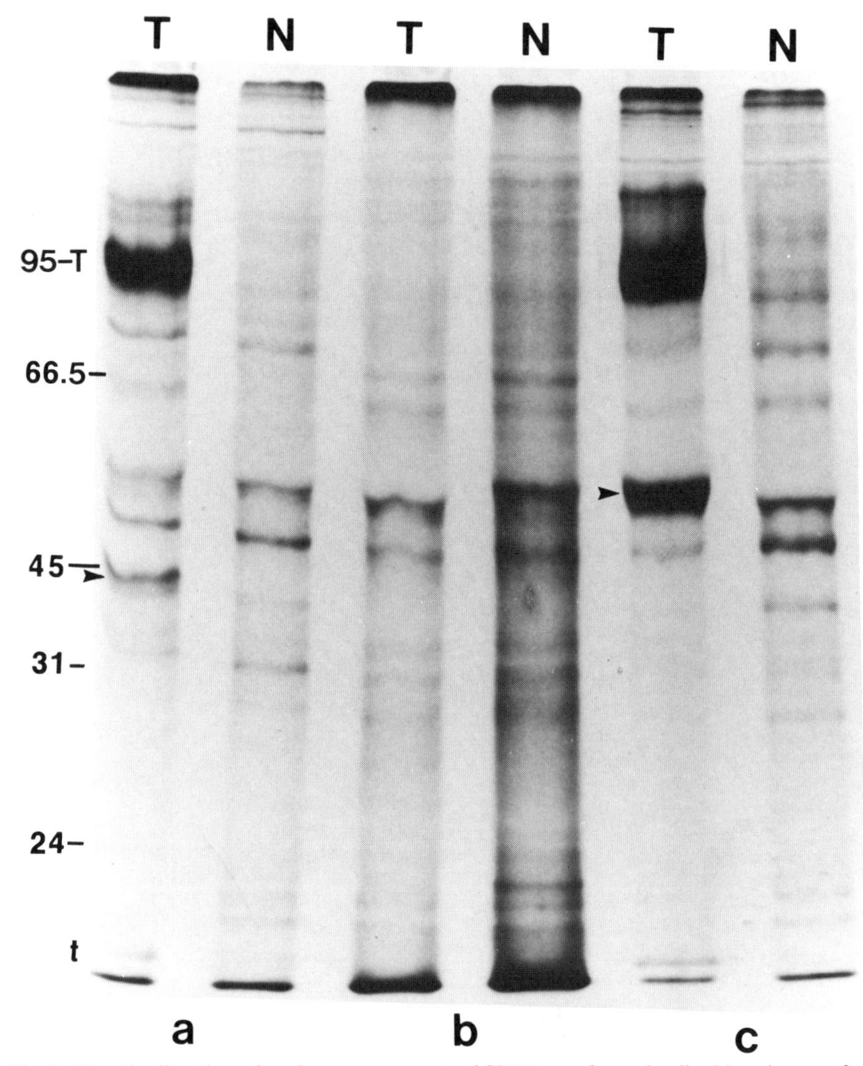

Fig. 3. Tryptic digestion of surface components of SV40-transformed cells. Monolayers of SV40-transformed clone 215 CSC were labeled with L-[^{35}S]methionine as in Fig. 2. The cell sheets were washed three times with Tris-buffered saline (TBS) and incubated with 2.7 ml trypsin (ICN 5105) (100 μg/ml) in TBS for 30 min at 37°C. Trypsinization was terminated by the addition of 0.3 ml soybean trypsin inhibitor (1 mg/ml). The cells were collected by centrifugation at 200*g* for 10 min. The cell pellet (a) and supernatant (b) were separated and the pellet was extracted in 1 ml of extraction buffer (Chang *et al.*, 1979b). Cells not treated with trypsin and subjected to similar procedures were used for control (c). The extracts of the control cells and the trypsinized cell pellet and also the supernatant were divided into

with the SV40-coded T and t antigens, precipitates specifically with the T serum in the SV40-transformed T antigen-positive 215 CSC and 214 CSC cells [for more examples see Chang *et al.* (1979b)].

The peptide pattern of the ^{35}S-methionine-labeled 55K protein as obtained after tryptic digestion by a cation-exchange chromatography showed that the labeled peptides of the 55K proteins from SV40-transformed mouse cells are not the same as those of the SV40 T or t antigens (Chang *et al.*, 1979b). As there are no substantially long regions of open frames in the SV40 DNA sequence besides those used for the T antigen (as described in the Introduction), the 55K protein is most unlikely to be coded by the SV40 genome. Our observations and conclusions were rapidly confirmed (Kress *et al.*, 1979; Melero *et al.*, 1979), and now it is generally accepted that the 55K protein is of cellular origin. It is induced [or stabilized possibly by interaction with the SV40 T antigen (Lane and Crawford, 1979)] in almost all SV40-transformed mouse cells which produce T antigen. Similar proteins are induced in rat, hamster, or human cells transformed or lytically infected by SV40 (Simmons *et al.*, 1980; Smith *et al.*, 1979). The protein, however, is species specific: the [^{35}S]-methionine tryptic peptides in two-dimensional fingerprints from various mouse cells are equal but show a hierarchy of relationship from different species paralleling the evolutionary relationship (Simmons *et al.*, 1980). As the 55K protein is an evolutionarily conserved protein, this indicates that it may have had or still has function. It is a phosphorylated protein (Chang *et al.*, 1979b; Kress *et al.*, 1979). The phosphorylated amino acids

---

two aliquots and 20 μl of either SV40 hamster tumor (T) or normal preimmune serum (N) was added and incubated for 30 min at 4°C. A 15% suspension of *S. aureus* protein A-Sepharose (120 μl) was then added and the solutions were shaken for 3 hr at 4°C. The immunoprecipitates were separated at the end and washed six times with a washing buffer (Chang *et al.*, 1979b). The immunoprecipitates were dissociated, separated on SDS–gel and flurographed, as described in Fig. 1. The positions of protein markers used have been indicated: phosphorylase *a* (94K), bovine serum albumin (66K), ovalbumin (45K), deoxyribonuclease I (31K), and trypsinogen (24K). The strong ~55K band in the untreated cell which precipitated with T serum is marked with an arrow in the control (c) cells. There is no such specific radioactive immunoprecipitate in the trypsinized cells (a) which remained substantially intact (but note a smaller specifically immunoprecipitating component, also marked by an arrow, which may be the partially digested 55K protein); neither could it be recovered from the supernatant fraction (b) probably because insufficient concentration during the immunoprecipitation. Most of the SV40 large T (nuclear) antigen and detectable amounts of small t (cytoplasmic) antigen remains with the trypsinized pellet. The removal of the 55K protein by such gentle trypsinization is in accord with the evidence on the presence of detectable amount of this protein on the plasma membrane, detected by surface iodination and by subcellular fractionation (Luborsky and Chandrasekaran, 1980; Chandrasekaran *et al.*, 1981a).

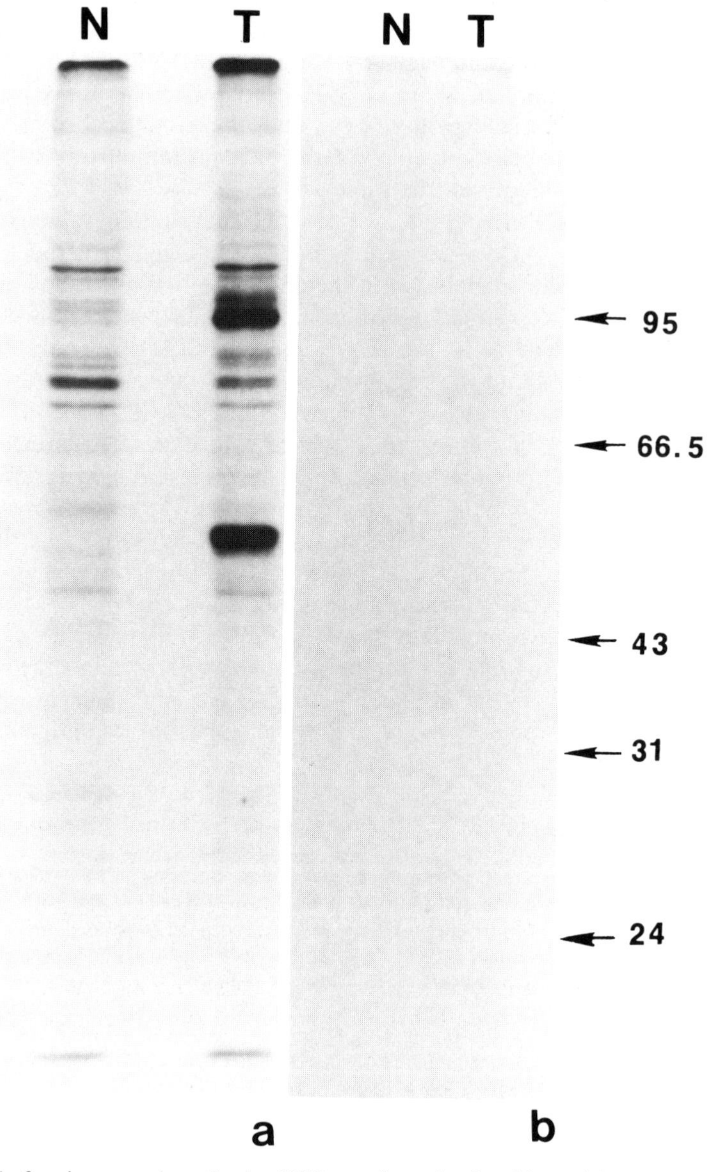

Fig. 4. Surface immunoadsorption by SV40-transformed cells with or without trypsin treatment, using normal hamster serum or hamster anti-SV40 tumor serum. SV40-transformed clone 215 CSC cells were grown in monolayers and labeled in methionine-free Eagle's minimum essential medium, 2% dialyzed fetal calf serum, and 50 μCi L-[^{35}S]methionine (800–1000 Ci/mmol)/ml for 3 hr at 37°C. Cell sheets were then washed three times with TBS, pH 7.4, and incubated with either normal hamster serum (N) or SV40 tumor-bearing hamster

include serine but apparently no tyrosine. The protein has an associated phosphotransferase activity (Jay *et al.*, 1981; also our experiments with Dr. T. Martensen and C. Parrott).

Interestingly, there are similar 53–56K proteins, with apparently partially similar peptide pattern and with immunological cross-reactivities with certain antisera, in cells which were transformed by many different agents: chemicals such as methylcholantrane (DeLeo *et al.*, 1979), RNA virus such as the Moloney or the Abelson leukemia virus (Jay *et al.*, 1979; Witte *et al.*, 1980), and the Epstein–Barr DNA virus (Luka *et al.*, 1980). Claims appeared that similar 53–55K proteins can be detected in cells which were denoted as spontaneously transformed (Jay *et al.*, 1979; Pollack *et al.*, 1979). However, significant amounts of the 55K protein is not detectable in many spontaneously transformed mouse fibroblast cells we tested (such as the 104 C or 219 CT cells) which were well characterized and possessed high tumorigenicity (Mora *et al.*, 1980). Furthermore, the 55K protein is detectable in normal cells (Dippold *et al.*, 1981; Oren *et al.*, 1981; Milner and Milner, 1981; see also Chapter 4 on embryo cells). In mouse cells, the presence, the amount, and the stability (half life) of the 55K protein is not generally correlated with cellular tumorigenicity (Mora *et al.*, 1982). It remains to be elucidated, however, why similar proteins are frequently found in all kinds of induced cellular transformations and in many (but not all) human tumor lines (Crawford *et al.*, 1981). We initiated studies (with Dr. C. Chang and C. Coll) on the frequency of its occurrence in human tumor biopsies.

In the SV40-transformed cells, a portion of the 55K protein, together with the SV40 T antigen can be detected in cell surface (plasma membrane)-enriched fractions (Soule and Butel, 1979; Luborsky and Chandrasekaran, 1980). The accessibility of these proteins on the cell surface was confirmed by radioactive iodination, as catalyzed by lactoperoxidase, from the outside of the cell (Chandrasekaran *et al.*, 1981a). Figures 3 and 4 present further evidence that this protein is present on the surface of the cells: it is susceptible to removal by gentle trypsinization. The fact that the protein is programmed to be expressed on the cell surface and that it is an evolutionary conserved protein brings up an interesting question on a possibly very general role in cell recognition.

---

serum (T), 25 μl/ml TBS for 30 min at 37°C. The cells were washed twice with TBS and extracted, precipitated with the protein A-Sepharose, and subjected to 10% SDS–polyacrylamide gel electrophoresis and fluorography as in Fig. 1. (a) Cells without trypsin treatment. (b) Cells pretreated with trypsin (100 μg/ml) in TD (TD is TBS, pH 7.7, without Ca^{2+} and Mg^{2+}) as in the legend of Fig. 2 before the immunoadsorption. Trypsin treatment, which did not lead to observable cell lysis, destroyed the capacity of the surface of the 215 CSC cells to absorb the antibodies from the SV40 tumor-bearing hamster serum.

IV. THE 55K-DALTON CELLULAR PROTEIN "INDUCED" BY SV40 IS AN EMBRYO PROTEIN

Simultaneously to our finding the 55K protein in SV40-transformed cells, others have found that a similar 55K protein is also present in mouse embryonal concarcinoma (EC) cells, such as in the F9 cells (Linzer and Levine, 1979). These EC cells, of course, never "saw" the SV40 virus. For the immunoprecipitation of the EC 53–55K protein, the same kind of SV40 tumor-bearing hamster serum was used that we used (the SV40 T antiserum). [We have used the SV40 T antiserum which we selected as the best antiserum to detect the 55 K protein in the F9 cells—and which thus contained independent antibodies (from that of the T antigen) to the 55K cellular protein—to show the *absence* of any detectable amount of the protein in the spontaneously transformed, highly tumorigenic mouse cells, as described in Section III.] The protein from the EC cells had partial peptide homology to the protein from the SV40-transformed cells. Certain EC cells can induce tumors and can be considered to be malignant cells, but they can be also considered as paradigms of embryo cells in early stages of differentiation (cf. Martin, 1980). The question arises then: is the appearance of the 55K protein a general correlate of normal embryonal differentiation? For this reason we began a careful search in mouse embryos and embryonal tissues for the 55K protein, using SV40 T antisera.

Indeed postimplantation mouse embryo cells when put in primary culture and labeled with [^{35}S]-methionine produce the 55K immunoprecipitable protein but only when the embryos are less than 14 days in age (Mora *et al.*, 1980). In the 16-day-old embryo primaries, or in the secondary cultures from 10- to 14-day-old embryos, we could not detect the 55K protein, either because the particular cells are not present anymore or the cells do not synthesize the protein in sufficient amount. Indeed, in established mouse embryo cell lines or clones the amount of the 55K protein is very small or it is undetectable after a standard labeling procedure which takes 2–3 hr in tissue culture (Mora *et al.*, 1980).

When the [^{35}S]-methionine-labeled 55K protein was isolated by immunoprecipitation from 12-day-old mouse embryo primary cultures, the embryo protein gave a very similar peptide pattern to the 55K protein isolated from an SV40-transformed mouse cell. The experiment was done both by using *S. aureus* V8 protease digestion followed by re-electrophoresis in one direction (Mora *et al.*, 1980) and also by using tryptic digestion followed by conventional two-dimensional fingerprinting (Chandrasekaran *et al.*, 1981b). A monoclonal antibody prepared against the SV40-induced murine 55K protein (Gurney *et al.*, 1980) (kindly supplied by Dr. E. Gurney) precipitates the mouse embryo 55K protein. This is, of course, expected on the basis of the peptide analyses. Thus, we conclude that the SV40-induced mouse proteins are normal mouse embryo proteins or are closely related to embryo proteins. Immunologic cross-reactivities between embryo cell surfaces and SV40-transformed cells have been observed previously (cf. Weppner and Coggin, 1980) and our finding probably provides the molecular explanation for such observations.

As mentioned in Chapter 3, the SV40-induced 55K proteins are species specific and show a hierarchy of structural relatedness paralleling evolutionary relationship (Simmons *et al.*, 1980). The [^{35}S] methionine-labeled 55K embryo protein tryptic peptide patterns from primary cells from midgestation mouse, rat, and hamster embryos, when compared to the SV40-"induced" 55K proteins from the same species, are very similar or indistinguishable (Chandrasekaran *et al.*,1981b). In all of these experiments, as in the case of the isolation of the mouse embryos, the same monoclonal antibody (Gurney *et al.*, 1980) was used for the immunoprecipitation. Thus, the 55K embryo proteins are indeed the SV40-induced species-specific proteins, and these proteins have been preserved in recent evolution encompassing the above species. We expect that they are also present in the human species.

The amount of the 55K protein in the primary embryo cells from midgestation murine embryos is about 50% of that which was found in the corresponding SV40-transformed cells. In the embryos, however, the amount of the 55K protein, detectable under the standard 3-hr. labeling procedure, declines as the embryo develops (Chandrasekaran *et al.*, 1981b). Thus, the amounts of the detectable 55K protein in embryo cells is a function of embryo age, and in turn it may be a function of embryonal differentiation or a function of the division (growth) rate of certain cells which can also decline during the embryonic differentiation. Clearly, primary embryo cultures are not suitable to make this distinction.

In recent collaborative work we used various embryonal carcinoma cells in an attempt to make such a distinction. The mouse embryonal carcinoma cells, such as undifferentiated nullipotent F9 and pluripotent OTT6050 cells, as well as differentiated parietal endodermallike PYS-2 cells, were employed. When the amount of the 55K protein was determined, it was found to correlate with the stage of differentiation rather than with cell growth rate (Chandrasekaran *et al.*, 1982).

The amount of the detectable 55K protein depends in part on the stability or on the turnover of this protein. In SV40-transformed cells the half-life of the 55K protein is long ($\gtrsim$24 hr). Apparently the 55K protein is protected from turnover (degradation) by being in a complex with the SV40 T antigen (Oren *et al.*, 1981). In established and cloned (non-SV40 transformed) mouse fibroblast cells, the half-life is short, ≤60 min (Oren *et al.*, 1981; Mora *et al.*, 1982b). The embryonal carcinoma cells appear to be intermediate in half-life, furthermore the length of the half-life seems to be inversely related to differentiation (Chandrasekaran *et al.*, 1982). While the above implies posttranslational regulation of the 55K protein, other control mechanisms, such as on the messenger RNA level, are of course not excluded.

Indeed herein lies an unusual opportunity to study molecular embryongenesis. For example, using the SV40-transformed cells, the mRNA to the DNA coding for the 55K protein can be isolated (cf. Kress *et al.*, 1979), then presumably one should be able to clone and amplify this embryo gene. It appears now that molecular approaches will bridge the two active and difficult fields: the study of cellular transformation especiallythrough tumor viruses and that of embryonic differentiation.

V. CONCLUSIONS AND CODA

In the mouse cells, spontaneous transformation is sufficient to explain acquisition of cellular tumorigenicity of SV40-transformed fibroblasts. When considering on balance the *in vivo* phenotypic changes in this species, the SV40 early gene-coded proteins are more "dominant" as cell surface antigens. They facilitate immunologic recognition and rejection, rather than "cause" tumorigenic transformation (Mora *et al.*, 1977a,b). There is no binding general correlation between a cell growth in viscous medium, of tumorigenicity *in vivo*, and of the SV40 antigen expression in the mouse fibroblast cells we have investigated. This of course does not mean that the SV40 early gene, its expression, and the role of the T antigen is not a suitable system to study the correlation between certain biochemical controls in cells and their phenotypic tissue culture growth, when such correlations can be made. Our research, however, emphasizes a growing recognition that for the acquisition of tumorigenic potential the transformation by SV40 is not a sufficient cause and not even a crucial contributory event. This is in mouse or rat cells. In hamster cells, key changes involve altered immune recognition. (See Lewis and Cook, 1980). For a recent review on the biology of the SV40 transformation in various species see Mora (1982).

We concentrated most of our attention on the antigen properties of the SV40 early gene-coded proteins, since these studies *may* be instructive as models for TSTA recognition and rejection, although these studies are admittedly artifactual and may not necessarily pertain to what occurs in naturally occurring tumors. During these studies, we discovered a non-SV40-coded cellular 55K-dalton phosphoprotein, which is induced in all SV40-infected cells. We found that these proteins are species specific and are partially expressed on the cell surface. Finally, we demonstrated that these cellular proteins are embryo proteins, present up to and somewhat beyond the midgestation state. It is very puzzling that the same protein (or very similar proteins) appear in most tumorigenic transformations, but it remains to be seen how general this phenomenon is, and whether it really pertains to tumorigenic transformation or just the consequence of a frequent, coinciding expression of a gene or/genes active in early embryogenesis.

It is instructive to observe that an unexpected and intriguing molecular approach to embryology became now accessible to the multifarious techniques so highly developed in the molecular biology of small tumor viruses (such as the SV40) by just following through patiently the serendipitous research leads, as they appeared to us, one by one.

ACKNOWLEDGMENTS. We thank Mrs. V. McFarland and Drs. S. Luborsky, C. Chang, J. Küster, D. Winterbourne, and S. Pancake of our laboratory, and others, especially Drs. R. Martin and D. Simmons, for their collaboration and advice, and L. Waters and J. Hoffman for skillful technical assistance.

VI. REFERENCES

Alstein, A. D., Vassiljeva, N. N., and Sarycheva, O. F., 1967, Neoplastic transformation of rat embryo cells by simian papovavirus SV40, *Nature* (*London*)**213:**931.

Anderson, J. L., Chang, C., Mora, P. T., and Martin, R. G., 1977a, Expression and thermal stability of SV40 tumor specific transplantation antigen and tumor antigen in wild type and ts A mutant transformed cells, *J. Virol.* **21:**459.

Anderson, J. L., Martin, R. G., Chang, C., Mora, P. T., and Livingston, D. M., 1977b, Nuclear preparations of SV40 transformed cells contain tumor-specific transplantation antigen activity, *Virology* **76:**420.

Berk, A. J., and Sharp, P. A., 1978, Spliced early mRNAs of SV40, *Proc. Natl. Acad. Sci. USA* **75:**1274.

Black, P. H., and Rowe, W. P., 1963, SV40-induced proliferation of tissue culture cells of rabbit, mouse and porcine origin, *Proc. Soc. Exp. Biol. Med.* **114:**721.

Chandrasekaran, K., Winterbourne, D. J., Luborsky, S. W., and Mora, P. T., 1981a, Surface proteins of simian virus 40 transformed cells, *Int. J. Cancer* **27:**397.

Chandrasekaran, K., McFarland, V. W., Simmons, D. T., Dziadek, M., Gurney, E., and Mora, P. T., 1981b, Quantitation and characterization of a species specific and embryo stage dependent 55-kilodalton phosphoprotein also present in cells transformed by simian virus 40, *Proc. Natl. Acad. Sci. USA* **78:**6953.

Chandraseharan, K., Mora, P. T., Nagarajan, L., and Anderson, W. B., 1982, The amount of a specific cellular protein (p 53) is a correlate of differentiation in embryonal carcinoma cells, *J. Cell Physiology*, in press.

Chang, C., Luborsky, S. W., and Mora, P. T., 1977a, Tumour-specific transplantation antigen from SV40 transformed cells binds to DNA, *Nature* (*London*) **269:**438.

Chang, C., Pancake, S. J., Luborsky, S. W., and Mora, P. T., 1977b, Detergent solubilization and partial purification of tumor specific surface and transplantation antigens from SV40 virus-transformed mouse cells, *Int. J. Cancer* **19:**258.

Chang, C., Martin, R. G., Livingston, D. M., Luborsky, S. W., Hu, C., and Mora, P. T., 1979a, Relationship between T-antigen and tumor specific transplantation antigen in SV40-transformed cells, *J. Virol.* **29:**69.

Chang, C., Simmons, D. T., Martin, M. A., and Mora, P. T., 1979b, Identification and partial characterization of new antigens from SV40 transformed mouse cells, *J. Virol.* **31:**463.

Chang, C., Chang, R., Mora, P. T., and Hu, C. P., 1982, Generation of cytotoxic lymphocytes by SV-40 induced antigens, *J. Immunol.* **128:**2160.

Crawford, L. V., Pim, D. C., Gurney, E. G., Goodfellow, P., and Taylor-Papadimitriou, 1981, Detection of a common feature in several human tumor cell lines—a 53,000-dalton protein, *Proc. Natl. Acad. Sci. USA* **78:**41.

DeLeo, A. B., Jay, G., Appella, E., Dubois, G. C., Law, L. W., and Old, L. J., 1979, Detection of a transformation-related antigen in chemically induced sarcomas and other transformed cells of the mouse, *Proc. Natl. Acad. Sci. USA* **76:**2420.

Dippold, W. G., Jay, G., DeLeo, A. B., Khoury, G., and Old, L. J., 1981, P53 transformation related protein-detection by monoclonal antibody in mouse and human cells, *Proc. Natl. Acad. Sci. USA* **78:**1695.

Fiers, W. R., Contreas, R., Haegeman, G., Rogiers, R., Van de Voorde, A., Van Heuverswyn, H., Herreneghe, J. V., Volcraert, G., and Ysebaert, M., 1978, Complete nucleotide sequence of SV40, *Nature* (*London*) **273:**113.

Gee, C. J., and Harris, H., 1979, Tumorigenicity of cells transformed by simian virus 40 and of hybrids between such cells and normal diploid cells, *J. Cell Science* **36:**223.

Gurney, E., Harrison, R. O., and Fenno, J., 1980, Monoclonal antibodies against simian virus 40 T antigens: Evidence for two distinct subclasses of large T antigen, *J. Virol.* **34:**752.

Jay, G., DeLeo, A. B., Appella, E., Dubois, G. C., Law, L. W., Khoury, G., and Old, L. J., 1979, A common transformation-related protein in murine sarcomas and leukemias, *Cold Spring Harbor Symp. Quant. Biol.* **44:**659.

Jay, G., Khoury, G., DeLeo, A. B., Dippold, W. G., and Old, L. J., 1981, p53 transformation-related protein: Detection of an associated phosphotransferase activity, *Proc. Natl. Acad. Sci. USA* **78:**2932.

Kit, S., Kurimura, T., and Dubbs, D. R., 1969, Transplantable mouse tumor line induced by injection of SV40 transformed mouse kidney cells, *Int. J. Cancer* **4:**384.

Kress, M. E., May, E., Cassingena, R., and May, P., 1979, Simian virus 40 transformed cells express new species of proteins precipitable by antisimian virus 40 tumor serum, *J. Virol.* **31:**472.

Küster, J. M., Mora, P. T., Brown, M., and Khoury, G., 1977, Immunologic selection against SV40 transformed cells: Concomitant loss of viral antigens and early viral gene sequences, *Proc. Natl. Acad. Sci. USA* **74:**4796.

Lane, D. P., and Crawford, L. V., 1979, T antigen is bound to a host protein in SV40-transformed cells, *Nature* (*London*) **278:**261.

Lewis, A. M., Jr., and Cook, J. L., 1980, Presence of allograft-rejection resistance in SV40 transformed hamster cells and its possible role in tumor development, *Proc. Natl. Acad. Sci. USA* **77**:2886.

Linzer, D. I. H., and Levine, A. J., 1979, Characterization of 54K dalton cellular SV40 tumor antigen present in SV40-transformed cells and uninfected embryonal carcinoma cells, *Cell* **17**:43.

Luborsky, S. W., and Chandrasekaran, K., 1980, Subcellular distribution of SV40 T antigen species in various cell lines; the 56K protein, *J. Cancer* **25**:517.

Luborsky, S. W., Chang, C., Pancake, S. J., and Mora, P. T., 1976, Detergent solubilized and molecular weight estimation of tumor specific surface antigen from SV40 virus transformed cells, *Biochem. Biophys. Res. Commun.* **71**:990.

Luborsky, S. W., Chang, C., Pancake, S., and Mora, P. T., 1978, Comparative behavior of simian virus 40 T-antigen and of tumor specific surface and transplantation antigens during partial purification, *Cancer Res.* **38**:2367.

Luka, J., Jörnvall, H., and Klein, G., 1980, Purification and biochemical characterization of the Epstein-Barr virus-determined nuclear antigen and an associated protein with a 53,000 dalton subunit, *J. Virol.* **35**:592.

Martin, G. R., 1980, Teratocarcinomas and mammalain embryogenesis, *Science* **209**:768.

Martin, R G., 1981, The transformation of cell growth and transmogrification of DNA synthesis by simian virus 40, *Adv. Cancer Res.* **26**:1.

McFarland, V. W., Mora, P. T., Schultz, A., and Pancake, S., 1975, Cell properties after repeated transplantation of spontaneously and of SV40 virus transformed mouse cell lines, *J. Cell. Physiol.* **85**:101.

Melero, J. A., Stitt, D. T., Mangel, W. F., and Carroll, R. B., 1979, Identification of new polypeptide species (48-55K) immunoprecipitable by antiserum to purified large T antigen and present in SV40-infected and transformed cells, *Virology* **93**:466.

Milner, J., and Milner, S., 1981, SV40 53K protein: a possible role for 53K in normal cells, *Virology* **112**:785.

Mora, P. T., 1982, The immunopathology of SV40-induced transformation, *Seminars in Immunopathology*, in press.

Mora, P. T., Brady, R. O., Bradley, R. M., and McFarland, V. W., 1969, Gangliosides in DNA virus-transformed and spontaneously transformed tumorigenic mouse cell lines, *Proc. Natl. Acad. Sci. USA* **63**:1290.

Mora, P. T., Chang, C., Couvillion, L., Küster, J., and McFarland, V. W., 1977a, Immunological selection of tumor cells which have lost SV40 antigen expression, *Nature (London)* **269**:36.

Mora, P. T., Chang, C., Khoury, G., Küster, J. M., Luborsky, S. W., and McFarland, V. W., 1977b, Antigenic expression of and *in vivo* selection against the early SV40 gene, *INSERM* **69**:327.

Mora, P. T., Chandrasekaran, K., and McFarland, V. W., 1980, An embryo protein induced by SV40 virus transformation of mouse cells, *Nature (London)* **288**:722.

Mora, P. T., Chandrasekaran, K., Hoffman, J., and McFarland, V. W., 1982, Quantitation of a 55K cellular protein: Similar amount and instability in normal and malignant mouse cells, *Mol. Cell. Biol.*, **2**:763.

Oren, M., Maltzman, W., and Levine, A. J., 1981, Post translational regulation of the 54K cellular tumor antigen in normal and transformed cells, *J. Mol. Cell Biol.* **1**:101.

Pancake, S. J., and Mora, P. T., 1974, Comparison of SV40 induced antigens on the surface of cultivated cells by a cytolytic microassay, *Virology* **59**:323.

Pancake, S. J., and Mora, P. T., 1976, Limitations and utility of a cytolytic assay for measuring simian virus 40-induced cell surface antigens, *Cancer Res.* **36**:88.

Paucha, E., Harvey, R., and Smith, A. E., 1978, Cell-free synthesis of SV40 T antigens, *J. Virol.* **28:**154.

Pollack, R., Lo, A., Steinberg, S., Smith, K., Shure, H., Blanch, G., and Venderane, M., 1979, SV-40 and cellular gene expression in the maintenance of the tumorigenic syndrome, *Cold Spring Harbor Symp. Quant. Biol.* **44:**681.

Prives, C., Gilboa, E., Revel, M., and Winocour, E., 1977, Cell-free translation of simian virus 40 early messenger RNA coding for viral T antigen, *Proc. Natl. Acad. Sci. USA* **74:**457.

Reddy, V. B., Thimmappaya, B., Dhar, R., Subramanian, K. N., Zain, B. S., Pan, J., Ghosh, P. K., Celma, M. L., and Weissman, S. M., 1978, The genome of SV40, *Science* **200:**494.

Shih, C., Padhy, L. C., Murray, M., and Weinberg, R. A., 1981, Transforming genes of carcinomas and neuroblastomas introduced into mouse fibroblast, *Nature (London)* **290:**261.

Simmons, D. T., Martin, M. A., Mora, P. T., and Chang, C., 1980, Relationship among Tau antigens isolated from various lines of SV40 transformed cells, *J. Virol.* **34:**650.

Smith, A. E., Smith, R., and Paucha, E., 1979, Characterization of different tumor antigens present in cells transformed by SV40, *Cell* **18:**335.

Smith, R. W., Morganroth, J., and Mora, P. T., 1970, SV40 virus-induced tumor specific transplantation antigen in cultured mouse cells, *Nature (London)* **227:**141.

Soule, H. R., and Butel, J. S., 1979, Subcellular localization of simian virus 40 large tumor antigen, *J. Virol.* **30:**523.

Tevethia, S. S., and McMillan, V. L., 1974, Acquisition of malignant properties by SV40 transformed mouse cells: Relationship to type-C viral antigen expression, *Intervirology* **3:**269.

Tevethia, S. S., Flyder, D. C., and Tjian, R., 1980, Biology of simian virus 40 (SV40) transplantation antigen (TrAg). VI. Mechanism of induction of SV40 transplantation immunity in mice by purified SV40 T antigen (D2 protein), *Virology* **107:**13.

Weppner, W. A., and Coggin, J. H., 1980, Antigenic similarity between plasma membrane proteins of fetal hamster cells and simian virus 40 tumor surface antigens, *Cancer Res.* **40:**1380.

Wesslen, T., 1970, SV40 tumorigenesis in mouse, *Acta Pathol. Microbiol. Scand. Sect. B* **78:**479.

Winterbourne, D. J., and Mora, P. T., 1978, Altered metabolism of heparan sulfate in simian virus 40 transformed mouse cells, *J. Biol. Chem.* **253:**5109.

Winterbourne, D. J., and Mora, P. T., 1981, Cells selected for high tumorigenicity or transformed by simian virus 40 synthesize heparan sulfate with reduced degree of sulfation, *J. Biol. Chem.* **256:**4310.

Winterbourne, D. J., and Salisbury, J. B., 1981, Heparan sulphate is a potent inhibitor of DNA synthesis *in vitro*, *Biochem. Biophys. Res. Commun.* **101:**30.

Witte, O. N., Dasgupta, A., and Baltimore, D., 1980, Abelson murine leukemia virus protein in phosphorylated *in vitro* to form phosphotyrosine, *Nature (London)* **283:**826.

Ziff, E. B., 1980, Transcription and RNA processing by the DNA tumour viruses, *Nature (London)* **287:**491.

Chapter 10

Monoclonal Antibody-Defined Antigens on Tumor Cells

R. W. Baldwin, M. J. Embleton, and M. R. Price

Cancer Research Campaign Laboratories
University of Nottingham
Nottingham, United Kingdom

I. INTRODUCTION

The definition of antigens expressed on tumor cells using monoclonal antibodies is the latest development following early studies showing that inbred animals often can be immunized against otherwise lethal inocula of tumor cells, such effects being attributed to host reactions against tumor-associated antigens (Byers and Baldwin, 1980; Woodruff, 1980). The original demonstration of these antigens involved pretreatment with tumor prevented from progressive growth by attenuation, for example, using γ irradiation or cytotoxic drugs. This was followed by a challenge with viable syngeneic tumor cells sufficient to produce growth in untreated animals but which were rejected in the immunized host. In this way, many chemically induced tumors (Embleton and Baldwin, 1980; Baldwin and Price, 1981) and most virally induced tumors (Lilly and Steeves, 1974; Levy and Leclerc, 1977) were found to have neo-antigens which are absent in normal tissues, at least those from the adult host. These antigens have been termed tumor-associated rejection antigens (TARA). Various *in vitro* assays have also been employed to identify antigens of equivalent specificity and distribution. There is, however, no formal proof that the antigens detected by such assays for cell-mediated immune reactivity (Bloom and David, 1976) or humoral antibody (Ting and Herberman, 1976) are identical with TARAs, and these are best referred to as tumor cell surface antigens (TCSA).

These early findings stimulated interest in the possibility that human tumors might also have antigens capable of eliciting immune reactions in the primary host, and the many recorded instances of spontaneous regression of metastases following removal of a primary tumor or the emergence of metastases several years after successful therapy of the primary tumor has been taken to support this view (Everson and Cole, 1966; Woodruff, 1980). Since such observations generally lack sufficient independent corroboration and, with the recognition that a number of factors including those of a nonimmunological nature may influence the clinical response, further supportive evidence is necessary to confirm the existence of human tumor-associated antigens. *In vivo*, the elicitation of cutaneous delayed-type hypersensitivity reactions following intradermal injection of tumor extract provides some indication of an antitumor immune response (Hollinshead *et al.*, 1974; Weese *et al.*, 1978; Vandebark *et al.*, 1979), although these observations have not been universally accepted or independently confirmed. Comparably, *in vitro* methods involving cellular immunity have often proved to be unreliable indicators of the antigenicity of human tumors (Baldwin and Embleton, 1977). However, from tests measuring cell-mediated cytotoxicity, leukocyte migration inhibition, leukocyte adherence inhibition, and mixed lymphocyte–tumor cell blastogenesis, the consensus of opinion that developed was that human tumors possess neo-antigens common to all tumors developing from a particular organ or tissue but not shared by histologically different tumors (Bloom and David, 1976; Baldwin and Embleton, 1977; Herberman, 1974).

Assays for humoral immunity have also proved to be far from satisfactory. The usual sources of antibody have been either the cancer patient or extensively absorbed xenogeneic antisera. In tests with patients' sera, both normal control and patients' sera frequently contain autoantibodies reactive with normal and tumor cells (Carey *et al.*, 1976; Shiku *et al.*, 1977; Rosenberg, 1980), whereas with xenogeneic antisera against tumor cells, it is virtually impossible to ensure the complete removal of antibodies against normal antigens, leaving reactivity exclusively directed against the tumor. Despite these reservations, techniques, such as membrane immunofluorescence (Morton *et al.*, 1968; Lewis and Phillipps, 1972; Embleton *et al.*, 1980), immune adherence (Cornain *et al.*, 1975), complement-dependent cytotoxicity (Canevari *et al.*, 1978), and antibody-dependent cellular cytotoxicity (Murray *et al.*, 1978), have been used to study antigens on melanomas and also to a lesser extent sarcomas and breast and colon carcinomas. The results tend to confirm the existence of histological-type specific neo-antigens as demonstrated in tests of cellular immunity (Rosenberg, 1980).

Against this somewhat confused background, the emergence of monoclonal antibody technology came as a breath of fresh air (Lennox, 1980). For the first time it was possible to envisage on a sound scientific basis the development of reagents of unique specificity for their target antigens and theoretically, in an unlimited supply. Thus, the objective of achieving an accurate analysis of tumor-associated antigens, particularly those on human tumors, became feasible.

II. MONOCLONAL ANTIBODY TECHNOLOGY

Köhler and Milstein (1975) developed the technique of immortalizing individual clones of antibody secreting cells by fusing them with cultured cells of a mouse myeloma. The tendency of two types of cells to fuse and form heterokaryons when brought into close proximity is vastly increased if the cells are exposed to fusing agents such as ultraviolet-inactivated Sendai virus, lysolecithin, or polyethylene glycol. The survival of hybrid progeny with preferred characteristics of both parental cell types can be promoted by means of appropriate selection.

In the case of antibody-producing hybrids, the objective is the fusion of B cells, which have the genetic information for producing immunoglobulin of defined specificity, and myeloma cells to produce hybrids which inherit from the parental cells both the capacity for unlimited growth and the ability to secrete large amounts of the specific immunoglobulin. Selective pressures are imposed against the survival of unfused parental cells or megakaryons formed by the fusion of like cells. With the lymphocyte population containing actively dividing activated B cells (typically a spleen cell preparation from an immunized mouse), this presents no problem since they are incapable of unlimited proliferation and will die out in prolonged culture. The myeloma cells require a more positive approach for their elimination. This is generally achieved by using mutants which lack an important enzyme such as thymidine kinase or hypoxanthine–guanine phosphoribosyl transferase (HGPRT), since these are susceptible to toxicity by tissue culture media containing hypoxanthine, aminopterin, and thymidine (HAT) (Littlefield, 1964) by virtue of the fact that they are no longer able to synthesize nucleic acid by salvage pathways available to normal cells after treatment with aminopterin. The hybrid cells inheriting the necessary enzyme from the parental lymphocytes can survive in HAT medium and are referred to as hybridomas.

Following the production of hybridoma cultures, the next stage is the screening of supernatants for the presence of antibody to the immunizing antigen. It should be recognized that not all hybridomas secrete antibody, and nonsecretors may be eliminated in primary screening using a sensitive radioimmunoassay for immunoglobulin. Then, after a secondary screen for the selection of those hybridomas producing antibody against the immunizing antigen, the chosen hybridoma is cloned. Clones producing the desired monoclonal antibody are then identified by testing culture supernatants, and these may then be propogated in culture. Alternatively, the selected clones may be grown in appropriate laboratory animals where they continue to secrete monoclonal antibody. In this case, often exceedingly high antibody titers are achieved as long as the chromosome complement of the hybridoma remains stable.

For the selection of hybridomas producing antibodies against TCSA, the most popular and versatile assay for antibody in culture supernatants is the radioisotopic antiglobulin test originally developed by Harder and McKhann (1968). Also available and in extensive use is a variant of this assay in which radioiodinated protein A (the Fc-reactive protein from *Staphylococcus aureus*) is employed as an alternative to labeled antiglobulin to identify cell-bound antibodies. Such assays are objective and reproducible and have been used in other systems for quantitative studies on the interaction of antibodies with cell surface-expressed antigens (Williams, 1977).In cases where target cells are not available in sufficient numbers or of an adequate quality (e.g., cells prepared from primary human tumors such as colon carcinoma), it is possible to use tumor membrane preparations adhered to plastic culture dishes as the targets in the antibody binding assay (Sikora and Phillips, 1981) although antibodies reactive with intracellular membranes will be detected using this assay.

With regard to the choice of myeloma, most monoclonal antibodies have been produced using murine cell lines developed from the Balb/c plasmacytoma MOPC-21 which was adapted to culture by Horibata and Harris (1970) and renamed P3K. An 8-azaguanine-resistant line which lacks HGPRT was selected, and this line (referred to as P3-X63-Ag-8) was employed in the original study by Köhler and Milstein (1975) to produce monoclonal antibodies against sheep red blood cells and later by Galfre *et al.* (1977) for the production of antibodies to rat histocompatibility antigens. This line secretes both the heavy and light chains of the original MOPC-21 tumor and a later variant, designated the P3-NSl-Ag4-1 line, was obtained which synthesizes but does not secrete the light chain and does not synthesize the heavy chain (Köhler *et al.*, 1976). Although this is currently the most extensively used line for hybridoma production, other lines which do not synthesize any immunoglobulin components are available, e.g., X63-Ag-8.653 (Kearney *et al.*, 1979) and Sp2/0-Ag-14 (Shulman *et al.*, 1978).

As already mentioned, spleen cells from mice immunized against the relevant antigen are typically used as a source of activated lymphocytes for fusion with myeloma cells and it is often desirable to employ mice syngeneic with the myeloma cell line (e.g., Balb/c in the case of P3-NS1-Ag4-1) since the resulting hybridoma can then be passaged in syngeneic hosts. The yields of antibody produced *in vivo* may greatly exceed those obtained from culture (Koprowski *et al.*, 1978).

Fusion products of mouse myeloma cells and lymphocytes from other species have also been prepared, but with variable success. Thus, while hybrids between mouse myelomas and rat lymphocytes tend to be stable products (Galfre *et al.*, 1977; Gunn *et al.*, 1980), those produced with human lymphocytes tend to be unstable since there is a preferential deletion of human chromosomes with a concomitant decrease in antibody production or loss in viability. Nevertheless, short-term mouse–human hybridomas have yielded antibodies against a human glioma (Sikora and Phillips, 1981), and Schlom *et al.* (1980) have described a interspecies hybridoma secreting antibody reactive against human breast tumor cells.

Highly stable rat–rat hybridomas have been produced using the rat myeloma, 210-TCY3-Agl (Galfre *et al.*, 1979) and antibodies to human colorectal tumors have been obtained (Berry *et al.*, 1981; Grant *et al.*, 1981). In addition, two human myelomas suitable for hybridoma production are currently receiving considerable attention. These are designated SK0-007 and U-266 (Olsson and Kaplan, 1980),and they have dual significance in the production of antibodies against human tumor-associated antigens. First, because rodent–human hybrids are generally short lived, stable human–human hybridomas can be prepared with lymphocytes from cancer patients, offering the opportunity of studying immunologically relevant tumor-associated antigens against which the lymphocyte donor has responded. Second, in the clinical situation, complications such as serum sickness or rapid elimination of foreign immunoglobulins would be minimized using entirely human antibodies. Already, anti-measles monoclonal antibodies have been prepared using human hybridomas (Croce *et al.*, 1980) so that the production of anti-human tumor antibodies may be anticipated.

III. MONOCLONAL ANTIBODIES DEFINING ANTIGENS ON EXPERIMENTAL TUMORS

The antigenic profiles of tumors in experimental animals are more clearly defined in comparison to those of human tumors, and it is therefore appropriate to consider the performance of monoclonal antibodies against animal tumors.

The diversity of antigens of chemically induced tumors, particularly in mice and rats, is well established although the molecular basis for their polymorphism is poorly understood. Since most tumors induced by carcinogens, such as 3-methylcholanthrene (MCA) or the azo dyes, express individually distinct tumor-associated antigens readily demonstrable, and rarely cross-reacting in tumor transplant rejection tests (Baldwin and Price, 1981), attempts have been made to prepare monoclonal antibodies against these targets. At the time of writing, no monoclonal antibodies have been identified which are exclusively reactive with these characteristic determinants of chemically induced tumors although it is most likely only a matter of time before such antibodies are obtained. In one of the first investigations reported on this topic, Simrell and Klein (1979) performed many fusions using spleen cells of C57BL/6 mice bearing, or immunized against, MCA-induced murine sarcomas, and the resulting monoclonal antibodies were analyzed in detail against a panel of MCA-induced sarcomas. In addition, these antibodies were screened for reactivity with murine leukemia virus (MuLV), since although not originally induced by virus, murine sarcomas are prone to infection with MuLV, particularly following repeated transplantation. Two of the antibodies were found to react with the immunizing tumor and with several virus preparations, including one from the tumor itself. One antibody reacting strongly with the immunizing tumor did not react with the MuLV preparations tested but reacted with several other sarcomas so that tumor specificity was not established (Simrell and Klein, 1979). These findings are comparable to those of Brown *et al.* (1978), who examined a series of MCA-induced murine sarcomas for their capacity to induce transplantation rejection and humoral antibody following immunization of syngeneic mice with irradiated tumor cells. The tumors all induced individually specific tumor rejection responses but only a proportion induced humoral antibody, and this was cross-reactive. Those tumors inducing this response or reactive with antibody were all MuLV positive, and although mice were able to react against individual TARAs, it was clear that the predominant humoral response was directed against viral antigens. In another study, Lostrom *et al.* (1979) raised monoclonal antibodies against MuLV by immunizing C57BL/6 or 129 strain mice with allogeneic AKR MuLV-producing leukemia cells and fusing their spleen cells with the P3NS1 mouse myeloma. Twenty-three clones were derived which displayed six patterns of antibody reactivity against a panel of MuLV or nonmurine retraviruses. One representative of each of the six patterns was selected, three of which identified as p15(E) on diverse MuLV preparations and three of which reacted with gp70 in selected isolates of MuLV while the other was separate and the same was found in the case of gp70 epitopes (Stone and Nowinski, 1980).

Lennox (1980) attempted to resolve the contribution of MuLV and its antigens in the expression of unique and cross-reactive antigens on carcinogen-induced murine tumors. Five of the monoclonal antibodies produced against MCA-induced sarcomas in B10 mice initially appeared specific for one particular tumor (MC6A). One of these (4B1) was selected for further study, and it was found that if the radioisotopic antiglobulin test for its detection was made very sensitive, this antibody bound to several other B10 tumors. Nevertheless, the same antibody also reacted with several B10 tumors grown from cells transformed *in vitro* with chemicals and viruses and it was determined that the antigens recognized by the anti-MC6A monoclonal antibody 4B1 corresponded with those recognized by an anti-gp70 monoclonal antibody (Stone and Nowinski, 1980). In addition, the 4B1 antibody was found to precipitate a protein of about 70,000 daltons from two 4B1-positive tumors and this comigrated in acrylamide gels with labeled gp 70. These observations taken together infer that the 4B1-defined antigen on the MC6A sarcoma is related to MuLV gp70, and since the original specificity of 4B1 made this antigen a candidate for the role of the MC6A TARA, Lennox (1980) has postulated that perhaps all rejection antigens on MCA-induced murine sarcomas are modified MuLV gp70 molecules. This contrasts with the findings of DeLeo *et al.* (1977), who have identified a specifically cytotoxic syngeneic antibody against the murine sarcoma Meth A, the reactivity of which appears not to be related to the MuLV phenotype. There is also some evidence to suggest that this serologically defined antigen is the TARA in this system although the case is not yet proven (Dubois *et al.*, 1980).

Monoclonal antibodies have also been prepared against nonviral antigens associated with rodent tumors. Antibodies against a mouse teratoma cell line, C86-51, were produced by fusing P3-X63-Ag-8 mouse myeloma cells with spleen cells from a rat immunized against the teratoma (Goodfellow *et al.*, 1979). One hybridoma produced secreted antibody reactive with all mouse teratoma cell lines tested, an endodermal cell line, and a neuroblastoma but not other target cell lines. In addition it was unreactive with pre-implantation embryos so it is not clear what antigen was being detected. In other investigations, a monoclonal antibody was prepared against rat spleen cell (Stern *et al.*, 1978; Willison and Stern, 1978). Fortuitously, it reacted with teratocarcinoma stem cells and trophoectoderm but it also reacted with sheep red blood cells and the antigen detected was identified as Forssman antigen.

At present, the only monoclonal antibody which appears to exhibit the classical individually distinct specificity against an experimental tumor is the product of an interspecies hybridoma prepared using spleen cells from a rat immunized against a spontaneously arising mammary carcinoma, Sp4 (Gunn *et al.*, 1980). Among tumors of unknown etiology, which are rarely immunogenic (Hewitt *et al.*, 1976; Middle and Embleton, 1981), this tumor is somewhat of an exception since it was possible to induce significant levels of immunoprotection against challenge with viable tumor and the TARA responsible was specific for Sp4 (Baldwin and Embleton, 1969). The humoral response to immunization of syngeneic rats was also specific for this tumor (Baldwin and Embleton, 1970). After cloning hybrids between spleen cells from a syngeneic Sp4 immune rat and cells of the mouse myeloma, P3-NS1, the products of one hybridoma were selected for further study. The monoclonal antibody produced (rat IgG2b, mouse κ light chain) reacted in cell binding assays exclusively with Sp4 cells and not with cells of other tumors including those derived from other spontaneously arising and chemically induced tumors or a variety of normal rat tissues (Gunn *et al.*, 1980). This antibody lacks complement-dependent cytotoxicity or ADCC reactivity for Sp4 target cells and has been purified for use in studies on the radiolocalization of tumors *in vivo* and the successful systemic treatment of Sp4 tumors following its conjugation to adriamycin (Pimm *et al.*, 1981). The target antigen would appear to be a protein susceptible to release from the cell surface by treatment with papain which liberates a single-chain polypeptide of 20,000 daltons (Price *et al.*, 1981).

IV. MONOCLONAL ANTIBODIES DEFINING ANTIGENS ON HUMAN TUMORS

A. Malignant Melanoma

Human malignant melanoma has received more attention from tumor immunologists than any other type of human tumor and there are a number of reports now describing the production of antimelanoma monoclonal antibodies by fusing spleen cells from mice immunized with cultured melanoma cells, with either the P3-NS1 or P3-X63 mouse myeloma. Koprowski *et al.* (1978) first immunized mice with a somatic cell hybrid between a human melanoma line (SW691) and mouse IT 22 cells in an attempt to reduce reactivity to nonrelevant human cell surface antigens. A variety of reactive hybridomas were produced, differing in their specificity and reactivity for the SW691 melanoma. From the pattern of reactivity of these antibodies with various tumor and nontumor target cells, the antibodies were categorized as follows: (1) human species-associated, (2) common melanoma-associated, and (3) individual SW691 melanoma-associated antibodies. Where cross-reactions occurred, they were different for the different hybridoma clones and in competitive binding assays, distinct monoclonal antibody-defined epitopes were identified (Koprowski *et al.*, 1978). Further investigations on the monoclonal antibodies showing melanoma cell line-related reactivity demonstrated that these antibodies also bound to nine freshly isolated cell preparations from primary or metastatic melanomas (Steplewski *et al.*, 1979). The same antibodies were, however, negative with fibroblasts from the donors or with cells from a giant cell hairy naevus. Not all antibodies bound to all melanomas and each patient's melanoma showed a different profile of reactivity, so it would appear that more than one common melanoma-associated antigenic determinant was involved.

The experiences of other groups who have raised monoclonal antibodies to human melanomas are broadly similar. Three monoclonal antibodies to the melanoma cell line M1804 were produced by Yeh *et al.* (1979), and of these two showed complement-dependent cytotoxic reactivity. These antibodies bound most strongly to cells of the immunizing line M1804 but also, albeit more weakly, to 2 of 11 other melanomas. Antibodies secreted by one clone reacted with a breast carcinoma cell line, but otherwise cross-tests involving other tumor cell lines, fibroblasts, lymphoblastoid cell lines, and normal leukocytes were negative.

The existence of common melanoma-associated antigens is further supported by studies by Carrel *et al.* (1980), who reported improved specificity using membrane fractions of the melanoma cell line Me-43, rather than intact cells, to immunize mice. Two hybridomas (Mel/5 and Mel/14) produced antibodies reactive with 15/16 different melanoma cell lines although neither reacted with 16 unrelated tumors or 14 lymphoblastoid cell lines and thus appeared to define common melanoma-associated antigens. However, in reciprocal binding inhibition tests using antibodies intrinsically labeled with [^{3}H]leucine and "cold" antibody as the inhibitor, both antibodies were found to define different antigenic determinants. Other monoclonal antibodies derived from the same fusion were, however, less discriminatory between melanoma and unrelated cell lines.

The pattern of reactivity of murine monoclonal antibodies to melanoma cell lines Mel/Juso and Mel/Wei when tested against other melanoma lines, unrelated tumors, and autologous lymphoblastoid cells allowed Johnson *et al.* (1981) to propose the following classes of reactivity of anti-melanoma antibodies: (1) broadly reactive, anti-human antibodies, (2) antibodies restricted to the immunizing and autologous cells, (3) antibodies reactive with melanoma and carcinoma cell lines but negative with lymphoblastoid cells and peripheral blood lymphocytes, and (4) antibodies to common antigens expressed only by melanoma cells from cultured lines or fresh tumors. Further studies with anti-melanoma monoclonal antibodies will reveal whether additional categories such as individually distinct melanoma-associated antigens (Lewis and Philipps, 1972) are necessary.

B. Osteogenic Sarcoma

As with human melanoma, there is considerable evidence that osteogenic sarcomas possess antigens which elicit a humoral antibody response in cancer patients (Morton and Malmgren, 1968; Bloom, 1972; Moore and Hughes, 1973; Byers and Johnston, 1977; Rosenberg, 1980). It was therefore considered appropriate in this laboratory to prepare monoclonal antibodies against osteogenic sarcoma, and of the several osteogenic sarcoma cell lines available, one, 791T, was selected and employed to immunize a Balb/c mouse. Spleen cells from this donor were then fused with the P3-NS1 myeloma cell line and hybridoma supernatants were examined for reactivity using a cell binding assay in which ^{125}I-protein A was employed to detect cell surface-bound antibodies. Two of forty-eight hybridoma cultures from a single fusion were unreactive against red cells and lymphocytes but reacted positively with four of ten osteogenic sarcoma lines. In order to determine whether these reactions were specific for osteogenic sarcoma, both antibodies were then tested against a panel of 22 cultured cell lines derived from different tumors. Here, some cross-reactions were observed. Anti-791T/36 clone 3 bound to a prostate carcinoma line, EB33, and HeLa cells and more weakly to lung carcinoma A549 and colon carcinomas HT29 and LS174. However, no binding was observed with two other lung carcinoma lines or two colorectal carcinoma lines. Antibody 791T/48 clone 15 bound to breast carcinoma SkBr3 and

Table I
Reactivity of Monoclonal Antibodies to Osteogenic Sarcoma 791T with a Panel of Human Target Cells

Target cells		Reactivity of monoclonal antibody[a]	
Cell line	Cell type	791T/36 clone 3	791T/48 clone 15
791T[b]	Osteogenic sarcoma	+ + +	+ +
788T[c]	Osteogenic sarcoma	+ + +	+ +
845T	Osteogenic sarcoma	+	NT
805T[d]	Osteogenic sarcoma	+	–
803T	Osteogenic sarcoma	–	–
836T	Osteogenic sarcoma	–	–
706T	Osteogenic sarcoma	–	NT
781T	Osteogenic sarcoma	–	+

Table I
(*Continued*)

Target cells		Reactivity of monoclonal antibody[a]	
Cell line	Cell type	791T/36 clone 3	791T/48 clone 15
888T	Osteogenic sarcoma	−	−
792T	Osteogenic sarcoma	−	NT
393T	Osteogenic sarcoma	+	−
T278	Osteogenic sarcoma	++	−
20S	Osteogenic sarcoma	+	−
791TSk[b]	Skin fibroblasts	−	−
788Sk[c]	Skin fibroblasts	−	−
803Sk	Skin fibroblasts	−	−
805Sk[d]	Skin fibroblasts	−	−
181Sk	Skin fibroblasts	−	−
836Sk	Skin fibroblasts	−	−
860[b]	Tumor-derived fibroblasts	−	−
870[b]	Tumor-derived fibroblasts	−	−
618 Lu	Lung fibroblasts	−	−
74BM	Fetal bone marrow	−	+
	Fresh erythrocytes	−	−
	Fresh blood lymphocytes	−	−
HT29	Colorectal carcinoma	++	−
HCT8	Colorectal carcinoma	−	−
HRT18	Colorectal carcinoma	−	−
LS174	Colorectal carcinoma	++	−
734B	Breast carcinoma	−	−
MCF7	Breast carcinoma	−	−
SkBr3	Breast carcinoma	−	+++
HS578T	Breast carcinosarcoma	−	−
MeWo	Melanoma	−	−
Mel-57	Melanoma	−	−
Mel-Swift	Melanoma	−	−
Mel 2a	Melanoma	−	−
NK1-4	Melanoma	−	−
RPM1 5966	Melanoma	−	−
A549	Lung carcinoma	+	−
A427	Lung carcinoma	−	−
9812	Lung carcinoma	−	−
HeLa	Cervix carcinoma	+++	−
EB33	Prostate carcinoma	+++	−
T24	Bladder carcinoma	−	−
PA-1	Ovarian carcinoma	−	−
RAJI	Burkitt lymphoma	−	−
K562	Erythroleukemia	−	−

[a] −, Binding ratio, <2; +, binding ratio, 2–5; ++, binding ratio, 5–10; +++, binding ratio, >10; where binding ratio = (mean cpm bound by monoclonal antibody) ÷ (mean cpm by P3NS1 spent medium control).
[b] Cells derived from same patient (M.U.).
[c] Cells derived from same patient (P.R.).
[d] Cells derived from same patient (Q.L.).

a fetal bone marrow line, 74BM, but not to other tumor target cells. It was clearly reactive with the immunizing cell line, 791T (Embleton *et al.*, 1981).

Following cloning, antibody produced by these two hybridomas was tested for reactivity against an extensive panel of cultured target cells, and also freshly prepared erythrocytes and blood lymphocytes from a panel of donors (Table I). One of the antibodies, 791T/36 clone 3, was found to react against 7 of 13 osteogenic sarcoma cell lines, including the original immunizing tumor, 791T. It did not react with a panel of fibroblast lines, including fibroblasts from the 791T tumor donor and the donors of two other sarcomas which reacted with the antibody. The antibody was negative with human erythrocytes and lymphocytes from normal donors and with sheep erythrocytes. The other antibody, 791T/48 clone 15, was similarly unreactive with normal adult fibroblasts, so that although both antibodies were reactive with an antigen associated with tumor cells rather than normal adult cells (Table I), they were not specific for osteogenic sarcoma. The difference in cross-reactivity indicates also that the two antibodies recognized different epitopes, which were both strongly expressed by the immunizing tumor, 791T, and another osteogenic sarcoma, 788T.

Both anti-791T/36 clone 3 and anti-791T/48 clone 15 antibodies belong to the IgG2b subclass and express the mouse κ light chain. As expected of a mouse IgG2b antibody, anti-791T/36 exhibits complement-dependent cytotoxicity for tumor cells using rabbit but not guinea pig serum as the source of complement. The anti-791T/48 antibody would appear, however, from the data in Table II to be noncytotoxic for tumor cells. The explanation for this is unknown although it is possible that in anti-791T/48 supernatants, the concentration of antibody is insufficient to induce lysis in the presence of complement. The yields of anti-791T/48 antibody following purification on Sepharose-protein A are indeed considerably less than for anti-791T/36 antibody. In addition, the density of epitopes reactive with anti-791T/48 is less than that for anti-791T/36 antibodies. The ^{51}Cr cytotoxicity release test is less sensitive than the cell binding assays for the detection of anti-791T/36 antibodies. Using the latter assay, antibody reactivity is clearly demonstrable in supernatants diluted to at least 1/1000. The cytotoxicity studies (Table II) also infer that anti-791T/36 is unreactive with EB33 and A549 target cells which are, respectively, positively and weakly positively reacting cell lines when employed in the ^{125}I-protein A binding assay (Embleton *et al.*, 1981). However, when Sepharose-protein A, affinity-purified anti-791T/36 antibodies are tested in the cytotoxicity assay at concentrations exceeding 100 $\mu g/\mu l$, their cytotoxicity for EB33 cells at least (e.g., 40.9 $\pm$ 6.3% at 125 μg antibody protein/ml) is readily detectable.

Table II
Complement-Dependent Cytotoxicity of Anti-Osteogenic Sarcoma 791T Monoclonal Antibodies

Antibody preparation tested	Dilution	Percentage cytotoxicity[a] (mean ± SD) against			
		Osteogenic sarcoma, 791T	Osteogenic sarcoma, 788T	Prostate carcinoma, EB33	Lung carcinoma, A549
Complement alone	—	0.0 ± 1.6	1.1 ± 0.4	−0.1 ± 0.5	0.8 ± 0.5
	—	(−0.1 ± 1.7)[b]	(0.4 ± 0.5)	(0.0 ± 0.2)	(0.0 ± 0.4)
Anti-791T/36 clone 3 hybridoma supernatant	1/10	34.8 ± 11.3	17.7 ± 2.0	2.5 ± 0.3	0.7 ± 0.2
	1/10	(−1.5 ± 0.4)	(0.5 ± 0.2)	(0.1 ± 0.4)	(0.2 ± 0.1)
	1/50	20.5 ± 3.7	4.7 ± 0.7	0.8 ± 0.6	0.8 ± 0.2
	1/250	7.4 ± 2.4	1.5 ± 0.6	0.1 ± 0.6	0.4 ± 0.3
	1/1250	−1.4 ± 0.6	2.1 ± 0.4	0.4 ± 0.3	0.4 ± 0.3
Anti-791T/48 clone 15 hybridoma supernatant	1/10	0.0 ± 1.6	1.7 ± 0.4	0.3 ± 0.2	0.9 ± 0.1
	1/10	(−1.7 ± 1.7)	(0.7 ± 0.2)	(0.1 ± 0.4)	(0.6 ± 0.3)
	1/50	−1.1 ± 0.6	2.1 ± 0.5	0.6 ± 0.5	0.7 ± 0.2
	1/250	−0.1 ± 1.4	1.9 ± 0.7	0.6 ± 0.2	0.8 ± 0.1
	1/1250	−1.5 ± 1.0	2.1 ± 0.4	1.2 ± 0.5	0.6 ± 0.2

[a] The cytotoxicity test was performed as described by Price (1978).
[b] Values in parentheses represent tests performed in the presence of heat-inactivated (56° C for 60 min) complement.

In addition to the indirect cell binding assay and complement-dependent cytotoxicity tests for the demonstration of anti-791T monoclonal antibody reactions against tumor target cells, a direct cell binding assay has recently been developed (Dawood *et al.*, 1981). With this test, affinity-purified anti-791T/36 and anti-791T/48 antibodies have been labeled with ^{125}I and the binding of antibody to tumor cells is determined. In the initial study (Dawood *et al.*, 1981), antibodies were labeled using the chloramine-T procedure of McConahey and Dixon (1966), although subsequently labeling has been achieved using the Iodogen (1, 3, 4, 6-tetrachloro-3α, 6α-diphenylglycoluril) reagent (Fraker and Speck, 1978). Antibodies labeled using the latter reagent have proved to be superior to those formerly prepared in that in excess of 60% of labeled antibody may rebind to 791T tumor target cells. Nevertheless, it was possible to establish a number of parameters with chloramine-T radioiodinated antibodies. The pattern of binding to target cells was equivalent to that found with the ^{125}I-protein A assay (Table I) but the weak cross-reactions (e.g., anti-791T/36 against HT29 and A549) were not easy to confirm. At saturation, and after extensive washing of antibody-treated 791T cells, it was determined that approximately 1.6×10^6 and 2×10^5 antibody molecules are bound per 791T target cell using anti-791T/36 and anti-791T/48 monoclonal antibodies, respectively. This is in accord with the weaker reaction of anti-791T/48 antibody with 791T targets as assessed in the indirect cell binding assay (Embleton *et al.*, 1981). Reciprocal binding inhibition tests using "cold" antibody and ^{125}I-labeled antibodies established conclusively that the two anti-791T monoclonal antibodies reacted with different epitopes on the 791T cell, in agreement with their differing profiles of reactivity with various tumors and the different numbers of labeled immunoglobulin molecules binding at saturation. In addition with ^{125}I-labeled anti-791T/36 antibody it was attempted to resolve whether this monoclonal antibody recognized antigens which are immunogenic in the tumor host. A number of serum samples from the 791T tumor donor and also several control sera were screened for reactivity with tumor cells in the ^{125}I-protein A cell binding assay. Some were found to be positive, both for 791T cells and for control fibroblasts, while others were negative for both tumor and control cells. When tested for their capacity to inhibit the binding of ^{125}I-labeled anti-791T/36 monoclonal antibody to 791T cells, neither positive nor negative autochthonous host sera showed any consistent inhibitory effect. Thus, there was no indication that the anti-791T/36 antibody recognized a 791T-associated antigen also recognized by the host.

Using the purified anti-791T monoclonal antibodies radioiodinated by the Iodogen reagent, it has been possible to attempt more quantitative studies upon the interaction of monoclonal antibodies with tumor cells (F. A. Dawood, M. R. Price, M. J. Embleton, to be published). Labeled

Table III
Quantitative Binding of Radioiodinated Anti-Osteogenic Sarcoma Monoclonal Antibodies to a Panel of Human Tumor Cells[a]

Target cells		Molecules bound per cell[b] ($\times 10^{-4}$) of antibodies from hybridoma	
Cell line	Cell type	791T/36 clone 3	791T/48 clone 15
791T	Osteogenic sarcoma	220	20
788T	Osteogenic sarcoma	160	11
278T	Osteogenic sarcoma	51	4
20S	Osteogenic sarcoma	53	4
393T	Osteogenic sarcoma	43	2
HeLa	Cervix carcinoma	190	7
EB33	Prostate carcinoma	35	4

[a] F. A. Dawood, M. R. Price, M. J. Embleton, to be published.
[b] Determined according to Fazekas de St. Groth (1979).

anti-791T/36 and anti-791T/48 monoclonal antibodies were reacted at various concentrations with a constant number of different tumor target cells and using the computations developed by Fazekas de St Groth (1979), the maximum number of antibodies binding per cell at equilibrium was determined for seven tumor cell lines (Table III). The figure calculated for the number of anti-791T/36 antibodies bound per 791T cell (2.2×10^6) is slightly higher than that determined in previous experiments (1.6×10^6) although the latter figure was calculated after extensive washing of target cells rather than by sampling supernatants of cells and antibody at equilibrium. In addition, high equilibrium constants, K of 10^9 M^{-1} or greater, were determined for the binding of these two monoclonal antibodies to 791T tumor cells and this information, together with the large number of epitopes available for antibody binding (at least with the anti-791T/36 monoclonal antibody), predicts that these labeled antibodies would be of value in *in vivo* tumor localization studies. Preliminary investigations using iodinated anti-791T/36 antibodies have established their preferential uptake into subcutaneous growths of 791T tumor in athymic or immunodeprived mice (M. V. Pimm, unpublished findings).

V. BIOCHEMICAL CHARACTERIZATION OF TUMOR CELL SURFACE-ASSOCIATED ANTIGENS DEFINED BY MONOCLONAL ANTIBODIES

Formal proof that any of the monoclonal antibodies already discussed define structures which are restricted exclusively to malignant cells is lacking. Thus, the biochemical characterization of antigens defined by these antibodies does not represent the study of classical "tumor-specific" antigens but rather at the present stage of development it should be viewed as the mapping of cell surface molecules which are at least elevated in neoplasia. This approach, with continued application of hybridoma technology, should result in a comprehensive picture of the cell surface antigenic structure of tumor cells.

Current findings have established that the antigens identified by monoclonal antibodies (and which appear to be preferentially expressed upon tumor cells) do not belong to a discrete group of surface molecules. Proteins and glycoproteins, either as single-chain polypeptides or more complex moieties, as well as glycolipids have been detected. The heterogeneity of these surface antigens is best exemplified by studies upon human malignant melanoma. Cultured melanoma cells were employed by Brown *et al.* (1980) as the initial immunogen and after fusion of spleen cells with mouse myeloma cells, somatic cell hybrids were grown in selective medium. Eight hybridomas which secreted antibodies to protein antigens of the melanoma cell line were identified by immunoprecipitation of ^{125}I-labeled melanoma cell lysates followed by sodium dodecyl sulfate (SDS)–polyacrylamide gel electrophoresis of immunoprecipitates and visualization of antigens by autoradiography. Five of the proteins, p23, p33, p40, p200, and p270, were also present at the surface of autologous skin fibroblasts but two, p80 and p97, were absent from the fibroblasts suggesting that they might be markers of differentiation or malignancy (Brown *et al.*, 1980). Comparably, with 18 mouse monoclonal antibodies against the human melanoma cell line SK-MEL-28, Dippold *et al.* (1980) defined six distinct antigenic systems using a panel of 41 cell lines. Again, SDS–gel electrophoresis of radioactively labeled immunoprecipitates with autoradiographic examination of gels was employed to determine molecular weights and two of the antigens were found to be 95,000 and 150,000.

The technique of immunoprecipitation of cell surface antigens by admixing detergent lysates of radiolabeled tumor cells or membranes with antibody followed by precipitation of the immune complexes with a suitable agent (*S. aureus* or an appropriate antiglobulin reagent) is without doubt a powerful tool for the characterization of cell surface antigens (Silver and Hood, 1974; Kessler, 1976). Gel electrophoresis of the immunoprecipitates after dissociation with SDS, coupled with autoradiography or gel slicing, allows apparent molecular weights to be assigned to the labeled species. Of course, analyses are not restricted to the study of protein antigens and the initial choice of labeling of tumor cells may be designed to identify structures of a specific composition (e.g., intrinsic labeling with radioactive amino acids for the detection of proteins or use of labeled sugars for the analysis of carbohydrate-containing structures). This was exploited by Dippold *et al.* (1980) to establish that the two antigens of molecular weight 95,000 and 180,000, associated with the melanoma cell line SK-Mel-28, were glycoproteins. This was determined by analysis of cells labeled extrinsically with ^{125}I (using the lactoperoxidase procedure) or by the metabolic incorporation of [^{35}S]methionine and [^{3}H]glucosamine. The finding that both antigens also bound to concanavalin A confirmed their identification as glycoproteins. Another important aspect of this investigation was that not all antigens assayed by conventional serological techniques could be precipitated with antibody and *S. aureus* although antigenic activity in solubilized cell extracts was demonstrable by their capacity to neutralize antibody. In addition, with one antigenic system defined by a monoclonal antibody termed O_5, no antigenic activity could be detected in either antibody inhibition tests or by immunoprecipitation, and the initial indications were that this antigen is a glycolipid (Dippold *et al.*, 1980). Comparably, Hellström *et al.* (1981b) have speculated that one of the two anti-melanoma antibodies they selected for further investigation might be a glycolipid. This antigen functioned as a target for both complement-dependent cytotoxicity and antibody-dependent cellular cytotoxicity in the presence of normal human leukocytes, and, of importance, it was expressed in tumor biopsy material. The second antibody was determined to be a glycoprotein of 97,000 daltons. Using a sensitive "double-determinant immunoassay," it was found that this gp97 was present on most tumors and, in small amounts, in most normal tissues, although certain melanomas tested expressed about 100 times more gp97 than normal tissues. In contrast, two other anti-melanoma cell line antibodies defining antigens (p45 and p70) preferentially on cultured melanomas and a proportion of human carcinomas (but not on normal lymphoblastoid and myeloid cell lines or peripheral blood lymphocytes) failed to react with cells derived from fresh melanoma specimens (Johnson *et al.*, 1981).

The molecular complexity of some surface structures identified on human melanoma cells using monoclonal antibodies is illustrated by an analysis of the target antigen for the antibody 691I5 Nu-4-B (Mitchell *et al.*, 1981; see also Koprowski *et al.*, 1978, for the derivation of this hybridoma). SDS–gel electrophoresis under reducing and nonreducing conditions indicated that the antigen was a macromolecular complex consisting of four associated polypeptide chains with molecular weights of 116,000, 95,000, 29,000, and 26,000. In the native state, the two smaller polypeptide chains were linked to the 116,000-dalton moiety by disulfide bonds, but the 95,000 unit was attached by noncovalent interactions. 2-Nitro-5-thiocyanobenzoic acid cleavage and peptide mapping of the two major polypeptides revealed distinct patterns of cleavage indicating that the two proteins have different primary amino acid sequences (Mitchell *et al.*, 1981).

Previous studies, particularly by Thomson and his colleagues using the leukocyte adherence inhibition assay (Thomson *et al.*, 1976, 1979) have indicated that human melanoma and other tumor-associated antigens may be related to, or even represent, modified histocompatibility antigens. This possible relationship has been the subject of much debate although with melanoma, the results of Carey *et al.*, (1979), McCabe *et al.* (1980), and Imai and Feronne (1980), using polyclonal allogeneic and xenogenic antisera, do not support the contention that melanoma-associated antigens, at least the serologically detectable antigens, are modified histocompatibility antigens. One of these melanoma-associated antigens defined by xenogeneic antisera was of molecular weight 94,000 and was found on melanoma cells and some tumors of nonlymphoid origin (McCabe *et al.*, 1980). Comparably, the same group reported the production of an anti-melanoma monoclonal antibody which again detects a 94,000-dalton glycoprotein on melanoma and carcinoma cells and their spent media (Galloway *et al.*, 1980). This single-chain polypeptide containing galactose (as suggested by lectin affinity) was immunoprecipitated from intrinsically radiolabeled cultured melanoma (M-14) and carcinoma (T-24) cell lines and tryptic peptide maps indicated that there was about a 70% homology in the 94,000-dalton components isolated from the two tumor sources. Fourteen of nineteen peptides were homologous when analyzed by high-pressure liquid chromatography, and two-dimensional gel electrophoresis revealed striking similarities together with distinct differences in their molecular profiles. The data in all suggest that this anti-melanoma monoclonal antibody recognizes two structurally similar glycoproteins with 94,000 molecular weight that also exhibit distinct differences in their molecular structure (Galloway *et al.*, 1980).

Finally a note of caution in consideration of studies upon human malignant melanoma: it was recently found that melanomas (both fresh tumors and cell lines) expressed Ia-like antigens as defined by their reaction with two monoclonal antibodies, Q 5/6 and Q 5/13, and a xenogeneic antiserum to these structures (Howe *et al.*, 1981). In contrast, a wide variety of other tumors, including other neuroectodermally derived tumors appeared not to do so. This observation was most apparent for neuroblastomas for which nine cell lines and ten noncultured tumors lacked Ia-like antigens; one medulloblastoma and three glioma cell lines were also deficient in these antigens. The practical implication from this investigation was that since only a few tumor cell types including melanoma, express Ia-like molecules, they may be mistaken for tumor-specific antigens. Therefore, careful serological and biochemical characterization of cell surface antigens is of considerable importance, particularly with melanoma cells (Howe *et al.*, 1981).

Few other tumors have received the concentrated attention devoted to human malignant melanoma with regard to antigen characterization. The target antigens defined by the monoclonal antibodies against the human osteogenic sarcoma cell line, 791T (Embleton *et al.*, 1981), appear to belong to that type of surface structure which is not easily precipitable with antibody from detergent lysates of radiolabeled cells, as was also found to be the case with the anti-melanoma SK-MEL-28 monoclonal antibodies J_{11}, M_{19}, O_5, or R_8, for example (Dippold *et al.*, 1980). However, the detergent lysates of 791T cells inhibit the binding of ^{125}I-labeled antibodies to tumor target cells so that the basis for an antigen assay is available and currently being applied to antigen purification. The heat lability of the antigen detected by the antibody 791T/36 suggests it to be a protein or glycoprotein and the observed differences in the equilibrium constants for the binding of ^{125}I-labeled 791T/36 antibody to several cross-reacting tumor cells indicate that there may be distinct structural differences in the target epitopes on the various tumors examined (F. A. Dawood and M. R., Price, unpublished findings). This situation is somewhat analogous to that found with the 94,000-dalton antigen commonly expressed upon melanoma (the immunizing tumor) and carcinoma cells which, as already mentioned, exhibit unique differences in their molecular structure (Galloway *et al.*, 1980).

Monoclonal antibodies have been prepared against antigens associated with human colon carcinomas, but, in this case, most reports have confined attention to the development of antibodies against the well-characterized carcinoembryonic antigen (CEA) (Accolla *et al.*, 1979; Mitchell, 1980; Rogers *et al.*, 1981a,b). It is recognized that although potent antisera are available to this component, monoclonal antibodies are of particular interest since they may overcome difficulties associated with the antigenic heterogeneity of this glycoprotein (Rogers, 1976). Comparably, monoclonal antibodies against other soluble tumor markers such as hCG (Stahli *et al.*, 1980) and α-fetoprotein (Tsung *et al.*, 1980) have recently been prepared.

Eleven monoclonal antibodies binding to colorectal carcinoma cells have been examined by Koprowski (1981), and of these, one reacted with 180,000-dalton CEA. Three others were found to bind to a monosialoganglioside that could be isolated from colorectal carcinoma and meconium but not from other tissues. Three reacted with neutral glycolipids extracted from colorectal carcinoma and two immunoprecipitated antigens with a molecular weight of 21,000–28,000. In contrast, the molecular weights of the antigens precipitated by anticolon carcinoma monoclonal antibodies, C15L and D20L, prepared by Arklie and Bodmer (1981), were and 40,000, respectively, although D20L was found to react with normal intestine.

The value of monoclonal antibodies defining tumor-associated antigens has yet to be realized. One of the next steps that may be anticipated with these antibodies in the characterization of their respective target cell surface antigens will be to examine their potential for immunoprecipitating the *in vitro*-translated products of mRNA. These antibodies may prove to be superior compared with conventional antisera because of their improved specificity: they are more likely to identify the correct protein rather than a contaminating one. However, not all monoclonal antibodies against membrane-bound or multimeric protein antigens recognize the nascent form of the translation product *in vitro*, and selection of antibodies for this purpose may have to be carried out. It should be further possible to isolate the relevant mRNA from the polysomes using monoclonal antibodies. Several of these approaches have already been initiated in studies on the HL-A system (Pleogh *et al.*, 1979; Lee *et al.*, 1980), and with protein antigens defined by monoclonal antibodies it may be expected that future investigations will follow similar pathways.

VI. APPLICATION OF MONOCLONAL ANTIBODIES DEFINING TUMOR-ASSOCIATED ANTIGENS, AND CONCLUDING REMARKS

Monoclonal antibodies must now be considered as the reagents of choice for the definition of tumor-associated antigens, particularly those expressed upon human tumors. This, together with the biochemical analyses of their target antigens, represents the first phase in their development by which the antigenic profiles of tumors may be evaluated. Their usage is not, however, confined to providing a means of mapping the antigenic structure of tumor cells. Their ability to mediate and participate in immune reactions *in vivo* anticipates their role in tumor therapy. Already, many of the monoclonal antibodies prepared have been shown to display complement-dependent cytotoxicity for tumor cells, the specificity of which parallels their cell binding reactions *in vitro*. This has been found to be the case with anti-human osteogenic sarcoma antibodies (Table I and Embleton *et al.*, 1981; Dawood *et al.*, 1981) and anti-human colon carcinoma antibodies (Herlyn and Koprowski, 1981; Koprowski *et al.*, 1979). Thus, the passive administration of antitumor monoclonal antibodies (ideally those from entirely human hybridomas) to cancer patients may represent an alternative to conventional therapy for the retardation or arrest of metastatic spread. The few reports on the *in vivo* response to inoculation of monoclonal antibodies are not at this stage particularly encouraging. For example, Herlyn and Koprowski (1981) prepared anti-human colon carcinoma specific antibodies of the IgM class which mediated specific complement-dependent cytotoxicity against human colon carcinoma cells from culture or obtained directly from patients. These antibodies, however, failed to inhibit the growth of tumor cells in nude mice. This may be attributable to the rapid clearance of IgM molecules or to an effective host complement system, but even when the antibody was administered with the potent source of complement, normal rabbit serum, there was no inhibition of growth. This contrasts with the complete susceptibility to tumor growth of leukemic mice treated with monoclonal anti-leukemia IgM antibodies (Bernstein *et al.*, 1980).

In one of the first human trials, Nadler *et al.* (1980a) prepared a cytotoxic murine monoclonal antibody to a lymphoma and attempted to use this antibody therapeutically in the patient from whom the lymphoma was obtained (Nadler *et al.*, 1980b). Two courses of the monoclonal antibody were administered to the patient after cytotoxic drugs had failed to produce any objective anti-tumor response. Following intravenous antibody infusion, flow cytometry indicated that the number of tumor cells decreased transiently, accompanied by the appearance of increased numbers of dead lymphoma cells. No significant toxic effects of the antibody were noted, but on the other hand, no clinical benefit was noted either and the patient eventually died of his lymphoma.

Some monoclonal antibodies have, in addition, been found to exhibit antibody-dependent cellular cytotoxicity for tumors. This has been adequately demonstrated with human colonic tumors (Herlyn *et al.*, 1980) and melanomas (Hellström *et al.*, 1981a). It remains obscure how effective this cytotoxic mechanism is for eliminating tumor cells *in vivo*, but there is some evidence to suggest that the suppression of growth of human colon tumors in nude mice by IgG monoclonal antibodies may be mediated by antibody-dependent cellular cytotoxicity (Herlyn *et al.*, 1980). However, although monoclonal antibodies specifically or preferentially reactive with human tumor cells may have limited value in the direct treatment of progressive disease, they do offer new approaches for therapy and diagnosis. Already, radiolabeled monoclonal antibodies have been shown to localize in tumor deposits with experimental animal tumors (Pimm *et al.*, 1981; Levine *et al.*, 1980) and with human tumors grown in immunodeprived mice (Moshakis *et al.*, 1981) so that radiodiagnostic imaging of small inaccessible tumors becomes a feasible objective. Comparably, since there is a degree of specificity in the localization of antibody to tumors, monoclonal antibodies may be employed as carriers for anticancer agents. As carriers for chemotherapeutic agents, this procedure enhances the selectivity of action of the drug against malignant cells and may reduce damage to normal cells. With the rat mammary carcinoma, Sp4, adriamycin has been coupled to the specific anti-Sp4 monoclonal antibody and conjugates significantly retarded Sp4 tumor at 1/20th of the effective dose of free drug and in some cases brought about total regression (Pimm *et al.*, 1981).

Based upon the above arguments it should also be possible to link highly toxic substances to monoclonal antibodies for application in therapy. In this respect, attention has been directed to the use of bacterial and plant toxins, such as diphtheria toxin, abrin, and ricin, since tumor cells can be killed by interaction with minute amounts, down to the level of single molecules per cell. To exploit these properties, diphtheria toxin has been enzymatically cleaved and the separated A chain (the toxic moiety) linked to anti-human colon carcinoma monoclonal antibody (Gilliland *et al.*, 1980). Such conjugates not only bound specifically to the relevant tumor targets *in vitro* but also markedly inhibited protein synthesis in colon carcinoma cells. Comparably, the A chain of ricin (one of the most powerful toxins known; Olsnes and Pihl, 1982) has been coupled to anti-human colon carcinoma monoclonal antibody and has been used *in vitro* as a specific reagent for killing tumor cells (Gilliland *et al.*, 1980).

Other examples of toxins and their active fragments linked to monoclonal antibodies are being reported (Blythman *et al.*, 1981; Youle and Neville, 1980) so that the therapeutic approach with these conjugates is not only theoretically appealing but presents itself as a practical alternative for the control of malignant disease. Future studies will resolve the extent of application of these new, specific, and potent reagents.

ACKNOWLEDGMENT. These studies were supported by a departmental grant from the Cancer Research Campaign.

VII. REFERENCES

Accolla, R. S., Carrel, S., Phan, M., Henmann, D., and Mach, J. -P., 1979, First report of the production of somatic cell hybrids secreting antibodies specific for carcinoembryonic antigen, *Protides Biol. Fluids* **27**:31.

Arklie, J., and Bodmer, W. F., 1981, Monoclonal antibodies to human carcinoma cell lines, *Br. J. Cancer* **43**:563.

Baldwin, R. W., and Embleton, M. J., 1969, Immunology of spontaneously arising rat mammary adenocarcinomas, *Int. J. Cancer* **4**:430.

Baldwin, R. W., and Embleton, M. J., 1970, Detection and isolation of tumour-specific antigen associated with a spontaneously arising rat mammary carcinoma, *Int. J. Cancer* **6**:373.

Baldwin, R. W., and Embleton, M. J., 1977, Assessment of cell-mediated immunity to human tumor-associated antigens, *Int. Rev. Exp. Path.* **17**:49.

Baldwin, R. W., and Price, M. R., 1981, Neoantigen expression in chemical carcinogenesis, *in* "Cancer: A Comprehensive Treatise" (F. F. Becker, ed.), 2nd ed., pp. 507–548, Plenum Press, New York.

Bernstein, I. D., Tam, M. R., and Nowinski, R. C., 1980, Mouse leukemia: Therapy with monoclonal antibodies against a thymus differentiation antigen, *Science* **107**:68.

Berry, J., Takei, F., and Lennox, E., 1981, Anti-colon carcinoma antibodies analysed with immunofluorescence in frozen sections, *Br. J. Cancer* **43**:562.

Bloom, B. R., and David, J. R., 1976, "In Vitro Methods in Cell Mediated and Tumor Immunity," Academic Press, New York.

Blythman, H. E., Casellas, P., Gros, O., Gros, P., Jansen, F. K., Paolucci, F., Pau, B., an Vidal, H., 1981, Immunotoxins: Hybrid molecules of monoclonal antibodies and a toxin subunit specifically kill tumor cells, *Nature* (*London*) **290**:145.

Brown, J. P., Klitzman, J. M., Hellström, I., Nowinski, R. C., and Hellström, K. E., 1978, Antibody response of mice to chemically induced tumors, *Proc. Natl. Acad. Sci. USA* **75**:955.

Brown, J. P., Wright, P. W., Hart, C. E., Woodbury, R. G., Hellström, K. E., and Hellström, I., 1980, Protein antigens of normal and malignant human cells identified by immunoprecipitation with monoclonal antibodies, *J. Biol. Chem.* **255**:4980.

Byers, V. S., and Baldwin, R. W., 1980, Tumor immunology, *in* "Basic and Clinical Immunology" (H. H. Fudenberg, D. P. Stites, J. L. Caldwell, and J. V. Wells, eds.), 3rd ed., pp. 296–312, Lange, Los Altos.

Byers, V. S., and Johnston, J. O., 1977, Antigenic differences among osteogenic sarcoma tumor cells taken from different locations in human tumors, *Cancer Res.* **37**:3173.

Canevari, S., Fossati, G., Della Porta, G., and Balzarini, G. P., 1978, Humoral cytotoxicity in melanoma patients and its correlation with the extent and course of the disease, *Int. J. Cancer* **16**:722.

Carey, R. E., Takahashi, T., Resnick, L. A., Öettgen, H. F., and Old, L. J., 1976, Cell surface antigens of human malignant melanoma: Mixed hemadsorption assays for humoral immunity to cultured autologous melanoma cells, *Proc. Natl. Acad. Sci. USA* **73**:3278.

Carey, T. E., Lloyd, K. E., Takahashi, T., Travasses, L. R., and Old, L. J., 1979, A cell surface antigen of human malignant melanoma: Solubilization and partial characterization, *Proc. Natl. Acad. Sci. USA* **76**:2898.

Carrel, S., Accolla, R. S., Carmagnola, A. L., and Mach, J. -P., 1980, Common human melanoma-associated antigen(s) detected by monoclonal antibodies, *Cancer Res.* **40**:2523.

Cornain, S., DeVries, J. E., Collard, J., Vennegoor, C., Van Wingerden, I., and Rümke, Ph. R., 1975, Antibodies and antigen expression in human melanoma detected by the immune adherence test, *Int. J. Cancer* **16**:981.

Croce, C. M., Linnenbach, A., Hall, W., Steplewski, Z., and Koprowski, H., 1980, Production of human hybridomas secreting antibodies to measles virus, *Nature* (*London*) **288**:488.

Dawood, F. A., Price, M. R., Embleton, M. J., and Baldwin, R. W., 1981, Detection of human osteogenic sarcoma cell surface antigens using radiolabelled anti-tumor monoclonal antibodies, *Br. J. Cancer*, **44**:310.

DeLeo, A. B., Shiku, H., Takahashi, T., John, M., and Old, L. J., 1977, Cell surface antigens of chemically induced sarcomas in the mouse, *J. Exp. Med.* **146**:720.

Dippold, W. G., Lloyd, K. O., Li, L. T. C., Ikeda, H., Öettgen, H. F., and Old, L. J., 1980, Cell surface antigens of human malignant melanoma: Definition of six antigenic systems with mouse monoclonal antibodies, *Proc. Natl. Acad. Sci. USA* **77**:6114.

Dubois, G. C., Appella, E., Law, L. W., DeLeo, A. B., and Old, L. J., 1980, Immunogenic properties of soluble cytosol fractions of Meth A sarcoma cells, *Cancer Res.* **40**:4204.

Embleton, M. J., and Baldwin, R. W., 1980, Antigenic changes in chemical carcinogenesis, *Br. Med. Bull.* **36**:83.

Embleton, M. J., Price, M. R., and Baldwin, R. W., 1980, Demonstration and partial purification of common melanoma-associated antigen(s), *Eur. J. Cancer* **16**:575.

Embleton, M. F., Gunn, B., Byers, V. S., and Baldwin, R. W., 1981, Antitumor reactions of monoclonal antibody against a human osteogenic-sarcoma cell line, *Br. J. Cancer* **43**:582.

Everson, T. C., and Cole, W. H., 1966, "Spontaneous Regression of Cancer," Saunders, Philadelphia, Pennsylvania.

Fazekas de St. Groth, S., 1979, The quality of antibodies and cellular receptors, *in* "Immunological Methods" (I. Lefkovits and B. Pernis, eds.), pp. 1–42, Academic Press, New York.

Fraker, P. J., and Speck, J. C., 1978, Protein and cell membrane iodinations with a sparingly soluble chloramide, 1, 3, 4, 6-tetrachloro - 3α, 6α, - diphenylglycoluril, *Biochem. Biophys. Res. Commun.* **80**:849.

Galfre, G., Howe, S. C., Milstein, C., Butcher, G. W., and Howard, J. C., 1977, Antibodies to major histocompatibility antigens produced by hybrid cells, *Nature* (*London*) **266**:550.

Galfre, G., Milstein, C., and Wright, B., 1979, Rat x rat hybrid myelomas and a monoclonal anti-Fd portion of mouse IgG, *Nature* (*London*) **277**:131.

Galloway, D. R., Walker, L. E., and Ferrone, S., 1980, Isolation and characterization of human tumor associated antigens with monoclonal antibody, *Proc. Am. Assoc. Cancer Res.* **21**:25.

Gilliland, D. G., Steplewski, Z., Collier, R. J., Mitchell, K. F., Chang, T. H., and Korprowski, H., 1980, Antibody-directed cytotoxic agents: Use of monoclonal antibody to direct the action of toxin A chains to colorectal carcinoma cells, *Proc. Natl. Acad. Sci. USA* **77**:4539.

Goodfellow, P. N., Lennson, J. R., Williams, V. E., and McDevitt, D. O., 1979, Monoclonal antibodies reacting with murine teratocarcinoma cells, *Proc. Natl. Acad. Sci. USA* **76**:377.

Grant, R. M., Finan, P. J., Lennox, E., and Bleehen, N. M., 1981, Demonstration of activity of monoclonal antibodies on colonic tissue using an indirect immunoperoxidase technique, *Br. J. Cancer*, **44**:298.

Gunn, B., Embleton, M. J., Middle, J., and Baldwin, R. W., 1980, Monoclonal antibody against a naturally occuring rat mammary carcinoma, *Int. J. Cancer* **26**:325.

Harder, F. H., and McKhann, C. F., 1968, Demonstration of cellular antigens on sarcoma cells by an indirect ^{125}I-labelled antibody technique, *J. Natl. Cancer Inst.* **40**:231.

Hellström, I., Hellström, K. E., and Yeh, M.-Y., 1981a, Lymphocytedependent antibodies to antigen 3.1, a cell-surface antigen expressed by a subgroup of human melanomas, *Int. J. Cancer* **27**:281.

Hellström, K. E., Hellström, I., Brown, J. P., Yeh, M-Y., Woodbury, R. G., and Nishiyama, K., 1981b, Two antigens identified on human melanoma cells by monoclonal antibodies, *Br. J. Cancer* **43**:560.

Herberman, R. B., 1974, Cell-mediated immunity to tumor cells, *Adv. Cancer Res.* **19**:207.

Herlyn, D. M., and Koprowski, H., 1981, Monoclonal anticolon carcinoma antibodies in complement dependent cytotoxicity, *Int. J. Cancer* **27**:769.

Herlyn, D., Herlyn, M., Steplewski, Z. and Koprowski, H., 1979, Monoclonal antibodies in cell-mediated cytotoxicity against human melanoma and colorectal carcinoma, *Eur. J. Immunol.* **9**:657.

Herlyn, D., Steplewski, Z., Herlyn, M., and Koprowski, H., 1980, Inhibition of growth of colorectal carcinoma in nude mice by monoclonal antibody, *Cancer Res.* **40**:717.

Hewitt, H. B., Blake, E. R., and Walder, A. S., 1976, A critique of the evidence for active host defence against cancer, based on personal studies of 27 murine tumors of spontaneous origin, *Br. J. Cancer* **33**:241.

Hollinshead, A. C., Stewart, T. H. M., and Herberman, R. B., 1974, Delayed-hypersensitivity reactions to soluble membrane antigens of human malignant lung cells, *J. Natl. Cancer Inst.* **52**:327.

Horibata, K., and Harris, A. W., 1970, Mouse myelomas and lymphomas in culture, *Exp. Cell. Res.* **60**:61.

Howe, A. J., Seeger, R. C., Molinaro, G. A., and Ferrone, S., 1981, Analysis of human tumor cells for Ia-like antigens with monoclonal antibodies, *J. Natl. Cancer Inst.* **66**:827.

Imai, K., and Ferrone, S., 1980, An indirect rosette microassay to characterize human melanoma associated antigens (MAA) recognized by operationally specific xenoantisera, *Cancer Res.* **40**:2252.

Johnson, J., Meo, T., Hadam, M., Reithmuller, G., 1981, Mouse monoclonal antibodies raised against human T lymphocytes and melanoma cells, *Protides Biol. Fluids* **28**: 533.

Kearney, J. F., Radbruch, A., Liesegang, B., and Rajewsky, K., 1979, A new mouse myeloma cell line that has lost immunoglobulin expression but permits the construction of antibody-secreting hybrid cell lines, *J. Immunol.* **123**:1548.

Kessler, S. W., 1976, Cell membrane antigen isolation with the Staphylococcal Protein A-antibody absorbent, *J. Immunol.* **117**:1482.

Köhler, G., and Milstein, C., 1975, Continuous cultures of fused cells secreting antibody of predefined specificity, *Nature (London)* **256**:495.

Köhler, G., Howe, C. S., and Milstein, C., 1976, Fusion between immunoglobulin -secreting and non-secreting myeloma cell lines, *Eur. J. Immunol.* **6**:292.

Koprowski, H., 1981, Monoclonal antibodies in the study of tumour antigens, *Br. J. Cancer* **43**:560.

Koprowski, H., Steplewski, Z., and Herlyn, D., 1978, Study of antibodies against human melanoma produced by somatic cell hybrids, *Proc. Natl. Acad. Sci. USA* **75**:3405.

Koprowski, H., Steplewski, Z., Mitchell, K., Herlyn, M., Herlyn, D., and Fuhrer, P., 1979, Colorectal carcinoma antigens detected by hybridoma antibodies, *Somat. Cell Genet.* **5**:957.

Lee, J. S., Trowsdale, J., and Bodmer, W. F., 1980, Synthesis of HLA antigens from membrane associated messenger RNA, *J. Exp. Med.* **152**:3s.

Lennox, E. S., 1980, The antigens of chemically induced tumours, *in* "Progress in Immunology IV" (M. Fougereau and J. Dausset, eds.), pp. 659–667, Academic Press, New York/London.

Levine, G., Ballou, B., Relland, J., Solter, D., Gumerman, L., and Hakala, T., 1980, Localization of I-131-labelled tumor specific monoclonal antibody in the tumor-bearing Balb/c mouse, *J. Nucl. Med.* **21**:570.

Levy, J. P., and Leclerc, J. C., 1977, The murine sarcoma virus-induced tumor: Exception or general model in tumor immunology, *Adv. Cancer Res.* **24**:1.

Lewis, M. G., and Philipps, T. M., 1972, Separation of two distinct tumor-associated antibodies in the serum of melanoma patients. *J. Natl. Cancer Inst.* **49:**915.

Lilly, F., and Steeves, R., 1974, Antigens of murine leukemia viruses, *Biochim. Biophys. Acta* **355:**105.

Littlefield, J. W., 1974, Selection of hybrids from matings of fibroblasts in vitro and their presumed recombinants, *Science* **145:**709.

Lostrom, M. E., Stone, M. R., Tam, M., Barnette, W. N., Pinter, A., and Nowinski, R. C., 1979, Monoclonal antibodies against murine leukemia viruses: Identification of six antigenic determinants on the p15 (E) and gp70 envelope proteins, *Virology* **98:**336.

McCabe, R. P., Indiveri, F., Galloway, D. R., Ferrone, S., and Reisfeld, R. A., 1980, Lack of association of serologically detectable human melanoma associated antigens with beta 2 microglobulin: Serologic and immunochemical evidence, *J. Natl. Cancer Inst.* **65:**703.

McConahey, P. J., and Dixon, F. J., 1966, A method of trace iodination of proteins for immunological studies, *Int. Arch. Allergy* **29:**185.

Middle, J. G., and Embleton, M. J., 1981, Immune responses to naturally occuring rat sarcomas, *Br. J. Cancer* **43:**44.

Mitchell, K. F., 1980, A carcinoembryonic antigen (CEA) specific monoclonal hybridoma antibody that reacts only with high-molecular-weight CEA, *Cancer Immunol. Immunother.* **10:**1.

Mitchell, K. F., Fuhrer, P., Steplewski, Z., and Koprowski, H., 1981, Structural characterization of the 'melanoma-specific' antigen detected by monoclonal antibody 69115 Nu-4-B, *Mol. Immunol.* **18:**207.

Moore, M., and Hughes, L. A., 1973, Circulating antibodies in human connective tissue malignancy, *Br. J. Cancer* **28:** (Suppl. I): 175.

Morton, D. L., and Malmgren, R. A., 1968, Human osteosarcomas: Immunologic evidence suggesting an associated infectious agent, *Science* **162:**1279.

Morton, D. L., Malmgren, R. A., Holmes, E. C., and Ketcham, A. S., 1968, Demonstration of antibodies against human melanoma by immunoflorescence, *Surgery* **64:**233.

Moshakis, V., McIlhinney, A. J., Raghavan, D., and Neville, A. M., 1981, Monoclonal antibodies to detect human tumours: An experimental approach, *J. Clin. Pathol.* **34:**314.

Murray, E. Raygrok, S., Milton, G. W., and Hersey, P., 1978, Analysis of serum blocking factors against leukocyte-dependent antibody in melanoma patients, *Int. J. Cancer* **21:**578.

Nadler, L. M., Stashenko, P., Hardy, R., and Schlossman, S. F., 1980a, A monoclonal antibody defining a lymphoma associated antigen in man, *J. Immunol.* **125:**570.

Nadler, L. M., Stashenko, P., Hardy, R., Kaplan, W. P., Button, L. D., Kufe, D. W., Antman, K. H., and Schlossman, S. F., 1980b, Serotherapy of a patient with a monoclonal antibody directed against a human lymphoma-associated antigen, *Cancer Res.* **40:**3147.

Olsnes, S., and Pihl, A., 1982, Chimaeric toxins, *in* "Pharmacology of Bacterial Toxins" (J. Drews and F. Dorner, eds.), Pergamon, New York, in press.

Olsson, L., and Kaplan, H. S., 1980, Human-human hybridomas producing monoclonal antibodies of predifined antigenic specificity, *Proc. Natl. Acad. Sci. USA* **77:**5429.

Pimm, M. V., Jones, J. A., Price, M. R., Middle, J. G., Embleton, M. J., and Baldwin, R. W., 1981, Tumour localization of monoclonal antibody against a rat mammary carcinoma and suppression of tumour growth with adriamycin-antibody conjugates, *Cancer Immunol. Immunother.* **12:**125.

Pleogh, H. L., Canon, L. E., and Strominger, J. L., 1979, Cell-free translation of the mRNAs for the heavy and light chains of HLA-A and HLA-B antigens, *Proc. Natl. Cancer Inst. USA* **76:**2273.

Price, M. R., 1978, A microassay for the detection of tumour-specific complement dependent serum cytotoxicity against a chemically induced rat hepatoma, *Transplantation* **25**:224.

Price, M. R., Dennick, R. G., Chen, Yu Fang, Hannant, D., Embleton, M. J., Gunn, B., and Baldwin, R. W., 1981, Characteristics of two anti-tumour monoclonal antibody preparations, *Arch. Geschwulstforsch.* **51**:302.

Rogers, G. T., 1976, Heterogeneity of carcinoembryonic antigen: Implications on its role as a tumour marker substance *Biochim. Biophys. Acta* **458**:355.

Rogers, G. T., Rawlins, G. A., and Bagshawe, K. D., 1981a, Somatic-cell hybrids producing antibodies against CEA, *Br. J. Cancer* **43**:1.

Rogers, G. T., Rawlins, G. A., and Bagshawe, K. D., 1981b, Monoclonal anti-CEA antibodies, *Br. J. Cancer* **43**:562.

Rosenberg, S., 1980, "Serologic Analysis of Human Cancer Antigens," Academic Press, New York/London.

Schlom, J., Wunderlich, D., and Teramoto, Y. A., 1980, Generation of human monoclonal antibodies reactive with human mammary carcinoma cells, *Proc. Natl. Acad. Sci. USA* **77**:6841.

Shiku, H., Takahashi, T., Reswick, L. A., Öettgen, H. F., and Old, L. J. 1977, Cell surface antigens of human malignant melanoma. III. Recognition of autoantibodies with unusual characteristics, *J. Exp. Med.* **145**:784.

Shulman, M., Wilde, C. D., and Köhler, G., 1978, A better cell line for making hybridomas secreting specific antibodies, *Nature* (*London*) **276**:269.

Sikora, K., and Phillips, J., 1981, Human monoclonal antibodies to glioma cells, *Br. J. Cancer* **43**:105.

Silver, J., and Hood, L., 1974, Detergent solubilised H-2 alloantigen is associated with a small molecular weight polypeptide, *Nature* (*London*) **249**:764.

Simrell, C. R., and Klein, P. A., 1979, Antibody responses of tumor-bearing mice to their own tumors captured and perpetuated as hybridomas, *J. Immunol.* **123**:2386.

Stahli, C., Staehelin, T., Miggiano, V., Schmidt, J., and Haring, P., 1980, High frequencies of antigen-specific hybridomas: Dependence on immunization parameters and prediction by spleen cell analysis, *J. Immunol. Methods.* **32**:297.

Steplewski, Z. Herlyn D., Clarke, W. H., and Koprowski, H., 1979, Reactivity of monclonal antibodies with melanoma cells freshly isolated from primary and metastatic melanoma, *Eur. J. Immunol.* **9**:94.

Stern, P. L., Willison, K. R., Lennox, E. S., Galfre, G., Milstein, C., Secher, D., Ziegler, A., and Springer, T., 1978, Monoclonal antibodies as probes for differentiation and tumor-associated antigen: A Forssman-specificity on teratocarcinoma stem cells, *Cell* **14**:775.

Stone, M. R., and Nowinski, R. C., 1980, Topological mapping of murine leukemia virus proteins by competition binding assays with monoclonal antibodies, *Virology* **100**:370.

Thomson, D. M. P., Gold, P., Freedman, S. O., and Shuster, J., 1976, The isolation and characterization of tumor-specific antigens of rodent and human tumors, *Cancer Res.* **36**:3518.

Thomson, D. M. P., Tataryn, D. N., O'Connor, R., Rauch, J., Friedlander, P., Gold, P., and Shuster, J., 1979, Evidence for the expression of human tumor-specific antigens associated with β-2 microglobulin in human cancer and in some adenomas and benign breast lesions, *Cancer Res.* **39**:604.

Ting, C. C., and Herberman, R. B., 1976, Humoral host defense mechanisms, *Int. Rev. Exp. Pathol.* **15**:93.

Tsung, Y -K., Milunsky, A., and Alpert, E., 1980, Secretion by a hybridoma of antibodies against α-fetoprotein, *N. Engl. J. Med.* **302**:180.

Vandebark, A. A., Greene, M. H., Burger, D. R., Vetto, R. M., and Reimer, R. R., 1979,

Immune response to melanoma extracts in three melanoma-prone families, *J. Natl. Cancer Inst.* **63:**1147.

Weese, J. L., Herberman, R. B., Hollinshead, A. C., Cannon, G. B., Keels, M., Kibrite, A., Morales, A., Char, D. H., and Oldham, R. K., 1978, Specificity of delayed cutaneous hypersensitivity reactions to extracts of human tumor cells, *J. Natl. Cancer Inst.* **60:**255.

Williams, A. F., 1977, Differentiation antigens of the lymphocyte cell surface, *Contemp. Top. Mol. Immunol.* **6:**83.

Willison, K. R., and Stern, P. L., 1978, Expression of a Forssman antigenic specificity in the pre-implantation mouse embryo, *Cell* **14:**785.

Woodruff, M. F. A., 1980, "The Interaction of Cancer and Host," Grune & Stratton, New York.

Yeh, M.-Y., Hellström, I., Brown, J. P., Warner, G. A., Hansen, J. A., and Hellström, K. E., 1979, Cell surface antigens of human melanoma identified by monoclonal antibody, *Proc. Natl. Acad. Sci. USA* **77:**5483.

Youle, R. J., and Neville, D. M., Jr., 1980, Anti-Thy 1.2 monoclonal antibody linked to ricin is a potent cell-type specific toxin, *Proc. Natl. Acad. Sci. USA* **77:**5483.

Chapter 11

Modulation of Immune Lysis of Tumor Cells by Interferon

A. K. Ng, K. Imai, M. A. Pellegrino, A. Vitiello, F. Indiveri, B. S. Wilson, and S. Ferrone

Departments of Pathology and Surgery
College of Physicians and Surgeons
Columbia University
New York, New York

I. INTRODUCTION

Interferons are a family of proteins produced by different cell types in response to virus infection or other stimuli, which can inhibit virus replication in cells, possess antiproliferative activity, and modulate a variety of immunological phenomena (Stewart, 1979). Three distinct human interferons have been described (Stewart *et al.*, 1980): (1) human leukocyte interferon or α-interferon (IFN-α),* which is induced in blood buffy-coat leukocytes infected by sendai virus and is a glycoprotein composed of two subunits of approximately 21,000 and 15,000 daltons; (2) human fibroblast interferon or β-interferon (IFN-β), which is induced in diploid fibroblast treated with a repeating polymer of inosine and cytosine polyribonucleotides and is a 20,000-dalton glycoprotein; and (3) immune interferon or γ-interferon (IFN-γ), which is produced primarily by T lymphocytes in response to mitogens or antigens and the biochemical nature

* Abbreviations used: ADCC, antibody-dependent cellular cytotoxicity; β_2-μ, β-2-microglobulin; HLA, human histocompatibility antigen; HMW-MAA, high-molecular-weight melanoma-associated antigen; Ia, immune response gene-associated antigen; IFN, interferon; K, killer cells that mediate ADCC; MAA, melanoma-associated antigen; MoAb, monoclonal antibody; NK, natural killer cells that mediate spontaneous cytotoxicity; SpA, purified protein A from *Staphylococcus aureus* Cowan I; SDS, sodium dodecyl sulfate.

of this type of interferon is still unclear. Human interferons were initially used in clinical trials as an antiviral agent, however, the more recent discovery that interferon shows an antitumor effect in animal studies has led to its use in pilot studies as an anticancer agent (Gresser, 1977; Merigan *et al.*, 1978; Strander *et al.*, 1978; Gutterman *et al.*, 1980; Osserman *et al.*, 1980). In view of the potential antitumor effect of human interferon, we have investigated its ability to modulate membrane antigen expression and immune lysis of human tumor cells *in vitro*. In this chapter we will summarize the results of our studies which have utilized human leukocyte interferon and monoclonal xenoantibodies to human melanoma-associated antigens (MAAs) and to histocompatibility antigens (HLA). Specifically, we will discuss the effect of interferon on the expression of MAA and HLA antigens as well as on three types of immune lysis of tumor cells: the latter includes complement-dependent antibody-mediated lysis, antibody-dependent, K cell-mediated cytotoxicity (ADCC), and natural killer (NK) cell-mediated cytotoxicity. We will not review the extensive literature on interferons and we refer the interested reader to recent publications on the subject (Hovanessian, 1979; Borden, 1979; Krim, 1980a,b; Stringfellow, 1981; Vilcek *et al.*, 1980; Weissman and Droller, 1980); we will mention the work by other investigators as it pertains to the discussion and interpretation of our own data.

II. MELANOMA-ASSOCIATED ANTIGENS AND HISTOCOMPATIBILITY ANTIGENS IDENTIFIED BY MONOCLONAL ANTIBODIES

Three types of plasma membrane-bound MAAs identified by monoclonal antibodies (MoAbs) will be discussed. One type of MAA is identified by MoAbs 225.28S, 653.40S, and 763.24T, which recognize distinct antigenic determinants on this structure. This antigen is composed of two high-molecular-weight glycopolypeptides, one at about 280,000 and the other at >440,000 (Fig. 1). The structure of this high-molecular-weight MAA (HMW-MAA) is, however, much more complex since analysis of ^{125}I surface-labeled melanoma cell extracts reveals several lower-molecular-weight polypeptides in addition to the 280K and >440K glycopolypeptides. None of the various components of the HMW-MAA are bridged by disulfide bonds since no higher-molecular-weight forms were seen when electrophoresis was performed in the absence of reducing agents (Fig. 1). Recent results from peptide-mapping experiments show that the

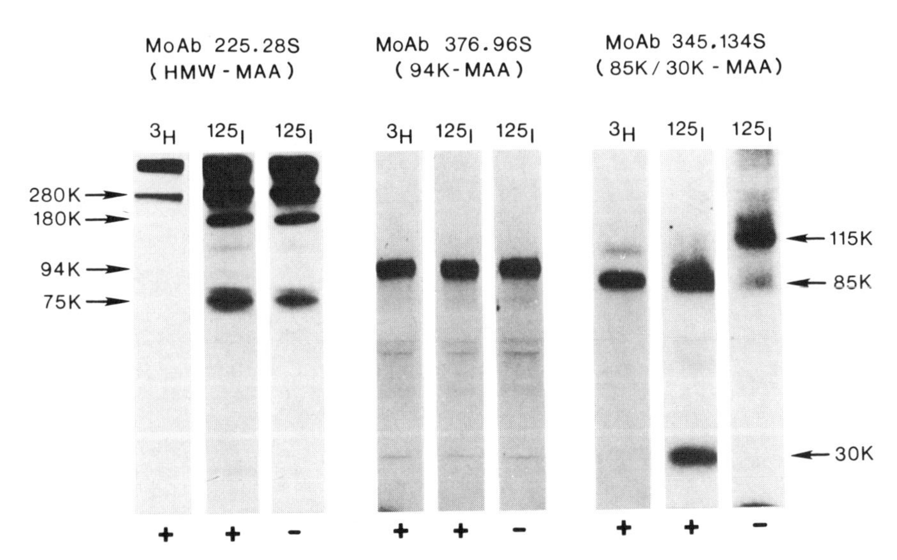

Fig. 1. Molecular profile of human MAAs identified with MoAbs. NP40 extracts of melanoma cells M21 either synthetically labeled with [^{3}H]glucosamine or surface labeled with ^{125}I were immunoprecipitated with insolubilized MoAbs 225.28S to high-molecular-weight MAA (left panel), 376.96S to 94K MAA (central panel), and 345.134S to 85K/30K MAA (right panel). The antigens were eluted with 2% sodium dodecyl sulfate (SDS) and then electrophoresed under reducing (2% 2-mercaptoethanol) and nonreducing conditions in a 7.5% polyacrylamide slab gel. Mobilities of molecular-weight markers are indicated by arrows.

various components of the HMW-MAA have a similar peptide composition and that the molecular-weight heterogeneity may result from various combinations of several subunits (B. S. Wilson, unpublished observations). Immunofluorescence analysis of cryostat thin sections of various normal human tissues, benign skin lesions, and tumor biopsies with the MoAb 225.28S showed that the HMW-MAA could be detected in melanomas, skin carcinomas, and nevi but were not detectable in any other normal tissues or benign skin lesions tested including skin melanocytes (Natali *et al.*, 1981a; Wilson *et al.*, 1981b).

The second type of MAA is a single-chain 94,000-dalton glycoprotein detected by the MoAb 376.96S (Fig. 1) (Wilson *et al.*, 1981a). This 94K MAA is not disulfide bridged in its native state and is similar to the HMW-MAA being readily solubilized in the absence of detergent, thus making it peripheral rather than integral to the plasma membrane (Wilson *et al.*, 1981a). Indirect immunofluorescence of cryostat thin sections of normal human tissues, benign skin lesions, and tumor biopsies with the MoAb

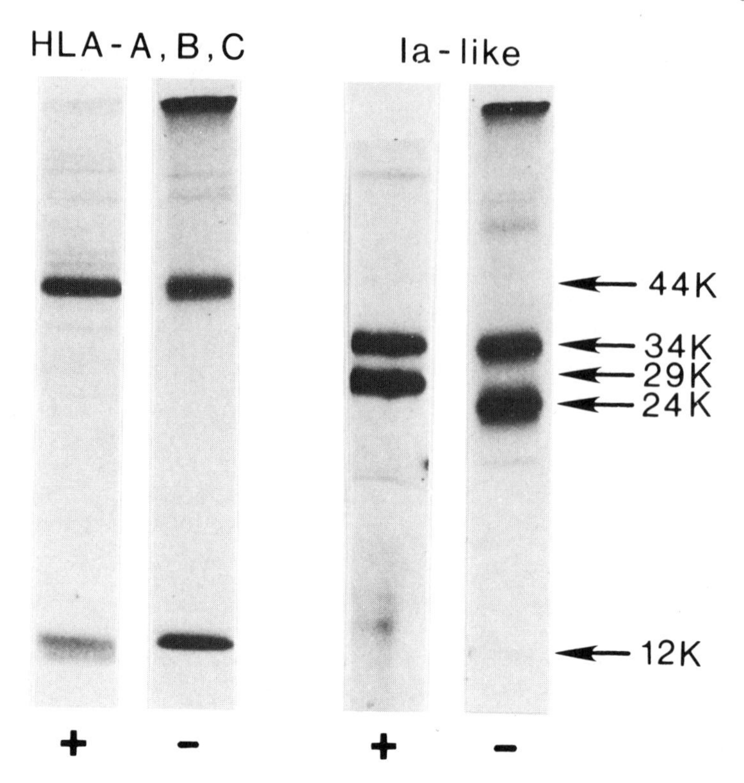

Fig. 2. Molecular profile of histocompatibility antigens synthesized by cultured human melanoma cells. NP40 extracts of [^{3}H]phenylalanine biosynthetically labeled melanoma cells M21 were immunoprecipitated with insolubilized MoAbs 6/31 to HLA-A, B antigens (left panel) and Q5/13 to Ia-like antigens (right panel). The antigens were eluted with 2% SDS and then electrophoresed under reducing (2% 2-mercaptoethanol) and nonreducing conditions in a 12.5% polyacrylamide slab gel. Mobilities of molecular-weight markers are indicated by arrows.

376.96S revealed that the 94K MAA is expressed in small amounts on stomach epithelium and basal layer of skin and in most cases of nevi, melanoma, and some cases of breast carcinoma (Wilson *et al.*, 1981a). The 94K MAA has not been detected on all other tissues and tumors tested including benign skin lesions and melanocytes.

The third type of MAA is identified by the MoAb 345.134S and is composed of an 85,000 dalton glycopolypeptide bridged by disulfide bonds to a 30,000 dalton polypeptide having little if any carbohydrate (Fig. 1). Like the other two MAA described before, this 85K/30K MAA is peripheral rather than integral in association with the plasma cell membrane

(Imai *et al.*, 1982a). Immunofluorescence analysis of cryostat thin sections of normal human tissues, benign skin lesions, and tumor biopsies with the MoAb 345.134S shows a complex pattern of reactions. Strong staining was seen in the epithelium of the sebaceous glands and the basal layer of pigmented skin while weak staining was seen in epithelium of gastrointestinal tract, urinary bladder, renal proximal tubules, parotid, and thyroid. Strong staining was also observed in melanoma, nevi, and carcinomas of the breast, lung, skin, gastrointestinal tract, and urinary bladder. Benign skin lesions and skin melanocytes were negative as were brain tumors and carcinomas of the mammary gland, lung, and liver (Imai *et al.*, 1982a).

HLA antigens are composed of two major classes of molecules, the HLA-A,B,C antigens, and the Ia-like antigens. The HLA-A,B,C antigen is a two-chain structure composed of a 44,000-dalton glycopolypeptide, referred to as heavy chain, associated noncovalently with β_2-microglobulin (β_2-μ), a 12,000-dalton polypeptide (Fig. 2). MoAb 6/31 to public determinants on the heavy chain (Russo *et al.*, 1981) and MoAb NAMB-1 to β_2-μ (Ng *et al.*, 1981a) have been used to identify HLA-A,B,C molecules on melanoma cells. Ia-like antigens, the human counterpart of the murine I–E subregion antigens (for review, see Ferrone *et al.*, 1978a), are more restricted in their tissue distribution than HLA-A,B,C antigens (Natali *et al.*, 1981b) and may unexpectedly appear on cells which undergo malignant transformation (Winchester *et al.*, 1978; Wilson *et al.*, 1979; Natali *et al.*, 1981c; Pellegrino *et al.*, 1981a). Ia-like antigens are composed of two noncovalently associated glycopolypeptides: one with an apparent molecular weight of 34,000 is referred to as α chain and the other one with an apparent molecular weight of 27,000 is referred to as β chain (Fig. 2). In the absence of reducing agents, the β chain migrates with an apparent molecular weight of 24,000 (Fig. 2), suggesting the presence of an intrachain disulfide bond which is broken by reduction with consequent slowing of migration due to some unfolding of the subunit. Human Ia-like antigens have been identified with the MoAb Q5/13 to a common determinant of this molecule (Quaranta *et al.*, 1981).

III. EFFECT OF INTERFERON ON THE EXPRESSION OF MELANOMA-ASSOCIATED ANTIGENS AND HISTOCOMPATIBILITY ANTIGENS

We have chosen not to use the classical complement-dependent cytotoxicity assay to measure alterations in the expression of antigens on interferon-treated cells, and instead we utilized two binding assays since

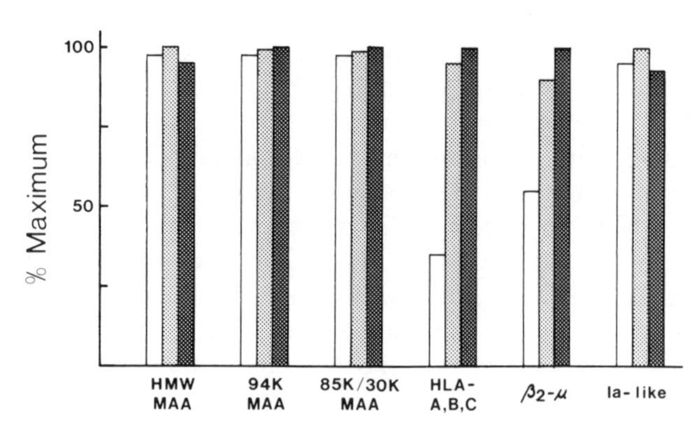

Fig. 3. Effect of human leukocyte interferon on the expression of MAAs and HLA on cultured human melanoma cells M21. The cells were incubated at 37°C for 16 hr in medium alone (□) or in medium containing interferon [final concentrations: 500 (▨) and 2000 U/ml (■)] and then reacted with an excess of MoAbs 225.28S to high-molecular-weight MAA, 376.96S to 94K MAA, 345.134S to 85K/30K MAA, 6/31 to HLA-A, B antigens, NAMB-1 to human β_2-μ, and Q5/13 to Ia-like antigens. The amount of antibody bound was measured by the uptake of ^{125}I-protein A.

the former assay has several drawbacks and may result in erroneous estimates of antigen density (for review, see Ferrone and Pellegrino, 1973). In the radioimmunometric binding assay, target cells are incubated with dilutions of antibody and the amount of bound antibody is determined by the uptake of ^{125}I-labeled protein A from *Staphylococcus aureus* Cowan I strain (^{125}I-SpA) (McCabe *et al.*, 1979) while in the indirect rosette microassay, the sensitized target cells are rosetted with sheep erythrocytes chemically coated with SpA (Imai and Ferrone, 1980). Incubation of melanoma cells with 2000 U/ml human leukocyte interferon for 16 hr did not cause any significant change in the levels of HMW-MAA, 94K MAA, 85K/30K MAA, and Ia-like antigens detectable on cell surface but it did increase the levels of HLA-A,B antigens and β_2-μ (Imai *et al.*, 1981a) (Fig. 3). The slope of the titration curves of the monoclonal antibodies tested in the indirect rosette assay with interferon-treated cells was similar to that with control target cells, suggesting that most if not all cells treated with interferon exhibit increased levels of HLA-A,B antigens. This interferon-induced augmentation of HLA-A,B antigen and β_2-μ expression is dependent on the concentration of interferon used and the incubation time (Imai *et al.*, 1981a). An increase in the level of HLA-A,B antigens on melanoma cells is first detected at an interferon concentration of 100 U/ml. Higher concentrations of interferon further increases the levels of HLA-A,B antigens on melanoma cells reaching a plateau at concentrations of 2000 U/ml. This effect becomes evident after

9 hr incubation with interferon and requires almost 16 hr before maximal increase in HLA-A,B and β_2-μ expression is reached.

The ability of interferon to selectively increase HLA antigen expression has also been observed with other types of human cells as well as murine cells. Thus, human peripheral blood or cultured lymphoid cells treated with interferon express an increased amount of HLA-A,B antigens (Heron *et al.*, 1978; Attallah and Strong, 1979; Fellous *et al.*, 1979; Perussia *et al.*, 1980; Imai *et al.*, 1981b) but do not change in the expression of Ia-like antigens (Attallah and Strong, 1979; Perussia *et al.*, 1980; Imai *et al.*, 1981b); murine cells treated with interferon show an increased expression of H-2 antigens but do not change in the expression of θ or Ia antigens (Lindahl *et al.*, 1974; Vignaux and Gresser, 1977). It is not known why interferon has a selective effect on HLA-A,B,C antigens and β_2-μ nor whether the mechanism behind this phenomenon involves enhanced biosynthesis, decreased degradation rate, mobilization of molecules from preformed intracellular pools, and/or uncovering of antigenic sites on cell membrane.

IV. EFFECT OF INTERFERON ON THE SUSCEPTIBILITY OF TUMOR CELLS TO IMMUNE LYSIS

A. Complement- and Cell-Dependent Antibody-Mediated Lysis

Melanoma cells which exhibited increased levels of HLA-A,B and β_2-μ expression following interferon treatment become more susceptible to lysis mediated by MoAbs to these antigens (Imai *et al.*, 1981a). The endpoint titers of MoAbs to both HLA-A,B and β_2-μ antigens in the cytotoxicity test with rabbit complement and the extent of ^{51}Cr release in the ADCC assay are significantly increased (Figs. 4 and 5). On the other hand, no significant change can be detected in the extent of immune lysis mediated by MoAbs to antigens which do not change in their cell surface expression following treatment with interferon, i.e., the HMW-MAA, the 94K MAA, the 85K/30K MAA, and the Ia-like antigens (Figs. 4 and 5). These results show that the increased susceptibility of interferon-treated melanoma cells to antibody-mediated lysis results from the increase in antigen density rather than from a nonspecific effect on the cell surface such as an alteration of the integrity of the plasma membrane. These data are in agreement with previous findings that the extent of lysis in ADCC is influenced by the number of antigenic determinants on the target cells (Wiedermann *et al.*, 1975; Lustig and Bianco, 1976; Ohlander

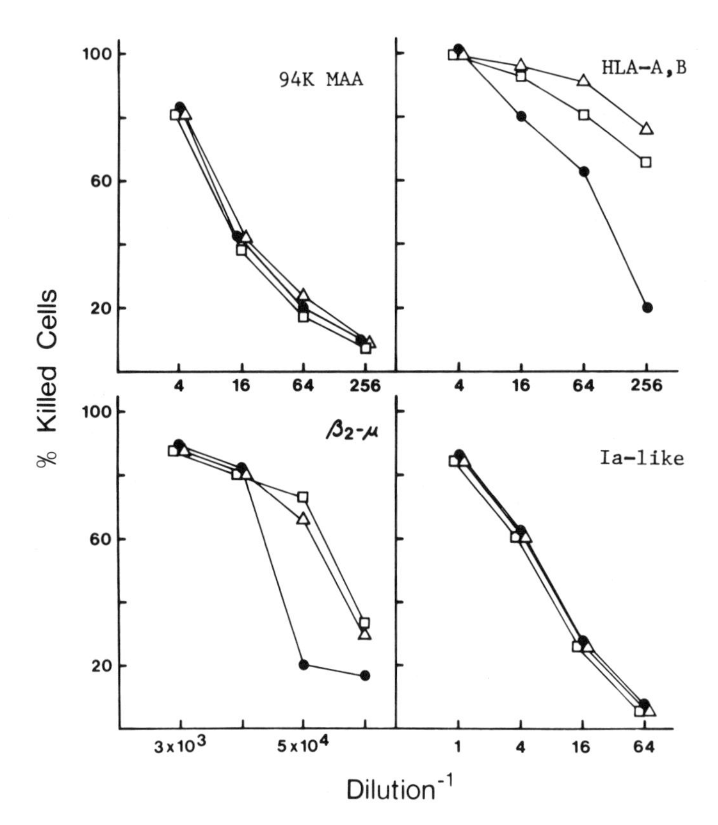

Fig. 4. Effect of human leukocyte interferon on the susceptibility of cultured human melanoma cells M21 to complement-dependent lysis mediated by MoAbs, 376.96S to 94K MAA, 6/31 to HLA-A, B antigens, NAMB-1 to human β_2-μ, and Q5/13 to Ia-like antigens. M21 cells were incubated at 37°C for 16 hr in medium alone (●) or in medium containing interferon [final concentrations: 500 (□) and 2000 U/ml (△)] and then used as targets in a complement-dependent microcytotoxicity assay.

et al., 1981). However, the relationship between density of antigens and extent of antibody-mediated lysis seen with interferon-treated cells is not a general phenomenon because significant variations in the density of antigens have been reported to occur without changing the susceptibility to immune lysis (for review, see Ferrone *et al.*, 1974b). For example, the expression of serologically detectable tumor-associated antigens is significantly higher on melanoma cells in late-log phase rather than melanoma cells in the initial stage of their growth cycle (Curry *et al.*, 1981), yet the endpoint titers of xenoantisera to these antigens did not differ for these two types of target cells when analyzed in the ADCC assays (Curry *et al.*, 1981; Imai *et al.*, 1981a). Furthermore, melanoma cells treated with the antibiotic tunicamycin, an inhibitor of *N*-asparagine-linked glycosylation (Struck and Lennarz, 1977), do not change in their susceptibility to ADCC mediated by MoAbs to MAAs, although the expression of these antigens is significantly reduced by the antibiotic (Imai *et al.*, 1981c). Finally, inhibitors of protein synthesis may increase the susceptibility of cultured melanoma cells to ADCC without affecting the expression of MAAs (Imai *et al.*, 1981c).

The species of complement used in the cytotoxicity assays mediated by MoAbs to HLA-A,B antigens is a critical factor in achieving lysis. Thus, human complement and MoAbs to HLA-A,B antigens are unable to effect lysis of melanoma cells treated with interferon even though this treatment increases the expression of HLA-A,B antigens. Human complement is also unable to effect lysis even when the melanoma cells are sensitized with combinations of antibodies directed to distinct determinants of HLA-A,B antigens or of antibodies directed to HLA-A,B antigens and to other antigenic structures (Imai *et al.*, 1981d). In contrast, rabbit complement is very effective in mediating lysis of melanoma cells treated with a single type of MoAb to HLA-A,B antigens. We believe that the difference in the lytic capabilities of rabbit and human complement is due to natural antibodies to human cell surface antigens (Ferrone *et al.*, 1974a) present in rabbit complement; these antibodies increase the

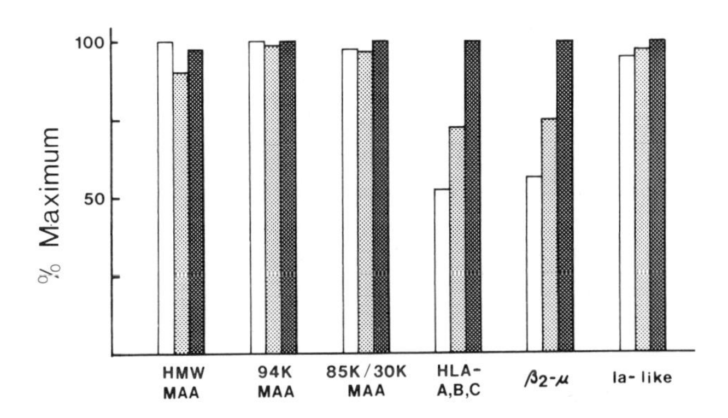

Fig. 5. Effect of human leukocyte interferon on the susceptibility of cultured human melanoma cells M21 to cell-dependent lysis mediated by MoAbs 225.28S to high-molecular-weight MAA, 376.96S to 94K MAA, 345.134S to 85K/30K MAA, 6/31 to HLA-A, B antigens, NAMB-1 to human β_2-μ, and Q5/13 to Ia-like antigens. M21 cells were incubated at 37°C for 16 hr in medium control (□) or in medium containing interferon [final concentrations: 500 (▨) and 2000 U/ml (■)] and then used as targets in a 16-hr ^{51}Cr release ADCC assay. The susceptibility of untreated and interferon-treated M21 cells to ADCC is expressed as the percentage of maximum lysis observed with individual MoAbs.

overall antigen–antibody interactions on the surface of cells sensitized with antibodies to HLA-A,B antigens (Wilson *et al.*, 1981c). If this interpretation is correct then the inability of human complement to mediate cytotoxicity of interferon-treated melanoma cells sensitized with MoAbs to HLA-A,B antigens is probably due to the failure of interferon to increase the concentration of these antigenic sites to the minimum required to achieve lysis.

It is noteworthy that even with rabbit complement we have not observed any synergistic effects of combinations of MoAbs to distinct determinants of HLA-A,B antigens, Ia-like antigens, and MAAs in lysis of target cells, even when the latter had been treated with interferon (unpublished observations). These results differ from those obtained by Howard *et al.* (1979) with MoAbs to rat histocompatibility antigens. These authors showed that combinations of antibodies to distinct determinants of a single antigenic structure are efficient in mediating synergistic lysis. The difference between our results and those by Howard *et al.* (1979) suggest that synergy does not only result from an increase in the total number of antibody molecules bound to cell surface but may also depend on other factors such as spatial relationship between the different determinants identified by the antibodies and between antigenic structures and areas of cell membrane susceptible to lysis.

B. Lysis by Natural Killer Cells

Melanoma cells treated with interferon become less susceptible to lysis by naturally occurring cytotoxic cells in human peripheral blood lymphocytes, i.e., the NK cells (Fig. 6). The properties of these NK cells will be described in the next section. A similar resistance to NK lysis of target cells treated with interferon has been reported with other types of human cells (Trinchieri and Santoli, 1978; Moore *et al.*, 1980; Trinchieri *et al.*, 1981) and with murine cells (Welsh *et al.*, 1981). This phenomenon is dose dependent and reversible, requires several hours of incubation with interferon, does not result from a lack of adherence to target cells, and does not occur with all cells tested (Trinchieri *et al.*, 1981; Welsh *et al.*, 1981). The protective effect of interferon on target cells requires nucleic acid and/or protein synthesis since it is blocked by inhibitors of RNA and protein synthesis (Trinchieri *et al.*, 1981; Welsh *et al.*, 1981). Furthermore, the inverse relationship between the effect of interferon on HLA expression (discussed in the previous section) and NK susceptibility of melanoma cells suggests that the cell surface recognition structures

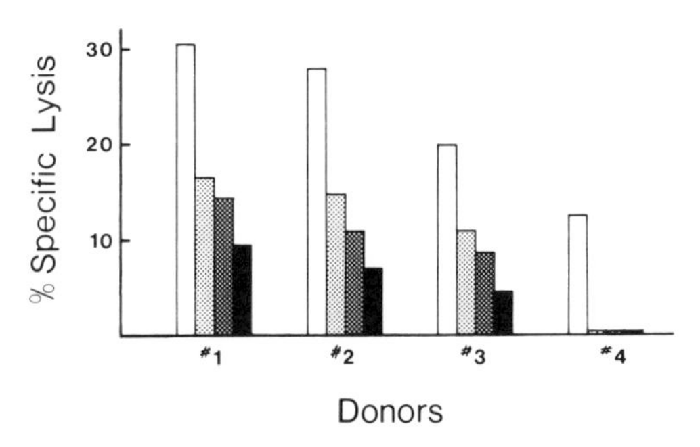

Fig. 6. Effect of human leukocyte interferon on the susceptibility of cultured human melanoma cells M21 to lysis by human NK cells. M21 cells were incubated at 37°C for 16 hr in medium alone (□) or in medium containing interferon [final concentrations: 500 (▨), 1000 (▩), and 2000 U/ml (■)] and then used as targets in a 16-hr ^{51}Cr release microcytotoxicity assay with human peripheral blood lymphocytes from four donors as a source of NK effector cells.

involved in natural killing are different from those of the major histocompatibility complex, in agreement with the finding by Roder *et al.* (1979).

It is not known whether interferon induces resistance of target cells to NK lysis by affecting (1) the expression of the target membrane structure recognized by NK cells, (2) the interaction between NK cells and targets in the lytic process, or (3) the susceptibility of cell membrane to lysis and/or ability of target cells to repair immune damage. The first possibility is based on the premise that NK cytotoxicity is a phenomenon requiring cell-to-cell contact and that exposure of cells to interferon results in an alteration of NK target antigen expression. This hypothesis is supported by the observation that interferon-treated cells compete poorly against the lysis of ^{51}Cr-labeled target cells in cold target competition experiments performed with both murine (Welsh *et al.*, 1981) and human cells (Moore *et al.*, 1980; Trinchieri *et al.*, 1981). The second possibility is supported by the recent observations that resistance of target cells to NK lysis is mediated by a mechanism that interferes with the lytic process of target cell-bound NK cells and is dependent on protein synthesis (Collins *et al.*, 1981; Kunkel and Welsh, 1981). The second hypothesis, however, would imply that the mechanism of NK cell-mediated lysis at the level of the plasma membrane differs from that of T cell lysis and ADCC, since the susceptibility of tumor cells to these latter activities are not affected by interferon treatment (Trinchieri *et al.*, 1981; Welsh *et al.*, 1981; see also Fig. 5). The third possibility is highly unlikely because interferon-treated cells remain highly susceptible to ADCC and to complement-mediated lysis. Additional information about the recognition and lytic process of NK activities is necessary before a mechanism for the protective effect of interferon can be established.

V. EFFECT OF INTERFERON ON THE LYTIC ACTIVITY OF CYTOTOXIC CELLS AGAINST TUMOR TARGETS

A. Effect on NK Cell Activity

Spontaneous cytotoxicity for a variety of tumor cells *in vitro* has been observed with lymphocyte preparations from normal mice (Herberman *et al.*, 1975; Kiessling *et al.*, 1975; Sendo *et al.*, 1975; Zarling *et al.*, 1975), rats (Nunn *et al.*, 1976; Shellam and Hogg, 1977; Oehler *et al.*, 1978), and man (McCoy *et al.*, 1973; Takasugi *et al.*, 1973; Rosenberg *et al.*, 1974; Pross and Jondal, 1975; Herberman, 1980). The cytotoxic effector cells are referred to as NK cells and may have an important role in host immunosurveillance against neoplasia (see review by Herberman, 1981). Although NK cells appear to be a heterogeneous population, there is general agreement that they lack surface immunoglobulins and Ia-like antigens but express receptors for the Fc portion of IgG (see reviews by Herberman *et al.*, 1978; Roder *et al.*, 1981). Our studies and those by others have repeatedly demonstrated that NK activities in human PBL can be enriched in lymphocyte subpopulations showing T cell characteristics, i.e., expression of receptors for sheep red blood cells (Bakacs *et al.*, 1977; Kay *et al.*, 1977; Ng *et al.*, 1980; Fast *et al.*, 1981) or in subpopulations lacking the surface characteristics of mature B cells and T cells, i.e., null cells (Kalden *et al.*, 1977; Ng *et al.*, 1981b). We find that the null cell population has a higher NK lytic activity than the T cell population. It was recently demonstrated that NK cell activity is associated with a subpopulation of lymphoid cells termed large granular lymphocytes; however, the proportion of active NK effector cells among large granular lymphocytes has remained undetermined (Timonen *et al.*, 1981).

Increased NK activity was originally observed when human PBL from normal donors were treated *in vitro* with culture supernatant known to contain human interferon (Skurkovich *et al.*, 1978; Trinchieri and Santoli, 1978); that the effect is mediated by interferon was later shown by experiments utilizing purified human interferon (Herberman *et al.*, 1979; Saksela *et al.*, 1979; Masucci *et al.*, 1980; Moore and Potter, 1980; Zarling 1980) or human recombinant interferon (Herberman *et al.*, 1982) and by blocking of the enhancement with antibodies specific for interferon (Herberman *et al.*, 1979; Zarling 1980). A similar increase in NK activity is also observed when peripheral blood lymphocytes from tumor-bearing patients are treated with interferon (Lucero *et al.*, 1981) and when lymphocytes are isolated from patients that have been treated with interferon (Einhorn *et al.*, 1978; Huddlestone *et al.*, 1979). The increased NK activity

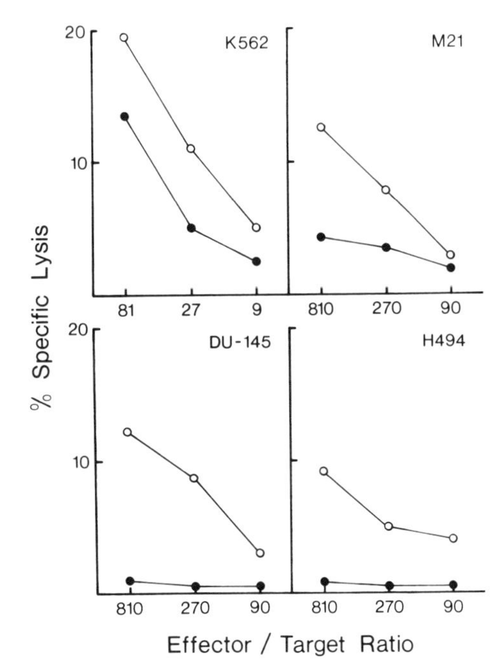

Fig. 7. Effect of human leukocyte interferon on the NK activity of human peripheral blood lymphocytes. Lymphocytes were incubated in medium alone (○) or medium containing interferon [final concentration: 500 U/ml (●)] at 37°C for 2 hr and subsequently tested for lytic activity against cultured erythromyeloid leukemia cells K562, melanoma cells M21, and prostate carcinoma cells DU-145 and H494 in a 4-hr ^{51}Cr release microcytotoxicity assay.

of interferon-treated cells is manifested by either a higher lytic capability of the interferon-treated cells (Figs. 7 and 8) or by the acquired ability of interferon-treated effector cells to lyse certain NK-resistant target cells (Fig. 7). The effect of interferon depends on the concentration used: concentrations as low as 20 U/ml interferon significantly boost NK activity and higher levels of enhancement can be achieved with larger

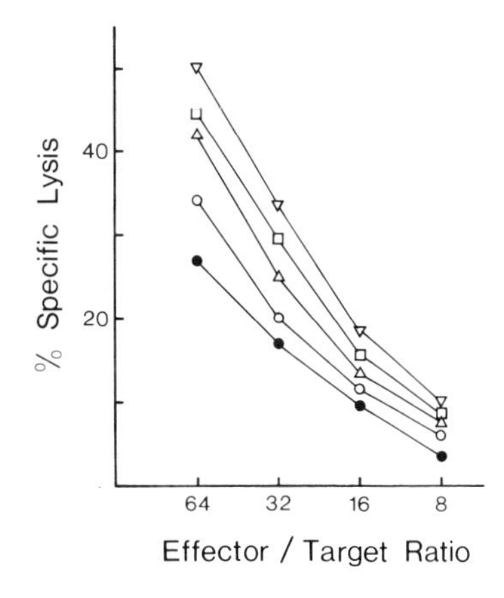

Fig. 8. Dose dependence of enhancing effect of human leukocyte interferon on NK activity of human peripheral blood lymphocytes. Lymphocytes were incubated at 37°C for 4 hr in medium alone (●) or in medium containing interferon [final concentrations: 20 (○), 100 (△), 500 (□), and 1000 U/ml (▽)] and then tested for lytic activity against cultured erythromyeloid leukemia cells K562 in a 4-hr ^{51}Cr release microcytotoxicity assay.

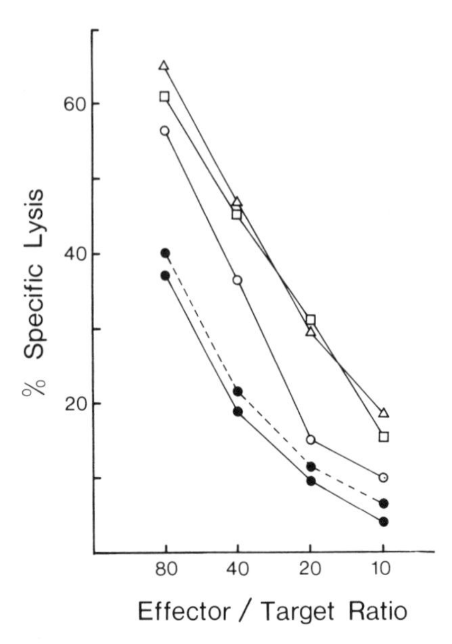

Fig. 9. Kinetics of enhancing effect of human leukocyte interferon on NK activity of human peripheral blood lymphocytes. Lymphocytes were incubated at 37°C in medium alone for 60 min (●-- ●) or in medium containing interferon (final concentration: 500 U/ml) for 0 (●——●), 15 (○), 30 (□), and 60 (△) min and then tested for lytic activity against K562 target cells in a 4-hr ^{51}Cr release microcytotoxicity assay.

amounts of interferon (Fig. 8). The effect of interferon is also time dependent: pretreatment of PBL with interferon for periods of time as short as 15 min markedly augments the natural cytotoxicity against target cells, while higher levels of augmentation are observed after longer periods of pretreatment (Fig. 9). The effect of interferon on null cells is not mediated by interactions with other lymphocyte subpopulations and does not require the presence of other cell types in the incubation mixtures: following incubation with interferon, null cells isolated from peripheral blood lymphocytes by depletion of B and T lymphocytes through rosetting with AET-sheep red blood cells and sheep red blood cells coated with anti-Ia-like antigen monoclonal antibodies (Ng *et al.*, 1981b) increase their NK activity to an extent similar to that of interferon-treated whole lymphocyte populations which contain B,T, and null cells and monocytes (Fig. 10). The type of target cells used in the NK assay does not appear to influence the effect of interferon on NK cells from various donors. As shown in Table I, interferon-treated lymphocytes from the donor F.F. showed a threefold higher NK activity than lymphocytes from the donor B.W. with both prostate carcinoma cell lines tested. On the other hand, lymphocytes from different donors differ in their sensitivity to the effect of interferon (Fig. 11), exhibiting increases in cytotoxicity ranging from 2- to 25-fold.

The limited information about the molecular nature of the cell surface structure which are the targets of NK lysis as well as about the mechanism

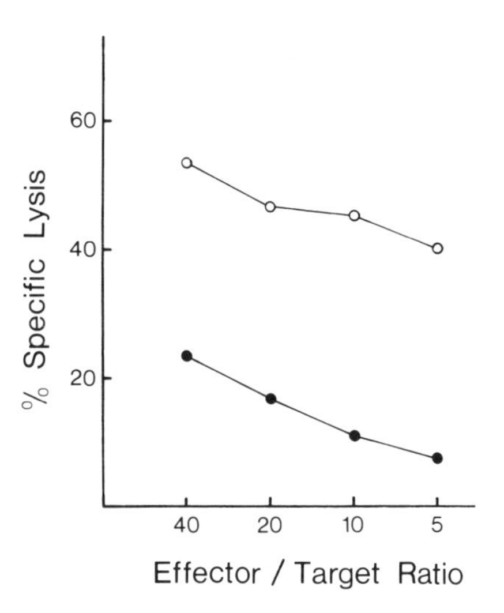

Fig. 10. Effect of human leukocyte interferon on the NK activity of human peripheral blood lymphocytes depleted of T and B lymphocytes by simultaneous rosetting with 2-aminoethylisothiourium bromide-treated sheep erythrocytes and sheep erythrocytes coated with monoclonal antibody Q5/13 to Ia-like antigens. Null cells were incubated in medium alone (●) or in medium containing interferon [final concentration: 500 U/ml (○)] at 37°C for 4 hr and then tested for lytic activities against K562 target cells.

of NK lysis complicates any interpretation of the effect of interferon on NK cell activity. A model to explain this finding has been formulated by Ortaldo *et al.* (1980a, 1981), and we refer the interested reader to their recent publications. We will conclude this section by listing findings which may be useful to plan additional experiments to investigate the phenomenon of enhancement of NK lysis by interferon. Boosting of NK activity by interferon is blocked by inhibitors of RNA or protein synthesis but is not blocked by inhibition of DNA synthesis either just before incubation of lymphocytes with interferon or 18 hr earlier (Ortaldo *et al.*, 1980b).

Table I
NK Lytic Activity of Interferon-Treated Peripheral Blood Lymphocytes[a] from Two Donors against Two Prostate Carcinoma Cell Lines

	LU/10^6 cells[c]			
	Donor F.F.		Donor B.W.	
NK targets[b]	Untreated	Interferon	Untreated	Interferon
DU-145	<1	10.5	<1	3.7
PC-3	<1	10.0	<1	3.3

[a] Human PBL were incubated at 37°C for 6 hr in medium alone or in medium containing interferon (final concentration, 500 U/ml).
[b] NK lysis of target cells was tested in a 16-hr ^{51}Cr release microcytotoxicity assay.
[c] One lytic unit (LU) is defined as the number of effector cells required to lyse 20% of total target cells, calculated from a dose–response curve.

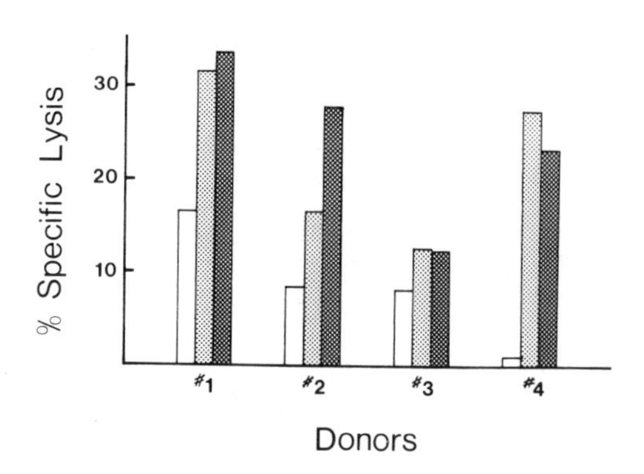

Fig. 11. Differential effect of human leukocyte interferon on the NK activities of human peripheral blood lymphocytes from various donors. Lymphocytes were incubated at 37°C for 16 hr in medium alone (□) or medium containing interferon [final concentrations: 500 (▨) and 2000 U/ml (▩)] and subsequently tested for NK activity against M21 melanoma cells in a 16-hr ^{51}Cr release microcytotoxicity assay.

Increases in expression of HLA-A,B antigens on NK cells is not a necessary prerequisite for their enhanced lytic activity, since Perussia *et al.* (1980) have reported that interferon-treated NK cells from a donor displayed enhanced lytic activity although the level of HLA-A,B antigens was not significantly different from that of control NK cells. Finally, experiments performed with a single-cell assay where NK killing of tumor cell targets is visualized by the reading of trypan blue uptake of NK cell–target cell conjugates plated in agarose gel suggest that the augmentation of NK cytotoxicity by interferon is due to recruitment of new effector cells and activation of both new and already active NK cells (Targan and Dorey, 1980). Using the same assay, it was further found that the effect of interferon depends on the type of target cells used. Thus, interferon increases the lytic rate of effector cells when the targets are cells in suspension but recruits new effector cells when the targets are adherent cells (Ortaldo *et al.*, 1981; Timonen *et al.*, 1982).

B. Effect on K Cell Activity

K cells that mediate ADCC are generally agreed to be closely related, if not identical, to NK cells (for review, see Herberman and Holden, 1978). The effect of interferon on K cell activity, is a controversial issue: Trinchieri and his colleagues (Trinchieri and Santoli, 1978; Trinchieri *et al.*, 1978) did not observe any increase in the lytic activity of human K cells for human target cells sensitized with anti-HLA alloantibodies. Similar findings have been reported by Heron *et al.* (1979) who used either human lymphoid cells sensitized with anti-HLA alloantisera or murine mastocytoma cells sensitized with a rabbit antiserum as targets. In contrast, Herberman *et al.* (1979), reported an increase of the lytic activity of K cells using Chang liver cells sensitized with rabbit IgG antibodies or murine lymphoma cells RBL-5 sensitized with rabbit Ig antibodies as targets. The latter findings are in agreement with other investigators who have used rabbit antisera as a source of antibody and human tumor cells or chicken erythrocytes as targets (Attallah and Folks, 1979; Droller *et al.*, 1979; Itoh *et al.*, 1980; Ortaldo *et al.*, 1980c). In our experiments, we found a reproducible increase in the lytic activity of interferon-treated cells for human melanoma cells sensitized with MoAbs to various MAA and for cultured B lymphoid cells sensitized with conventional xenoantisera to HLA-A,B and Ia-like antigens (Fig. 12 and Table II).

The reason for the conflicting results reported in the literature is not known; the properties of the antibodies and of the target cells are not likely to play a major role in this regard since in our experience a similar increase of K cell lysis occurs with antibodies of different specificity and Ig subclass (Imai *et al.*, 1982b) and with target cells of different histological origin. Furthermore, the increase of lysis in ADCC reported by some investigators cannot be explained by increased NK cell activity since the phenomenon occurs with target cells resistant to interferon-treated NK cells (Atallah and Folks, 1979; Itoh *et al.*, 1980; Ortaldo *et al.*, 1980c).

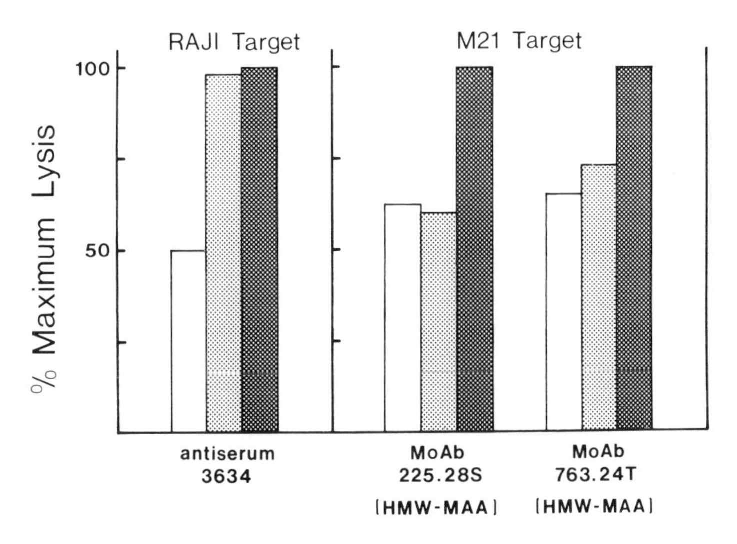

Fig. 12. Effect of human leukocyte interferon on the lytic activity of human peripheral blood lymphocytes in ADCC. Lymphocytes were incubated at 37°C for 16 hr in medium alone (□) or in medium containing interferon [final concentrations: 500 (▒) and 2000 U/ml (▩)] and subsequently tested as effectors in a 16-hr ^{51}Cr release microcytotoxicity assay. Targets were human B lymphoblastoid cells Raji sensitized with the anti-HLA xenoantiserum 3634 (left panel) and human melanoma cells M21 sensitized with MoAbs 225.28S or 763.24T to high-molecular-weight MAA.

Table II
Dose Dependence of the Effect of Human Leukocyte Interferon[a] on the ADCC Activity[b] of Human Peripheral Blood Lymphocytes

Interferon concentration (U/ml)	LU/10^6 cells[c]
0	11.1
20	21.1
100	22.3
500	23.3
1000	24.7

[a] Human PBL were incubated at 37°C for 4 hr in medium alone or in medium containing interferon.
[b] ADCC activity was tested in a 4-hr ^{51}Cr release microcytotoxicity assay using Raji cells coated with xenoantiserum 3634 as targets. This xenoantiserum is from a rabbit immunized with partially purified HLA antigens (Ferrone *et al.*, 1977). Serologic and immunochemical assays have shown that it contains antibodies to common determinants of human Ia-like antigens and to the heavy chain of HAL-A,B,C antigens (Ferrone *et al.*, 1978a,b).
[c] One lytic unit (LU) is defined as the number of cells needed to lyse 20% of total target cells, calculated from a dose–response curve.

Interferon boosted the lytic activity in ADCC of leukemic cells from a patient with chronic lymphocytic leukemia which lacked any detectable NK activity (Ortaldo *et al.*, 1980c). The conflicting results are also not due to variations in the level of a K cell-activating factor contaminating some of the partially purified interferon preparations of interferon used in these studies since purified interferon can boost the lytic activity of K cells in ADCC (Herberman *et al.*, 1982b). These investigators suggested that the discrepancy may result from interferon preparations containing factors which selectively interfere with the augmentation of ADCC. The method of isolation of effector cells may also be important since our preliminary results suggest that the effect of interferon is more pronounced on human null cells than on other lymphocyte subpopulations, and Kimber and Moore (1981) have reported that enhancement of ADCC does not occur if the effector cell preparation does not contain monocytes.

The mechanism by which interferon enhances K cell activity has not been investigated as extensively as the enhancement of NK lysis. It is of interest that the expression of receptors for Fc portions of IgG is increased on interferon-treated lymphocytes (Fridman *et al.*, 1980; Itoh *et al.*, 1980); whether this plays a role in the increased lytic activity of K cells is, however, not known.

VI. CONCLUSION

In the previous pages we have summarized a series of experimental data which indicates that interferon can modulate different mechanisms of immune lysis of human tumor cells. Interferon can increase the extent of immune lysis by augmenting the lytic activity of NK and K effector cells or by enhancing the susceptibility of the target cells to antibody-mediated lysis through an increase in the expression of cell surface antigens. On the other hand, interferon can decrease the extent of lysis by reducing the susceptibility of target cells to NK effectors. In this final section we will compare some of our results with those published in the literature and we will discuss the biological implications of these findings.

The dual and antagonistic effects of interferon on NK activity, i.e., enhancement of lytic activity of NK effector cells and protection of target cells from NK lysis, complicates the evaluation of the role of interferon–NK system in antitumor activity. Trinchieri *et al.* (1981) have hypothesized that during the process of immune surveillance, interferon enhances NK lysis of virus-infected cells or tumor cells while at the same time protecting normal tissue from NK-mediated damage. The hypothesis, however, appears to be oversimplified because studies by others (Moore *et al.*, 1980; Welsh and Kiessling, 1980) indicate that the protective effect of interferon does not always seem to correlate with normal versus transformed cells. At any rate, protection of tumor cells by interferon ought to be an important consideration when attempting to apply interferon as antitumor agent in clinical trials.

There is very limited information about the effects of interferon on the expression of tumor-associated antigens. We are aware of only one report by Liao *et al.* (1980) who published that incubation of melanoma cells with interferon increased the expression of tumor-associated antigens recognized by a conventional xenoantiserum. These data are at variance with our own findings and wc cannot easily identify the reason(s) for these differences because of the limited information available on the experiments performed by Liao *et al.* (1980). The most likely explanation for the different experimental results is that the xenoantiserum used by Liao *et al.* (1980) identifies tumor-associated antigens different from those recognized by the MoAbs we have developed. If so, then interferon may have a differential effect on the various types of tumor-associated antigens which are expressed by human melanoma cells. An exchange of reagents will be necessary to compare their specificity and to prove or disprove this possibility. An alternative, although less likely, explanation is that the conflicting results reflect either the different incubation times, the interferon preparations, or the individual characteristics of melanoma cells used by Liao *et al.* (1980) and by ourselves. Liao *et al.* (1980) did not describe any effect of interferon on immune lysis of melanoma cells by antisera to MAAs. We did not detect any significant change in the susceptibility of interferon-treated melanoma cells to immune lysis mediated by MoAbs to tumor-associated antigens (Imai *et al.*, 1981a). Our finding suggests that the MAAs we have identified with our MoAb are not likely to play a significant role in the beneficial effect interferon may have on the clinical course of melanoma if the mechanism behind this phenomenon is immunologic in nature.

There is general agreement in the literature on the increased expression of HLA-A,B antigens on interferon-treated cells (Heron *et al.*, 1978; Attallah and Strong, 1979; Fellous *et al.*, 1979; Imai *et al.*, 1981a,b). Our own results show for the first time that the enhanced expression of HLA-A,B antigens is associated with an increase in the extent of complement-dependent and cell-dependent lysis of target cells sensitized with anti HLA-A,B antibodies. Although no change in the HLA-A,B phenotype defined by conventional antisera has been reported for melanoma cells (Pellegrino *et al.*, 1977, 1981b; Pollack *et al.*, 1980), it is our working hypothesis that this finding may have clinical relevance and this lytic mechanism may contribute to the accelerated destruction of melanoma cells in patients treated with interferon, since the following lines of evidence suggest that malignant transformation of cells may be associated with changes in the immunological profile of HLA-A,B,C molecules. At the workshop on Clinical Application of Monoclonal Antibodies to Tumor Associated Antigens held at the National Institutes of Health, Bethesda, Maryland, in February, 1981, Dr. E. Lennox (MRC Laboratory of Molecular Biology, University Postgraduate Medical School, Cambridge, U.K.) reported that the MoAb W6/32 to a framework determinant of HLA-A,B,C antigens (Parham *et al.*, 1979) cross-reacts with H-2 molecules synthesized by murine tumor cells but not with those synthesized by their normal counterparts, although the expression of H-2 alloantigenic determinants on these two types of cells was the same. These findings show for the first time that changes of HLA on tumor cells may affect not only the determinants defining the conventional serological polymorphism (Parmiani *et al.*, 1979) but also other antigenic sites of the molecule. Furthermore, our recent observation that the carbohydrate side chain of HLA-A,B,C heavy chains can effect their antigenic profile (Wilson *et al.*, 1981d) raises the possibility that abnormalities of glycosylation which are frequent in tumor cells may result in changes of the antigenic profile of HLA-A,B,C molecules synthesized by melanoma cells. If through these and/or other mechanisms, alien determinants appear on HLA-A,B,C antigens of melanoma cells and are immunogenic in the tumor host, then interferon treatment may amplify their expression and render melanoma cells more susceptible to immune lysis specific for these alien determinants. Proof of this hypothesis will depend on the availability of a large library of MoAbs to distinct epitopes of HLA-A,B,C molecules, since these reagents will allow an extensive characterization of the immunologic profile of HLA-A,B,C antigens synthesized by tumor cells.

ACKNOWLEDGMENTS. Special thanks are due to Dr. Ronald B. Herberman for his helpful comments and generosity in providing us unpublished data from his laboratory. This work was supported by the National Institutes of Health Grants AI 13154, CA 16069, CA 16071, CA 24329, and CA 29897, a Research Career Development (M.A.P.), a Columbia University Biomedical Research Support Grant (A.K.N.) and a Special Fellowship of the Leukemia Society of America (B.S.W.). K. Imai is a visiting investigator from Sapporo Medical College (Japan). The authors wish to thank Ms. Ellen Schmeding for expert secretarial assistance.

VII. REFERENCES

Attallah, A. M., and Folks, T., 1979, Interferon enhanced human natural killer and antibody-dependent cell-mediated cytotoxic activity, *Int. Arch. Allergy Appl. Immunol.* **60:**377.

Attallah, A. M., and Strong, D. M., 1979, Differential effects of interferon on the MHC expression of human lymphocytes, *Int. Arch. Allergy Appl. Immunol.* **60:**101.

Bakacs, T., Gergely, P., and Klein, E., 1977, Characterization of cytotoxic human lymphocyte subpopulations. The role of Fc-receptor carrying cells, *Cell. Immunol.* **32:**317.

Borden, E. C., 1979, Interferons: Rationale for clinical trials in neoplastic disease, *Ann. Intern. Med.* **91:**472.

Collins, L. J., Patek, P. O., and Cohn, M., 1981, Tumorigenicity and lysis by natural killer cells, *J. Exp. Med.* **153:**89.

Curry, R. A., Quaranta, V., Pellegrino, M. A., and Ferrone, S., 1981, Lysis of cultured human melanoma M10 cells by polyclonal xenoantibodies to melanoma-associated antigens, *Cancer Res.* **41:**463.

Droller, M. J., Borg, H., and Perlmann, P., 1979, *In vitro* enhancement of natural and antibody-dependent lymphocyte-mediated cytotoxicity against tumor target cells by interferon, *Cell. Immunol.* **47:**248.

Einhorn, S., Blomgren, H., and Strander, H., 1978, Interferon and spontaneous cytotoxicity in man. II. Studies in patients receiving exogenous leukocyte interferon, *Acta Med. Scand.* **204:**477.

Fast, L. D., Hansen, J. A., and Newman, W., 1981, Evidence for T cell nature and heterogeneity with natural killer (NK) and antibody-dependent cellular cytotoxicity (ADCC) effectors: A comparison with cytolytic T lymphocytes (CTL), *J. Immunol.* **127:**448.

Fellous, M., Komoun, M., Gresser, I., and Bono, R., 1979, Enhanced expression of HLA antigens and β_2-microglobulin on interferon treated human lymphoid cells, *Eur. J. Immunol.* **9:**446.

Ferrone, S., and Pellegrino, M. A., 1973, HLA antigen, antibody and complement in the lymphocytotoxic reaction, *in* "Contemporary Topics in Molecular Immunology," Vol. 2 (R. A. Reisfeld and W. J. Mandy, eds.), pp. 185–235, Plenum Press, New York.

Ferrone, S., Cooper, N. R., Pellegrino, M. A., and Reisfeld, R. A., 1974a, The role of complement in the HLA antibody-mediated lysis of lymphocytes, *Transplant. Proc.* **6:**13.

Ferrone, S., Pellegrino, M. A., Dierich, M. P., and Reisfeld, R. A., 1974b, Expression of histocompatibility antigens during the growth cycle of cultured lymphoid cells, *Curr. Top. Microbiol. Immunol.* **55:**1.

Ferrone, S., Pellegrino, M. A., and Reisfeld, R. A., 1977, Immunogenicity of human B cell antigens solubilized from cultured human lymphoid cells, *J. Immunol.* **118:**1036.

Ferrone, S., Allison, J. P., and Pellegrino, M. A., 1978a, Human DR (Ia-like) antigens: Biological and molecular profile, *Contemp. Top. Mol. Immunol.* **7:**239.

Ferrone, S., Naeim, F., Indiveri, F., Walker, L. E., and Pellegrino, M. A., 1978b, Xenoantisera to human DR antigens: Serological and immunochemical characterization, *Immunogenetics* **7:**349.

Fridman, W. H., Gresser, I., Bandu, M. T., Aguet, M., and Neauport-Sautes, C., 1980, Interferon enhances the expression of Fc receptors, *J. Immunol.* **124:**2436.

Gresser, I., 1977, Antitumor effects of interferon, *in* "A Comprehensive Treatise" Vol. 5 (F. Becker, ed.), pp. 521–571, Plenum Press, New York.

Gutterman, J. U., Blumenschein, G. R., Alexanian, R., Yah, H. Y., Buzdar, A. V., Cabanilas, F., Hortobagy, G. N., Hersh, E. M., Rasmussen, S. L., Harmon, M., Kramer, M., and Pestka, S., 1980, Leukocyte interferon induced tumor regression in human metastatic breast cancer, multiple myeloma and malignant lymphoma, *Ann. Int. Med.* **93:**399.

Herberman, R. B. (ed.), 1980, "Natural Cell-Mediated Immunity against Tumors," Academic Press, New York.

Herberman, R. B., 1981, Natural killer (NK) cells and their possible roles in resistance against disease, *in* "Clinical Immunology Reviews," Vol. 1 (R. Rocklin, ed.), pp. 1–65, Dekker, New York.

Herberman, R. B., and Holden, H. T., 1978, Natural cell-mediated immunity, *in* "Advances in Cancer Research," Vol. 27 (G. Klein and S. Weinhouse, eds.), pp. 305–377, Academic Press, New York.

Herberman, R. B., Nunn, M. E., and Lavrin, D. H., 1975, Natural cytotoxic reactivity of mouse lymphoid cells against syngeneic and allogeneic tumors. I. Distribution of reactivity and specificity, *Int. J. Cancer* **16:**216.

Herberman, R. B., Djeu, J. Y., Kay, H. D., Ortaldo, J. R., Riccardi, C., Bonnard, G. D., Holden, H. T., Fagnani, R., Santoni, A., and Puccetti, P., 1978, Natural killer cells: Characteristics and regulation of activity, *Immunol. Rev.* **44:**43.

Herberman, R. B., Ortaldo, J. R., and Bonnard, G. P., 1979, Augmentation by interferon of human natural and antibody-dependent cell-mediated cytotoxicity, *Nature* (*London*) **277:**221.

Herberman, R. B., Ortaldo, J. R., Mantovani, A., Hobbs, D. S., Kung, H. S., and Pestka, S., 1982a, Effect of human recombinant interferon on cytotoxic activity of natural killer (NK) cells and monocytes, *Cell Immunol.* **67:**160.

Herberman, R. B., Ortaldo, J. R., Rubinstein, M., and Pestka, S., 1982b, Augmentation of natural and antibody-dependent cell-mediated cytotoxicity by pure human leukocyte interferon, *J. Clin. Immunol.* **1:**149.

Heron, I., Hokland, M., and Berg, K., 1978, Enhanced expression of β_2-microglobulin and HLA antigens on human lymphoid cells by interferon, *Proc. Natl. Acad. Sci. USA* **75:**6215.

Heron, I., Hokland, M., Moller-Larsen, A., and Berg, K., 1979, The effect of interferon on lymphocyte-mediated effector cell functions: Selective enhancement of natural killer cells, *Cell. Immunol.* **42:**183.

Hovanessian, A. G., 1979, Intracellular events in interferon-treated cells, *Differentiation* **15:**139.

Howard, J. C., Butcher, G. W., Galfre, G., Milstein, C., and Milstein, C. P., 1979, Monoclonal antibodies as tools to analyze the serological and genetic complexities of major transplantation antigens, *Immunol. Rev.* **47:**139.

Huddlestone, J. R., Merigan, T. C., Jr., and Oldstone, M. B. A., 1979, Induction and kinetics of natural killer cells in humans following interferon therapy, *Nature (London)* **282:**417.

Imai, K., and Ferrone, S., 1980, Indirect rosette microassay to characterize human melanoma-associated antigens (MAA) recognized by operationally specific xenoantisera, *Cancer Res.* **40:**2252.

Imai, K., Ng, A. K., Glassy, M. C., and Ferrone, S., 1981a, Differential effect of interferon on the expression of tumor associated antigens and histocompatibility antigens on human melanoma cells: Relationship to susceptibility to immune lysis mediated by monoclonal antibodies, *J. Immunol.* **127:**505.

Imai, K., Pellegrino, M. A., Ng, A. K., and Ferrone, S., 1981b, Role of antigen density in immune lysis of interferon treated human lymphoid cells: Analysis with monoclonal antibodies to the HLA-A,B antigenic molecular complex and to Ia-like antigens, *Scand. J. Immunol.* **14:**529.

Imai, K., Ng, A. K., Glassy, M. C., and Ferrone, S., 1981c, ADCC of cultured human melanoma cells: Analysis with monoclonal antibodies to human melanoma associated antigens, *Scand. J. Immunol.* **14:**369.

Imai, K., Wilson, B. S., Ruberto, G., Nakanishi, T., Yachi, A., and Ferrone, S., 1981d, Molecular heterogeneity of a high molecular weight human melanoma associated antigen (MAA) detected by monoclonal antibodies, *in* "Protides of the Biological Fluids: 29th Colloquium 1981" (H. Peters, ed.), pp. 8993–8997, Pergamon, New York.

Imai, K., Natali, P. G., Kay, N. E., Wilson, B. S., and Ferrone, S., 1982a, Tissue distribution and molecular profile of a differentiation antigen detected by monoclonal antibody (345.134S) produced against human melanoma cells, *Cancer Immunol. Immunother.* **12:**159.

Imai, K., Pellegrino, M. A., Wilson, B. S., and Ferrone, S., 1982b, Lytic activity of an IgGl and an IgG2a monoclonal antibody recognizing the same determinant on a human melanoma associated antigen, *Cell Immunol.,* in press.

Itoh, K., Inoue, M., Kataoka, S., and Kumagai, K., 1980, Differential effect of interferon on expression of IgG- and IgM-Fc receptors on human lymphocytes, *J. Immunol.* **124:**2589.

Kalden, J. F., Peter, H. H., Roubin, R., and Cesarini, J. P., 1977, Human peripheral null lymphocytes. I. Isolation, immunological and functional characterization, *Eur. J. Immunol.* **7:**537.

Kay, H. D., Bonnard, G. D., West, W. H., and Herberman, R. B., 1977, A functional comparison of human Fc-receptor-bearing lymphocytes active in natural cytotoxicity and antibody-dependent cellular cytotoxicity, *J. Immunol.* **118:**2058.

Kiessling, R., Klein, E., and Wigzell, H., 1975, "Natural" killer cells in the mouse. I. Cytotoxic cells with specificity for mouse moloney leukemia cells. Specificity and distribution according to genotype, *Eur. J. Immunol.* **5:**112.

Kimber, I., and Moore, M., 1981, Selective enhancement of human mononuclear leucocyte cytotoxic function by interferon, *Scand. J. Immunol.* **13**:375.

Krim, M., 1980a, Towards tumor therapy with interferons. I. Interferons: Production and properties, *Blood* **55**:711.

Krim, M., 1980b, Towards tumor therapy with interferons. II. Interferons: *In vivo* effects, *Blood* **55**:875.

Kunkel, L. A., and Welsh, R. M., 1981, Metabolic inhibitors render "resistant" target cells sensitive to natural killer cell-mediated lysis. *Int. J. Cancer* **27**:73.

Liao, S-K., Kwong, P. C., Khosravi, M., and Dent, P. B., 1982, Enhanced expression of melanoma-associated antigens and beta-2 microglublin on cultured human melanoma cells by interferon. *J. Natl. Cancer Inst.* **68**:19.

Lindahl, P., Leary, P., and Gresser, I., 1974, Enhancement of the expression of histocompatibility antigens of mouse lymphoid cells by interferon *in vitro, Eur. J. Immunol.* **4**:779.

Lucero, M. A., Fridman, W. H., Provost, M. A., Billardin, C., Pouillart, P., Dumont, J., and Falcoff, E., 1981, Effect of various interferons on the spontaneous cytotoxicity exerted by lymphocytes from normal and tumor-bearing patients, *Cancer Res.* **41**:294.

Lustig, H. J., and Bianco, C., 1976, Antibody-mediated cell cytotoxicity in a defined system: Regulation by antigen, antibody and complement, *J. Immunol.* **116**:253.

Masucci, M. G., Masucci, G., Klein, E., and Berthold, W., 1980, Target selectivity of interferon-induced human killer lymphocytes related to their Fc receptor expression, *Proc. Natl. Acad. Sci. USA* **77**:3620.

McCabe, R. P., Quaranta, V., Frugis, L., Ferrone, S., and Reisfeld, R. A., 1979, A radioimmunometric antibody binding assay for the evaluation of xenoantisera to melanoma associated antigens, *J. Natl. Cancer Inst.* **62**:455.

McCoy, J. L., Herberman, R. B., Rosenberg, E. B., Donnelly, F. C., Levine, P. H., and Alford, C., 1973, ^{51}Chromium-release assay for cell-mediated cytotoxicity of human leukemia and lymphoid tissue-culture cells, *Natl. Cancer Inst. Monogr.* **37**:59.

Merigan, T. C., Sikora, K., Breeden, J. H., Levy, R., and Rosenberg, S. A., 1978, Preliminary observations on the effect of human leukocyte interferon in non-Hodgkin's lymphoma, *N. Engl. J. Med.* **299**:1449.

Moore, M., and Potter, M. R., 1980, Enhancement of human natural cell-mediated cytotoxicity by interferon, *Br. J. Cancer* **41**:378.

Moore, M., White, W. J., and Potter, M. R., 1980, Modulation of target cell susceptibility to human natural killer cells by interferon. *Int. J. Cancer* **25**:565.

Natali, P. G., Imai, K., Wilson, B. S., Bigotti, A., Cavaliere, R., Pellegrino, M. A., and Ferrone, S., 1981a, Structural properties and tissue distribution of the antigens recognized by the monoclonal antibody 653.40S to human melanoma cells, *J. Natl. Cancer Inst.* **67**:591.

Natali, P. G., De Martino, C., Quaranta, V., Nicotra, M. R., Frezza, F., Pellegrino, M. A., and Ferrone, S., 1981b, Expression of Ia-like antigens in normal human non lymphoid tissues, *Transplantation* **31**:75.

Natali, P. G., Quaranta, V., Nicotra, M. R., Apollonj, C., Pellegrino, M. A., and Ferrone, S., 1981c, Tissue distribution of Ia-like antigens: Analysis with monoclonal antibodies, *Transplant. Proc.* **13**:1026.

Ng, A. K., Indiveri, F., Pellegrino, M. A., Molinaro, G. A., Quaranta, V., and Ferrone, S., 1980, Natural cytotoxicity and antibody-dependent cellular cytotoxicity of human lymphocytes depleted of HLA-DR bearing cells with monoclonal HLA-DR antibodies, *J. Immunol.* **124**:2336.

Ng, A. K., Pellegrino, M. A., Imai, K., and Ferrone, S., 1981a, HLA-A,B antigens, Ia-like antigens and tumor associated antigens on prostate carcinoma cell lines: Serological and immunochemical analysis with monoclonal antibodies, *J. Immunol.* **127**:443.

Ng, A. K., Indiveri, F., Russo, C., Quaranta, V., Pellegrino, M. A., and Ferrone, S., 1981b, Characterization of human null cells isolated from peripheral lymphocytes by a simultaneous double rosetting procedure, *Scand. J. Immunol.*, **14**:225.

Nunn, M. E., Djeu, J. Y., Glaser, M., Lavrin, D. H., and Herberman, R. B., 1976, natural cytotoxic reactivity of rat lymphocytes against syngeneic Gross virus-induced lymphoma, *J. Natl. Cancer Inst.* **56**:393.

Oehler, J. R., Lindsay, L. R., Nunn, M. E., and Herberman, R. B., 1978, Natural cell-mediated cytotoxicity in rats. I. Tissue and strain distribution and demonstration of a membrane receptor for the Fc portion of IgG, *Int. J. Cancer* **21**:204.

Ohlander, C., Larsson, A., and Perlmann, P., 1981, Regulation of effector functions of human K-cells and monocytes by antigen density of the target cells, *Scand. J. Immunol.* **13**:503.

Ortaldo, J. R., Herberman, R. B., and Djeu, J. Y., 1980a, Characterization of augmentation by interferon of cell-mediated cytotoxicity, *in* "Natural Cell-Mediated Immunity against Tumors" (R. B. Herberman, ed.), pp. 593–607, Plenum Press, New York.

Ortaldo, J. R., Phillips, W., Wasserman, K., and Herberman, R. B., 1980b, Effects of metabolic inhibitors on spontaneous and interferon-boosted human natural killer cell activity, *J. Immunol.* **125**:1839.

Ortaldo, J. R., Pestka, S., Slease, R. B., Rubinstein, M., and Herberman, R. B., 1980c, Augmentation of human K-cell activity with interferon, *Scand. J. Immunol.* **12**:365.

Ortaldo, J. R., Timonen, T., Mantovani, A., and Pestka, S., 1981, The effect of interferon on natural immunity, *in* "The Biology of the Interferon System" (E. de Mayer, G. Gralasso, and H. Schellekens, eds.), Elsevier/North-Holland, Amsterdam, in press.

Osserman, E. G., Sherman, W. H., Alexanian, R., Gutterman, J. U., and Humphrey, R. L., 1980, Preliminary results of the American Cancer Society (ACS)-sponsored trial of human leukocyte interferon (IF) in multiple myeloma (MM), *Abstr. Proc. Am. Assoc. Cancer Res.* **21**:161.

Parham, P., Barnstable, C. J., and Bodmen, W. F., 1979, Use of a monoclonal antibody (W6/32) in structural studies of HLA-A,B,C antigens, *J. Immunol.* **123**:342.

Parmiani, G., Carbone, G., Invernizzi, G., Pierotti, M. A., Sensi, M. L., Rogers, M. J., and Appella, E., 1979, Alien histocompatibility antigens on tumor cells, *Immunogenetics* **9**:1.

Pellegrino, M. A., Ferrone, S., Reisfeld, R. A., Irie, R. F., and Golub, S. H., 1977, Expression of histocompatibility (HLA) antigens on tumor cells and normal cells from patients from melanoma, *Cancer* **40**:36.

Pellegrino, M. A., Weaver, J. F., Nelson-Rees, W. A., and Ferrone, S., 1981, Ia-like and HLA-A,B antigens on tumor cells in long term culture, *Transplant. Proc.* **13**:1935.

Pellegrino, M.A., Natali, P.G., Ng, A.K., Imai, K., Russo, C., Bigotti, A., and Ferrone, S., 1982, Unorthodox expression of Ia-like antigens on human tumor cells of non lymphoid origin: Structural properties and biological significance, *in* "Recent Progress in Diagnostic Laboratory Immunology" (R.M. Nakamura, ed.), pp.157–173, Masson, New York.

Perussia, B., Santoli, D., and Trinchieri, G., 1980, Interferon modulation of natural killer cell activity, *Ann. N.Y. Acad. Sci.* **350**:55.

Pollack, M. S., Livingston, P. O., Fogh, J., Carey, T. E., Oettgen, H. F., and Dupont, B., 1980, Genetically appropriate expression of HLA and DR(Ia) alloantigens on human melanoma cell lines. *Tissue Antigens* **15**:249.

Pross, H. F., and Jondal, M., 1975, Cytotoxic lymphocytes from normal donors. A functional marker of human non-T lymphocytes, *Clin. Exp. Immunol.* **21**:226.

Quaranta, V., Pellegrino, M. A., and Ferrone, S., 1981, Serologic and immunochemical characterization of the specificity of four monoclonal antibodies to distinct antigenic determinants expressed on subpopulations of human Ia-like antigens, *J. Immunol.* **126:**548.

Roder, J. C., Ahrlund-Richter, L., and Jondal, M., 1979, Target-effector interaction in the human and murine natural killer system. Specificity and xenogeneic reactivity of the solubilized natural-killer-target structure complex and its loss in a somatic cell hybrid, *J. Exp. Med.* **150:**471.

Roder, J. C., Karre, K., and Kiessling, R., 1981, Natural killer cells. *Prog. Allergy* **28:**66.

Rosenberg, E. B., McCoy, J. L., Green, S. S., Donnelly, F. C., Siwarski, D. F., Levine, P. H., and Herberman, R. B., 1974, Destruction of human lymphoid tissue culture cell lines by human peripheral lymphocytes in ^{51}Cr-release cellular cytotoxicity assays, *J. Natl. Cancer Inst.* **52:**345.

Russo, C., Pellegrino, M. A., Ng, A. K., Wilson, B. S., and Ferrone, S., 1982, Antigenic constitution of HLA-A,B,C molecules: Immunochemical analysis with monoclonal xenoantibodies, *in* "HLA-Typing: Methodology and Clinical Aspects" (S. Ferrone and B. Solheim, eds.), pp. 77–100, CRC Press, Boca Raton, Florida.

Salsela, E., Timonen, T., Ranki, A., and Hayry, P., 1979, Fractionation, morphological and functional characterization of effector cells responsible for human natural killer activity to fetal fibroblasts and to culture cell line targets, *Immunol. Rev.* **44:**71.

Sendo, F., Aoki, T., Boyse, E. A., and Buofo, C. K., 1975, Natural occurrence of lymphocytes showing cytotoxic activity to BALB/c radiation-induced leukemia RL♂ 1 cells, *J. Natl. Cancer Inst.* **55:**603.

Shellam, G. R., and Hogg, N., 1977, Gross-virus-induced lymphoma in the rat. IV. Cytotoxic cells in normal rats, *Int. J. Cancer* **19:**212.

Skurkovich, S. V., Skorikova, A. S., and Eremkina, E. I., 1978, Enhancement by interferon of lymphocyte cytotoxicity in normal individuals to cells of human lymphoblastoid lines, *J. Immunol.* **121:**1173.

Stewart, W. E., II, 1979, "The Interferon System," Springer-Verlag, Vienna.

Stewart, W. E., Blalock, J. E., Burke, D. C., Chany, C., Dunnick, J. K., Falcoff, E., Friedman, R. M., Galasso, G. J., Joklik, W. K., Vilcek, J. T., Youngner, J. S., and Zoon, K. C., 1980, Interferon nomenclature: Letter to editor. *J. Immunol.* **125:**2353.

Strander, H., Cantell, K., Carlstrom, G., Ingimarsson, S., Jakobsson, P., and Nilsonne, V., 1978, Acute infections in interferon-treated patients with osteosarcoma: A preliminary report of a comparative study. *J. Infect. Dis.* **133**(Suppl. A)**:**245.

Stringfellow, D. A. (ed.), 1980, "Interferons and Interferon Inducers: Clinical Applications," Dekker, New York.

Struck, D. R., and Lennarz, W. J., 1977, Evidence for the participation of saccharide-lipids in the synthesis of the oligosaccharide chain of ovalbumin, *J. Biol. Chem.* **252:**1007.

Takasugi, M., Mickey, M. R., and Terasaki, P. I., 1973, Reactivity of lymphocytes from normal persons on cultured tumor cells, *Cancer Res.* **33:**2898.

Targan, S., and Dorey, F., 1980, Interferon activation of "pre-spontaneous killer" (pre-Sk) cells and alteration in kinetics of lysis of both "Pre-Sk" and active Sk cells, *J. Immunol.* **124:**2157.

Timonen, T., Ortaldo, J. R., and Herberman, R. B., 1981, Characteristics of human large granular lymphocytes and relationships to natural killer and K cells, *J. Exp. Med.* **153:**569.

Timonen, T., Ortaldo, J. R. and Herberman, R. B., Analysis by a single cell cytotoxicity assay of natural killer (NK) cell frequencies among human large granular lymphocytes and of the effects of interferon on their activity, *J. Immunol.* **128:**2514.

Trinchieri, G. D., and Santoli, D., 1978, Antiviral activity induced by culturing lymphocytes with tumor-derived or virus-transformed cells. Enhancement of human natural killer cell activity by interferon and antagonistic inhibition of susceptibility of target cells to lysis. *J. Exp. Med.* **147:**1314.

Trinchieri, G., Santoli, D., and Koprowski, H., 1978, Spontaneous cell-mediated cytotoxicity in humans: Role of interferon and immunoglobulins, *J. Immunol.* **120:**1849.

Trinchieri, G., Granto, D., and Perussia, B., 1981, Interferon-induced resistance of fibroblasts to cytolysis mediated by natural killer cells: Specificity and mechanism, *J. Immunol.* **126:**335.

Vignaux, F., and Gresser, I., 1977, Differential effects of interferon on the expression of H-2K, H-2D and Ia antigens on mouse lymphocytes, *J. Immunol.* **118:**721.

Vilcek, J., Gresser, I., and Merigan, T. C. (eds.), 1980, Regulatory functions of interferons, *Ann. N.Y. Acad. Sci.* **350.**

Weissman, R. M., and Droller, M. J., 1980, Interferon: A perspective, *Invest. Urol.* **18:**189.

Welsh, R. M., Jr., and Kiessling, R. W., 1980, Modification of target susceptibility to activated mouse NK cells by interferon and virus infections, *in* "Natural Cell-Mediated Immunity against Tumors" (R. B. Herberman, ed.), pp. 963–972, Plenum Press, New York.

Welsh, R. M., Karre, K., Hanson, M., Kunkel, L. A., and Kiessling, R. W., 1981, Interferon-mediated protection of normal and tumor target cells against lysis of mouse natural killer cells, *J. Immunol.* **126:**219.

Wiedermann, G., Denk, H., Stemberger, H., Eckerstorfer, R., and Tappeiner, G., 1975, Influence of the antigenicity of target cells on the antibody-mediated cytotoxicity of nonsensitized lymphocytes, *Cell. Immunol.* **17:**440.

Wilson, B. S., Indiveri, F., Pellegrino, M. A., and Ferrone, S., 1979, DR (Ia-like) antigens on human melanoma cells: Serological detection and immunological characterization, *J. Exp. Med.* **149:**658.

Wilson, B. S., Imai, K., Natali, P. G., Kay, N. E., Cavaliere, R., Pellegrino, M. A., and Ferrone, S., 1981a, Immunochemical analysis of the antigenic profile of human melanoma cells with monoclonal antibodies, *in* "Melanoma Antigens and Antibodies" (R. A. Reisfeld and S. Ferrone, eds.), Plenum Press, New York.

Wilson, B. S., Imai, K., Natali, P. G., and Ferrone, S., 1981b, Distribution and molecular characterization of a cell surface and a cytoplasmic antigen detectable in human melanoma cells with monoclonal antibodies, *Int. J. Cancer* **28:**293.

Wilson, B. S., Ng, A. K., Quaranta, V., and Ferrone, S., 1981c, HLA polyclonal and monoclonal antibodies: Production, characterization and application to the study of HLA antigens, *in* "Current Trends in Histocompatibility," Vol. 1 (R. A. Reisfeld and S. Ferrone, eds.), pp. 307–345. Plenum Press, New York.

Wilson, B. S., Glassy, M. C., Quaranta, V., Ng, A. K., and Ferrone, S., 1981d, Effect of tunicamycin on the assembly and antigenicity of HLA antigens: Analysis with monoclonal antibodies, *Scand. J. Immunol.* **14:**201.

Winchester, R. J., Wang, C. Y., Gibofsky, A., Kunkel, H. G., Lloyd, K. O., and Old, L. J., 1978, Expression of Ia-like antigens on cultured human malignant melanoma cell lines, *Proc. Natl. Acad. Sci. USA* **75:**6235.

Zarling, J. M., 1980, Augmentation of human natural killer cell activity by purified interferon and polyribonucleotides. *in* "Natural Cell-Mediated Immunity against Tumors" (R. B. Herberman, ed.), pp. 687–719, Plenum Press, New York.

Zarling, J. M., Nowinski, R. C., and Bach, F. H., 1975, Lysis of leukemia cells of normal mice. *Proc. Natl. Acad. Sci.* **72:**2780.

Chapter 12

Experimental Systems for Analysis of the Surface Properties of Metastatic Tumor Cells

George Poste

Smith Kline and French Laboratories
and
Department of Pathology and Laboratory Medicine
University of Pennsylvania
Philadelphia, Pennsylvania

and

Garth L. Nicolson

Department of Tumor Biology
The University of Texas System Cancer Center
M.D. Anderson Hospital and Tumor Institute
Houston, Texas

I. INTRODUCTION

Understanding how malignant tumor cells can spread from the primary tumor to establish metastases at other body sites is one of the major challenges in cancer research. Although the phenomenon of metastasis has long attracted the interest of both clinicians and experimentalists, the last few years have witnessed a dramatic increase in experimental studies on the basic mechanisms responsible for this life-threatening disease.

In this chapter we present a brief survey of experimental tumor systems that are currently being used to study the properties of metastatic tumor cells and we review recent findings on the contribution of cell surface changes to the altered behavior of metastatic tumor cells.

II. CHOICE OF TUMOR SYSTEMS

To study metastasis, it is necessary to work with a tumor that is metastatic. This may seem to be stating the obvious to the point of absurdity. However, the most cursory examination of the literature reveals numerous examples of papers in which the stated purpose of the work is to identify the cellular properties associated with "malignancy" or "metastasis," yet the tumor cells being studied do not metastasize!

With the stated prerequisite of a metastatic tumor, the next question concerns whether spontaneous or transplanted tumors are preferable. No single tumor system stands out as providing an all-inclusive experimental model, and the choice will depend on the question(s) being asked. However, the relative rarity of spontaneous neoplasms in most laboratory animal species (review, Poste and Fidler, 1980) dictates that transplanted tumors have of necessity been used extensively in experimental studies of the metastatic phenotype. The transplantable, metastatic animal tumors that have gained considerable popularity for experimental studies on metastasis are summarized in Table I.

The metastatic potential of transplantable tumors can be assayed in two ways. The more traditional approach is to implant tumor cells subcutaneously (s.c.) or intramuscularly (i.m.) and monitor formation of spontaneous metastases in various organs. The second approach is to inject tumor cells directly into the circulation and measure formation of so-called "experimental" metastases in different organs. By introducing tumor cells directly into the circulation, this method introduces the risk that noninvasive tumor cells could form metastases when injected i.v. but be unable to metastasize spontaneously when implanted s.c. or i.m. This question was addressed by Kripke *et al.* (1978) who evaluated the behavior of 21 clones of the UV2237 fibrosarcoma in three different assays: (1) formation of experimental metastases when injected i.v.; (2) formation of spontaneous metastases when implanted s.c.; and (3) survival time of animals injected i.v. with tumor cells. When the clones were ranked in order of increasing metastatic behavior from the s.c. site an excellent correlation was found with their ranking in the other two assays. This indicates that formation of experimental metastases by i.v. injection of tumor cells is a valid assay of cellular metastatic potential and satisfactorily predicts the outcome of more tedious and time-consuming assays of spontaneous metastatic activity.

Table I
Commonly Used Solid Metastasizing Transplantable Tumors of Rodents

Species	Syngeneic strain	Designation	Tumor type	Usual site(s) of implantation	Major site(s) of metastasis
Mouse	C57/BL6	Lewis lung carcinoma	Poorly differentiated epidermoid carcinoma	s.c. (axillary region), i.m.	Lung
Mouse	C57/BL6	B16 melanoma	Melanoma	s.c.	Lung + extrapulmonary
Mouse	Balb/c, CDF1	No. 26	Undifferentiated colon carcinoma	i.p., s.c. (axillary region)	Liver, lung
Mouse	Balb/c, CDF1	No. 51	Colon mucinous adenocarcinoma	i.p., s.c. (axillary region)	Liver, lung
Mouse	C3H/He, B6C3F1	16/C	Mammary adenocarcinoma	s.c.	Lung (lymph nodes)
Mouse	CD8F1	—	Mammary adenocarcinoma	Spontaneous mammary	Lung
Mouse	Balb/c	—	Renal cortical adenocarcinoma	Intrarenal	Lung
Mouse	CBA/Rij	2661	Mammary carcinoma	s.c. (footpad)	Lymph nodes, lung
Mouse	C57/BL6	RL-67	Lung adenocarcinoma	i.m. (hindleg)	Lung
Mouse	Balb/c	MS-2	Murine sarcoma virus-Moloney induced tumor	i.m. (hindleg)	Mainly lung, also spleen liver, kidney, brain
Mouse	Swiss	Ehrlich carcinoma	—	Marrow cavity of tibia	Lymph nodes
Mouse	Swiss	S180	—	Marrow cavity of tibia	Lymph nodes
Rat	Donryu	Yoshida sarcoma	—	s.c. (thigh, ear, footpad)	Lymph nodes (lung)
Rat	Donryu	AH-109A	Ascites hepatoma	s.c. (penis)	Lymph node
Rat	Lobund-Wistar	—	Prostate adenocarcinoma	s.c. (multiple sites)	Lung, visceral organs, bones
Rat	Buffalo	Line I, line II	Colon adenocarcinoma, signet ring cell, adenocarcinoma of colon	s.c. (abdomen) s.c. (abdomen) }	Lymph nodes, lung, parietal pleura

III. PHENOTYPIC HETEROGENEITY IN TUMOR CELL POPULATIONS: IMPLICATIONS FOR EXPERIMENTAL ANALYSIS OF THE METASTATIC PHENOTYPE

From a conceptual standpoint, one of the most important findings to have been made in the study of metastasis has been the demonstration that metastasis is not a random process in which every cell from the primary tumor is capable of causing a metastatic lesion but is instead caused by specialized subpopulations of tumor cells that are endowed with specific properties that befit them to complete each potentially destructive step in the metastatic process. This important concept has emerged from studies done initially by Fidler and Kripke (1977) with the mouse B16 melanoma and now confirmed by others using different tumors of diverse histotypic origin (Table II). These studies demonstrate that primary tumors contain subpopulations of cells of widely differing metastatic potential, including cells that are tumorigenic but have little or no ability to metastasize. These observations provide experimental confirmation of the hypothesis advanced by Nowell (1976) that progressive tumor growth will result in the emergence of variant cell subpopulations with increasingly diverse phenotypes and that selection pressures imposed by the host will favor survival of those subpopulations with increasing metastatic potential. Regardless of whether a neoplasm is monoclonal or polyclonal in origin, by the time it can be diagnosed clinically it will exhibit substantial cellular heterogeneity and contain subpopulations of cells with diverse phenotypic properties.

At any given time during tumor progression the number of subpopulations present, and the extent of their phenotypic diversity, will depend on the selection pressures encountered during the lifetime of the tumor.

Table II
Solid Transplantable Animal Tumors Containing Cell Subpopulations with Differing Metastatic Properties

Tumor	Species	Subpopulation	Reference
Melanoma	Mouse	Clones	Fidler and Kripke (1977); Fidler and Nicolson (1981); Poste *et al.* (1981)
Fibrosarcoma	Mouse	Clones	Kripke *et al.* (1978)
Fibrosarcoma	Mouse	Clones	Suzuki *et al.* (1978)
Mammary tumor	Mouse	Clones	Dexter *et al.* (1978)
Carcinoma	Mouse	Sublines	Heppner *et al.* (1978)
Carcinoma	Rat	Clones	Talmadge *et al.* (1979)
Lymphosarcoma	Rat	Clones	Reading *et al.* (1980b)

Selection pressures can be natural (e.g., assault from host defense mechanisms, limiting nutritional conditions) or artificial (e.g., clinical therapy).

Efforts to demonstrate similar heterogeneity in the metastatic properties of cells from human tumors have been hindered by the lack of suitable experimental systems for analyzing the *in vitro* behavior of human cells. Transplantation into athymic nude mice has been used widely to assay the tumorigenic potential of human tumor cells. However, a consistent finding has been that in most instances the resulting tumors fail to invade or metastasize, including cells isolated from metastases in the patient. The same situation applies to experimental animal tumors implanted into nude mice, including tumors that are invasive and metastatic in immunocompetent hosts. Recently, however, it has been shown that the absence of invasion and metastasis in tumors implanted in nude mice is an age-dependent phenomenon, and tumors which are nonmetastatic in nude mice 10 weeks of age or older will metastasize when inoculated into 3-week-old animals (Hanna and Fidler 1981). This important finding opens the way for experimental analysis of the important question of whether human tumors contain subpopulations of cells with differing invasive and metastatic capacities in analogous fashion to experimental animal tumors.

Identification of significant variation in the invasive and metastatic properties of cells isolated from the same tumor has important implications for experimental attempts to define which cellular properties correlate with invasiveness. Hitherto, the question has been investigated by analyzing tumor cells isolated at random from cultured tumor cell lines or from biopsies of tumor tissue *in vivo*. This is valid only if the tumor cell population being studied is uniform and the cells are phenotypically homogenous. If, however, extensive cellular heterogeneity exists, and cell subpopulations endowed with invasive properties are only a minor fraction of the total population, random analysis of the entire population will mean that noninvasive subpopulations may present a "background noise" which obscures features unique to invasive subpopulations.

Continued experimental efforts to characterize the invasive phenotype using heterogeneous tumor cell populations may therefore be less productive than studies in which invasive/metastatic subpopulations are isolated and compared with noninvasive/nonmetastatic subpopulations from the same tumor.

Whereas the experimental strategy adopted for identifying tumor cell properties that correlate with tumorigenicity involves comparison of nonneoplastic and neoplastic cells, analysis of the metastatic phenotype is more complicated and will require comparison of the following cell pop-

Table III
Selection of Tumor Cell Variants with Preferential Anatomic Sites for Metastatic Colonization

Species	Tumor	Major site(s) of metastasis	Reference
Mouse	Carcinoma	Lung	Koch (1939)
Mouse	Ascites carcinoma	Lung	Klein (1955)
Mouse	MDAY undifferentiated tumor	Liver, lung	Kerbel *et al.* (1978)
Mouse	Lewis lung carcinoma	Lung	Fogel *et al.* (1979)
Mouse	B16 melanoma	Lung	Fidler (1973)
Mouse	B16 melanoma	Brain (forebrain)	Brunson *et al.* (1978)
Mouse	B16 melanoma	Brain	Raz and Hart (1980)
Mouse	B16 melanoma	Ovary	Brunson and Nicolson (1979)
Mouse	B16 melanoma	Liver	Tao *et al.* (1979)
Mouse	EB lymphoma	Liver	Schirrmacher *et al.* (1979b)
Mouse	SV3T3 sarcoma	Lung	Nicolson *et al.* (1978)
Mouse	SV3T3 sarcoma	Lung	Yogeeswaran *et al.* (1980)
Mouse	RAW117 lymphosarcoma	Liver	Brunson and Nicolson (1978)
Mouse	UV2237 fibrosarcoma	Lung	Raz *et al.* (1981)
Mouse	M5076 reticulum cell sarcoma	Liver, spleen	Hart *et al.* (1981)
Rat	13762 adenocarcinoma	Lung, RLN	Neri *et al.* (1982b)
Rat	1ARG-1-RT7 hepatocarcinoma	Lung	Talmadge *et al.* (1979)
Guinea pig	Line 10 hepatocarcinoma	Liver	Fidler (1978)
Chicken	HV-transformed lymphoma	Liver	Shearman and Longenecker (1980)
Human	Poorly differentiated adenocarcinoma	Ascites, lung	Takahashi *et al.* (1978)

ulations (preferably isolated from the same parent tumor cell population):

tumorigenic, noninvasive
versus
tumorigenic, invasive, nonmetastatic
versus
tumorigenic, invasive, metastatic

Three different experimental approaches have been used to date to isolate tumor cell subpopulations with differing invasive and/or metastatic abilities.

The first involves enrichment of the fraction of invasive/metastatic subpopulations present in heterogeneous tumor cell populations and subsequent comparison with the original parent population. Koch (1939) was the first to attempt to study the cellular characteristics of organ metastasis by selecting for tumor cell variants with enhanced malignant properties. Later Klein (1955) selected repeatedly a carcinoma cell line by ascites injection and found that the highly selected variant cells were more malignant. One of the more useful techniques for obtaining malignant variant sublines of common genetic origin but differing in metastatic properties was initiated by Fidler (1973). By sequential selection of B16 melanoma sublines for enhanced abilities to form blood-borne experimental pulmonary metastases in syngeneic mice, he was able to develop a series of melanoma cell sublines that showed increasing preference for lung colonization with selection. Several types of *in vivo* selections have now been performed utilizing the B16 melanoma and B16 sublines are available that show enhanced abilities to colonize different organs (Table III). Analogous selection techniques have also been successful in isolating sublines with enhanced metastatic potential from other animal tumors (Table III).

A related approach is to select for (or against) cells which exhibit properties considered important for successful metastasis (e.g., production of tissue lytic enzymes, locomotion). As in the enrichment method, the selection procedures are applied to heterogeneous (uncloned) tumor cell populations and variants displaying (or lacking) the property of interest are recovered and assayed to determine whether their invasiveness differs significantly from that of the original parent cell population. Selection *in vitro* for tumor cell variants with reduced adhesive properties, resistance to lysis by cytotoxic lymphocytes, resistance to lectin-mediated toxicity, and altered attachment to collagen have been used successfully to isolate variant tumor cell lines with altered metastatic capabilities (Table IV).

The third approach involves cloning of heterogeneous tumor cell populations to identify clones with desired invasive/metastatic properties which can be compared with clones from the same population that fail

Table IV
Selection of Tumor Cell Variants with Altered Metastatic Properties from Heterogeneous Parent Tumor Cell Populations via Selection for or against Nonmetastatic Properties

Cell type	Property selected	Reference
B16 melanoma	Resistance to lymphocyte killing	Fidler *et al.* (1976)
B16 melanoma	Detachment from plastic	Briles and Kornfeld (1978)
B16 melanoma	Resistance to lectin toxicity	Tao and Burger (1977)
B16 melanoma	Attachment to collagen	Liotta *et al.* (1977)
B16 melanoma	Increased invasiveness	Poste *et al.* (1980)
B16 melanoma	Increased invasiveness	Hart (1979)
RAW117 lymphosarcoma	Loss of lectin-binding sites	Reading *et al.* (1980a)

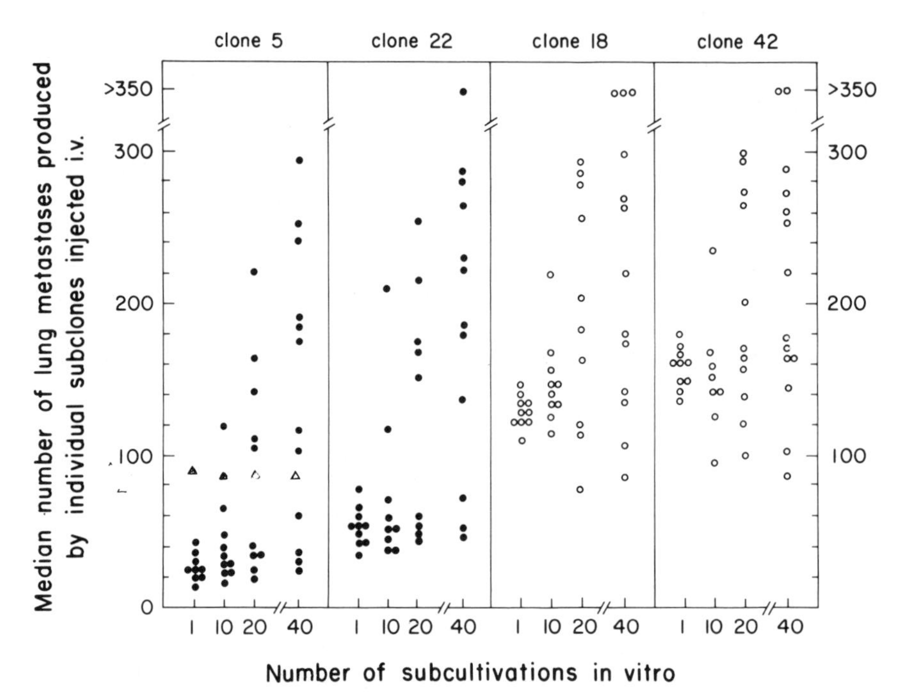

Fig. 1. The incidence of lung metastases produced by subclones (●, ○) isolated after serial subcultivation of four clones derived from the murine B16-F10 cell line. The metastatic activity of the uncloned parental B16–F10 cell line is also shown (▲). Metastasis formation was assayed by i.v. injection of 2.5×10^4 viable cells as a single-cell suspension in 0.2 ml Hank's balanced salt solution. Mice were killed 18 days later, the lungs removed, rinsed in water, and fixed in formalin, and the number of melanotic lung tumor colonies (experimental metastases) counted under a dissecting microscope. Reproduced with permission from Poste *et al.* (1981).

to express these properties. This approach was used by Fidler and Kripke (1977) to isolate B16 melanoma clones with highly different metastatic properties from the same parent cell population. Similar demonstrations of the presence of clones with widely differing invasive and/or metastatic properties within the same tumor have been made in several other animal tumors (Table II).

Phenotypic analysis of tumor cell clones is the most direct and satisfactory of the three approaches. In the other two, selection to enrich the subpopulations showing the desired phenotype, though useful, still carries the difficulty that several subpopulations with differing phenotypes persist within the population and may complicate interpretation of experimental data.

The value of tumor cell clones in analyzing any aspect of tumor cell behavior requires that the phenotypic character(s) of interest are stable during serial passage of the clones whether *in vivo* or *in vitro*. Although many cellular properties are highly stable in cloned cells in the absence of specific selection pressures, recent work in our laboratories has revealed that the invasive and metastatic properties of B16 melanoma clones are highly unstable during serial passage and subclones with different invasive and metastatic properties are generated rapidly on serial passage *in vitro* or *in vivo* (Fidler and Nicolson, 1981; Poste *et al.*, 1981). The instability of these properties in clones propagated in isolation contrasts with the apparent stability of these traits in the heterogeneous polyclonal parent cell populations from which they were isolated (Fig. 1). Observations on individual B16 clones carrying a variety of stable biochemical markers have revealed that whereas the metastatic phenotype is unstable when clones are grown singly (Fig. 1), mixing and cocultivation of clones to create artificially a polyclonal population eliminates this phenotypic instability and formation of variant subclones with altered properties is reduced dramatically (Fig. 2). This suggests that some form of "interaction" is occurring between the various cellular subpopulations in polyclonal populations which somehow "stabilizes" not only their invasive/metastatic properties but also their relative proportions within the total population. This type of interaction would conserve clonal diversity within a tumor cell population and prevent domination of the population by a few, or even a single, subpopulation.

Introduction of a new selection pressure can alter the "equilibrium" between different clonal subpopulations and restrict subpopulation diversity by eliminating unfit clones. We have shown that if the majority of the subpopulations are eliminated, the "stabilizing" interaction between subpopulations is lost and the surviving subpopulations become phenotypically unstable and quickly generate a new panel of variant subpopulations with different invasive/metastatic properties. This was demonstrated by showing that in polyclonal populations produced by mixing three wild type (*wt*) and three drug-resistant B16 melanoma clones subsequent treatment with drugs to eliminate *wt* clones stimulated phenotypic instability within the surviving drug-resistant clones and new panel of subclones with different invasive/metastatic properties emerged rapidly (Fig. 2).

This phenomenon is not unique to the B16 melanoma. Similar instability of the metastatic phenotype has been found in the mouse UV2237 fibrosarcoma (Cifone and Fidler, 1981), mouse RAW117 lymphosarcoma, 13762 mammary adenocarcinoma (G. L. Nicolson, unpublished), and the rat 1AR6 hepatocarcinoma (Talmadge *et al.*, 1979).

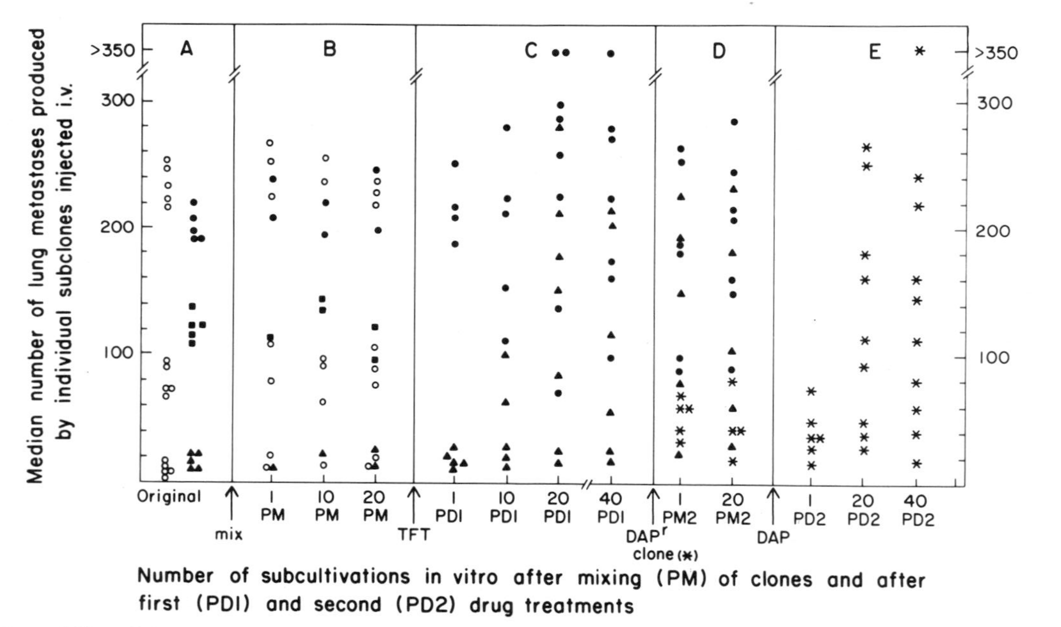

Fig. 2. The stability of the metastatic phenotype in polyclonal cultures prepared by cocultivation of a series of clones isolated from the murine B16-F10 melanoma cell line. Wild-type, drug-sensitive B16-F10 clones (*wt*; ○) were mixed with clones resistant to trifluorothymidine (TFT^r; ▲), ouabain (Oua^r; ■) or to both drugs (TFT^r/Oua^r; ●) (A). Subclones were isolated after 10 or 20 subcultivations (B) and assayed for their metastatic properties and drug sensitivities. After 20 subcultivations, the cultures were treated with TFT. The surviving cells were then passaged (C) and subclones isolated and tested for metastatic properties and for resistance to 2 μg/ml TFT (▲) or TFT and 1×10^{-3} M ouabain (●). After 40 subcultivations, a new clone of DAP^r cells (*) were added. Subclones were isolated from this mixed cell population after 20 subcultivations (D), and their metastatic properties and susceptibility to TFT (2 μg/ml), ouabain (1×10^{-3} M), and DAP (4.7×10^{-5} M) evaluated. Replicate cultures were treated with DAP (4.7×10^{-5} M). The surviving cells were passaged (E), and subclones isolated at the indicated intervals and tested for their metastatic properties and their ability to grow in HAT medium [DAP^r variants (*) grow, TFT^r variants fail to grow]. Reproduced with permission from Poste *et al.* (1981).

The identification of interactions occurring between clonal subpopulations within the same tumor introduces an additional level of complexity to the experimental analysis of tumor cell properties since it suggests that a full understanding of the behavior of tumors may not be possible by analysis of individual component cells but may require more sophisticated analyses of subpopulation interactions.

Cloning provides the most direct and efficient method for the identification of tumor cells with differing invasive and/or metastatic properties. However, if the data described above concerning the phenotypic instability of clones from several murine tumors are applicable to tumor cells in general, then cloning alone will not guarantee uniform experimental material. By eliminating the interaction between subpopulations, cloning provides a potent stimulus for emergence of variant subclones with differing phenotypes. If this proves to be a general feature of malignant tumor cells, it will be essential to practice the additional step of recloning tumor cell clones at regular intervals to ensure that cell preparations with uniform properties are being studied. A possible alternative might be to attempt to "stabilize" the metastatic phenotype in clones by artificially creating polyclonal populations via cocultivation of a series of clones. In this case, the most logical approach would seem to be first to identify a series of clones with similar invasive properties and mix them. This population could then be compared with a polyclonal population prepared by mixing a similar number of noninvasive clones isolated from the same parent cell population.

IV. THE SURFACE PROPERTIES OF METASTATIC TUMOR CELLS

The central role of cell surface components in the interaction of cells with their environment has stimulated a massive research effort to identify differences in the surface properties of normal and tumor cells. This has been successful to the extent that a vast catalog of differences have been detected. Fundamental deficiencies remain, however, in our understanding of how specific surface changes may contribute to complex traits such as tumorigenicity, invasiveness, and metastatic behavior. Indeed, surprisingly few of the enormous number of publications on this topic pertain to metastatic cells. The majority of studies reported to date have used continuous cell lines "transformed" *in vitro* by oncogenic viruses or chemical carcinogens. Most of these lines are not metastatic, and, worse still, in many cases their *in vivo* behavior has not even been examined in detail. In far too many publications, cells have been classified as "transformed" merely because they exhibit altered morphology and/or growth properties following *in vitro* exposure to a putative oncogenic stimulus and experiments to determine tumorigenicity and metastatic potential have not been undertaken. The shortcomings inherent in reliance on *in vitro* parameters of transformation is illustrated by the widespread use of heteroploid established cell lines such as BHK-21, NIL-B, and NIL-8 hamster cells as so called "normal" control cells to compare with cells of the same type after "transformation" by tumor viruses, even though all of these lines are tumorigenic before "transformation"!

With the recent recognition that metastases are caused by specialized tumor cell subpopulations, an increasing number of studies are being done to compare the surface properties of nonmetastatic and metastatic tumor cell subpopulations isolated from the same tumor. Recent findings on surface changes in metastatic cells obtained from comparisons of this kind are reviewed in the following sections.

A. Cell Surface Properties and the Behavior of Metastatic Tumor Cells

During transport in the blood malignant cells undergo cellular interactions with other circulating host cells, soluble blood components, and the vascular endothelium. Heterotypic cell adhesion of malignant cells with platelets (review, Nicolson *et al.*, 1980) is known to be important in blood-borne implantation. For example, it has been shown that experimental metastasis can be reduced by inducing thrombocytopenia, in animals, as well as administration of antiplatelet agents (review, Giraldi and Sava, 1981). Platelet aggregation activities have been found in a variety of tumor cell lines, and in isolated cell membrane vesicles from tumor and transformed cells (see Pearlstein *et al.*, 1980, for references). Metastatic cells also aggregate with host lymphocytes. Fidler *et al.* (1976) found that metastatic B16-F10 cells adhere to host lymphocytes at higher rates than B16-F1 cells of lower metastatic potential. Using sequential selection procedures based on lymphocyte killing of B16 melanoma cells, Fidler *et al.* (1976) were able to obtain after six *in vitro* selections sublines that were no longer susceptible to lymphocyte-mediated cytotoxicity. These sublines no longer bound lymphocytes to the same degree as the parent line, and, interestingly, they produced significantly fewer experimental lung metastases *in vivo*.

The fact that malignant cells enter the blood does not necessarily indicate that arrest and metastatic colonization will follow (see Fisher and Fisher, 1976). Stable attachment or adhesion to the capillary endothelium is necessary to prevent detachment and recirculation and the larger the circulating tumor cell emboli the greater the chance for successful arrest (see Poste and Fidler, 1980, for references). Adhesion of malignant cells to host organ cells has been used to demonstrate a role for specific cell adhesion in determining specificity of organ arrest and metastasis. Nicolson *et al.* (1980) found that B16 melanoma sublines selected for enhanced lung colonization and metastasis adhere at much faster rates to lung cells than to cells obtained from other organs not involved in metastasis. Similarly, Phondke *et al.* (1981) found that leukemia cells which colonize spleen, but not lung, adhere to isolate spleen cells but fail to adhere to suspended lung cells, and Schirrmacher *et al.* (1980) have demonstrated that liver colonizing lymphoma variants adhere preferentially to hepatocytes. Since these interactions normally take place at the level of the vascular endothelium, adhesion experiments should take this into account by using endothelial cells. By establishing endothelial cells from brain and using B16 melanoma sublines selected for enhanced lung or brain implantation, the relationship between target organ endothelial cells and organ-selected B16 melanoma subpopulations has been explored further (Nicolson, 1982). These experiments showed that brain-selected B16 melanoma cells adhere at faster rates to brain endothelial cell monolayers than lung-selected melanoma cells.

Once tumor cell arrest has occurred, malignant cells must escape into the extravascular compartment by invading the underlying basement membrane or basal lamina. The sequence of events in this process has been studied using malignant cell binding to monolayers of vascular endothelial cells (review, Nicolson, 1982). These studies indicate that adhesion of malignant cells to vascular endothelial cells stimulates endothelial cell retraction and exposure of the underlying basal lamina, whereupon the malignant cells migrate to the basal lamina and adhere firmly to this structure. Basal lamina components that are believed to be important in adhesion of tumor cells to the basal lamina are the glycoproteins fibronectin, laminin, type IV collagen, and possibly sulfated proteoglycans (see Nicolson, 1982). Once the basal lamina has been solubilized, malignant cells can enter extravascular tissues and establish new tumor colonies.

Another intriguing aspect of metastasis concerns the nonrandom distribution of metastasis caused by certain tumors.

Clinical and experimental observations showing a predilection of certain neoplasms to metastasize to particular organs (for references, see Poste and Fidler, 1980) cannot be explained on mechanical grounds alone and these nonrandom patterns of cell arrest and growth have been variously attributed to properties of the tumor cells, the vascular bed and/or the environment of the organ involved. Evidence that tumor cell properties can in part determine the pattern of metastasis has come from studies in which cell variants that localize preferentially in specific "target organs" have been isolated from a common parental B16 melanoma population (Table I). The term *target organ* is used here only with respect to the initial "arrest" behavior of cells and does not imply that cells are incapable of growing in other organs. Indeed, as discussed later, this is known not to be so. However, the specificity of the organ-localization patterns exhibited by these cell lines has been demonstrated by the finding that they will localize in pieces of their appropriate target organ implanted ectopically (Hart and Fidler, 1980) and in parabiosis experiments in which tumor cells released from the lungs of tumor-bearing animals localized and grew in the lungs of the non-tumor-bearing parabiotic recipients (Fidler, 1978).

The possible role of cell surface components in determining such localization patterns was first suggested by experiments showing that perturbation of cell surface organization by proteinases (Fidler, 1978) and by agents that disrupt plasma membrane-associated cytoskeletal elements (Raz *et al.*, 1980) resulted in altered organ-localization patterns, together with metastasis formation in the organs in which they arrested. Additional evidence implicating cell surface components in the arrest behavior of circulating B16 melanoma cells has come from work in our laboratories in which plasma membrane components were transferred between B16 sublines of differing lung-colonizing potential. B16 melanoma cells (and many others) growing *in vivo* and *in vitro* spontaneously shed closed vesicles of intact plasma membrane that can be harvested and fused with other cells by using polyethylene glycol (Poste and Nicolson, 1980). Membrane vesicles from the B16-F10 subline, which has a high capacity for lung colonization, were used with B16-F1 cells, which produce few pulmonary metastases. When the vesicle-modified cells were injected intravenously into C57BL/6 mice, they were found to arrest in the lung capillary bed in significantly greater numbers than untreated B16-F1 cells. Equally important, increased cell arrest resulted in increased production of pulmonary metastases. The possibility that the increased arrest of vesicle-modified B16-F1 cells in the lung was due simply to increased cell volume produced by vesicle uptake was excluded by showing that the volume of vesicle-treated cells (determined from cells isolated by centrifugal elutriation) did not differ significantly from that of control B16-F1 cells.

In related experiments, lymphocyte-resistant sublines of B16-F1 were rendered susceptible to killing *in vitro* by cytotoxic lymphocytes by treatment with vesicles from lymphocyte-sensitive B16 sublines plus polyethylene glycol. Since lymphocyte-mediated cytolysis occurs only when the target antigen resides within the plasma membrane, this experiment also shows that the vesicles fuse with the cellular plasma membrane and are not merely adsorbed to the cell surface.

The effect of B16-F10 vesicles in modifying the arrest of B16-F1 cells was highly specific. Veiscles from B16-F10^{lr} cells (a variant of the B16-F10 subline that is resistant to killing by syngeneic immune lymphocytes but which resembles B16-F1 in its arrest and metastatic behavior) had no effect on the arrest of B16-F1 cells. Interestingly, fusion of B16-F1 vesicles with B16-F10 cells did not alter the arrest or metastatic behavior of these cells, but this may have been because the vesicles were not introduced in sufficient amounts.

Studies with B16-F1 vesicles surface-labeled with ^{125}I or ferritin reveal that vesicle components persist in the plasma membrane of B16-F10 acceptor cells for only 18–24 hr. How then does vesicle treatment enhance metastasis formation when proliferation of extravasated cell to form metastases does not begin until 1–2 days after injection, when vesicle components are no longer present? B16 cells invade the blood vessel wall within 2–4 hr after their arrest, and once within the vessel wall they are presumably less susceptible to damage by hemodynamic forces or circulating host cells and thus have a greater opportunity of continuing the process of metastasis. Therefore, simply by increasing the efficiency of cell arrest and invasion in the lung capillary bed, vesicle-induced modification of B16-F1 cells increases the likelihood of metastasis formation, even though vesicle components per se make no direct contribution to any events occurring 24 hr after initial cell arrest.

B. Cell Surface Proteins and Glycoproteins on Metastatic Cells

Differences between cell surface glycoproteins on sublines of low or high metastatic potential have been detected by the use of lectins. Quantitative lectin binding to B16 melanoma sublines of differing metastatic potential has shown that some low metastatic sublines bind less ^{125}I-concanavalin A (such as the lymphocyte-resistant subline B16-F10$^{Lr\text{-}6}$), while the highly metastatic subline B16-F10 was found to have fewer wheat germ and soy bean agglutinin-binding sites (Raz *et al.*, 1980). Differences in the binding of ^{125}I-labeled lectins have also been documented in metastatic RAW117 cell variants. The most dramatic difference found was the sequential loss of concanavalin A receptors with increasing malignancy and metastatic potential (Nicolson *et al.*, 1980). Reading *et al.* (1980a) have examined a number of cell lines and clones selected *in vivo* for enhanced metastasis or *in vitro* for loss of adherence to immobilized lectins. In almost every cell line and clone examined, there was a correlation between the loss of concanavalin A-binding sites and metastatic potential, while there was no correlation between the number of wheat germ agglutinin-binding sites and malignancy. Similarly, Talmadge *et al.* (1981) found that the metastatic potential of a hepatocarcinoma cell line correlated with the number of concanavalin A-binding sites but did not correlate with a number of other *in vitro* parameters such as cell growth rate, saturation densities, cell shedding and adhesion characteristics, plasminogen activator production, and procoagulant activities.

The lectin-binding components involved in metastatic cell interactions have also been studied by analyzing the binding of radioactive lectins to solubilized, separated cellular glycoproteins on sodium dodecylsulfate (SDS)–polyacrylamide slab gels. Although differences are often not found using this technique (e.g., see Raz *et al.*, 1980), lectin-binding glycoprotein alterations have been found which sometimes correlate nicely with the potential to metastasize. For example, Irimura *et al.* (1981) used ^{125}I-labeled *Ricinus communis* agglutinin to detect differences on tunicamycin-treated B16 melanoma sublines. Tunicamycin treatment inhibited experimental metastasis and resulted in loss of high-molecular-weight sialogalactoproteins. The reappearance of these sialogalactoproteins correlated with the reappearance of metastatic properties after drug removal.

Changes in galactoproteins have been identified in metastatic cells using oxidation with galactose oxidase followed by reduction with [^{3}H]-borohydride with or without pretreatment of cells with neuraminidase to remove terminal sialic acid. Using this procedure, Raz *et al.* (1980) found that expression of a major sialogalactoprotein on lung-colonizing B16 melanoma sublines correlated with the degree of lung implantation, survival, and growth. Using similar techniques, other workers have found other changes in sialogalactoproteins in B16 melanoma glycoproteins (Yogeeswaran *et al.*, 1980). These findings correlate with those using ^{125}I-labeled galactose-specific lectins to stain SDS–gels containing detergent-solubilized, neuraminidase-treated B16 melanoma sublines (Irimura *et al.*, 1981; Nicolson, 1982).

One of the generalized changes found on many "transformed" cells compared to untransformed cells is the loss of certain cell surface glycoproteins identifiable by cell surface labeling utilizing lactoperoxidase-catalyzed iodination (review, Hynes, 1976). Not all metastatic systems show distinct cell surface changes using lactoperoxidase-catalyzed iodination techniques. While differences were not found in iodinatable proteins exposed on the surfaces of lung-colonizing B16 melanoma cells (Nicolson *et al.*, 1980; Raz *et al.*, 1980), differences have been found on brain-selected sublines (Brunson *et al.*, 1978), ovary-selected B16 melanoma sublines (Brunson and Nicolson, 1979), and whcat germ agglutinin-resistant clones (Finne *et al.*, 1980). Modifications in lactoperoxidase-catalyzed iodinatable surface proteins have also been seen in other metastatic tumor systems. Highly metastatic liver-colonizing RAW117 lymphosarcoma sublines were found to progressively lose a 70,000-dalton surface glycoprotein during *in vivo* selection for enhanced liver colonization, while simultaneously gaining a component of 135,000 daltons (Nicolson *et al.*, 1980). Alterations have also been noted on virus-transformed 3T3 cell sublines of high and low lung colonization potential (Yogeeswaran *et al.*, 1980) and mammary adenocarcinoma clones differing in their ability to spontaneously metastasize to lymph nodes and lungs (G. L. Nicolson, unpublished).

Cell surface glycopeptide fragments removed from tumor cells by trypsinization and rendered free of oligopeptides by Pronase treatment have been studied on metastatic tumor cells. Although similar studies have revealed that trypsin-released glycopeptides from virally transformed cells (of approx. 4000 mol. wt.) are of higher apparent molecular weight when chromatographed on Sephadex G-50 columns, examination of peptides from low and high lung-colonizing sublines of B16 melanoma has revealed no significant differences (Nicolson *et al.*, 1980).

One of the more consistent cell surface changes accompanying transformation of cells *in vitro* is the loss or decrease in exposure of cell surface fibronectin (review, Hynes, 1976). Smith *et al.* (1979) examined several human tumor-derived cell lines for the presence of fibronectin and found this glycoprotein was expressed on cell lines obtained from primary mammary carcinomas but not on cell lines from metastatic carcinomas. On the other hand, Neri *et al.* (1982a,b) failed to find a correlation between fibronectin expression *in vivo* or *in vitro* and the potential of rat mammary adenocarcinoma cell lines and clones to metastasize spontaneously from mammary sites to lung. Although cell lines obtained from secondary sites generally expressed less fibronectin compared to cell lines from primary sites, there was no absolute relationship between loss of fibronectin and ability to metastasize.

V. GLYCOLIPIDS OF METASTATIC CELLS

Several investigators have examined the glycolipids of tumor cell sublines of defined metastatic potential. A very simple ganglioside pattern is found in B16 melanoma cells. The most prominent components are the sialoganglioside G_{M3} [Cer-Glc-Gal-(NANA)-GalNAc] and G_{M1} [Cer-Glc-Gal-(NANA)-GalNAc-Gal] (NANA, *N*-acetylneuraminic acid; GalNAc, *N*-acetylgalactosamine). The most distinctive change in glycolipid pattern found in selected B16 metastatic variants has been an increase in G_{M3} and a reduction in the ganglioside GD_{1a} [Cer-Glc-Gal-(NANA)-Gal-NAc-Gal-(NANA)] in lung-colonizing sublines of high metastatic potential (Nicolson *et al.*, 1980; Raz *et al.*, 1980). Yogeeswaran and Stein (1980) have examined the glycosphingolipids on several metastatic variant sublines of virus-transformed 3T3 lines and found differences between various sublines depending on cell growth characteristics, state of transformation, type of transforming virus, and metastatic potential. In the latter case, only the enhanced external labeling of G_{M2} by galactose oxidase-NaB^3H_4 correlated with ability to metastasize to the lung.

Immunologic Alterations in Metastatic Cells

There are numerous reports in the literature in which tumor cells isolated from metastases exhibit different antigenicities and/or immunogenicities to cells in the primary tumor (see Gorelik *et al.*, 1979; Shirrmacher *et al.*,1979a; Fidler and Kripke, 1980, for references).

Although sufficient data indicate that the immunologic properties of metastatic tumor cells are altered compared to nonmetastatic tumor cells, the role of malignant cell antigens and host immune antitumor response in the metastatic process has not been defined clearly (review, Fidler and Kripke, 1980). In some systems depression of host antitumor immunity has led to an increase in the incidence of metastasis, while in other systems abrogation of host immunity results in prevention or suppression of metastasis. These apparent conflicts, and his own data, led Prehn (1977) to propose that host immunity can have dual or even simultaneous effects on metastasis—in some cases stimulating tumor growth and aiding in metastatic cell survival, while in other instances causing immune-mediated inhibition or cytotoxic destruction of metastases. Strong support for the dual concept of antitumor immunity has come from the studies of Fidler *et al.* (1979), where histologically similar fibrosarcomas of differing immunogenicities were compared for their abilities to form spontaneous and experimental metastases in normal, immunosuppressed, and immunologically reconstituted animals. They found that a highly immunogenic tumor produced more metastases in immunosuppressed animals compared to the other groups. Fibrosarcomas of intermediate immunogenicity formed more metastases in immunosuppressed animals, but this could be reversed by immunological reconstitution, and the least immunogenic tumor formed fewer metastases in immunosuppressed or reconstituted mice compared to normal animals. Thus, as Prehn predicted, host antitumor immunity can be both inhibitory or stimulatory with regard to tumor metastasis.

In immunologically competent hosts, it is thought that immunoselection may occur at the level of the primary or secondary tumor resulting in the eventual emergence of tumor cell subpopulations which lack strong antigens that could lead to their recognition and destruction. Evidence for this type of selection has come from experiments where the selected, highly malignant cells have been shown to display reduced antigen levels or to have antigen deletions on their cell surfaces. For example, Reading *et al.* (1980b) analyzed a number of *in vivo*- and *in vitro*-selected murine RAW117 lymphosarcoma cell lines (and clones derived from these cells lines) for their metastatic properties and cell surface antigen contents and found that the ability to metastasize to the liver showed an inverse correlation with their content of the antigenic RNA tumor virus envelope glycoprotein gp70 determined by competition radioimmune assay. In this system, successful metastasis apparently requires escape from host immune surveillance via antigen deletion on the highly metastatic lymphosarcoma cells. However, in other metastatic systems such as B16 melanoma, there is no relationship between metastasis and viral antigens such as gp70 (Fidler and Nicolson, 1981), and in other tumors particular antigens may even be increased on metastatic cells. Shearman and Longenecker (1980) reported an increase in cellular antigen content that correlated with ability of Marek's disease virus-transformed chick lymphoma cells to metastasize to the liver. In this system, the amount of a cell surface antigen detectable with monoclonal antibody increased concomitant with increased ability to colonize the liver. Thus, there is no simple relationship between the display of cell surface antigens, immunogenicity, and metastasis.

VI. CONCLUDING REMARKS

Two of the most difficult problems facing researchers involved in studying tumor cell properties and metastasis are (1) the inherent phenotypic instabilities of malignant cells and (2) the continuous *in vivo* selection pressures that discriminate against portions of the tumor cell population. These characteristics of the malignant cell and its environment lead to tumor progression and the emergence of new rare tumor cell variants. This, coupled with natural selection pressures, dictates that tumors are undergoing continuous phenotypic change and contain subpopulations of phenotypically different cells. Careful comparison of tumor cell subpopulations with differing metastatic potentials has allowed identification of cell surface characteristics which may be important in different steps in the metastatic process. However, malignant cell subpopulations are often unstable, even in the absence of host selective pressures, and this fact has made it more difficult to identify cell surface properties that may correlate with metastatic ability. Most studies on this topic have used cell populations that have been expanded by long-term growth *in vitro* or obtained from tumors contaminated with dead or dying cells and containing unknown numbers of normal host cells. Although frequent assays of metastatic potential, performed in parallel with biochemical and immunological experiments, have eliminated some of the problems associated with phenotypic divergence of tumor cell properties during growth *in vivo* or *in vitro*, this has not been practical when using malignant human cells. Strict utilization of short-term tissue culture may or may not exclude some of the problems of phenotypic instability. Finally, it should be mentioned that in certain tumor cell systms growth *in vivo* can lead to even more rapid phenotypic divergence than that found in tissue culture, so the stringent growth of a tumor cell population *in vivo* does not necessarily remove problems associated with phenotype instability. Phenotypic instability (due to genotypic instability?) and the rapid generation of tumor cells with altered properties could be the unique characteristics that are common to all highly malignant tumor cell populations.

Reliable correlations between specific phenotypic traits studied *in vitro* and metastatic behavior *in vivo* will not only require detailed comparison of a number of sublines with different abilities to metastasize but also that experiments *in vivo* be undertaken in parallel to evaluate the behavior of the particular cells being studied *in vitro*. The search for a property uniform to all metastatic cells may be unproductive. Work carried out with isolated variant tumor cell lines must consider that the development of metastases is dependent on a complex interplay between host factors and the intrinsic properties of the tumor cells. To establish metastases, tumor cells must complete all steps involved in the metastatic process. Enhanced performance of a tumor cell in one step of the process cannot compensate for an inability to complete a subsequent step. The failure of most tumor cells to produce metastases need not be due to a single, common factor. Interruption of the metastatic sequence at any step in the process will prevent the production of clinical metastasis.

The increasing availability of tumor cell populations with differing metastatic capabilities should facilitate experimental efforts to identify the properties that enable malignant cells to metastasize. With the availability of this knowledge it might then be possible to design more effective procedures for the treatment of this feared disease.

ACKNOWLEDGMENTS. The personal research cited in this article was supported by Grants CA13393, CA18260, and CA17609 (G.P.) and CA28867, CA29571 and CA28844 (G.L.N.) from the National Cancer Institute.

VII. REFERENCES

Briles, E. B., and Kornfeld, S., 1978, Isolation and metastatic properties of detachment variants of B16 melanoma cells, *J. Natl. Cancer Inst.* **60:**1217.

Brunson, K. W., and Nicolson, G. L., 1978, Selection and biologic properties of malignant variants of a murine lymphosarcoma, *J. Natl. Cancer Inst.* **61:**1499.

Brunson, K. W., and Nicolson, G. L., 1979, Selection of malignant melanoma variant cell lines for ovary colonization, *J. Supramol. Struct.* **11:**517.

Brunson, K. W., Beattie, G., and Nicolson, G. L., 1978, Selection and altered tumour cell properties of brain-colonising metastatic melanoma, *Nature (London)* **272:**543.

Cifone, M. A., and Fidler, I. J., 1981, Increasing metastatic potential is associated with increasing genetic instability of clones isolated from murine neoplasms, *Proc. Natl. Acad. Sci. USA* **78:**6949.

Dexter, D. L., Kowalski, H. M., Blazar, B. A., Fligiel, A., Vogel, R., and Heppner, G. H., 1978, Heterogeneity of tumor cells from a single mouse mammary tumor, *Cancer Res.* **38:**3174.

Fidler, I. J., 1973, Selection of successive tumor lines for metastasis, *Nature New Biol.* **242:**148.

Fidler, I. J., 1978, General considerations for studies of experimental cancer metastasis, *Methods Cancer Res.* **15:**399.

Fidler, I. J., and Kripke, M. L., 1977, Metastasis results from pre-existing variant cells within a malignant tumor, *Science* **197:**893.

Fidler, I. J., and Kripke, M., 1980, Tumor cell antigenicity, host immunity and cancer metastasis, *Cancer Immunol. Immunother.* **7:**201.

Fidler, I. J., and Nicolson, G. L., 1981, The immunobiology of experimental metastatic melanoma, *Cancer Biol. Rev.* **2:**1.

Fidler, I. J., Gersten, D. M., and Budmen, M. B., 1976, Characterization *in vivo* and *in vitro* of tumor cells selected for resistance to syngeneic lymphocyte-mediated cytotoxicity, *Cancer Res.* **36:**3160.

Fidler, I. J., Gersten, D. M., and Kripke, M. L., 1979, The influence of immunity on the metastasis of three murine fibrosarcomas of differing immunogenicity, *Cancer Res.* **39:**3816.

Finne, J., Tao, T-W., and Burger, M. M., 1980, Carbohydrate changes in glycoproteins of a poorly metastasizing wheat germ agglutinin-resistant melanoma clone, *Cancer Res.* **40:**2580.

Fisher, E. R., and Fisher, B., 1976, Circulating cancer cells and metastasis, *Int. J. Radiat. Oncol. Biol. Phys.* **1:**87.

Fogel, M., Gorelik, E., Segal, S., and Feldman, M., 1979, Differences in cell surface antigens of tumor metastases and those of the local tumor, *J. Natl. Cancer Inst.* **62:**585.

Giraldi, T., and Sava, G., 1981, Selective antimetastatic drugs, *Anticancer Res.* **1:**163.

Gorelik, E., Fogel, M., Segal, S., and Feldman, M., 1979, Tumor-associated antigenic differences between the primary and the descendant metastatic tumor cell population, *J. Supramol. Struct.* **12:**385.

Hanna, N., and Fidler, I. J., 1981, Expression of metastatic potential of allogeneic and xenogeneic neoplasma in young nude mice, *Cancer Res.* **41:**438.

Hart, I. R., 1979, The selection and characterization of an invasive variant of B16 melanoma, *Am. J. Pathol.* **97:**587.

Hart, I. R., and Fidler, I. J., 1980, Role of organ selectivity in the determination of metastatic patterns of B16 melanoma, *Cancer Res.* **40:**2282.

Hart, I. R., Talmadge, J. E., and Fidler, I. J., 1981, Metastatic behavior of a murine reticulum cell sarcoma exhibiting organ-specific growth, *Cancer Res.* **41:**1281.

Heppner, G. H., Dexter, D. L., DeNucci, T., Miller, F. R., and Calabresi, P., 1978, Heterogeneity in drug sensitivity among tumor cell subpopulations of a single mammary tumor, *Cancer Res.* **38:**3758.

Hynes, R. O., 1976, Cell surface proteins and malignant transformation, *Biochim. Biophys. Acta* **458:**73.

Irimura, T., Gonzalez, R., and Nicolson, G. L., 1981, Effects of tunicamycin on B16 metastatic melanoma cell surface glycoproteins and blood-borne arrest and survival properties *Cancer Res.* **41:**3411.

Kerbel, R. S., Twiddy, R. R., and Robertson, D. M., 1978, Induction of a tumor with greatly increased metastatic growth potential by injection of cells from a low-metastatic H-2 heterozygous tumor cell line into an H-2 incompatible parental strain, *Int. J. Cancer* **22:**583.

Klein, E., 1955, Gradual transformation of solid into ascites tumors. Evidence favoring the mutation-selection theory. *Exp. Cell Res.* **8:**188.

Koch, F. E., 1939, Iur Fragi der Metastasenbildung bei Impftumour, *Krebforsch.* **28:**495.

Kripke, M. L., Gruys, E., and Fidler, I. J., 1978, Metastatic heterogeneity of cells from an ultraviolet light-induced murine fibrosarcoma of recent origin, *Cancer Res.* **38:**2962.

Liotta, L. A., Kleinerman, J., Catanzaro, P., and Rynbrandt, D., 1977, Degradation of basement membrane by murine tumor cells, *J. Natl. Cancer Inst.* **58:**1427.

Neri, A., Ruoslahti, E., and Nicolson, G. L., 1982a, The distribution of fibronectin on clonal cell lines of a rat mammary adenocarcinoma growing in vitro and in vivo at primary and metastatic sites, *Cancer Res.* **41:**5082.

Neri, A., Welch, D., Kawaguchi, T., and Nicolson, G. L., 1982b, The development and biologic properties of malignant cell sublines and clones of a spontaneously metastasizing rat mammary adenocarcinoma, Submitted, for publication.

Nicolson, G. L., 1982, Metastatic tumor cell attachment and invasion assay utilizing vascular endothelial cell monolayers, *J. Histochem. Cytochem.* **30:**214.

Nicolson, G. L., Brunson, K. W., and Fidler, I. J., 1978, Specificity of arrest, survival and growth of selected metastatic variant cell lines, *Cancer Res.* **38:**4105.

Nicolson, G. L., Reading, C. L., and Brunson, K. W., 1980, Blood-borne tumor metastasis: Some properties of selected tumor cell variants of differing malignancies, *in* "Tumor Progression," (R. G. Crispen, ed.), p. 31, Elsevier/North-Holland, Amsterdam.

Nowell, P. C., 1976, The clonal evolution of tumor cell populations, *Science* **194:**23.

Pearlstein, E., Salk, P. L., Yogeeswaran, G., and Karpatkin, S., 1980, Correlation between spontaneous metastatic potential, platelet-aggregating activity of cell surface extracts, and cell surface sialylation in 10 metastatic-variant derivatives of rat renal sarcoma cell line, *Proc. Nat. Acad. Sci. USA* **77:**4336.

Phondke, G. P., Madyastha, K. R., Madyastha, P. R., and Barth, R. F., 1981, Relationship between Concanvalin A-induced agglutin-ability of murine leukemia cells and their propensity to form heterotypic aggregates with syngeneic lymphoid cells, *J. Natl. Cancer Inst.* **66:**643.

Poste, G., and Fidler, I. J., 1980, The pathogenesis of cancer metastasis, *Nature* (*London*) **283:**139.

Poste, G., and Nicolson, G. L., 1980, Arrest and metastasis of blood-borne tumor cells are modified by fusion of plasma membrane vesicles from highly metastatic cells, *Proc. Natl. Acad. Sci. USA* **77:**399.

Poste, G., Doll, J., Hart, I. R., and Fidler, I. J., 1980, In vitro selection of murine B16 melanoma variants with enhanced tissue invasive properties, *Cancer Res.* **40:**1636.

Poste, G., Doll, J., and Fidler, I. J., 1981, Interactions between clonal subpopulations affect the stability of the metastatic phenotype in polyclonal populations of B16 melanoma cells. *Proc. Natl. Acad. Sci. USA* **78:**6226.

Prehn, R. T., 1977, Immunostimulation of the lymphodependent phase of neoplastic growth, *J. Natl. Cancer Inst.* **59:**1043.

Raz, A., and Hart, I. R., 1980, Murine melanoma. A model for intracranial metastasis, *Br. J. Cancer* **42:**331.

Raz, A., McLellan, W. L., Hart, I. R., Bucana, C. D., Hoyer, L. C., Sela B-A., Dragsten, P., and Fidler, I. J., 1980, Cell surface properties of B16 melanoma variants with differing metastatic potential, *Cancer Res.* **40:**1645.

Raz, A., Hanna, N., and Fidler, I. J., 1981, In vivo isolation of a metastatic tumor cell variant involving selective and nonadaptive processes. *J. Natl. Cancer Inst.* **66:**183.

Reading, C. L., Belloni, P. N., and Nicolson, G. L., 1980a, Selection and in vivo properties of lectin-attachment variants of malignant murine lymphosarcoma cell lines, *J. Natl. Cancer Inst.* **64:**1241.

Reading, C. L., Brunson, K. W., Torrianni, M., and Nicolson, G. L., 1980b, Malignancies of metastatic murine lymphosarcoma cell lines and clones correlate with decreased cell surface display of RNA-tumor virus envelope glycoprotein gp70, *Proc. Natl. Acad. Sci. USA* **77:**5943.

Schirrmacher, V., Bosslet, K., Shantz, G., Claver, K., and Hubsch, D., 1979a, Tumor metastasis and cell-mediated immunity in a model system in DBA/2 mice. IV. Antigenic differences between a metastasizing variant and the parental tumor line revealed by cytotoxic T lymphocytes, *Int. J. Cancer* **23:**245.

Schirrmacher, V., Shantz, G., Clauer, K., Komitowski, D., Zimmermann, H-P., and Lohmann-Matthes, M-L., 1979b, Tumor metastases and cell-mediated immunity in a model system in DBA/2 mice. I. Tumor invasiveness in vitro and metastasis formation in vivo, *Int. J. Cancer* **23:**233.

Schirrmacher, V., Cheinsong-PoPov, R., and Arheiter, H., 1980, Hepatocyte-tumour cell interaction in vivo. I. Conditions for rosetta formation and inhibition by anti-H2 antibody, *J. Exp. Med.* **151:**984.

Shearman, P. J., and Longenecker, B. M., 1980, Selection for virulence and organ-specific metastasis of herpes virus-transformed lymphoma cells, *Int. J. cancer* **24:**363.

Shearman, P. J., and Longenecker, B. M., 1981, Clonal variation and functional correlation of organ-specific metastasis and an organ-specific metastasis-associated antigen, *Int. J. Cancer* **27:**387.

Smith, H. S., Riggs, J. L., and Mosesson, M. W., 1979, Production of fibronectin by human epithelial cell lines, *Cancer Res.* **39:**4138.

Suzuki, M., Withers, H. R., and Koehler, M. W., 1978, Heterogeneity and variability of artificial lung colony-forming ability among clones from a mouse fibrosarcoma, *Cancer Res.* **38:**3349.

Takahashi, S., Yoichi, K., Nakatani, I., Inui, S., Kojima, K., and Shiratori, T., 1978, Conversion of a poorly differentiated human adenocarcinoma to ascites form with invasion and metastasis in nude mice, *J. Natl. Cancer Inst.* **60:**925.

Talmadge, J. E., Starkey, J. R., Davis, W. C., and Cohen, A. L., 1979, Introduction of metastatic heterogeneity by short-term in vivo passage of a cloned transformed cell line, *J. Supramol. Struct.* **12:**227.

Talmadge, J. E., Starkey, J. R., and Stanford, D. R., 1981, In vitro characteristics of metastatic variant subclones of restricted genetic origin, *J. Supramol. Struct.,* **15:**139.

Tao, T-W., and Burger, M. M., 1977, Non-metastasising variants selected from metastasising melanoma cells, *Nature* (*London*) **270:**437.

Tao, T-W., Matter, A., Vogel, K., and Burger, M. M., 1979, Liver-colonizing melanoma cells selected from B16 melanoma, *Int. J. Cancer* **23:**854.

Yogeeswaran, G., and Stein, B. S., 1980, Glycosphingolipids of metastatic variants of RNA virus-transformed nonproducer Balb/3T3 cell lines: Altered metabolism and cell surface exposure, *J. Natl. Cancer Inst.* **65:**967.

Yogeeswaran, G., Stein, B. S., and Sebastian, H., 1980, Characterization of tumorigenic and metastatic properties of murine sarcoma virus-transformed non-producer BALB/3T3 cell lines, *J. Natl. Cancer Inst.* **64:**951.

Chapter 13

Antigen-Specific Suppressor ("Blocking") Factors in Tumor Immunity

Karl Erik Hellström, Ingegerd Hellström, and Karen Nelson

Division of Tumor Immunology
Fred Hutchinson Cancer Research Center
Seattle, Washington

I. INTRODUCTION

Many neoplasms have antigens, which are either not expressed or expressed in minute amounts in normal tissues. One class of tumor antigens, tumor-specific transplantation antigens (TSTA), is detectable in some experimentally induced animal tumors, e.g., on most 3-methylcholanthrene (MCA)-induced sarcomas. These antigens are immunogenic in syngeneic hosts and can be demonstrated by the ability of animals immunized with a syngeneic tumor to reject a graft of that tumor (Prehn and Main, 1957; Old and Boyse, 1964; Sjögren, 1965; Hellström and Brown, 1979). However, expression of TSTA does not normally prevent neoplasms from growing progressively in nonimmune but immunocompetent animals. Also, immunity to TSTA appears to be relatively weak; increasing the number of tumor cells in the challenge graft results in progressive growth of the tumor in immunized mice.

Other neoplasms, including many that are "spontaneous" or have been induced by a low dose of chemical carcinogen or by the implantation of plastic films, are nonimmunogenic, that is, they fail to induce tumor-specific transplantation resistance (Old and Boyse, 1964; Sjögren, 1965; Hewitt *et al.*, 1976). These neoplasms may either lack cell surface structures that can be recognized as foreign by syngeneic hosts (Klein and Klein, 1977), or they may be particularly capable of evading an immune response. Definition of the nature of the nonimmunogenic tumors should be facilitated by a better understanding of the regulation of immune responses to tumor antigens, an understanding easiest gained from studying immunogenic tumors.

One of the most challenging findings in tumor immunology is that of "concomitant tumor immunity" (Gershon *et al.*, 1967): animals bearing a growing, immunogenic neoplasm often reject a second transplant of cells from the same neoplasm while the original tumor continues to grow. Lymphocytes from tumor-bearing animals are commonly specifically cytotoxic to cells from their tumors *in vitro* (Hellström *et al.*, 1968), and the outgrowth of transplanted tumor cells is sometimes prevented when a mixture of the tumor cells and lymphocytes from animals bearing the respective neoplasm is injected into immunologically compromised syngeneic hosts (Mikulska *et al.*, 1966). Therefore, at least part of the concomitant tumor immunity phenomenon results from an immunological response to tumor antigens. An important question is then why small numbers of transplanted cells can be destroyed by that response, while large numbers of cells in the form of a tumor mass can not. An answer to this question may help us to understand why tumor-specific transplantation immunity is generally weak, and it may point toward ways of improving this immunity and make it therapeutically useful.

The fact that an immunogenic tumor can grow in an animal which has the capacity to destroy its cells implies that there must be escape mechanisms which prevent this destruction from occurring *in vivo*. The existence of concomitant tumor immunity further suggests that these mechanisms are more effective at the local than at the systemic level. Various escape mechanisms have, indeed, been identified (reviewed by Baldwin and Robins, 1977; Herberman, 1977; Hellström and Brown, 1979). They include the masking of tumor antigens by material produced by either the host or the tumor, the modulation of tumor antigens in the presence of specific antibody, inability of effector cells and molecules to penetrate into a growing tumor, and the loss of tumor antigens. There are good reasons to believe, however, that the release of antigen from tumor cells followed by the activation of suppressor cell-mediated responses provides one of the most crucial of the escape mechanisms. In this article we shall review the evidence for this view.

II. THE CONCEPT OF SPECIFIC BLOCKING FACTORS (SBF)

Some of the first evidence that tumor-bearing animals can mount an immune response to their neoplasms was obtained in the late 1960s from studies on lymphocyte-mediated reactivity against cultured tumor cells (reviewed by Hellström and Hellström, 1969). This evidence argued against the then prevailing view that those lymphocyte clones which were specific for a given tumor's antigens had been deleted in the tumor-bearing animal, and it led to searches for mechanisms which could render the immune response to a growing tumor ineffective *in vivo*. Specific blocking factors (SBF) were identified as a result of this search (Hellström *et al.*, 1969, 1977) and were among the first antigen-specific factors described which "block" (inhibit, suppress) cell-mediated immune responses. When discussing SBF here, we shall begin with a short historical background in order to remove the still common misconception that SBF and suppressor cells provide alternative ways for a tumor's escape from immunological control. Since, as discussed below, SBF constitute molecules which can activate suppressor cells as well as molecules which are the products of such cells, the two mechanisms are interdependent.

Most studies of SBF have used colony inhibition (Hellström, 1967) and microcytotoxicity (Hellström and Hellström, 1971) techniques. Both of these involve long exposure of effector and target cells to each other (4–6 days for the colony inhibition and about 40 hr for the microcytotoxicity test). Therefore, these assays probably measure a combination of lymphocyte activation and the effects of the activated lymphocytes on the target cells, such as killing, inhibition of growth, and loss of adherence. This should be borne in mind when interpreting the data to be discussed, and one must be aware that short-term assays, such as 3 to 4 hr ^{51}Cr release tests, measure target cell killing by already activated lymphocytes and may reflect a different aspect of the antitumor response.

The experiments which led to the discovery of SBF were inspired by the demonstration of "enhancing antibodies" (Kaliss, 1958; Hellström and Möller, 1965), present in certain hyperimmune sera. Such antibodies had been shown to prevent both the induction of cell-mediated immune responses ("afferent" and "central" enhancement) and the effectiveness of these responses once activated ("efferent" enhancement). Furthermore, there was evidence that the rejection response to tumor transplants was decreased in mice given appropriate hyperimmune serum (G. Möller, 1964). Also, alloreactivity of immune lymphocytes *in vitro* was inhibited if the target cells were incubated with specific hyperimmune serum (E. Möller, 1965). It was tested, therefore, whether sera from mice with either Moloney sarcoma virus (MSV)-induced sarcomas or mammary tumor virus (MTV)-associated mammary carcinomas, or from rabbits with persistent Shope papillomas, could inhibit ("block") the reactivity of specifically immune lymphocytes to the appropriate tumor cells (Hellström *et al.*, 1969, 1970; Hellström and Hellström, 1969; Heppner, 1969). Such inhibition was, indeed, detected, and it was found to be specific for the antigens of each tumor type. Subsequent work in our own and other laboratories confirmed and extended these findings for a variety of animal and human neoplasms (for review, see Sjögren *et al.*, 1971; Hellström and Hellström, 1974; Baldwin and Robins, 1977; Hellström *et al.*, 1977; Hellström and Brown, 1979). The serum factor responsible for the inhibitory effects has been called by us (Hellström *et al.*, 1977) *specific blocking factors*.

The degree of blocking serum activity varies for different antigens. A particularly strong activity has been detected for some oncofetal (MuLV related?) antigens shared by cells from MCA-induced sarcomas and normal embryos from BALB/c mice (Hellström and Hellström, 1975). This may contribute to the fact that oncofetal antigens only rarely induce a tumor rejection response (Castro *et al.*, 1974; Hellström and Hellström, 1975, Hellström and Hellström, 1976).

In a study of transplanted MCA-induced sarcomas it was found that the concentration of SBF is much higher within growing tumors than it is systemically (Hellström and Hellström, 1979). Only SBF obtained directly from tumors in the form of cell-free fluid, but not those obtained from serum, enhanced tumor outgrowth *in vivo*. Therefore, it seems likely that a high local concentration of SBF provides at least part of the explanation why a tumor can grow in an animal displaying concomitant immunity.

As interpretations of scientific data often have to change as more findings emerge, our understanding of the blocking phenomenon today differs in some important points from the view which we held when the phenomenon was discovered about 12 years ago. At that time, it seemed most likely that the blocking effect was similar to the postulated "masking" of target cell antigens by alloantibodies (E. Möller, 1965) and represented a case of "efferent" immunological enhancement (Hellström and Hellström, 1969, 1970). The finding that blocking factors can be removed by absorption with cultured cells from the respective tumors (Hellström and Hellström, 1969) provided strong support for this view, as did the observation that target cells incubated with tumor-bearer sera are protected from destruction by immune lymphocytes (Hellström and Hellström, 1969). We argued that free tumor antigen should not behave this way, and molecules other than antibodies with such characteristics were unknown. It was natural, therefore, to attribute the serum effects to *blocking antibodies* (Hellström and Hellström, 1969, 1970), a term borrowed from the allergy field. However, this interpretation had to be abandoned soon in the view of emerging, contradictory findings. First, we found that the blocking serum effect disappeared within 3–4 days of tumor removal (Hellström *et al.*, 1970), an observation which was difficult to reconcile with the idea of blocking antibodies. Conventional IgG antibodies of the typ which appeared to be responsible for immunological enhancement were not cleared so quickly. Second, antibodies from tumor-immunized animals and from animals whose tumors had spontaneously regressed (MSV sarcomas in mice, Shope papillomas in rabbits) not only lacked blocking activity (Hellström and Hellström, 1969; Hellström *et al.*, 1969) but could cancel ("unblock") the blocking activity of sera from animals bearing the same tumor (Hellström and Hellström, 1970).

A new hypothesis had to be introduced to take into account both the evidence that the blocking phenomenon could not be explained on the basis of circulating "blocking antibodies" and the fact that the blocking activity of tumor-bearer serum can be removed by absorption with the specific tumor cells (Hellström and Hellström, 1971). The hypothesis stated that antigen–antibody complexes were the most important SBF, and experiments were set up to test this hypothesis (Sjögren *et al.*, 1971). Sera from mice with growing MSV sarcomas were absorbed onto MSV-induced syngeneic sarcoma cells and eluted from the surface of these cells at low pH (3.1). The eluted material was then subjected to ultrafiltration by using Amicon filters which isolated material having molecular weights of 10,000–100,000 ("E10") or more than 100,000 ("E100"). After ultrafiltration, the fractions were tested for biological activity, using the colony inhibition technique. Molecules in the E10 or E100 fractions, when tested alone, could not prevent specifically immune lymphocytes from reacting against MSV-induced sarcoma cells. However, pretreatment of the lymphocytes or the target cells with a combination of the E10 and E100 fractions blocked reactivity. Furthermore, if the lymphocytes were pretreated with the E10 fraction and this was then allowed to remain in contact with the lymphocytes for the remainder of the assay, reactivity was blocked. Sjögren *et al.* (1971) tentatively concluded that the E10 fraction contained tumor antigen, the E100 fraction antitumor antibodies, and that the combination of the two contained antigen–antibody complexes. The "blocking effect" which is detectable *in vitro* was ascribed to an effect on the effector lymphocytes and not to "masking" of target cell antigens, and it has since never been attributed, by our group, to any "blocking antibodies."

One of our collaborators, Brawn, showed shortly thereafter that cell-free tumor homogenates could block the *in vitro* reactivity of lymphocytes from multiparous mice in a system where reactivity appeared to be directed against oncofetal antigens (Brawn, 1973).

Baldwin *et al.* (1972, 1973) extended these findings by systematicaclly analyzing the blocking of cell-mediated antitumor immunity by sera, cell-free tumor homogenates, and artificially prepared antigen–antibody complexes. Inhibition of reactivity was detected following incubation of the lymphocytes with tumor homogenates (presumed to contain tumor antigen), tumor-bearer sera or "antigen–antibody complexes" (mixtures of tumor cell preparations and antitumor antibodies). Suppressed killing could be also observed upon incubation of the target cells with sera or complexes, followed by washing, while the tumor homogenates were not inhibitory under these conditions. During different phases of tumor growth, a suppressive serum effect at the target cell level (called *blocking* by Baldwin *et al.*) or suppression of the reactivity of pre-incubated and washed lymphocytes (referred to as *inhibition* by Baldwin's group) could be detected, with an effect being observed at both the lymphocyte and the target cell level most of the time (Baldwin and Robins, 1977).

Hayami *et al.* (1973, 1974) studied Rous virus-induced sarcomas in Japanese quails and detected factors in the sera of quails with progressively growing tumors that inhibited the reactivity of specifically immune lymphocytes on plated sarcoma cells. No such effects were seen with sera from quails whose Rous sarcomas had spontaneously regressed ("regressors"); they were, instead, "unblocking," similar to regressor sera studied in other systems (Hellström and Hellström, 1970). However, when regressor sera were incubated with Rous sarcoma cells overnight, the supernatant of the tumor cultures was found to inhibit the killing of Rous sarcoma cells. It was speculated that antibodies in the regressor sera combined with antigen from the tumor cells and formed a complex with blocking activity (Hayami *et al.*, 1973).

A substantial amount of evidence has since accumulated which shows that tumor-bearing animals and human patients have higher levels of circulating immune complexes than do healthy individuals or patients with some diseases other than cancer, these levels decrease after tumor removal, and they possibly are of prognostic significance (see, e.g., Theophilopolous *et al.*, 1977; Hoffken *et al.*, 1978; Brandeis *et al.*, 1978; Amlot *et al.*, 1978). Although it has not been studied in all systems that circulating complexes do, indeed, consist of tumor antigens and their specific antibodies, this concept is supported by evidence from work done with certain animal tumors (Oldstone, 1975; Jennette and Feldman, 1977) and with patients with Burkitt lymphoma (Oldstone *et al.*, 1975; Heimer and Klein, 1976).

If circulating tumor antigens and their complexes are the major SBF, it should be possible to isolate the antigens from tumor-bearer serum using antibody with proper specificity as immunoadsorbent and formally prove it to be involved in the *in vitro* suppression of cell-mediated antitumor reactivity. Antibodies which are responsible for the "unblocking" (Hellström and Hellström, 1970) effect of sera from tumor-immunized donors would be expected to be ideal for this purpose. Based on this assumption, our group started in the middle of the 1970s to isolate SBF from the sera of BALB/c mice carrying transplanted, MCA-induced sarcomas, studying in parallel two tumors expressing different unique antigens. The sera were passaged over immunoadsorbent columns. These were prepared by using "unblocking" sera from BALB/c mice which had received a graft of the respective sarcoma followed by excision of the tumor and challenge with irradiated cells from the same sarcoma. When tumor-bearer serum was passed through such a column, its blocking activity was removed. The blocking activity could be recovered by elution of the column at pH 3.1 (Tamerius *et al.*, 1976).

The SBF purified on the columns were found to have the same specificities as those detected in the corresponding tumor-bearer sera (Nepom *et al.*, 1976). The "unblocking" sera used for their isolation had to be specific: only serum from mice immunized against the same MCA sarcomas could be used to make an immunoadsorbent on which to isolate a given SBF (Nepom *et al.*, 1976, 1977).

The finding that SBF could be purified by immunoadsorption made it possible to study their molecular nature. Nepom *et al.* (1976, 1977) found them to be glycoproteins with a molecular weight of approximately 56,000. Isolates which had specific blocking activity were consistently found to contain this protein, while isolates lacking the activity did not contain it. Peptide-mapping experiments on SBF purified from the sera of mice carrying two antigenically different MCA sarcomas failed to detect any differences between the two, suggesting that a class of closely related molecules was involved (Nepom, 1977; Hellström *et al.*, 1977).

One of the most important findings was that the 56,000-dalton SBF could be specifically removed by absorption with cultivated cells from the respective tumor. This was studied, in parallel, with two antigenically distinct MCA sarcomas and their respective SBF (Nepom *et al.*, 1977). This observation was in agreement with earlier results using whole tumor-bearer sera (Hellström and Hellström, 1969), and established that molecules other than antibody can have a receptor for a marker unique to a tumor. Because of work on the role of T cells in immune regulation (Gershon, 1974; Tada *et al.*, 1975; Asherson and Zembala, 1976) and of Nelson *et al.* (1975a,b,c; see discussion below), the concept of antigen-specific molecules other than immunoglobulins, which would have been heretical in 1969, was acceptable in 1977. The possibility that the unique tumor marker recognized was TSTA, or a molecule related to that, deserves consideration.

There are two reasons why it is unlikely that the 56,000-dalton SBF is free tumor antigen. First, there is no precedent that such antigen should bind selectively to tumor cells expressing it. Second, it has not been possible, in spite of intense attempts, to demonstrate antibodies in either the sera used to prepare the "unblocking" columns or in sera of similarly immunized mice which specifically bind to any cell surface antigens unique to the respective MCA-induced sarcomas (Nepom *et al.*, 1977, and unpublished findings; Klitzman *et al.*, 1980). Antibodies in such sera which bound to tumor cells were directed against MuLV antigens, particularly gp70 (Klitzman *et al.*, 1980). The most attractive hypothesis was, therefore (Nepom *et al.*, 1977; Hellström *et al.*, 1977), that the 56,000-dalton SBF is a host-derived immunoregulatory molecule similar to the thymus-derived lymphocyte (T cell) suppressor factors found in nontumor systems (Tada *et al.*, 1975; Taniguchi *et al.*, 1979). As a corollary to this hypothesis we have postulated that the "unblocking" sera contain antibodies to idiotypic determinants expressed on the suppressor molecules (Hellström *et al.*, 1977).

Other evidence that circulating SBF with specificity for tumor antigens can be host-derived, antigen-binding, immunosuppressive molecules comes from work which showed that SBF can be formed *in vitro* by T cells from tumor-bearing mice (Nelson *et al.*, 1975a,b,c; Nelson, 1975). Spleen cells from BALB/c mice bearing either a progressively growing MSV-induced sarcoma or a transplanted MCA-induced sarcoma were found to produce factors *in vitro* which could block the *in vitro* cytotoxic effect of specifically tumor-immune T cells, as assessed by a 30- to 40-hr microcytotoxicity test. The SBF could bind specifically to tumor cells carrying the appropriate antigens (Nelson *et al.*, 1975c), and its molecular weight was substantially smaller than that of IgG antibodies (Nelson, 1975; Nelson and Brown, unpublished findings). Inhibition of protein synthesis in the spleen cultures with cyclohexamide prevented SBF production (Nelson, 1975c). Removal of Thy-1-positive cells from the cultures stopped their production of SBF, while this production was restored if Thy-1-positive cells from spleens of tumor-bearing or normal syngeneic mice were added back to the cultures (Nelson *et al.*, 1975b). Removal of adherent cells or of immunoglobulin carrying cells did not interfere with SBF synthesis. More recently, thymus cells from mice carrying MCA-induced sarcomas have also been found to produce SBF *in vitro* (Nelson *et al.*, 1980).

A puzzling finding, for which we do not have a good explanation, is that the SBF we have studied, in whole serum or following purification as done by Nepom *et al.* (1976, 1977), or in the supernatants of spleen cultures (Nelson *et al.*, 1975a,b,c), have been retained by immunoadsorbents prepared against mouse immunoglobulins (Hellström *et al.*, 1977). There is no experimental evidence that the SBF studied by Nepom *et al.*, or by Nelson *et al.* contain immunoglobulins (e.g., both molecules are smaller than immunoglobulins). We tend to believe, therefore, that the antiimmunoglobulin preparations used contained antibodies to some cross-reactive determinant present on the SBF. However, this needs to be studied.

To conclude this discussion of SBF, we like to note that the most direct evidence from our own group is for circulating SBF to be immunosuppressive factors which are different from antigens, antibodies, and immune complexes (Nepom *et al.*, 1976, 1977; Nelson *et al.*, 1975a,b,c). Our evidence that free antigens and complexes acts as SBF is indirect (Sjögren *et al.*, 1971; Hayami *et al.*, 1973). Furthermore, one should bear in mind that in both the studies of Sjögren *et al.* (1971) and of Hayami *et al.* (1973) virally induced neoplasms were used, in which tumor-specific circulating antibodies are known to occur, while Nelson *et al.* (1975a,b,c) and Nepom *et al.* (1976, 1977) worked on chemically induced mouse sarcomas, where such antibodies have not been detected.

We also like to emphasize, however, that the findings of Baldwin's group, in particular (Baldwin *et al.*, 1972, 1973) have shown, beyond a reasonable doubt, that circulating antigens and complexes are important SBF, in agreement with the suggestive evidence of Sjögren *et al.* (1971). However, one cannot exclude that the effects which have been measured primarily by colony inhibition and long-term microcytotoxicity tests and ascribed to antigens or complexes (Sjögren *et al.*, 1971; Baldwin *et al.*, 1972, 1973) are the result of these activating a suppressor cell response in the lymphocyte populations used as the source of effector cells. It is interesting then that a tumor-specific SBF which was studied by a short-term assay of cellular immunity, the leukocyte adherence inhibition test, was found to be a suppressor factor with characteristics similar to those of T suppressor factors in nontumor systems (Koppi *et al.*, 1981).

III. TUMOR ANTIGEN AS INDUCER OF A SUPPRESSOR CELL RESPONSE

There is evidence that tumor antigen, alone or complexed with antibody, can inhibit tumor immunity *in vivo*. Vaage (1972) was one of the first to demonstrate this by showing that the rejection response to TSTA of chemically induced mouse tumors was abolished if cells containing the antigens of the respective tumors were given at the proper time; heavily irradiated tumor cells similar to those regularly used to induce a rejection response were employed as the source of the tumor antigen. Likewise, delayed hypersensitivity reactions to tumor cells have been shown to be abolished upon injection of soluble antigen prepared from the respective tumor (Paranjpe *et al.*, 1976).

Tumor antigen may suppress tumor immunity in several ways: it may inhibit induction of immunity or the activity of effector cells once induced. This suppression may be mediated by a direct effect on helper or effector cells or by an indirect mechanism, for example by induction of suppressor cells. The best evidence that tumor antigen can have a direct effect comes from the demonstration of "cold target cell inhibition" *in vitro*, in which the mechanism is known to be a competitive inhibition for available effector cells (Cerrotini and Brunner, 1974). However, this occurs mainly when the antigen is provided in the form of intact cells, and there is conflicting evidence from properly controlled short-term experiments that subcellular antigens directly block effector cells (reviewed by Berke, 1980). Long-term experiments such as colony inhibition or 20- to 40-hr microcytotoxicity tests are not informative as to the mechanism of inhibition, due to the fact that secondary events may take place during the assay.

We have investigated whether the inhibition ("blocking") of tumor immunity by antigen *in vivo* is due to a direct effect of the antigen on the effector cells or whether it is indirect, resulting, for example, from an activated suppressor cell response (Hellström and Hellström, 1978). As source of antigen, heavily irradiated cells were employed, similar to those used to induce a tumor rejection response (Klein *et al.*, 1960; Sjögren, 1965) or, in Vaage's (1972) experiments, to inhibit the effectiveness of such a response. The experiments used MCA-induced BALB/c sarcomas, transplanted into syngeneic hosts. First it was found (as expected) that small numbers of sarcoma cells (10^3 or 10^4) were rejected by mice which had been sensitized by subcutaneous injection of heavily irradiated cells from the respective sarcoma 10 days earlier. Second, also as expected, this effect was found to be abolished, if the tumor cells used for challenge were injected with an excess (5×10^6) of heavily irradiated cells from the respective sarcoma; cells from an antigenically different MCA sarcoma were used as a control. This confirmed that excess tumor antigen can inhibit an immune response to itself *in vivo*.

Since suppressor cell-mediated reactions include a radiosensitive component (reviewed in Katz, 1977), while transplantation immunity to antigenic tumors is not diminished following whole body radiation (Klein *et al.*, 1960), we then investigated whether inhibition of tumor rejection by excess antigen was possible in immune mice which had received whole body irradiation (400–450 rad). The answer was no. However, tumor outgrowth was observed if the irradiated immune mice received both excess tumor antigen and Thy-1-carrying normal spleen lymphocytes.

This indicated that tumor antigens can block the development of a rejection response by a mechanism which requires a population of radiosensitive T cells. These cells were postulated to be T suppressor lymphocytes (Hellström and Hellström, 1978).

We next asked whether inactivation or removal of the postulated radiosensitive T suppressor cells from tumor-bearing animals would result in an inhibition of tumor growth (Hellström *et al.*, 1978). BALB/c mice carrying small transplants of MCA-induced syngeneic sarcomas were given either 400 rad whole body radiation or kept as nonirradiated controls; 400 rad is too low a dose to directly affect the growth of these MCA sarcomas. Some of the irradiated mice received no further treatment. Others were reconstituted by an intraveneous injection of 10^7 BALB/c spleen cells, which had been depleted either of Thy-1-bearing cells by treatment with specific antibody and complement or of immunoglobulin carrying B cells and other cells adhering to a nylon wool column. Inhibition of tumor growth, and even regression of some established sarcomas, was observed in mice which either received radiation without reconstitution or which were irradiated and received spleen suspensions from which the Thy-1-bearing cells had been removed. Reconstitution with Thy-1-bearing cells, on the other hand, led to continued growth of the sarcomas. These data, then, as well as data from subsequent confirmatory studies (Enker and Jacobitz, 1980; Tilkin *et al.*, 1981) show that radiosensitive T cells are important inhibitors of tumor rejection and that their removal can be therapeutically beneficial.

There is by now an abundance of evidence that suppressor T cells play an important role in regulating tumor immunity (Gershon, 1974; Umiel and Trainin, 1974; Kall *et al.*, 1975; Rotter and Trainin, 1975; Kripke *et al.*, 1977; Fujimoto and Tada, 1978; Daynes and Spellman, 1977; Hellström and Brown, 1979; Naor, 1979; Greene, 1980), and it is beyond the scope of this review to discuss that evidence. We would like, however, to single out, as one of the most important of the contributions to the field, the demonstration by Greene *et al.* (1977) that injection of an antiserum to the IJ marker of suppressor T cells inhibits tumor growth *in vivo*. Antisera to the IJ markers of different mouse strains than that in which the tumor was growing had no effect, and neither had the proper anti-IJ serum if first absorbed with lymphocytes carrying the relevant IJ antigens.

So to learn more about the cells involved in the tumor antigen-induced suppressor mechanisms, experiments have been done in which lymphoid cells were adoptively transferred to BALB/c mice which carried progressively growth MSV-induced sarcomas (Hellström *et al.*, 1979). These sarcomas had been induced when the recipient mice were approximately

20 days old and would normally regress in more than 90% of cases. It was found that the adoptive transfer of Thy-1-positive lymphocytes from syngeneic mice, which had carried either MSV sarcomas that had spontaneously regressed ("regressors") or progressively growing sarcomas, ("progressors") inhibited the ability of the recipients to reject their tumors. Control lymphocytes from normal untreated mice, from mice immunized against any of several antigenically different neoplasms, or from mice bearing an antigenically different neoplasm had no such effect. Neither did lymphocytes from MSV regressors or progressors affect the growth of tumors carrying different antigens. Thy-1-positive cells from spleens and thymuses were found to be more effective than those from lymph nodes. The conclusion was that mice exposed to MSV sarcoma antigens have T cells which can suppress the immune response to antigens of MSV-induced sarcomas.

The next step was to study whether the Thy-1-positive cells which could transfer diminished resistance to MSV sarcomas were radiosensitive. Irradiation of the cell donors had no effect. However, the radioresistant T cells needed to interact with a radiosensitive population of Thy-1-positive cells in the recipient mice for suppression to occur (Hellström *et al.*, 1979). These radiosensitive cells could come from thymus or spleen of normal, untreated, syngeneic mice. Using the terminology of Cantor and Gershon (1979), we have referred to the antigen-specific, radioresistant cells as *suppressor activator cells* and the radiosensitive, naive cells as *suppressor acceptor cells*. On subsequent studies of mice with transplanted MCA-induced sarcomas, we demonstrated that the concept of suppressor activator and acceptor cells is also valid for tumors other than MSV-induced sarcomas (Hellström *et al.*, 1979). Similar findings have been independently obtained by Rao *et al.* (1980), who also found that the suppressor activator cells were induced by antigen–antibody complexes and that they were T cells expressing Lyt 1.

The conclusion is then that tumor antigen induces antigen-specific suppressor activator cells, which have immunological memory and are relatively radioresistant. To suppress tumor immunity, these cells need to interact with a population of suppressor acceptor cells. The latter cells can come from naive animals, and they are radiosensitive. Whether the tumor-derived molecules inducing suppression are the same as those inducing rejection remains to be studied.

The extent to which an immunological response leading to suppression is induced, rather than one leading to rejection, probably depends on both the nature of the antigen and the mode and timing of its presentation. Perry and Greene (1981) have recently shown that intravenously injected tumor cell membrane preparations decrease reactivity as measured by delayed hypersensitivity assays and that they do so by inducing a suppressor cell response. Those tumor-associated antigens, including some that are "oncofetal" (MuLV associated?), which are more prone to induce SBF activity than others (Hellström and Hellström, 1975), are likely to be also more prone to induce suppressor cells.

The suppressor activator cells are highly antigen specific, and they require contact with the specific antigen for suppression to occur. However, after this contact has occurred; the immune response to different antigens may also be suppressed (Hellström and Hellström, 1981a,b,; Mule *et al.*, 1981). This implies that if a tumor cell expresses an antigen that preferentially induces a suppressor cell response, immunity may be inhibited also to other antigens on the tumor cell surface. Such a mechanism may contribute to the fact that some tumors are "nonimmunogenic."

Suppression of the response to third-party antigens, following specific activation, has been previously demonstrated in nontumor systems (Truit *et al.*, 1978). One approach to search for specific tumor immunity, e.g., in man, can be based on the fact that such suppression occurs, confronting lymphocytes from a tumor patient with the putative tumor antigen and by measuring the suppressed response to a third-party antigen, evidence for tumor-specific immunity can be gained (Bean *et al.*, 1979).

In addition to suppressor T cells, macrophages in tumor-bearing animals can suppress immune reactions. As a rule, this suppression is seen only when the tumors are relatively large. It is antigenically nonspecific (reviewed by Herberman *et al.*, 1979; Garrigues *et al.*, 1981). The relationship between T cell and macrophage-mediated suppression of tumor immunity deserves to be investigated.

IV. THE USE OF HYBRIDOMA TECHNOLOGY TO OBTAIN TUMOR-SPECIFIC T CELL SUPPRESSOR FACTORS

The introduction by Köhler and Milstein (1975) of a technique by which antibody-forming B cells can be hybridized with murine myeloma cells has had a tremendous impact on immunology. Using an adaptation of this technique, T suppressor cells have been hybridized with murine thymoma cells and hybridomas obtained which form specific suppressor factors (Kontianen *et al.*, 1978; Taniguichi *et al.*, 1979). Isolation of clones of T hybridoma cells should enable biological and molecular characterization of the respective factors.

The finding that Thy-1-positive cells from the spleens (Nelson *et al.*,

1975a) or thymuses (Nelson *et al.*, 1980) of tumor-bearing animals form antigen-specific suppressor ("blocking") factors *in vitro* suggested that such cells might be used to form hybridomas producing tumor-specific suppressor factors. Therefore, Nelson *et al.* (1980) hybridized thymocytes from BALB/c mice carrying a syngeneic transplanted MCA-induced sarcoma, MCA-1490, with cells from the AKR thymoma BW5147. Supernatants from the resultant hybridomas were screened for suppressive activity using an assay in which ^{51}Cr-labeled MCA-1490 cells were exposed to immune lymphocytes from BALB/c mice specifically immunized to MCA-1490. Of the original 100 hybridoma cultures, supernatant of one hybridoma, 182, suppressed lysis of MCA-1490 cells but not lysis of cells from allogeneic tumor. This hybridoma was cloned; the clone, 182K54, was expanded, and its supernatant was further tested *in vitro* as well as *in vivo*. The culture supernatant of 182K54 did not suppress the killing of other syngeneic tumors by specifically immune lymphocytes. Neither did it affect the primary or secondary immune response of BALB/c lymphoyctes to antigens shared by MSV-induced sarcoma cells and cells of a BALB/c leukemia line, LSTRA. Furthermore, mixed-leukocyte responses of BALB/c lymphocytes cocultivated with C57BL/6 spleen cells were not affected, and neither was the response of BALB/c lymphocytes to three mitogens: phytohemaglutinin, concanavalin A, or LPS. In addition, Nelson *et al.* (to be published) showed that blastogenic response of specifically immune lymphocytes to irradiated MCA-1490 cells in mixed lymphocyte tumor cell assays, was specifically inhibited by the supernatant of 182K54.

182K54 supernatants have also been tested for effects on tumor growth and immunity to tumors *in vivo* (Nelson *et al.*, 1980; Cory *et al.*, 1981; Hellström *et al.*, 1981). This has been done in several ways. First, supernatant has been given to BALB/c mice during immunization with heavily irradiated MCA-1490 or control cells, or mixed with the tumor cells given as test challenge. The outgrowth of MCA-1490 cells in specifically immunized hosts was significantly facilitated by supernatants of 182K54. Likewise, outgrowth of MCA-1490 cells mixed with immune lymphoid cells in Winn assays was enhanced by supernatants of 182K54. 182K54 supernatants have also been assayed for their effect on delayed-type hypersensitivity (DTH) reactions to MCA-1490 or control tumor cells. In these assays, mice were immunized subcutaneously with irradiated tumor cells, and 5 days later, irradiated cells from the immunizing tumor were injected into one hind footpad. Twenty-four hours later, the swelling of the injected footpad, as compared to the contralateral pad, was measured and compared with the reaction in similarily injected, but not previously immunized, control mice. It was found that 182K54 supernatants suppressed the induction, and to a lesser extent the elicitation, of DTH reactions to MCA-1490 cells. This suppression was specific, no effects being seen against the reactions to cells from other MCA-induced BALB/c sarcomas.

Nelson *et al.* subsequently studied whether the suppressor effect of 182K54 supernatants could be removed by absorption with the respective tumor cells (Cory *et al.*, 1981; Hellström *et al.*, 1981; Nelson *et al.*, to be published). This was found to be the case: cultivated MCA-1490 cells, but not cells from other MCA-induced BALB/c sarcomas, or normal fibroblasts, removed the suppressive activity of 182K54 supernatants. This activity was recovered by elution, at pH 3, from the MCA-1490 cells. Eluates of MCA-1490 cells treated with control supernatants were not suppressive. Therefore, the data implied that the suppressor factor had a receptor for some membrane component of MCA-1490. Whether or not this molecule is a TSTA unique to MCA-1490 remains to be studied. In addition, the 182K54 suppressor factor was shown to be a glycoprotein by its binding to *Lens culinaris* lectin and its sensitivity to proteolytic digestion (Nelson *et al.*, in preparation).

Thus, the 182K54 suppressor factor specifically suppresses several aspects of lymphocyte-mediated reactivity to MCA-1490 cells *in vitro* and *in vivo*, and it has a receptor for a molecule expressed on MCA-1490 cells and not detectably present on normal fibroblasts or cells from other MCA-induced BALB/c sarcomas. We are now further studying the molecular nature of the 182K54 suppressor factor and its possible relationship to factors isolated from tumor-bearer serum (Nepom *et al.*, 1977).

V. GENERAL DISCUSSION

We have presented evidence that antigens released from tumor cells can suppress (block, inhibit) the host's ability to reject a tumor, and that this occurs in many cases, through a mechanism which involves suppressor T cells. Soluble factors produced by T cells appear to play an important part in these events. However, as is clear from the preceding sections, little is known about the cellular and molecular interactions leading to the suppressed antitumor reactivity.

There are several questions which need to, and which probably can, be answered, by using available techniques. One of the more important of these is whether the molecule on the neoplastic cell that induces a suppressor cell response is the same one that induces and/or is the target for tumor rejection. If not, one needs to know how the two are related, if at all. Recent work on chemically induced murine tumors indicates that different, tumor-associated molecules are inducing the two different responses (Greene and Perry, 1978; Yamauchi *et al.*, 1979). These differences need to be established at the molecular level, however, and to be shown for tumors of well-defined antigenicity.

One approach toward this is to use the 182K54 suppressor factor (Nelson *et al.*, 1980) to make "immunoadsorbents" on which to purify those tumor cell surface molecules to which the factor can bind and then to study their biological activity. However, an even better approach may be to concentrate more on the suppressor work on a DNA tumor virus-induced mouse model, e.g., polyoma or SV40 virus-induced mouse tumors. In such a model, one has a spectrum of tumors with shared TSTA, and one can obtain virus-infected cells in which the degree of expression of the viral DNA is known. Although the natures of the SV40 and polyoma TSTA are still controversial (Weil, 1978), it should be possible by using the monoclonal antibody technique to characterize the TSTA of SV40 and polyoma tumors and compare them with tumor cell surface molecules recognized by suppressor cells. Using a model with shared tumor antigens should also make it easier to investigate whether a modification of the production and/or action of suppressor cells and/or soluble suppressor factors can have tumor preventive and therapeutic possibilities.

Another important question concerns the molecular nature of the various T cell-derived suppressor factors so far identified. In spite of the fact that our own group has, for more than a year, had a hybridoma, 182K54, producing such a factor, we have still not been able to develop serological and biochemical assays for the factor. This is most likely so because the factor is active at very low concentrations and represents a small fraction of all the proteins produced by the hybridoma. A few facts about tumor-specific T cell-derived suppressor molecules are, nevertheless, known. First, they have an idiotypic determinant, responsible for their binding to tumor cells of the respective antigenicity, as demonstrated by Nelson *et al.* (1975c) and Nepom *et al.* (1977), and subsequently shown for other factors as well (reviewed by Tada and Okumura, 1979; Taussig, 1980; Germain and Benacerrat, 1980; Greene, 1980). Second, they may in part be encoded by I-J subregion genes (Greene, 1980).

The nature of the T cells producing the various factors, and the nature of the cells with which the factors interact, need to be further investigated. The T cell-derived suppressor factor described by Perry and Greene (reviewed by Greene, 1980) is produced by Lyt $1^{-}23^{+}$ cells and appears to play its major role as a recruiter of antigen-specific suppressor cells. On the other hand, the thymus cells which produce suppressor factors in the system of Nelson *et al.* are Lyt $1^{+},2^{-}$ (Nelson, unpublished findings).

The 182K54 factor can inhibit the cytotoxic response of Lyt $1^-,2^+$ T cells in a 40-hr microcytotoxicity assay. It does not appear capable of recruiting suppressor cells *in vivo* (Nelson *et al.*, unpublished findings).

One can only speculate about the possibilities of manipulating the immune response in favor of the host by selectively turning off the suppressor response to a given tumor. However, since the release of soluble antigen appears to play an important role for this to happen, the use of agents which can decrease this release may be considered. Procedures by which antigen already released may be removed also deserve consideration. Such removal might be one of the mechanisms by which plasmapheresis has an antitumor effect *in vivo* (Bansal *et al.*, 1978; Terman *et al.*, 1980; Jones *et al.*, 1980).

Another approach is to try to interfere with the activity and/or production of suppressor cells and their factors. Our own studies in this direction aim to raise monoclonal antibodies to various tumor antigen-specific suppressor factors, starting with the one which we have, the one produced by 182K54, and to analyze the extent to which antibodies to such factor can inhibit the activity of the factor and interfere with tumor growth *in vivo*. However, one has to be aware of the fact that the immunity to a tumor may be suppressed, even if the suppressor cell response to one set of tumor antigens is inhibited. This is illustrated by the work (Hellström and Hellström, 1981a,b) discussed in a preceding section and suggesting that a noninhibited suppressor cell response to one set of tumor antigens is sufficient for suppression to occur also to a different set of antigens on the same tumor. Therefore, such therapeutical procedures which aim to obtain effector cells at the stage where they are beyond sensitivity to suppression may be more rewarding.

"Nonspecific" ways of interfering with suppressor cell activity, including irradiation, treatment with drugs such as cyclophosphamide, or by antisera to markers of suppressor cells or suppressor factors (such as IJ) have already been found to have some antitumor effects, as discussed above. In spite of the fact that the effects obtained have been seen only against small tumors, this approach deserves consideration. Clinically rewarding effects may well be seen after good protocols for treating animals have been developed, probably based on a combination of drugs and antisera.

As stated in the Introduction, many tumors are "nonimmunogenic." Although many believe that this is because they do not express any antigens which can be recognized as nonself by the immune system, we doubt that this explanation is correct for the majority of the nonimmunogenic neoplasms. We do so in view of the fact that even many normal cells have surface structures which can, in proper circumstances, be recognized as nonself (resulting in autoimmune diseases), and it would be surprising, therefore, if the same could not be true for neoplastic cells. One must also be aware of the fact that one does not yet know whether the much discussed TSTA of chemically induced mouse sarcoma are, indeed, nonself molecules. Therefore, to set these tumors aside as immunogenic is only based on an operational distinction: they have been found to induce an immune response that leads to tumor rejection. We find it worthwhile, therefore, to consider the possibility that many tumors do not induce a rejection response (are "nonimmunogenic") because they express antigens which easily induce suppression. Finally, the finding of "nonimmunogenic" animal tumors should not allow one to forget the fact that a relatively large number of human tumors induce both cell-mediated and humoral responses in their host (Hellström and Brown, 1979; Shiku *et al.*, 1976; Ueda *et al.*, 1979) and are, therefore, immunogenic, even if the nature of the target antigen (tumor specific, oncofetal, tissue type specific) is often not known.

VI. CONCLUSIONS

Antigens released from tumor cells can thwart the host's immune reactivity against immunogenic tumors in favor of tumor growth. This occurs through a mechanism in which suppressor T cells play a major role. Although much remains to be learned about the cellular and molecular aspects of the suppression, a guarded optimism is warranted that better knowledge of the antigen-induced suppression of tumor immunity will lead to procedures by which the immune response can be modified in favor of the host.

ACKNOWLEDGMENTS. This work was supported by Grants 19148, 19149, 25558, and 26116 from the National Institutes of Health and IM 43M from the American Cancer Society. This article was written while two of us (K.E.H. and I.H.) were Humboldt Awardees and worked as guest scientists at Deutsches Krebsforschungszentrum in Heidelberg, West Germany.

VII. REFERENCES

Amlot, P. L., Pussel, B., Slaney, J. M., and Williams, B. D., 1978, Correlation between immune complexes and prognostic factors in Hodgkin's disease, *Clin. Exp. Immunol.* **31**:166.

Asherson, G. L., and Zembala, M., 1976, Suppressor T-cells in cell mediated immunity. *Br. Med. Bull.* **32:**158.

Baldwin, R. W., and Robins, R. A., 1977, Induction of tumor-immune responses and their interaction with the developing tumor, *in* "Contemporary Topics in Molecular Immunology," Vol. 6 (R. R. Porter and G. L. Ada, eds.), pp. 177–207, Plenum Press, New York.

Baldwin, R. W., Price, M. R., and Robins, R. A., 1972, Blocking of lymphocyte-mediated cytotoxicity for rat hepatoma cells by tumor-specific antigen-antibody complexes, *Nature New Biol.* **238:**185.

Baldwin, R. W., Price, M. R., and Robins, R. A., 1973, Inhibition of hepatoma-immune lymph node cell cytotoxicity by tumor-bearer serum, and solubilized hepatoma antigen, *Int. J. Cancer* **11:**527.

Bansal, S. C., Bansal, B. R., Thomas, H. L., Siegel, P. D., Shoads, J. E. Cooper, D. R., Terman, D. S., and Mark, R., 1978, *Ex vivo* removal of serum IgG in a patient with colon carcinoma, *Cancer* **42:**1.

Bean, M. A., Akiyama, M., Kodera, Y., Dupont, B., and Hansen, J. A., 1979, Human blood T lymphocytes that suppress the mixed leukocyte culture reactivity of lymphocytes from HLA-B14-bearing individuals, *J. Immunol.* **123:**1610.

Berke, G., 1980, Interaction of cytotoxic T lymphocytes and target cells, *Prog. Allergy* **27:**69.

Brandeis, W. E., Nelson, L., Wang, Y., Good, R. A., and Day, N. K., 1978, Circulating immune complexes in sera of children with neuroblastoma. Correlation with stage of disease, *J. Clin. Invest.* **62:**1201.

Brawn, R. J., 1973, "Evidence for association of embryonic antigen(s) with several 3 methylcholanthrene-induced murine sarcomas, *in* "Proceedings of the First Conference and Workshop on Embryonic and Fetal Antigens in Oak Ridge, Tenn. p. 143.

Cantor, H., and Gershon, R. K., 1979, Immunological circuits: cellular composition, *Fed. Proc., Fed. Am. Soc. Exp. Biol.* **38:**2058.

Castro, J. E., Hunt, R., Lance, E. M., and Medawar, P. B., 1974, Implication of the fetal antigen theory for fetal transplantation, *Cancer Res.* **34:**2055.

Cerrotini, J.-C., and Brunner, K. T., 1974, Cell-mediated cytotoxicity, allograft rejection and tumor immunity, *Adv. Immunol.* **18:**67.

Cory, J., K. Nelson, Forstrom, J. W., Hellström, I., and Hellström, K. E., 1981, A cell hybridoma suppressor factor which binds tumor antigen, *in* "Monoclonal Antibodies and T Cell Hybridomas" (G. J. Hammerling, W. Hammerling and J. F. Kearney, eds.), pp. 503–508, Elsevier/North-Holland, Amsterdam.

Daynes, R. A., and Spellman, C. W., 1977, Evidence for the generation of suppressor cells by ultraviolet radiation, *Cell. Immunol.* **31:**182.

Enker, W. E., and Jacobitz, J. L., 1980, *In vivo* splenic irradiation eradicates suppressor T cells causing the regression and inhibition of established tumor, *Int. J. Cancer* **25:**819.

Fujimoto, S., and Tada, T., 1978, I Region expression on cytotoxic and suppressor T cells against synteneic tumors in the mouse, *in* "Cancer Immunotherapy and Its Immunological Basis" (Y. Yamamura, M. Kitagawa, I. Azuma, eds.), pp. 11–10, University Park Press, Baltimore, Maryland.

Garrigues, J., Romero, P., Hellström, I., and Hellström, K. E., 1981, Adherent cells (macrophages?) in tumor-bearing mice suppress MLC responses, *Cellular Immunol.* **60:**109.

Germain, R. N., and Benacerraf, B., 1980, Helper and suppressor T cell factors, Springer Semlin, *Immunopathology* **3:**93.

Gerhson, R. K., 1974, T cell control of antibody production, *in* "Contemporary Topics in

Immunobiology" (M. D. Cooper and N. L. Warner, eds.), pp. 1–40, Plenum Press, New York.

Gershon, R. K., Carter, R. L., and Kondo, K., 1967, On concomitant immunity in tumor-bearing hamsters, *Nature (London)* **213**:674.

Gershon, R. K., Mokyr, M. B., and Mitchell, M. S., 1974, Activation of suppressor T cells by tumor cells and specific antibody, *Nature (London)* **250**:594.

Greene, M. L., 1980, The genetic and cellular basis for regulation of the immune response tumor antigens, *in* "Contemporary Topics in Immunobiology," Vol. 11 (N. L. Warner, ed.), pp. 81-116, Plenum Press, New York.

Greene, M. I., and Perry, L. L., 1978, Regulation of the immune response to tumor antigen. VI. Differential specificities of suppressor T cells or their products and effector T cells, *J. Immunol.* **121**:1263.

Greene, M. I., Dorf, M. E., Pierres, M., and Benacerraf, B., 1977, Reduction of syngeneic tumor growth by an anti-IJ-alloantiserum, *Proc. Natl. Acad. Sci. USA* **74**:5118.

Hayami, M., Hellström, I., and Hellström, K. E., 1973, Serum effects on cell-mediated destruction of Rous sarcomas, *Int. J. Cancer* **12**:667.

Hayami, M., Hellström, I., Hellström, K. E., and Lannin, D. R., 1974, Further studies on the ability of regressor sera to block cell-mediated destruction of Rous sarcomas, *Int. J. Cancer* **13**:43.

Heimer, R., and Klein, G., 1976, Circulating immune complexes in sera of patients with Burkitt's lymphoma and nasopharyngeal carcinoma, *Int. J. Cancer* **18**:310.

Hellström, I., 1967, A colony inhibition (CI) technique for demonstration of tumor cell destruction by lymphoid cells *in vitro, Int. J. Cancer* **2**:65.

Hellström, I., and Hellström, K. E., 1969, Studies on cellular immunity and its serum mediated inhibition in Moloney virus induced mouse sarcomas, *Int. J. Cancer* **4**:587.

Hellström, I., and Hellström, K. E., 1971, Colony inhibition and cytotoxicity assays, *in* "*In Vitro* Methods in Cell-Mediated Immunity" (B. R. Bloom and P. R. Glade, eds.), pp. 409–414, Academic Press, New York.

Hellström, K. E., and Hellström, I., 1976, Cell-mediated immunity to mouse tumors. Some recent findings. *The New York Academy of Sciences* **276**:176.

Hellström, I., and Hellström, K. E., 1981a, Cell-mediated suppression of tumor immunity has a nonspecific component. II. Evidence from cell culture experiments, *Int. J. Cancer* **27**:487.

Hellström, I., Hellström, K. E., and Pierce, G., 1968, *In vitro* studies of immune reactions against autochthonous and syngeneic mouse tumors induced by methylcholanthrene and plastic discs, *Int. J. Cancer* **3**:467.

Hellström, I., Hellström, K. E., Evans, C. A., Heppner, G., Pierce, G. E., and Yang, J. P. S., 1969, Serum mediated protection of neoplastic cells from inhibition by lymphocytes immune to their tumor specific antigens, *Proc. Natl. Acad. Sci. USA* **62**:362.

Hellström, I., Hellström, K. E., and Sjögren, H. O., 1970, Serum mediated inhibition of cellular immunity to methylcholanthrene induced murine sarcomas, *Cell. Immunol.* **1**:18.

Hellström, I., Hellström, K. E., and Bernstein, I. D., 1979, Tumor enhancing suppressor activator T cells in spleens and thymuses of tumor immune mice, *Proc. Natl. Acad. Sci. USA* **76**:52.

Hellström, K. E., and Brown, J. P., 1979, Tumor antigens, *in* "The Antigens," Vol. 5 (M. Sela, ed.), pp. 1–82, Academic Press, New York.

Hellström, K. E., and Hellström, I., 1970, Immunological enhancement as studied by cell culture techniques, *Annu. Rev. Microbiol.* **24**:373.

Hellström, K. E., and Hellström, I., 1974, Lymphocyte-mediated cytotoxicity and blocking serum activity to tumor antigens, *in* "Advances in Immunology," Vol. 18 (F. J. Dixon, ed.), p. 209, Academic Press, New York.

Hellström, K. E., and Hellström, I., 1975, Studies on the mechanism of tumor immunity: Some recent data on cellular immunity to common, possibly embryonic antigens in mouse sarcomas, *in* "Fundamental Aspects of Neoplasia" (A. A. Gottleib, ed.), Chap. 6, Springer-Verlag, New York.

Hellström, K. E., and Hellström, I., 1978, Evidence that tumor antigens enhance tumor growth *in vivo* by interacting with a radiosensitive (suppressor?) cell population, *Proc. Natl. Acad. Sci. USA* **75:**436.

Hellström, K. E., and Hellström, I., 1979, Enhancement of tumor outgrowth by tumor-associated blocking factors, *Int. J. Cancer* **23:**366.

Hellström, K. E., and Hellström, I., 1981b, Cell-mediated suppression of tumor immunity has a nonspecific component. I. Evidence from transplantation tests, *Int. J. Cancer.* **27:**481.

Hellström, K. E., and Möller, G., 1965, Immunological and immunogenetic aspects of tumor transplantation, *Prog. Allergy* **9:**158.

Hellström, K. E., Hellström, I., and Nepom, J. T., 1977, Specific blocking factors—are they important?, *in* "Biochimica et Biophysica Acta, 473" (C. Weissman and M. M. Burger, eds.) pp. 121–148, Elsevier/North-Holland Biomedical Press, Amsterdam.

Hellström, K. E., Hellström, I., Kant, J. A., and Tamerius, J. D., 1978, Regression and inhibition of sarcoma growth by interference with a radiosensitive T cell population, *J. Exp. Med.* **148:**799.

Hellström, K. E., Nelson, K., Cory, J., Forstrom, J. W., and Hellström, I., 1982, A tumor specific suppressor factor produced by a murine T cell hybridoma, *in* "Hybridomas in Cancer Diagnosis and Treatment" (M. M. Mitchell and H. F. Oettgen, eds.), pp. 47–50, Raven Press, New York.

Heppner, G. H., 1969, Studies on serum-mediated inhibition of cellular immunity to spontaneous mouse mammary tumor, *Int. J. Cancer* **4:**608.

Herberman, R. B., 1977, Immunogenicity of tumor antigens, *Biochim. Biophys Acta* **473:**93.

Herberman, R. B., Holden, H. T., Djeu, J. Y., Jerrels, T. R., Varesi, L., Tagliabue, A., White, S. L., Oheler, J. R., and Dean, J. H., 1979, Macrophages as regulators of immune responses against tumors, *Adv. Exp. Med. Biol.* **121:**361.

Hewitt, H. B., Blake, E. R., and Walder, A. S., 1976, A critque of the evidence for active host defense against cancer based on personal studies of 27 murine tumors of spontaneous origin, *Br. J. Cancer* **33:**241.

Hofken, H., Meredith, I. D., Robins, R. A., Baldwin, R. W., Davies, C. J., and Blamey, R. W., 1978, Immune complexes and prognosis of human breast cancer, *Lancet* **1:**672.

Jennette, J. C., and Feldman, J. D., 1977, Sequential quantitation of circulating immune complexes in syngeneic and allogeneic rats bearing Moloney sarcomas, *J. Immunol.* **118:**2269.

Jones, F. R., Yoshida, L. H., Ladiges, W. C., and Kenney, M. A., 1980, Treatment of feline leukemia and reversal of FeLV by *ex vivo* removal aof IgG: A preliminary report, *Cancer* **46:**675.

Kaliss, N., 1958, Immunological enhancement of tumor homografts in mice. A recent review, *Cancer Res.* **18:**992.

Kall, M. A., Hellström, I., and Hellström, K. E., 1975, Different responses of lymphoid cells from tumor-bearing as compared to tumor-immunized mice when sensitized to tumor specific antigens *in vitro, Proc. Natl. Acad. Sci. USA* **72:**5086.

Katz, D. H., 1977, *in* "Lymphocyte Differentiation, Recognition, and Regulation" Academic Press, New York.

Klein, G., and Klein, E., 1977, Rejectability of virus-induced tumors and nonrejectability of spontaneous tumors: A lesson in contrasts, *Transplant. Proc.* **9:**1095.

Klein, G., Sjögren, H. O., Klein, E., and Hellström, K. E., 1960, Demonstration of resistance against methylcholanthrene-induced sarcomas in the primary autochthonous host, *Cancer Res.* **20:**1561.

Klitzman, J. M., Brown, J. P., Hellström, K. E., and Hellström, I., 1980, Antibodies to murine leukemia virus gp70 and p15(E) in sera of BALB/c mice immunized with syngeneic chemically induced sarcomas, *J. Immunol.* **124:**2552.

Köhler, G., and Milstein, C., 1975, Continuous culture of fused cells secreting antibodies of predefined specificity, *Nature* (*London*) **256:**495.

Kontiainen, S., Simpson, E., Bohrer, E., Beverley, P. C. L., Herzenberg, L. A., Fitzpatrick, W. C., Vogt, P., Tarano, A., McKenzie, I. F. C., and Feldman, M., 1978, T cell lines producing antigen-specific suppressor factor. *Nature* (*London*) **274:**477.

Koppi, T. A., Halliday, W. J., and McKenzie, I. F. C., 1971, Regulation of cell-mediated immunologic reactivity to Moloney murine sarcoma virus-induced tumors. II. Nature of blocking and unblocking factors in serum, *J. Natl. Cancer Inst.* **66:**1097.

Kripke, M. L., Lofgreen, J. S., Beard, J., Jessup, J. M., and Fisher, M. S., 1977, *In vivo* immune responses of mice during carcinogenesis by ultraviolet irradiation, *J. Natl. Cancer Inst.* **59:**1227.

Mikulska, Z. B., Smith, C., and Alexander, P., 1966, Evidence for an immunological reaction of the host directed against its own actively growing primary tumor, *J. Natl. Cancer Inst.* **36:**29.

Möller, E., 1965, "Antagonistic effects of humoral isoantibodies on the *in vitro* cytotoxicity of immune lymphoid cells, *J. Exp. Med.* **122:**11.

Möller, G., 1964, Effect of tumor growth in syngeneic recipients of antibodies against tumor-specific antigens in methylcholanthrene-induced mouse sarcomas, *Nature* (*London*) **204:**846.

Mule, J. J., Stanton, T. H., Hellström, I., and Hellström, K. E., 1981, Suppressor pathways in tumor immunity: A requirement for Qa-1 positive tumor-bearer spleen T cells in suppression of the afferent immune response to tumor antigens, *Int. J. Cancer* **28:**353.

Naor, D., 1979, Suppressor cells: Permitters and promotors of malignancy?, *in* "Advances in Cancer Research," Vol. 29 (G. Klein and S. Weinhouse, eds.), pp. 45–125, Academic Press, New York.

Nelson, K., 1975, Products of Murine Spleen Cells That Specifically Modulate Cell-Mediated Immunity to Syngeneic Tumor Cells, Ph.D. Thesis, University of Washington, Seattle.

Nelson, K., Pollack, S. B., and Hellström, K. E., 1975a, Specific anti-tumor responses by cultured immune spleen cells. I. *In vitro* culture method and initial characterization of factors which block immune cell-mediated cytotoxicity *in vitro, Int. J. Cancer* **15:**806.

Nelson, K., Pollack, S. B., and Hellström, K. E., 1975b, Specific anti-tumor responses by cultured immune spleen cells. III. Further characterization of cells which synthesize factors with blocking and antiserum dependent cellular cytotoxic (ADC) activities, *Int. J. Cancer* **16:**539.

Nelson, K., Pollack, S. B., and Hellström, K. E., 1975c, *In vitro* synthesis of tumor-specific factors with blocking and antibody-dependent cellular cytotoxicity (ADC) activities, *Int. J. Cancer* **16:**932.

Nelson, K., Cory, J., Hellström, I., and Hellström, K. E., 1980, T-T hybridoma product specifically suppresses tumor immunity *Proc. Natl. Acad. Sci. USA* **77:**2866.

Nepom, J. T., 1977, Ph.D. Thesis, Serum blocking factors: Purifications and Properties, University of Washington, Seattle.

Nepom, J. T., Hellström, I., and Hellström, K. E., 1976, Purification and partial characterization of a tumor-specific blocking factor from sera of mice with growing chemically induced sarcomas, *J. Immunol.* **117:**1846.

Nepom, J. T., Hellström, I., and Hellström, K. E., 1977, Antigen-specific purification of blocking factors from sera of mice with chemically induced tumors, *Proc. Natl. Acad. Sci. USA* **74:**4605.

Old, L. J., and Boyse, E., 1964, Immunology of experimental tumors, *Annu. Rev. Med.* **15:**167.

Oldstone, M. B. A., 1975, Immune complexes in cancer: Demonstration of complexes in mice bearing neuroblastoma tumors, *J. Natl. Cancer Inst.* **54:**223.

Oldstone, M. B. A., Theofilopoulos, A. N., Gunven, P., and Klein, G., 1975, Immune complexes associated with neoplasia: Presence of Epstein-Barr virus antigen-antibody complexes in Burkitt's lymphoma, *Intervirology* **4:**292.

Paranjpe, M. S., Boone, C. W., and Takeichi, N., 1976, Specific paralysis of the anti-tumor cellular immune response produced by growing tumors studied with a radioisotope footpad assay, *Ann. N.Y. Acad. Sci.* **276:**254.

Perry, L. L., and Greene, M. I., 1981, T cell subset interactions in the regulation of syngeneic tumor immunity, *Fed. Proc.* **40:**39.

Prehn, R., and Main, D., 1957, Immunity to methylcholanthrene induced sarcomas, *J. Natl. Cancer Inst.* **18,** 768.

Rao, V. S., Bennet, J. A., Shen, F. W., Gershon, R. K., and Mitchell, M. S., 1980, Antigen-antibody complexes generate Lyt-1 inducers of suppressor cells, *J. Immunol.* **125:**63.

Rotter, V., and Trainin, N., 1975, Inhibition of tumor growth in syngeneic chimeric mice mediated by a depletion of suppressor T cells, *Transplantation* **20:**68.

Shiku, H., Takahashi, T., Oettgen, H. F., and Old, L. J., 1976, Cell surface antigens of human malignant melanoma. II. Serological typing with immune adherence assays and definition of two new surface antigens, *J. Exp. Med.* **144:**873.

Sjögren, H. O., 1965, Transplantation methods as a tool for detection of tumor-specific antigens, *Prog. Exp. Tumor Res.* **6:**289.

Sjögren, H. O., Hellström, I., Bansal, S. C., and Hellström, K. E., 1971, Suggestive evidence that the 'blocking antibodies' of tumor-bearing individuals may be antigen-antibody complexes, *Proc. Natl. Acad. Sci. USA* **68:**1372.

Tada, T., and Okumura, K., 1979, The role of antigen-specific T cell factors in the immune response, *Adv. Immunol.* **28:**1.

Tada, T., Taniguichi, M., and Takemori, T., 1975, T: Properties of primed suppressor T cells and their products, *Transplant. Rev.* **26:**106.

Tamerius, J., Nepom, J., Hellström, I., and Hellström, K. E. 1976, Tumor-associated blocking factors: Isolation from sera of tumor-bearing mice, *J. Immunol.* **116:**724.

Taniguichi, M., Saito, T., and Tomio, T., 1979, Antigen-specific suppressive factor produced by a transplantable I-J bearing T cell hybridoma, *Nature* **278:**555.

Taussig, M. J., 1980, Antigen-specific T cell factors, *Immunology* **41:**759.

Terman, D. S. Yamamota, Y., Mattioli, M., Cook, G., Tillquist, R., Henry, J., Poser, R., and Daskal, Y., 1980, Extensive necrosis of spontaneous canine mammary adenocarcinoma after extracorporeal perfusion over *Staphylococcus aureus* Cowan I *J. Immunol.* **124:**795.

Theofilopoulos, A. N., Andrews, B. S., Urist, M. M., Morton, D. L., and Dixon, F. J., 1977, The nature of immune complexes in human cancer sera, *J. Immunol.* **199:**657.

Tilkin, A. F., Schaaf, L. A., Fontaine, N., Van Acker, A., Boccadoro, M., and Urbain, J., 1981, Reduced tumor growth after low-dose irradiation or immunization against blastic suppressor cells, *Proc. Natl. Acad. Sci. USA* **78:**1809.

Truit, G. A., Rich, R. R., and Rich, S. S., 1978, Suppression of cytotoxic lymphocyte responses *in vitro* by soluble products of alloantigen-activated spleen cells, *J. Immunol.* **121:**1045.

Ueda, R., Shiku, H., Pfreundschuh, M., Takahashi, T., Li, L. T. C., Whitmore, W. F., Oettgen, H. F., and Old, L. J., 1979, Cell surface antigens of human renal cancer defined by autologous typing, *J. Exp. Med.* **150:**564.

Umiel, T., and Trainin, N., 1974, Immunological enhancement of tumor growth by syngeneic thymus-derived lymphocytes, *Transplantation* **18:**244.

Vaage, J., 1972, Specific desensitization of resistance against a syngeneic methylcholanthrene-induced sarcoma in C3HF mice, *Cancer Res.* **32:**193.

Weil, R., 1978, Viral 'tumor antigens': a novel type of mammalian regulator protein, *Biochim. Biophys. Acta* **516:**301.

Yamauchi, K., Fujimoto, S., and Tada, T., 1979, Differential activation of cytotoxic and suppressor T cells against syngeneic tumors in the mouse, *J. Immunol.* **123:**1653.

Chapter 14

Estrogen Regulation of Specific Proteins as a Mode of Hormone Action in Human Breast Cancer

David J. Adams, Dean P. Edwards, and William L. McGuire

Department of Medicine
University of Texas Health Science Center
San Antonio, Texas

I. INTRODUCTION

What effects do estrogens have on mammary tumors? Researchers have been pondering this question ever since George Beatson (1896) first reported breast tumor regression in premenopausal patients following oophorectomy. Over 70 years passed before Jensen *et al.* (1967) and Terenius (1968) were able to show that radioactive estradiol was bound specifically by certain human breast tumor biopsies and that this binding correlated with the clinical response to endocrine therapy. This observation and the discovery of a high-affinity receptor protein for estradiol in the cytoplasm of target cells (Toft and Gorski, 1966) lead to the widespread use of an estrogen receptor (ER) assay to identify hormone-responsive breast tumors (McGuire *et al.*, 1975). Subsequent studies have shown that roughly two-thirds of all breast tumor biopsies contain ER and of these tumors, about half respond to ablative and additive endocrine therapies (Edwards *et al.*, 1979). Furthermore, ER analysis of the primary tumor can predict response to endocrine therapy if inaccessible metastatic disease develops. Receptor-rich primary tumors display a lower probability and rapidity of recurrence (Knight *et al.*, 1977; Jensen, 1981), while absence of ER in the primary tumor is prognostic of a higher rate of recurrence and shorter survival (Osborne *et al.*, 1980). These clinical findings have further intensified efforts to understand at the molecular level why some breast tumors exhibit estrogen-dependent growth while other tumors do not, despite retention of ER activity.

The development of breast tumor cell lines which retain most, if not all, the receptors for steroid hormones (Horwitz *et al.*, 1978) has provided a convenient model system in which to study hormone action in breast cancer. We now know, for example, that estrogen can stimulate the synthesis of specific proteins in cultured breast tumor cells (described later), a recognized mode of hormone action. However, it is also apparent that estradiol *fails* to evoke responses in cell culture that are observed *in vivo*. Perhaps the most perplexing observation is that estrogens display little, if any, mitogenic activity toward breast tumor cells *in vitro*. This paradox probably results from our incomplete knowledge of culture requirements for hormone-regulated growth. Indeed, only recently have we come to realize the importance of hormonal effects at the cell surface, particularly membrane–substratum interactions. On the other hand, there is a distinct possibility that estrogens may not act directly on breast tumor cells at all (Sirbasku and Benson, 1980; Shafie, 1980).

In this review, we first consider normal mammary tissue as a target for estrogen action to anticipate estrogenic effects in neoplastic breast tissue. We then focus on a specific mode of hormone action: estrogen regulation of specific protein synthesis in cultured breast tumor cells. Finally, we discuss future directions in models for and concepts of hormone action in human breast cancer.

II. BREAST TUMORS AS TARGETS OF ESTROGEN ACTION

It is generally agreed that a steroid hormone environment is required for mammary tumor growth in many species. However, a direct and pivotal role for estrogens in regulation of tumor growth remains a controversial question. Essentially, we wish to know if breast tumors resemble rat uterus, chick oviduct, or avian liver, the classical targets for estrogen action. These tissues possess a well-characterized ER system and exhibit dramatic responses to estrogen either in cellular proliferation (uterus) or in production of large amounts of secretory proteins (egg proteins in oviduct and liver). Estrogen stimulation of ovalbumin synthesis in chick oviduct has been a particularly useful model for our understanding of the structure and regulation of hormone-dependent gene expression (Chambon *et al.*, 1979; McKnight and Palmiter, 1979; Chan and O'Malley, 1976). Here, estrogen appears to act exclusively by increasing transcription of the ovalbumin gene. A similar mechanism holds for induction of vitellogenin synthesis in *Xenopus laevis* and avian liver (Tata, 1979; Deeley *et al.*, 1977; Baker and Shapiro, 1977). In the rat uterus, estrogenic regulation of glucose-6-phosphate dehydrogenase occurs at both the transcriptional and translational levels. The hormone may also act indirectly by affecting the levels of $NADP^+$, the enzyme cofactor (Barker *et al.*, 1981).

By what mechanism do breast tumors respond to estrogen? Before addressing this question, it might be helpful to ask if *normal* mammary epithelial tissue displays target tissue properties. Unlike uterus, oviduct, or liver, the role of estrogen in mammary development and function is much less clear, probably because hormone regulation of this tissue is so diverse and complex. To briefly summarize a recent review of this subject (Topper and Freeman, 1980), estradiol is the one of several hormones required for mammary growth during both adolescence and pregnancy, although a direct effect is uncertain. Estradiol appears to have a permissive role in differentiated breast tissue, making the epithelial cells competent to respond to other hormones. For example estrogen alone does not induce synthesis of milk proteins *in vitro* but is necessary for optimal expression of lactogenic enzymes in response to thyroid hormone, insulin, glucocorticoid, and prolactin (Bolander and Topper, 1979).

Thus, it is reasonable to expect that breast tumors may not exhibit classical target responses to estrogen. In fact, evidence from rat, mouse, and human breast tumors maintained in organ culture all suggest that estrogens are weak mitogens at best and do not appear to cause pronounced changes in specific protein synthesis (Sirbasku and Benson, 1980). In a provocative paper, King (1979) has concluded that steroids are modulating agents in breast tumor cells and not the switch operators we have come to expect in target tissues. He argues that hormones need only be weak mitogens to produce the small change in tumor growth rate necessary to account for clinical responses to endocrine therapy. Furthermore, steroids may also exert indirect influence by affecting blood supply, cell–cell interaction, and the immune response. Sirbasku and coworkers (Sirbasku and Benson, 1980; Sirbasku, 1980) have extended this idea to propose that estrogens need not act directly on breast tumors at all but can induce production of specific polypeptide growth factors, or estromedins, in traditional target tissues that subsequently stimulate tumor growth *in vivo*. As yet, no such estromedin has been identified, although growth factor activity has been partially purified from rat uterine, plasma, and mammary tumor extracts and is apparently associated with rat serum albumin (Sirbasku, 1980). Shafie (1980) has also found evidence for an indirect mechanism of estrogen action in breast tumors. The MCF-7 human breast cancer cell line does not require estrogen for growth in culture. Yet, when these cells are inoculated into ovariectomized nude mice, cell growth and solid tumor formation are dependent on concomitant estrogen administration. If these tumors are then cultured *in vitro*, they revert to an estrogen-independent pattern of growth. These results are interpreted as evidence that breast tumor cell growth is normally inhibited *in vivo* and that estrogen stimulates synthesis of a gene product that blocks this inhibition. Whether this gene product comes from the host or from the tumor cell itself is unknown.

The scenario of estrogen action in normal and neoplastic breast tissue is therefore superficially different from that of classical target tissues. The hormone may act indirectly in concert with other hormones or growth factors, while direct effects may be subtle rather than dramatic. The latter point is illustrated when total cellular poly(+)RNA from DMBA-induced rat mammary adenocarcinomas is compared to that from normal mammary glands of midpregnant rats. Molecular hybridization and cell-free translation analysis do not reveal a major class of tumor-specific sequences (Supowit and Rosen, 1980). Thus, levels of specific proteins may be changed by regulation of the relative abundancies of certain mRNA species rather than by true suppression or induction of specific genes. If induction of new gene expression does occur, only a small number of proteins representing a small fraction of total protein synthesis may be involved. For example, the glucocorticoid "domain" in hepatoma cells includes perhaps 10 proteins out of over 1000 revealed by two-dimensional gel analysis (Ivarie and O'Farrell, 1978). Yet, subtle changes in protein synthesis may lead to profound effects. An analogy may be the estrogenic modulation of rat uterine induced protein (IP) (Notides and Gorski, 1966). This protein, which represents less than 0.10% of new protein synthesis has been proposed as a single "key intermediary protein" through which estrogen triggers synthesis of a protein cascade which in turn brings about the diverse uterine responses (Baulieu *et al.*, 1972). More recent evidence indicates that IP is synthesized constitutively in the rat uterus and is present in other tissues (Kaye and Reiss, 1980; Skipper *et al.*, 1980) but is induced by estrogen only in target cells. IP has also been shown to possess creatine kinase and enolase activities (Kaye *et al.*, 1980; Reiss and Kaye, 1981). As yet, the precise role of IP in the overall uterotrophic response to estrogen remains a mystery although a metabolic function appears likely.

III. ESTROGEN-REGULATED PROTEIN SYNTHESIS IN HUMAN BREAST CANCER

To better understand the mechanisms of estrogen action in mammary tumor development and to identify clinical markers for hormone-sensitive breast cancer, several laboratories have studied estrogen regulation of specific proteins in human breast tumor cell lines. This work may be roughly divided into two categories: estrogen-regulated functions or activities in which actual quantitation of the protein itself is lacking and estrogen regulation of the amounts of specific proteins whose function is unknown.

A. Estrogen Regulation of Specific Biological Activities

1. Progesterone Receptor

Progesterone receptor (PgR) was the first protein shown to be regulated specifically by estradiol in human breast cancer cells (Horwitz and McGuire, 1977a). MCF-7 cells grown on medium containing calf serum (stripped of endogenous estrogen by charcoal treatment) contain a low but consistent basal level of PgR. Addition of 1 nM estradiol stimulates PgR three- to four-fold by 4 days. The response is dose dependent and closely parallels accumulation and processing of ER complex in the nucleus. It has been postulated that estrogen stimulation of PgR indicates presence of a functional pathway of estrogen action in breast tumor cells (Horwitz *et al.*, 1975). Clinical measurement of both ER and PgR are consistent with this proposal, identifying a subset of ER-positive patients who have response rates to endocrine therapy approaching 80% (Osborne *et al.*, 1980). Although progesterone receptor has proven to be valuable in clinical diagnosis, it is present in such small amounts and ligand binding is so labile that purification of PgR is difficult and therefore its use as a research tool is limited. Consequently, other estrogen-regulated proteins have been sought.

2. Growth-Associated Enzymes

Our laboratory has analyzed two enzymes in MCF-7 cells that may be related to tumor growth. Lactate dehydrogenase (LDH), an enzyme thought to be related to the degree of tumor malignancy (Goldman *et al.*,

1964; Hilf *et al.*, 1976) and to be involved in metabolic functions crucial to growth, has been shown to be elevated two-fold by 10 nM estradiol treatment (Burke *et al.*, 1978). Curiously, only the fifth isozyme of LDH could be detected. Maximal stimulation occurs after 10 days making the LDH response, like that of PgR, a late effect of estrogen stimulation. We have also examined DNA polymerase α activity in MCF-7 cells (Edwards *et al.*, 1980b). This enzyme is known to increase dramatically during the S phase of the cell cycle (Lockwood *et al.*, 1967) and is therefore a logical choice as a marker for estrogen-regulated growth. Initial experiments did not reveal any estrogenic stimulation of cell growth or polymerase activity above control levels measured out to 8 days. Cell growth and enzyme activity could, however, be inhibited by growing cells on the anti-estrogen nafoxidine (1 μ M). If nafoxidine-pretreated cells are subsequently switched to medium containing estradiol (10 nM), cell growth and DNA polymerase activity increase four-fold after 4 days. These results suggest that MCF-7 cells may be replicating at a maximal rate in the absence of exogenous estrogen due to other growth factors in serum-containing medium. However, estrogen receptors are apparently able to regulate MCF-7 cell growth, since inhibition and subsequent "rescue" of growth and DNA polymerase activity are events specifically associated with the estrogen receptor.

Lippman and co-workers (Bronzert *et al.*, 1981) have measured another enzyme closely correlated with DNA synthesis in MCF-7 cells. Cytoplasmic thymidine kinase, an enzyme in the salvage pathway of deoxynucleotide biosynthesis, was found to increase in specific activity two-fold when assayed 24 hr after estradiol addition. Stimulation of enzyme activity was dose dependent and paralleled the dose curve for thymidine incorporation into DNA. Enzyme activity was inhibited by the anti-estrogen tamoxifen which also inhibited cell growth. Kinetic studies on thymidine kinase from MCF-7 cytosols suggest that estrogen may act by increasing the *V*max of the enzyme rather than the K_m and thus indicate the presence of more active enzyme. However, care must be taken when analyzing enzyme kinetics in crude preparations. Furthermore, estrogen effects on thymidine kinase (and for that matter on PgR, LDH, and DNA polymerase) do not necessarily reflect changes in intracellular concentration of the protein, only changes in protein activity.

3. *Proteases*

A characteristic feature of breast tumors is the ability to secrete proteolytic enzymes. These proteases may have normal roles in tissue remodeling, such as ovulation, blastocyst implantation, and in involution of mammary gland after lactation (Poole *et al.*, 1980). The fact that these enzymes are mitogens for normal cells in culture suggests that proteases may also have a role in growth control and malignant transformation (Cunningham *et al.*, 1979; Quigley *et al.*, 1980).

Studies of breast tumor explants cultured *in vitro* have shown that malignant adenocarcinomas, nonmalignant fibroadenomas, and normal breast specimens all release similar amounts of neutral proteases such as collagenase and plasminogen activator into culture medias (Poole *et al.*, 1980). Similar results obtain for cathepsin D activity. However, a thiol protease has been detected selectively in carcinomas. This protease resembles cathepsin B in substrate specificity but has distinct physical characteristics. More important, estradiol can stimulate secretion of the thiol protease in certain adenocarcinoma specimens.

Secretion of proteolytic activity can also be observed in breast tumor cell lines. Hakim (1980) has reported estrogen stimulation of thiol protease activity in human mammary carcinoma cells. Furthermore, a correlation was found between estrogen stimulation of several types of protease activity and the levels of ER and PgR in the cells. The highest basal and induced levels of protease activity were found in ER+ PgR+ lines, while normal mammary epithelial cells (ER− PgR−) had low endogenous levels that were not affected by hormone treatment.

The MCF-7 cell line secretes elastinolytic (Honbeck *et al.*, 1980), plasminogen activator (Butler *et al.*, 1979), and collagenase activities (Shafie and Liotta, 1980). Doses of estradiol that do not stimulate growth are nevertheless able to increase plasminogen activator activity 1.5-fold in as early as 8 hr. The response is dose dependent (increasing to 2.6-fold) and is inhibited by tamoxifen and by inhibitors of RNA and protein synthesis. These results suggest that estrogen regulation of plasminogen activator is due to receptor-mediated increases in specific protein synthesis. One consequence of plasminogen activator synthesis may be subsequent activation of latent collagenase via conversion of the zymogen plasminogen to plasmin. This process apparently occurs in the ZR-75-1 cell line among others (Paranjpe *et al.*, 1980) and may be responsible for the protease activity against Type I (stoma and bone) and Type II (basement membrane) collagen seen in MCF-7 cells by Shafie and Liotta (1980). These authors report a two- to three-fold increase in Type I collagenase activity by estradiol and a similar stimulation of both Types I and II collagenase by insulin. Since castration or diabetes prevents metastasis formation by MCF-7 cells injected into athymic nude mice, hormone regulation of collagenase activity may be involved in the metastatic potential of these cells. In fact, estrogenic stimulation of Type I collagenase may account for the ability of MCF-7 cells to erode bone *in vitro*, independent of osteoclast action (Martin *et al.*, 1980). Finally, plasminogen activation in breast tumor cytosols generates various cleavage products of estrogen receptor (Sherman *et al.*, 1980; Miller *et al.*, 1981). Observation of different species of steroid receptors due to protease action may imply a role for these enzymes in receptor activation and processing.

B. Estrogen-Regulated Proteins of Unknown Function

It is obvious that estrogen stimulation of human breast tumors or tumor cell lines can result in modulation of protein activities that could regulate tumor growth and development. What is needed now is to translate these effects on activities into quantitative changes in the levels of specific proteins. Such a transition usually requires purification of the protein of interest and production of a monospecific antibody for use in a quantitative immunological assay. As we have stated previously, this is often no easy task, especially if one has to deal with limited amounts of labile protein. An alternate approach to this problem is to identify estrogen-regulated proteins using some high-resolution technique for protein separation. With the advent of two- (O'Farrell, 1975) and even three (Skipper *et al.*, 1980) dimensional gel electrophoresis systems, this approach has become fruitful.

1. The 46K Glycoprotein

Westley and Rochefort (1979) were the first to successfully use two-dimensional gel analysis of estrogen-regulated proteins in human breast cancer. They discovered a glycoprotein of 46,000 daltons (46K), p*I* 5.5–6.5, that is secreted into the culture medium as early as 12 hr after hormone treatment. Induction of 46K is specific for ER+ breast cancer lines and is not detected in receptor-negative malignant or nonmalignant lines or in human milk or cystic disease fluid (Westley and Rochefort, 1980). The 46K protein is induced only by steroids known to interact with the estrogen receptor; similarly, the protein is repressed by the anti-estrogen tamoxifen and hydroxytamoxifen which inhibit growth in MCF-7. Since these anti-estrogens are partly estrogenic and can induce PgR in MCF-7 cells, induction of 46K may be more related to growth than is PgR and thus may be a better marker for estrogen-responsive breast tumors if an antibody to 46K can be raised. As with progesterone receptor, this could be a problem since intracellular levels of 46K cannot be measured, possibly because it is rapidly secreted and accounts for only 0.15% of total [^{35}S]methionine incorporation into soluble protein. However, it may be that 46K exists intracellularly in a different form. For example, 46K may be leaked into the media following estrogen-regulated changes in the cell surface. Once exposed to media, 46K may be modified sufficiently that two-dimensional gel analysis makes the protein appear distinct from the intracellular species. This idea is consistent with data from Mairesse *et al.* (1980), who found an intracellular 46K protein stimulated by 3-hr estrogen treatment of MCF-7 cells. Although the size and charge of their protein is similar to the secretary 46K, no definite conclusion regarding the identity of the two proteins could be drawn. A more recent report from this group (Mairesse *et al.*, 1981) now indicates the intracellular 46K may be larger than first thought and may be related to the 54K protein identified by our laboratory (described later).

2. *The 24K Cytosol Protein*

In our search for estrogen-regulated proteins in human breast cancer, we have chosen to use the double-label ratio method of Notides and Gorski (1966) that identifed rat uterine IP. In this method, control cells are pulse-labeled with [^{14}C]leucine, while experimental cells are labeled with [^{3}H]leucine. The cells are then mixed, proteins extracted, and aliquots analyzed on single- or double-dimensional polyacrylamide gels. An estradiol-stimulated increase in the rate of synthesis of a specific protein relative to other cellular proteins is reflected by an increase in the ^{3}H/^{14}C ratio in a particular band or spot on the gel. Although high resolution of proteins can be obtained, this method does have some notable drawbacks. First, only mixed samples from control and experimental groups can be analyzed, thus "nontarget" tissues are not easily assayed. Second, the method determines *relative* increases in specific proteins so that a large increase in general protein synthesis can mask individual increases. Ratio peaks can also be masked if the protein population in a limited molecular-weight range is too complex. This is a particular problem if single-dimension gels are used. On the other hand, the sample must contain enough proteins with sifficiently high incorporation so that a ratio baseline can be established.

Our initial attempts to identify estrogen-regulated proteins in MCF-7 cells by the double-labeling technique did not reveal any prominent ($\geq$ two-fold) increases in ratio. Considering our experience with estrogenic stimulation of DNA polymerase, we decided to use the nafoxidine rescue protocol, a method by which we can consistently observe estrogen stimulation of cell growth. Rescue from anti-estrogen growth inhibition is specific for estrogens (Edwards *et al.*, 1980a; Zava and McGuire, 1978), and likely results from estrogen displacement of anti-estrogen bound to

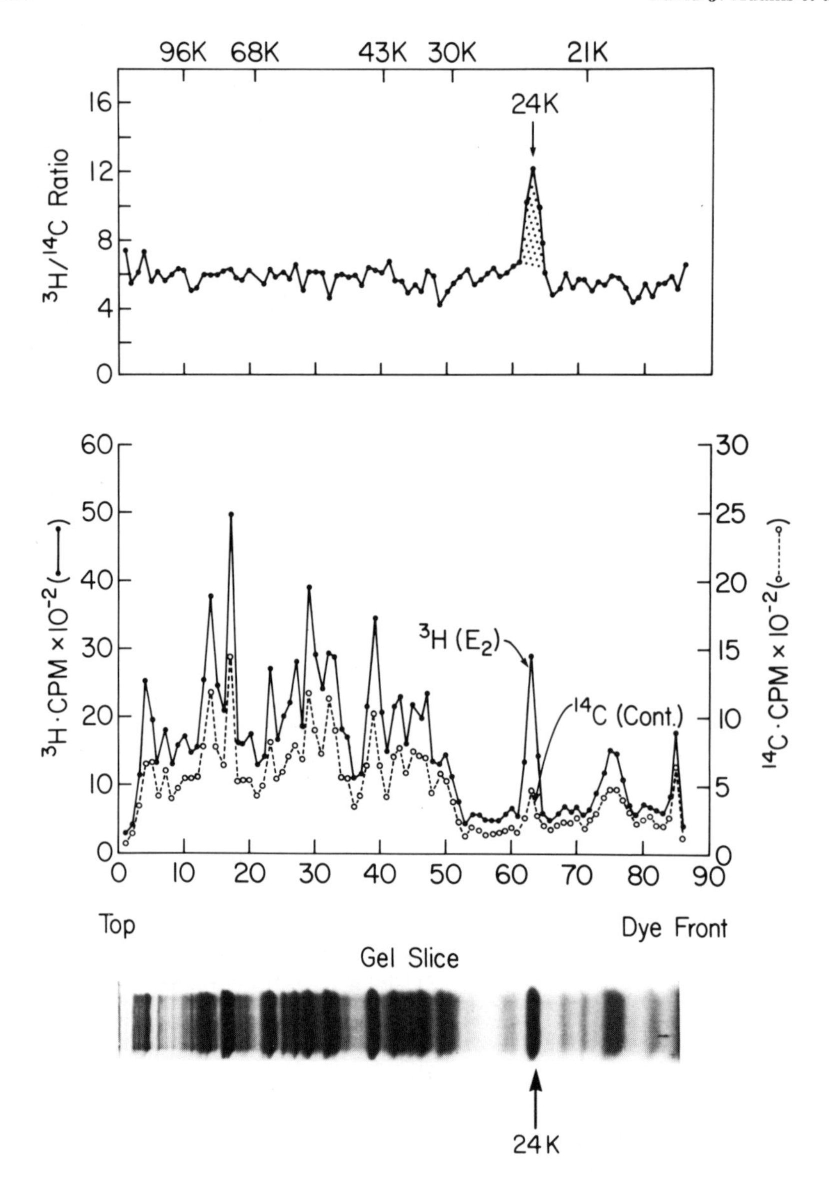

Fig. 1. Co-electrophoresis in SDS–polyacrylamide gel of cytosol proteins synthesized by nafoxidine-treated cells ([^{14}C]leucine) and cells "rescued" from nafoxidine inhibition by incubation with estradiol ([^{3}H]leucine). Cells were pretreated for 6 days with 1.0 μM nafoxidine and either continued on nafoxidine for another 6 days (reference cells) or changed to medium with 10 nM estradiol for 6 days. Equal numbers of control and estrogen-stimulated cells were combined and a mixed cytosol was prepared and analyzed by SDS–polyacrylamide

receptor. Horwitz *et al.*(1981) found that estrogen rescue of anti-estrogen-treated MCF-7 cells actually amplified the induction of progesterone receptor above levels obtained by treating with estradiol alone. The mechanism of this "superinduction" of PgR is unknown but suggests that the rescue protocol may maximize cell sensitivity to estrogen, perhaps by displacing residual endogenous estrogen and, in effect, creating a truly estrogen-withdrawn cell. This explanation is supported by evidence that even charcoal-stripped serum still retains conjugated estrogens which can be cleaved to biologically active hormones by MCF-7 cells (Vignon *et al.*, 1980). Alteratively, anti-estrogen pretreatment may allow enhanced expression of certain differentiated functions simply by slowing the cellular growth rate. Whatever the exact mechanism, we reasoned that the rescue protocol would provide a consistent basal level of protein synthesis and magnify estrogen stimulation of specific proteins. Consequently, in the studies that follow, we have compared synthesis rates of specific proteins between nafoxidine-treated (control) and estrogen-"rescued" (experimental) cells. Full details of our procedures will not be included here as they are described in previous publications (Edwards *et al.*, 1980a, 1981; Adams *et al.*, 1980).

Figure 1 shows a double-label analysis of cytoplasmic proteins after 6 days of estradiol stimulation. This is a single-dimension sodium dodecyl sulfate (SDS)–gel fractionating in the 94,000–14,000 molecular-weight range. The profile of the total ^{3}H and ^{14}C counts in each gel slice is indicated in the middle panel, while the upper panel gives the corresponding $^{3}H/^{14}C$ ratios. Under these gel conditions, a single, prominent ratio peak is found at 24,000 molecular weight coincident with a major radioactive and Coomassie blue-staining band. We refer to this protein as 24K. Stimulation of the 24K ratio peak occurs when the isotopes are reversed and does *not* occur if cells are rescued with ethanol vehicle alone (data not shown). Thus, stimulation of 24K is due to estrogen treatment and is not due to an isotope effect.

Because we have used an anti-estrogen rescue protocol, some important control experiments are necessary. We first examined the generality of the anti-estrogen pretreatment. As Fig. 2 illustrates, 24K stimulation does not require nafoxidine as anti-estrogen. Tamoxifen and CI-628 are equally effective. A more difficult question to address is whether

---

gel electrophoresis. The lower panel is a profile of the total ^{3}H and ^{14}C counts in each gel slice and the top is a plot of the corresponding $^{3}H/^{14}C$ ratios. A photograph of the Coomassie blue-stained gel is positioned at the bottom of the figure. An arrow indicates the molecular weight (24K) and position of the increased $^{3}H/^{14}C$ ratio. Mobilities of molecular-weight standards are indicated along the top margin.

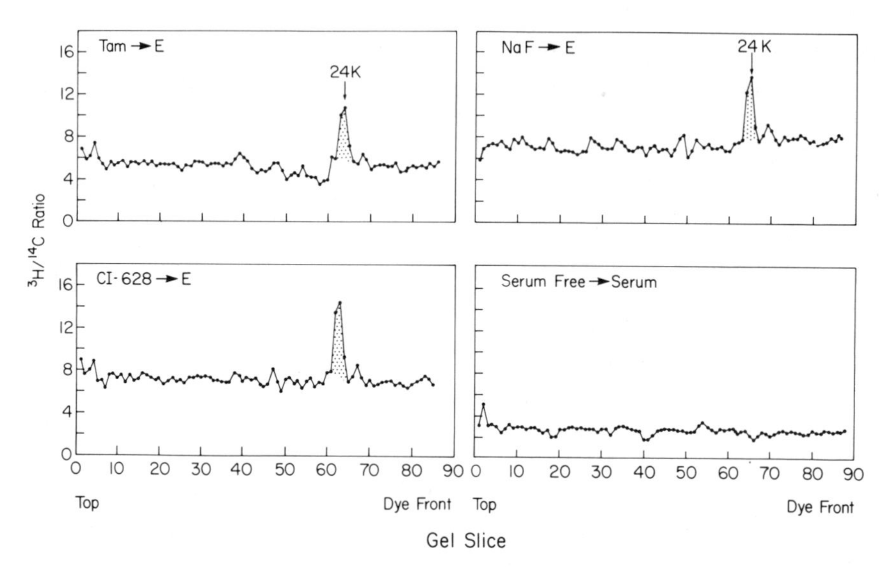

Fig. 2. Comparison of specific protein synthesis between serum-stimulated and estrogen-stimulated cells. MCF-7 cells were preincubated with three different anti-estrogens tamoxifen (Tam), CI-628, and nafoxidine (Naf), followed by incubation with estradiol (→E), and double-label cytosols were prepared and analyzed by SDS electrophoresis as described in Fig. 1. Double-label cytosols prepared from serum-deprived cells (labeled with [^{14}C]leucine) and cells restimulated by addition of serum (labeled with [^3H]leucine) were also analyzed by SDS electrophoresis (serum free → serum).

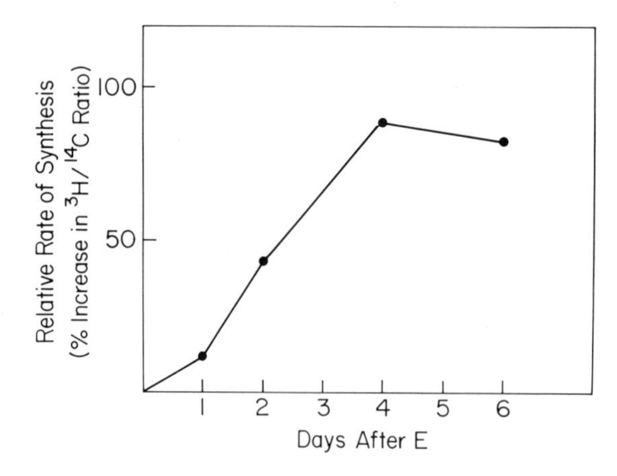

Fig. 3. Time course of stimulation of the 24K protein. Cells were pretreated with nafoxidine and $^3H/^{14}C$ cytosols were prepared and analyzed as described in Fig. 1, except at the times of estrogen stimulation indicated. The degree of stimulation of the 24K protein was estimated by calculating the percentage increase in the $^3H/^{14}C$ ratio at 24K over the baseline ratio.

increased synthesis of 24K is merely a nonspecific result of growth stimulation. Thus, the estrogen effect may be an indirect one exerted through a return to cellular proliferation and not via protein synthesis. We therefore did a serum rescue experiment to mimic nonspecific growth stimulation. After plating, MCF-7 cells were switched to serum-free medium for 6 days. Under serum-free conditions, growth is arrested but cells remain viable. On Day 6, control cells were continued on serum-free medium, while the experimental group was returned to medium containing serum. Although this serum rescue protocol results in a growth stimulation comparable to estrogen rescue (Edwards *et al.*, 1981), no ratio change at 24K (or anywhere else on the gel) was apparent (Fig. 2, right panels), despite the fact that *both* groups were pulse-labeled in medium containing serum. Furthermore, no relative change in 24K synthesis occurred at earlier time points of serum rescue (data not shown). Finally, 24K could represent a specific "nafoxidine-suppressed" protein. To test this possibility, cells treated with ethanol vehicle alone and pulse-labeled with [^{14}C]leucine were mixed with cells exposed to nafoxidine for 6 days and labeled with [^{3}H]leucine. Under these conditions, an anti-estrogen-sup-

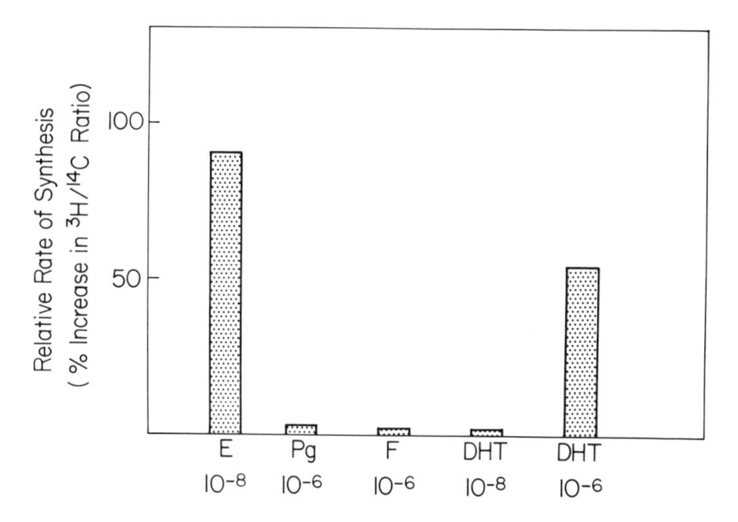

Fig. 4. Hormone specificity of stimulation of the 24K protein. Nafoxidine-pretreated cells were incubated for 6 days with various hormones at the concentrations indicated. Cells incubated with other hormones were labeled with [^{3}H]leucine in the same manner as estradiol-treated cells and were then mixed with ^{14}C-labeled nafoxidine-maintained reference cells. ^{3}H/^{14}C cytosols were analyzed by SDS electrophoresis. The degree of stimulation of the 24K protein was estimated by calculating the percentage increase in the ^{3}H/^{14}C ratio at 24K over the baseline ratio. E, Estradiol; Pg, progesterone; F, cortisol; DHT, dihydrotestosterone.

pressed protein should exhibit a *negative* ratio peak, but again no ratio change at 24,000 daltons was found (D. Edwards, unpublished observation).

Confident that we were observing an estrogenic effect on specific protein synthesis, we set out to characterize the 24K response. We first examined the time course of 24K stimulation in nafoxidine-pretreated cells rescued with 10 nM estradiol. A progressive increase in the relative rate of 24K synthesis was observed beginning at Day 1 and reaching maximum levels between Days 4 and 6 (Fig. 3). The 24K response is therefore considered a late effect of estradiol, although early general increases in protein synthesis observed under rescue conditions could obscure a rapid effect.

Next, we tested the hormone specificity of 24K stimulation. Nafoxidine-pretreated cells were rescued with high doses of progesterone, cortisol, and a physiological dose of dihydrotestosterone (DHT). As shown in Fig. 4, these hormones had no effect on synthesis of 24K. However, incubation with micromolar levels of DHT did stimulate 24K as might be

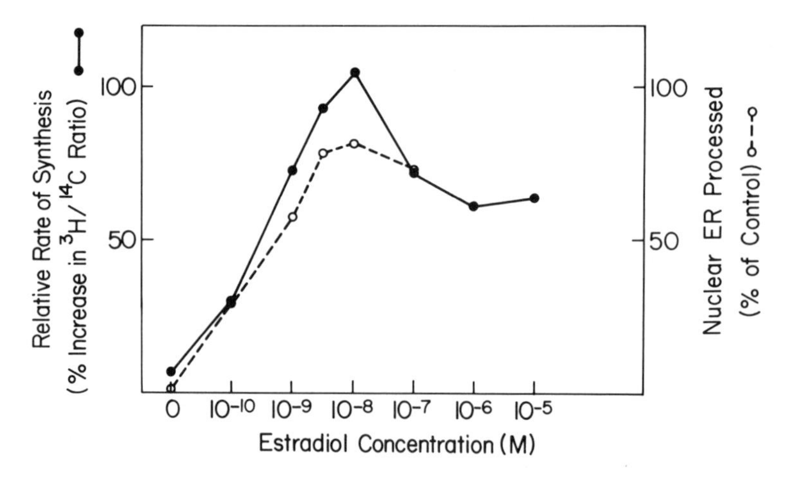

Fig. 5. Effect of different estradiol concentrations on synthesis of the 24K protein. Nafoxidine-pretreated cells were incubated for 6 days with the concentrations of estradiol indicated and double-label cytosols were prepared and analyzed by SDS electrophoresis as previously described. The degree of stimulation of the 24K protein (●) was estimated by calculating the percentage increase in the $^3H/^{14}C$ ratio at 24K over the baseline ratio. In a parallel experiment, cells were also preincubated with nafoxidine and then with the concentrations of estradiol indicated and 24 h later, nuclear ER was measured by protamine sulfate exchange assay (○). The amount of ER processed was estimated by taking the decrease in ER content at each dose of estradiol (compared with the level of ER in cells treated with nafoxidine only) and expressing this as a percentage of control; the control (or 100% processing level) being the processed level of nuclear ER in cells incubated for the entire period of the experiment with 10 nM estradiol.

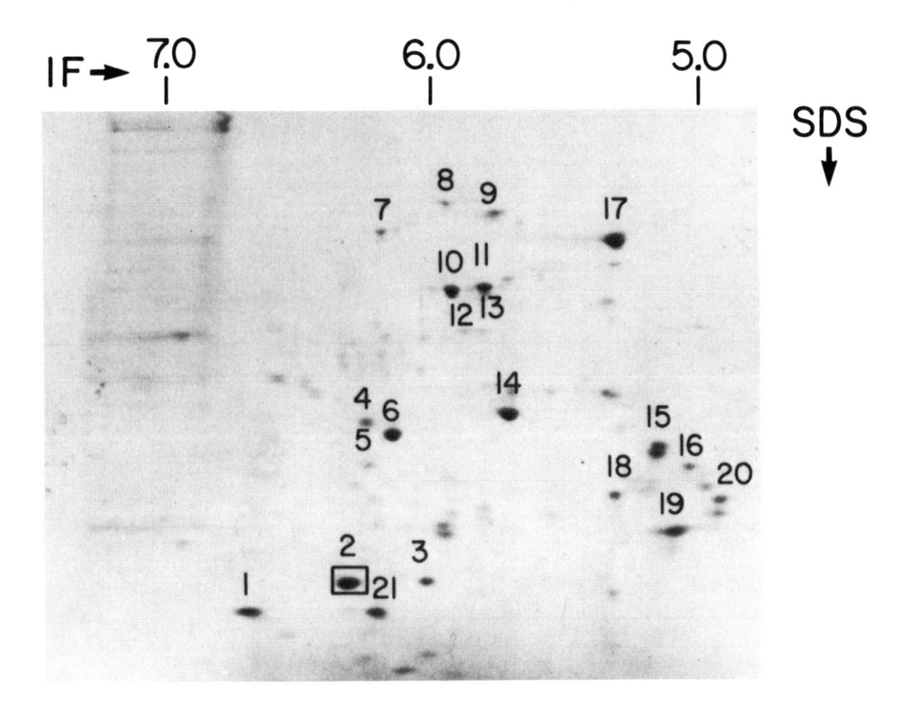

Fig. 6. Two-dimensional electrophoresis of $^{3}H/^{14}C$ cytosols from 6 days stimulation with 10 nM estradiol. An isoelectric focusing gel was applied directly to further separation by SDS–slab gel electrophoresis. Selected Coomassie blue-stained spots on the second-dimension SDS–gel were numbered, cut out, and counted for ^{3}H and ^{14}C radioactivity. Numbers on the SDS–gel indicate the stained spot below it and the enclosed spot (2) represents the estrogen-stimulated 24K protein.

expected, since we have previously shown that high doses of DHT elicit estrogenic responses in MCF-7 via binding to the estrogen receptor (Zava and McGuire, 1978). We then determined the estradiol dose response for 24K stimulation and found a progressive increase in the response throughout the physiological range of hormone with maximal stimulation at 10 nM estradiol (Fig. 5).

Previous work from our laboratory with estrogen stimulation of PgR has shown that this effect is dependent on nuclear processing of ER (Horwitz and McGuire, 1978). Although nafoxidine binds and translocates ER, nuclear processing does not occur so that nuclear ER levels remain elevated. Subsequent treatment with estradiol displaces nafoxidine from receptor, processing ensues, and the PgR response occurs. We were therefore interested in knowing whether the 24K response was also related to nuclear ER processing. Figure 5 also shows that in cells treated with

Table I
Incorporation and $^3H/^{14}C$ Ratio for Individual Proteins Resolved by Two-Dimensional Electrophoresis

Gel spot[a]	^{14}C-dpm (control)	^3H-dpm (estradiol)	$^3H/^{14}C$ ratio	Percentage of baseline ratio[b]
1	54	897	*16.57*	250
2 (24K)	538	7900	*14.67*	222
3	237	2237	9.41	142
4	65	274	4.21	64
5	153	1027	6.72	101
6	628	3907	6.22	94
7	43	267	6.26	95
8	67	385	5.76	87
9	219	1313	6.00	91
10	235	1428	6.06	92
11	349	2831	8.11	122
12	81	633	7.85	118
13	72	534	7.38	111
14	658	4028	6.12	92
15	145	914	6.28	95
16	286	2052	7.17	108
17	672	5185	7.72	116
18	116	595	5.13	77
19	256	1604	6.27	95
20	186	919	4.94	75
21	111	995	8.99	135
Blank	25	153	6.19	94
			Ave. $6.64^3 \pm 0.29$ (S.E.)	

[a] The numbered spots shown in Fig. 6 were cut out and counted for ^3H and ^{14}C radioactivity.
[b] Baseline ratio is the average ratio of all numbered spots, excluding spots 1 and 2.
[c] Average of ratios excluding spots 1 and 2.

estradiol alone the processed or steady-state level of nuclear ER is about 25% of that in cells treated with nafoxidine. Cytosol levels of ER remain low in both groups and are less than 10% of the total. If nafoxidine-pretreated cells (containing nonprocessed nuclear ER) are subsequently incubated with increasing doses of estradiol, nuclear ER levels decrease progressively. No change in nuclear ER concentration occurs if cells are incubated with ethanol vehicle alone. Analysis of the relative rate of 24K synthesis under these conditions indicates that estrogen stimulation of 24K parallels nuclear ER processing, again suggesting that regulation of the 24K protein is a receptor-mediated event.

A central problem in detection of estrogen-regulated proteins using gel analysis is that little information is gained regarding protein identity or function. The protein can, however, be characterized by its molecular weight and charge. Figure 6 is a two-dimensional gel of MCF-7 double-labeled cytosol after 6 days of estradiol treatment. A major Coomassie blue-stained spot is present at 24,000 daltons with a p*I* of 6.4. When this spot is excised and counted along with other prominent spots on the gel, a significant amount of radioactivity is detected with a $^{3}H/^{14}C$ ratio 2.2-fold above background (Table I). We estimate from these data that 24K in estrogen-stimulated cells represents about 1.6% of the total incorporation into cytoplasmic protein, making 24K a major intracellular protein regulated by estrogen.

In a further attempt to identify the 24K protein, we have compared the mobility of 24K on SDS–gels with that of the human milk proteins, casein, and α-lactalbumin and with the principal proteins found in cyst fluid obtained from patients with gross cystic breast disease (Haagensen *et al.*, 1979). Figure 7 indicates that 24K does not comigrate on the gel with any of these proteins; furthermore, the size, cellular location, and

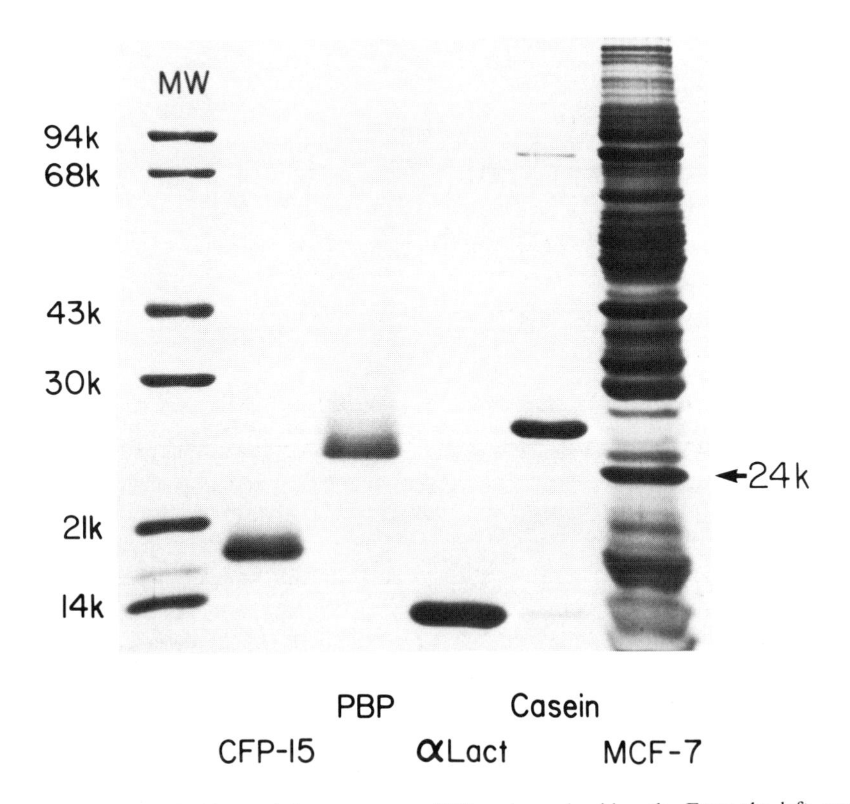

Fig. 7. Coomassie blue staining pattern on SDS–polyacrylamide gels. From the left, molecular-weight standards; gross cystic disease fluid protein (CFP-15); progesterone binding protein (PBP) from gross breast cystic disease; human α-lactalbumin; human casein; MCF-7 double-label cytosol from 6 days estradiol stimulation.

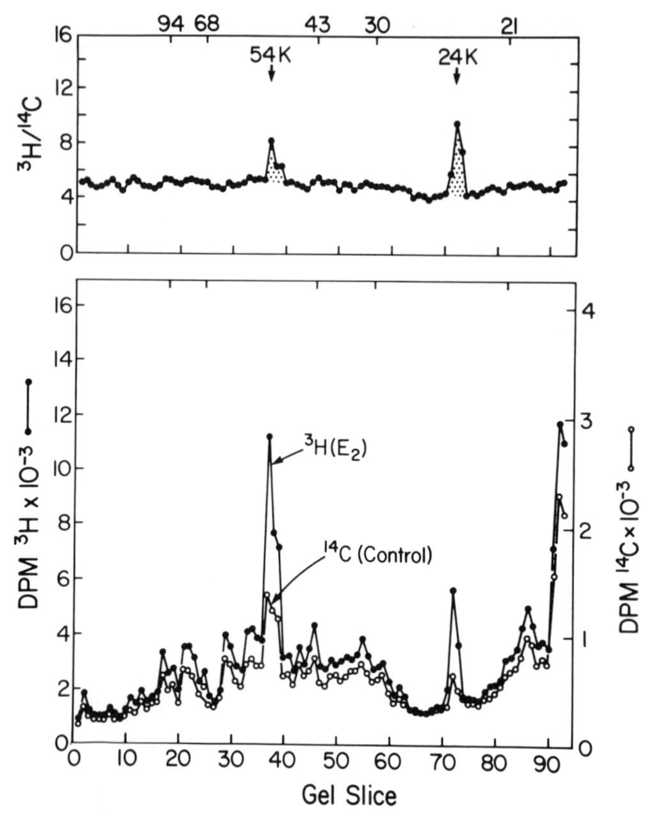

Fig. 8. Cell-free translation of MCF-7 mRNA. Two days after plating, MCF-7 cells were treated with nafoxidine (1 μM) for 6 days and then either continued on nafoxidine or treated with estradiol (10 nM) for 6 days. Messenger RNA was isolated from equal cell numbers in each treatment group, and equilvalent amounts of mRNA were translated in the reticulocyte lysate system using [^{14}C]leucine (nafoxidine) or [^{3}H]leucine (estradiol) as tracer. Cell-free translation products were mixed and co-electrophoresed on SDS–gels. The upper profile shows the ratio of ^{3}H/^{14}C in each gel slide; the lower profile indicates the total disintegrations per minute for ^{3}H (●) and ^{14}C (○).

amount of 24K suggest that it is not related to any of the other estrogen-regulated proteins reported for MCF-7 cells. It should be noted, however, that the size of 24K on SDS–gels may be misleading. Under nondenaturing conditions, 24K appears in the excluded volume on Sephacryl S-300 columns indicating a much greater molecular weight ($\geq 10^6$, D. Adams, unpublished observation). This result could reflect a high degree of aggregation of the 24K molecule with itself or with another high-molecular-weight protein. The 24K protein could also be a degradation product produced during cell fractionation, although our preliminary experiments using various homogenization buffers and protease inhibitors, together with the cell-free translation data described below argue against this possibility.

Like the identity of 24K, little is known about the mechanism of estrogenic regulation of this protein. Initial experiments suggest that transcriptional control may be involved. Using a double-label translation assay, we have shown that messenger RNA derived from estrogen-rescued MCF-7 cells encodes a major translation product of 24,000 molecular weight (Fig. 8). This radioactive peak coincides with a prominent ratio peak, suggesting that estrogen-rescued cells have an increased rate of 24K synthesis. Whether this effect is due to an increase in the specific mRNA for 24K is not yet known. However, double-label analysis of 24K synthesis *in vivo* (Fig. 6) and *in vitro* (Fig. 8) both imply that a significant amount of 24K is produced in anti-estrogen-treated control cells. Therefore, estrogenic regulation of 24K does not represent induction of a dormant gene; rather, the hormone probably increases the relative abundance of 24K mRNA.

3. The 54K Nuclear Protein

A surprising result of the double-label translation assay shown in Fig. 8 was the appearance of another putative estrogen-regulated protein at 54,000 daltons. The seemingly large amount of messenger RNA for 54K observed in this experiment was puzzling because our *in vivo* results did not reveal a major ratio change at this molecular weight. One explanation for this anomoly is that 54K is not a cytoplasmic protein. Accordingly, our recent data indicate that 54K is the major newly synthesized protein present in MCF-7 nuclei (D. Adams, manuscript in preparation).

Observation of an estrogen-regulated 54K nuclear protein in human breast tumor cells is significant for several reasons. First, 54K may be related to a nucleolar 54K protein isolated by Chan *et al.* (1980) and detected in a broad range of human malignancies, including 94% of malignant breast tumors (Busch *et al.*, 1979). The protein is not found in benign tumors or normal tissue. In another model system, Crawford *et al.* (1981) have prepared a monoclonal antibody against a 53K phosphoprotein (p53) associated with SV40 viral T antigen. Again, this antibody detects a 53K protein in all cell lines derived from spontaneous tumors or from normal cells transformed by SV40 but not in normal cells, including mammary epithelial cells cultured from human milk. Similar results have been reported by Dippold *et al.* (1981), who also showed that p53 has kinase activity capable of phosphorylating serine and threonine but not tyrosine (Jay *et al.*, 1981). Interestingly, a number of viral "oncogenes" appear to code for 53–60K protein kinases that specifically phosphorylate tyrosine residues.

Although expression of a 53–54K protein now appears to be a common feature of malignant cells, it is probably characteristic of rapidly proliferating cells in general. For example, treatment of nondividing T lymphocytes with the mitogen, concanavalin A, also induces expression of a 53K protein (Milner and McCormick, 1980). Furthermore, normal kidney epithelium and fetal brain cells express high levels of 53K phosphoprotein during exponential growth but promptly shut down p53 synthesis after reaching contact inhibition of cell division (Dippold *et al.*, 1981). This result has obvious implications for our observation of a nuclear 54K ratio peak in MCF-7 cells undergoing growth rescue compared to anti-estrogen-inhibited cells. Dippold *et al.* (1981) also find p53 antigenicity in the BT-20(ER-,PgR-) breast tumor cell line in addition to MCF-7, indicating that expression of this protein is not receptor dependent. Our nuclear 54K ratio peak could then conceivably be due to estrogen-regulated protein translocation rather than *de novo* protein synthesis or could be a nonspecific effect of growth rescue. The exact relation of our nuclear 54K protein with the similar proteins described by others is unknown, but the possibilities for 54K function in growth control and estrogenic regulation of this protein are fascinating and currently are under investigation in our laboratory.

IV. CONCLUSIONS

Although we have made significant advances in our understanding of estrogen-sensitive breast tumors, we are far from unraveling hormone effects at the molecular level. Perhaps we have been too anxious to fit estrogen regulation of breast tumors into the mold established by classical effects of this hormone on target cells. Diversity is a hallmark of natural processes and it is quite possible that new modes of estrogen action will surface in breast tumor cells. We are now realizing that estrogen may not have a straightforward mechanism that impinges directly on breast tumor cells. Estrogen may act in concert with other hormones affecting a variety of cell types *in vivo* that respond in a coordinated fashion.

How then, shall we interpret estrogen regulation of specific protein synthesis in breast tumor cells under *in vitro* conditions? If we accept that current cell culture conditions do not reproduce the *in vivo* tumor environment (as is suggested by lack of estrogenic effects on tumor growth in culture), it is perhaps surprising that any estrogen-regulated protein observed in cultured tumor cells has relevance *in vivo*. Clearly, however, estrogen-stimulated protease and progesterone receptor activities in MCF-7 cells have correlates *in vivo*. For example, tamoxifen treatment can increase PgR in ER+ cutaneous metastatic nodules taken from postmenopausal breast cancer patients. This estrogenic property of a clinically important anti-estrogen may improve progestagen effectiveness in counteracting the growth-promoting effects of estradiol in sensitive tumors (Namer *et al.*, 1980). As yet, no similar correlation can be made for the other hormone-regulated proteins described in breast tumor cell lines. Since biological activity-based assays are not available, development of highly specific antibodies will be required to detect the 46, 24, and 54K proteins *in vivo*. The advent of these immunological assays could very well bring a new level of quality control and sensitivity to the detection of hormone-dependent breast tumors.

Study of estrogenic effects in cultured breast tumor cells has therefore proved most valuable in identifying estrogen-regulated tumor markers. However, the old question of how estrogen stimulates tumor growth still remains unanswered. One might then ask whether estrogenic effects on tumor growth are necessarily related to hormone regulation of protein synthesis. A recent study has shown that estrogen regulates PgR levels in the MTW-9B rat mammary tumor but has no influence on tumor growth (Ip *et al.*, 1979). Similar observations were made in our laboratory using the DMBA-induced rat tumor model where a small number of tumors were autonomous in their growth but dependent upon estradiol for maintenance of PgR (Horwitz and McGuire, 1977b).

To sort out this question and others concerning the diverse effects and pathways of estrogen action in breast tumors will almost certainly require improvements in our cell culture models. A significant advance has already been made by development of defined media for the MCF-7 and ZR-75-1 cell lines (Barnes and Sato, 1979; Barnes, 1980; Allegra and Lippman, 1980). Removal of serum from culture medium permits an assessment of hormone function and interaction unclouded by the many growth factors known (and yet to be discovered) in serum. Furthermore, serum generally appears to stimulate cell proliferation and suppress differentiated functions which could lead to selection of cell populations that may not reflect those *in vivo* (Barnes and Sato, 1980; Orly *et al.*, 1980). Use of serum-free media will also permit detailed study of substratum and attachment factor requirements. There is increasing evidence that cellular morphology is determined by the extracellular matrix and that cellular shape is responsible for control of growth and function, including hormone responsiveness of mammary tumor cells (Gospodarowicz *et al.*, 1979; Yates and King, 1981). Advances in our understanding of cellular matrices could lead to successful culturing of normal mammary epithelial cells. Absence of hormone-responsive normal breast cell lines as control cultures is a notable deficiency in this field. Finally, a basic understanding of hormone-dependent tumor growth may require even more complex model systems. We may need to exploit tumor growth in athymic nude mice or coculture breast tumor cells with other cell lines. Indeed, at least one group has demonstrated a requirement for inoculation of pituitary tumor cells in addition to estradiol treatment to consistently produce growth of MCF-7 cells in nude mice (Leung and Shiu, 1981). Ideally, a system will be developed where estrogen-dependent tumor growth can be correlated with specific changes in the control of a hormone-dependent gene. A shift from inducible to constitutive expression of a gene product involved in growth control is one possible mechanism that may apply to breast cancer. Clearly, the future holds many interesting possibilities for defining the role of estrogen and other steroid hormones in regulating both normal and neoplastic human mammary tissue.

V. REFERENCES

Adams, D. J., Edwards, D. P., and McGuire, W. L., 1980, Estrogen regulation of specific messenger RNA's in human breast cancer cells, *Biochem. Biophys. Res. Commun.* **97:**1354.

Allegra, J. C., and Lippman, M. E., 1978, Growth of a human breast cancer cell line in serum-free hormone-supplemented medium, *Cancer Res.* **38:**3823.

Baker, H. J., and Shapiro, D. J., 1977, Kinetics of estrogen induction of *Xenopus laevis* vitellogenin messenger RNA as measured by hybridization to complimentary DNA, *J. Biol. Chem.* **252:**8428.

Barker, K. L., Adams, D. J., and Donohue, T. M., 1981, Regulation of the levels of mRNA for glucose-6-phosphate dehydrogenase and its rate of translation in the uterus by estradiol, *in* "Cellular and Molecular Aspects of Implantation" (S. Glasser and D. Bullock eds.), pp. 269–281, Plenum Press, New York.

Barnes, D., 1980, Factors that stimulate proliferation of breast cancer cells *in vitro* in serum-free medium, *in* "Cell Biology of Breast Cancer" (C. McGrath, M., Brennan, and M. Rich, eds.), pp. 227–287, Academic Press, New York.

Barnes, D., and Sato, G., 1979, Growth of a human mammary tumor cell line in a serum-free medium, *Nature* (*London*) **281:**388.

Barnes, D., and Sato, G., 1980, Serum-free cell culture: A unifying approach, *Cell* **22:**649.

Baulieu, E. E., Alberga, A., Raynaud-Jammet, C., and Wira, C. R., 1972, New look at the very early steps of oestrogen action in uterus, *Nature New Biol.* **236:**236.

Beatson, G. T., 1896, On the treatment of inoperable cases of carcinoma of the mamma: Suggestions for a new method of treatment, with illustrative cases, *Lancet* **2:**104.

Bolander, F. F., Jr., and Topper, Y. J., 1979, Stimulation of lactose synthetase activity and casein synthesis in mouse mammary explants by estradiol, *Endocrinology* **106:**490.

Bronzert, D. A., Monaco, M. E., Pinkus, L., Aitken, S., and Lippman, M. E., 1981, Purification and properties of estrogen-responsive cytoplasmic thymidine kinase from human breast cancer, *Cancer Res.* **41:**604.

Burke, R. E., Harris, S. C., and McGuire, W. L., 1978, Lactate dehydrogenase in estrogen-responsive human breast cancer cells, *Cancer Res.* **38:**2773.

Busch, H., Gyorkey, F., Busch, R. K., Davis, F. M., Gyorkey, R., and Smetna, A., 1979, A nucleolar antigen found in a broad range of human malignant tumor specimens, *Cancer Res.* **39:**3024.

Butler, W. B., Kirkland, W. L., and Jorgensen, T. L., 1979, Induction of plasminogen activator by estrogen in a human breast cancer cell line (MCF-7), *Biochem. Biophys. Res. Commun.* **90:**1328.

Chambon, P., Benoist, C., Breathnach, R., Cochet, M., Gannon, F., Gerlinger, P., Krust, A., Lemeur, M., LePennec, J. P., Mandel, J. L., O'Hare, K. U., and Perrin, F. 1979, Structural organization and expression of ovalbumin and related chicken genes, *in* "From Gene to Protein: Information Transfer in Normal and Abnormal Cells" (T. R. Russell, K. Brew, H. Faber, and J. Schultz, eds.), pp. 55–83, Academic Press, New York.

Chan, L., and O'Malley, B. W., 1976, Mechanism of action of the sex steroid hormones, *N. Engl. J. Med.* **294:**1322.

Chan, P. -K., Feyerabend, A., Busch, R. K., and Busch, H., 1980, Identification and partial purification of human tumor nucleolar antigen 54/6.3, *Cancer Res.* **40:**3194.

Crawford, L. V., Pim, D. C., Gurney, E. G., Goodfellow, P., and Taylor-Papadimitriou, J., 1981, Detection of a common feature in several human tumor cell lines—A 53,000 dalton protein, *Proc. Natl. Acad. Sci. USA* **78:**41.

Cunningham, D. D., Carney, D. H., and Glenn, K. C., 1979, A cell-surface component involved in thrombin-stimulated cell division, *in* "Hormones and Cell Culture, Book A" (G. H. Sato and R. Ross, eds.), pp.199, Cold Spring Harbor Laboratory, Cold Spring Harbor, New York.

Deeley, R. G., Gordon, J. I., Burns, A. T. H., Mullinix, K. P., Bina-Stein, M., and Goldberger, R. F., 1977, Primary activation of the vitellogenin gene in the rooster, *J. Biol. Chem.* **252:**8310.

Dippold, W. G., Jay, G., DeLeo, A. B., Khoury, G., and Old, L. J., 1981, p53 transformation-related protein: Detection by monoclonal antibody in mouse and human cells, *Proc. Natl. Acad. Sci. USA* **78:**1695.

Edwards, D. P., Chamness, G. C., and McGuire, W. L., 1979, Estrogen and progesterone receptor proteins in breast cancer, *Biochim. Biophys. Acta* **560:**457.

Edwards, D. P., Adams, D. J., and McGuire, W. L., 1980a, Estrogen induced synthesis of specific proteins in human breast cancer cells, *Biochem. Biophys. Res. Commun.* **93:**804.

Edwards, D. P., Murthy, S. R., and McGuire, W. L., 1980b, Effects of estrogen and antiestrogen on DNA polymerase in human breast cancer, *Cancer Res.* **40:**1722.

Edwards, D. P., Adams, D. J., and McGuire, W.L., 1981, Estradiol stimulated synthesis of a major intracellular protein in human breast cancer cells (MCF-7), *Breast Cancer Ther. Res.* **1:**209.

Goldman, R. D., Kaplan, N. O., and Hall, T. C., 1964, Lactate dehydrogenase in human neoplastic tissues, *Cancer Res.* **24:**389.

Gospodarowicz, D., Vlodavsky, I., Greenburg, G., and Johnson, L. K., 1979, Cellular shape is determined by the extracellular matrix and is responsible for the control of cellular growth and function, *in* "Hormones and Cell Culture, Book B" (G. H. Sato and R. Ross, eds.), pp. 561–592, Cold Spring Harbor Laboratory, Cold Spring Harbor, New York.

Haagensen, D. E., Mazoujian, G., Dilley, W. G., Pedersen, C. E., Kister, S. J., and Wells, S. A., 1979, Breast gross cystic disease fluid analysis. I. Isolation and radioimmunoassay for a major component protein, *J. Natl. Cancer Inst.* **62:**239.

Hakim, A. A., 1980, Estradiol-induced biochemical changes in human neoplastic cells: Estradiol-mediated protease, *Cancer Biochem. Biophys.* **4:**173.

Hilf, R., Rector, W. D., and Orlando, R. A., 1976, Multiple molecular forms of lactate dehydrogenase and glucose 6-phosphate dehydrogenase in hormonal and abnormal human breast tissues, *Cancer* **37:**1825.

Hornebeck, W., Brechemier, D., Bellon, G., Adnet, J. J., and Robert, L., 1980, Biological significance of elastase-like enzymes in arteriosclerosis and human breast cancer, *in* "Proteinases and Tumor Invasion" (A. J. Barrett, A. Baici, and P. Strauli eds.), pp. 117–141, Raven Press, New York.

Horwitz, K. B., and McGuire, W. L., 1977a, Estrogen control of progesterone receptor in human breast cancer, *J. Biol. Chem.* **253:**2223.

Horwitz, K. B., and McGuire, W. L., 1977b, Progesterone and progesterone receptors in experimental breast cancer, *Cancer Res.* **37:**1733.

Horwitz, K. B., and McGuire, W. L., 1978, Nuclear mechanisms of estrogen action: Effects of estradiol and antiestrogens on estrogen receptors and nuclear receptor processing, *J. Biol. Chem.* **253:**8185.

Horwitz, K. B., McGuire, W. L., Pearson, O. H., and Segaloff, A., 1975, Predicting response to endocrine therapy in human breast cancer, *Science* **189:**726.

Horwitz, K. B., Zava, D. T., Thilagar, A. K., Jensen, E. M., and McGuire, W. L., 1978, Steroid receptor analyses of nine human breast cancer cell lines, *Cancer Res.* **38:**2434.

Horwitz, K. B., Aiginger, P., Kuttenn, F., and McGuire, W. L., 1981, Nuclear estrogen receptor release from antiestrogen suppression: Amplified induction of progesterone receptor in MCF-7 human breast cancer cells, *Endocrinology* **108:**1703.

Ip, M., Milholland, R. J., and Rosen, F., 1979, Mammary cancer: Selective action of the estrogen receptor complex, *Science* **203:**361.

Ivarie, R. D., and O'Farrell, P. H., 1978, The glucocorticoid domain: Steroid-mediated changes in the rate of synthesis of rat hepatoma proteins, *Cell* **13:**41.

Jay, G., Khoury, G., DeLeo, A. B., Dippold, W. G., and Old, L. J., 1981, p53 transformation-related protein: Detection of an associated phosphotransferase activity, *Proc. Natl. Acad. Sci. USA* **78:**2932.

Jensen, E. V., 1981, Hormone dependency of breast cancer, *Cancer* **47:**2319.

Jensen, E. V., DeSombre, E. R., and Jungblut, P. W., 1967, Estrogen receptors in hormone-responsive tissues and tumors, *in* "Endogenous Factors Influencing Host-Tumor Balance" (R. W. Wissler, T. L. Dao, and S. Wood, Jr., eds.), pp. 15–30, Univ. Of Chicago Press, Chicago.

Kaye, A. M., and Reiss, N., 1980, The uterine "estrogen induced protein" (IP): Purification, distribution and possible function, *in* "Steroid Induced Uterine Proteins" (M. Beato, ed.), pp. 3–19, Elsevier/North-Holland, New York.

Kaye, A. M., Reiss, N., Iacobelli, S., Bartoccioni, E., and Marchetti, P., 1980, The "estrogen-induced protein" in normal and neoplastic cells, *in* "Hormones and Cancer" (S. Iacobelli, H. R. Lindner, R. J. B. Kino, M. E. Lippman, and eds.), pp. 41–51, Raven Press, New York.

King, R. J. B., 1979, How important are steroids in regulating the growth of mammary tumors?, *in* "Biochemical Actions of Hormones," Vol. VI (G. Litwack, ed.) pp. 247–264, Academic Press, New York.

Knight, W. A., Livingston, R. B., Gregory, E. J., and McGuire, W. L., 1977, Estrogen receptor as an independent prognostic factor for early recurrence in breast cancer, *Cancer Res.* **37:**4669.

Leung, C. K. H., and Shiu, R. P. C., 1981, Required presence of both estrogen and pituitary factors for the growth of human breast cancer cells in athymic nude mice, *Cancer Res.* **41**:546.

Lockwood, D. H., Boytovich, A. E., Stockdale, F. E., and Topper, Y. J., 1967, Insulin-dependent DNA polymerase and DNA synthesis in mammary epithelial cells *in vitro*, *Proc. Natl. Acad. Sci. USA* **58**:658.

Mairesse, N., Devleeschouwer N., Leclercq, G., and Galand, P., 1980, Estrogen-induced protein in the human breast cancer cell line MCF-7, *Biochem. Biophys. Res. Commun.* **97**:1251.

Mairesse, N., Devleeschouwer, N., Leclercq, G., and Galand, P., 1981, Estrogen-Induced Protein in the Human Breast Cancer Cell Line MCF-7: Further Characterization, Fifth International Symposium of the Journal of Steroid Biochemistry, Puerto Vallarta, Jalisco, Mexico.

Martin, T. J., Findlay, D. M., MacIntyre, I., Eisman, J. A., Michelangeli, V. P., Moseley, J. M., and Partridge, N. C., 1980, Calcitonin receptors in a cloned human breast cancer cell line (MCF-7), *Biochem. Biophys. Res. Commun.* **96**:150.

McGuire, W. L., Carbone, P. P., Sears, M. E., and Escher, G. C., 1975, Estrogen receptors in human breast cancer: An overview, *in* "Estrogen Receptors in Human Breast Cancer" (W. L. McGuire, P. P. Carbone, and E. P. Vollmer, eds.), pp. 1–7, Raven Press, New York.

McKnight, S. G., and Palmiter, R. D., 1979, Transcripitional regulation of the ovalbumin and conalbumin genes by steroid hormones in the chick oviduct, *J. Biol. Chem.* **254**:9050.

Miller, L. K., Tuazon, F. B., Niu, E.-M., and Sherman, M. R., 1981, Human breast tumor estrogen receptor: effects of molybdate and electrophoretic analyses, *Endocrinology* **108**:1369.

Milner, J., and McCormick, F., 1980, Lymphocyte stimulation: concanavalin A. Induces the expression of a 53K protein, *Cell Biol. Int. Rep.* **4**:663.

Namer, M., Lalanne, C., and Baulieu, E. E., 1980, Increase of progesterone receptor by tamoxifen as a hormonal challenge test in breast cancer, *Cancer Res.* **40**:1750.

Notides, A., and Gorski, J., 1966, Estrogen-induced synthesis of a specific uterine protein, *Proc. Natl. Acad. Sci. USA* **56**:230.

O'Farrell, P. J., 1975, High resolution two-dimensional electrophoresis of proteins, *J. Biol. Chem.* **250**:4007.

Orly, J., Sato, G., and Erickson, G. F., 1980, Serum suppresses the expression of hormonally induced functions in cultured granulosa cells, *Cell* **20**:817.

Osborne, C. K., Yockmowitz, M. G., Knight, W. A., and McGuire, W. L., 1980, The value of estrogen and progesterone receptors in the treatment of breast cancer, *Cancer* **46**:2884.

Paranjpe, M., Engel, L., Young, N., and Liotta, L. A., 1980, Activation of human breast carcinoma collagenase through plasminogen activator, *Life Sci.* **26**:1223.

Poole, A. R., Recklies, A. D., and Mort, J. S., 1980, Secretion of proteinases from human breast tumors: excessive release from carcinomas of a thiol proteinase, *in* "Proteinases and Tumor Invasion" (P. Strauli, A. J. Barrett, A. Baici, and eds.), pp. 81–95, Raven Press, New York.

Quigley, J. P., Goldfarb, R. H., Scheiner, C., O'Donnell-Tormey, J., and Yeo, T. K., 1980, Plasminogen activator and the membrane of transformed cells, *in* "Tumor Cell Surfaces and Malignancy" (R. O. Hynes and C. F. Fox, eds.), pp. 773–796, Liss, New York.

Reiss, N., and Kaye, A. M., 1981, Identification of the major component of the estrogen-induced protein of rat uterus as the BB isoenzyme of creatine-kinase, *J. Biol. Chem.* **256**:5741.

Shafie, S. M., 1980, Estrogen and the growth of breast cancer: New evidence suggests indirect action, *Science* **209:**701.

Shafie, S. M., and Liotta, L. A., 1980, Formation of metastasis by human breast carcinoma cells (MCF-7) in nude mice, *Cancer Lett.* **11:**81.

Sherman, M. R., Tuazon, F. B., and Miller, L. K., 1980, Estrogen receptor cleavage and plasminogen activation by enzymes in human breast tumor cytosol. *Endocrinology* **106:**1715.

Sirbasku, D. A., 1980, Estromedins: uterine-derived growth factors for estrogen-responsive tumor cells, *in* "Control Mechanisms in Animal Cells" (A. Shields, R. Levi-Montalcini, S. Iacobelli, and L. Jimenez de Asua, eds.), pp. 293–298, Raven Press, New York.

Sirbasku, D. A., and Benson, R. H., 1980, Proposal of an indirect (estromedin) mechanism of estrogen-induced mammary tumor cell growth, *in* "Cell Biology of Breast Cancer" (C. McGrath, M. J. Brennan, and M. A. Rich, eds.), pp. 289–314, Academic Press, New York.

Skipper, J. K., Eakle, S. D., and Hamilton, T. H., 1980, Modulation by estrogen of synthesis of specific uterine proteins, *Cell* **22:**69.

Supowit, S. C., and Rosen, J. M., 1980, Gene expression in normal and neoplastic mammary tissue, *Biochemistry* **19:**3432.

Tata, J. R., 1979, Control by oestrogen of reversible gene expression: The vitellogenin model, *J. Steroid Biochem.* **11:**361.

Terenius, L., 1968, Selective retention of estrogen isomers in estrogen-dependent breast tumors of rats demonstrated by *in vitro* methods, *Cancer Res.* **28:**328.

Toft, D., and Gorski, J., 1966, A receptor molecule for estrogens: Isolation fro the rat uterus and preliminary characterization, *Proc Natl. Acad. Sci USA* **55:**1574.

Topper, Y. J., and Freeman, C. S., 1980, Multiple hormone interactions in the developmental biology of the mammary gland, *Physiol. Rev.* **60:**1049.

Vignon, F. V., Terqui, M., Westley, B., Ducoq, D., and Rochefort, H. 1980, Effects of plasma estrogen sulfate in mammary cancer cells, *Endocrinology* **106:**1079.

Westley, B., and Rochefort, H., 1979, Estradiol induced proteins in the MCF-7 human breast cancer cell line, *Biochem. Biophys. Res. Commum.* **90:**410.

Westley, B., and Rochefort, H., 1980, A secreted glycoprotein induced by estrogen in human breast cancer cell lines, *Cell* **20:**353.

Yates, J., and King, T. J. B., 1981, Correlation of growth properties and morphology with hormone responsiveness of mammary tumor cells in culture, *Cancer Res.* **41:**258.

Zava, D. T., and McGuire, W. L., 1978, Androgen action through the estrogen receptor in a human breast cancer cell line, *Endocrinology* **103:**624.

Chapter 15

Molecular Characteristics of Brain Opiate and Nicotine Receptors

Jean M. Bidlack and Leo G. Abood

Center for Brain Research and Department of Biochemistry
University of Rochester Medical Center
Rochester, New York

I. INTRODUCTION

A considerable amount of current research on receptors has focused on elucidating the molecular mechanisms responsible for their pharmacological and psychological effects. Under investigation in the authors' laboratory are the molecular characteristics of the brain opiate and nicotine receptors. The opiates and their endogenous ligands (the enkephalins) are known to play a crucial role in analgesia (Terenius, 1978) and have been implicated in the pathophysiology of hypovolemic shock (Faden and Holoway, 1979).

The psychopharmacological effects of tobacco smoking, which are largely due to nicotine, include mixed depression and facilitation, EEG and behavioral arousal, and skeletal muscle relaxation (Domino, 1973; Kawamura and Domino, 1969). The action of nicotine, particularly in the autonomic nervous system, is known to be mediated via nicotinic cholinergic synapses; however, recent evidence from this laboratory suggests that nicotine may also be acting on noncholinergic sites in the brain (Abood *et al.*, 1978a, 1979, 1980c). This review is directed toward outlining an assessment of current knowledge of the molecular basis of the brain opiate and the nicotine receptor.

II. OPIATES AND ENKEPHALINS

A. Introduction

The search for a hypothetical opiate receptor commenced by measuring the binding of radioactive opiates to neural membranes with the expectation that such binding would involve the pharmacologically relevant opiate receptor. Goldstein *et al.* (1971) showed that neural membranes bound opiates sterospecifically and at relatively low drug concentrations. Since this time, the binding of opiates to neural membranes has been extensively studied and reviewed (Snyder *et al.*, 1979; Goldstein and Cox, 1978; Snyder and Simantov, 1977). Terenius (1973a,b) showed that high-affinity stereospecific opiate binding sites were present in the synaptic plasma membrane fraction of rat brain homogenates. A differential study of the subcellular distribution of opiate receptors by Pert *et al.* (1976) and Smith and Loh (1976) revealed the highest specific activity to be localized in the synaptosomal fraction, but some stereospecific opiate binding occurred in all membranous fractions. Opiate binding is found in all regions of the brain with the exception of the cerebellum. Hiller *et al.* (1973) and Goldstein and Cox (1978) present summaries of the regional distribution of sterospecific opiate binding in rat brain.

The close resemblance of the opiate receptor to receptors for other known neurotransmitters led researchers to look for an endogenous ligand for the opiate receptor. Terenius and Wahlstrom (1974) first suggested the presence of an endogenous ligand by measuring inhibition of opiate binding by a brain membrane preparation. Shortly thereafter, Hughes (1975a,b) identified an endogenous ligand for the opiate receptor as a substance in brain which mimics the naloxone-reversible inhibitory actions of morphine. This ligand was an opioid peptide from pig brain and identified as a mixture of two pentapeptides, methionine- (Met-enk, H-Tyr–Gly–Gly–Phe–Met–OH) and leucine-enkephalin (Leu-enk, H-Tyr–Gly–Gly–Phe–Leu-OH), Met-enk being in four times greater than *Leu-enk*. Some precursors of the enkephalins have opioid activity and much work is currently directed toward the mechanisms involved in the generation and regulation of the enkephalins as well as its neuroanatomical localization.

Not all opiates bind to neural membranes at the same site. Synder *et al.* (1979), Lord *et al.* (1977), Simantov *et al.* (1978), and Simantov and Snyder (1976) have summarized extensive studies on the ability of certain opiates to compete with other opiates for binding to neural membranes. Briefly, they found that all drugs, whether agonists, antagonists, mixed agonist–antagonists, or opioid peptides, display similar affinities in competing for binding sites labeled by [^{3}H]dihydromorphine and [^{3}H]naloxone. Some drugs, such as morphine, dihydromorphine, normorphine, oxymorphone, and fentanyl, are 20 to 50 times more potent in competing for [^{3}H]opiate than [^{3}H]enkephalin binding. On the other hand, however, etorphine, levorphanol, phanazocine, and the opioid peptides have more similar affinities to [^{3}H]Met-enk, [^{3}H]naloxone, and [^{3}H]dihydromorphine. Because not all opiates have the same binding characteristics, Martin *et al.* (1976) have proposed subclasses of opiate receptors. The presence of more than one endogenous ligand may imply that there are subclasses of receptors, but for the present, this remains problematical.

B. Agents That Affect the Opiate Receptor

1. Proteases

Sterospecific binding of opiates is destroyed by treating the neural membranes with proteolytic enzymes (Pasternak and Snyder, 1974; Simon *et al.*, 1973); a finding which suggests that the opiate receptor is comprised of a protein moiety. It still remains to be determined, however, whether these enzymes are directly attacking the receptor molecule, producing changes in membrane structure that indirectly alter receptor properties, or interacting with the receptor in a nonenzymatic manner.

It has recently been shown that when trypsin and chymotrypsin were enzymatically inactivated by a variety of chemical and physical procedures, the inhibitory effect on [^{3}H]dihydromorphine binding to neural membranes was about one-quarter that of the original intact enzymes (Abood *et al.*, 1980b). A number of peptides formed by autolysis of trypsin were found to be responsible for the nonenzymatic inhibition of opiate binding. The more active of these peptides had about 5% the inhibitory potency of Met-enk and each contained at least one tyrosine.

Klee (1979) has shown that peptides derived from pepsin digestion of dietary proteins, such as wheat gluten and α-casein, possess opiatelike properties. A number of small peptides derived from pepsin digests of wheat gluten produced a naloxone-reversible inhibition of adenyl cyclase in neuroblastoma cultures. Such studies indicate the complexity and diversity of peptides capable of interacting with opiate receptors and further serve to emphasize the possibility that pharmacologically active peptides are derivable from normal digestion.

While a protein moiety is still believed to be a component of the opiate receptor, these studies demonstrate that the use of proteases to demonstrate the involvement of a protein in the opiate, and possibly other types of receptors, requires that a distinction be made between the direct proteolytic destruction of the receptor and secondary effects resulting from autolytic and other peptides formed during proteolytic digestion.

2. *Phospholipases, Phosphatidylserine Decarboxylase, and Ascorbic Acid—Requirement for Phosphatidylserine*

Phospholipases have been used as a tool to determine whether phospholipids are an intrinsic component of the opiate receptor. Pasternak and Snyder (1974) demonstrated that opiate binding to nervous tissue was sensitive to phospholipases, particularly phospholipase A_2 from *Vipera russelli*. The addition of oleic acid or lysophosphatidylcholine, which they assumed were the major by-products of hydrolysis, was not inhibitory to binding. Using lower concentrations of phospholipase A_2 from *V. russelli*, Lin and Simon (1978) reported that the inhibition caused by phospholipase A_2 could be restored to within 15% of control values by washing the membranes with 1% (w/v) bovine serum albumin which binds the released fatty acids and lysophospholipids. We have shown that the addition of unsaturated fatty acids to neural membranes is highly inhibitory to opiate binding, with the degree of inhibition being related to the degree of unsaturation (Abood *et al.*, 1977). These results suggest that the by-products of phospholipase A_2 hydrolysis are inhibitory to opiate binding but do not rule out the possibility that inhibition could also be due in part to the destruction of a phospholipid component of the opiate receptor.

The addition, by homogenization, of phospholipids to neural membranes also affects opiate binding. Table I shows the effect on stereospecific opiate binding of adding various phospholipids with varying hydrocarbon chains. Phosphatidylserine produced nearly a 40% increase in stereospecific dihydromorphine binding. Both the saturated forms of phosphatidic acid and phosphoinositides also produced an enhancement in opiate binding. In contrast, phosphatidylethanolamine derived from brain produced a 20% inhibition in opiate binding.

Exposure of neural membranes to phosphatidylserine decarboxylase results in about a 30% inhibition in stereospecific dihydromorphine binding (Abood *et al.*, 1978b), while the addition of phosphatidylserine to membranes that have been treated with the decarboxylase restores opiate binding to control levels. Phosphatidylethanolamine is the product formed by the action of phosphatidylserine decarboxylase. The inhibitory effect that of the enzyme on the opiate receptor has at least two explanations: (1) Phosphatidylserine may be an integral component of the opiate, and when it is destroyed, binding is reduced, or (2) the inhibition may result

Table I
Effect of Phospholipids with Varying Unsaturation on Stereospecific Opiate Binding[a]

Phospholipid	Source	Acyl composition	[^{3}H]Dihydromorphine bound (% control)
None			100
Phosphatidylserine	Myelin	18:0(40), 18:1(43:3)	139
	Synaptic membrane	18:0(39.1), 18:1(13.1) 22 un(14.0), 22:6(25.8)	138
	White matter	18:0(37.6), 18:1(37.6) 22 un(4.5), 22:6(7.6)	136
	Gray matter	18:0(41.6), 18:1(15.3) 22 un(8.6), 22:6(28.7)	137
Phosphatidic acid (from egg phosphatidylcholine)	Synthetic	16:0, 16:0	130
	Egg	16:0(37.7), 18:0(9.2) 16:1(3.1), 18:1(32.9) 18:2(17.0)	75
Phosphoinositides	Yeast	16:0(23), 16:1(32) 18:0(8), 18:1(33)	128
	Brain	18:0(42), 20:4(40)	63
Phosphatidylcholine	Synthetic	18:0, 18:0	102
	Brain	16:0(35.3), 18:0(16.3) 18:1(34.6), 20:4(4.4) 22:6(7.4)	95
Ethanolamine phosphatides	Synthetic	16:0, 16:0	90
	Brain	16:0(8.3), 18:0(24.4) 18:1(12.5), 20:4(13.6) 22:6(29.2)	80

[a] Percentage of principal acyl groups given in parentheses. The control value for [^{3}H]dihydromorphine was 720 cpm. The values represent an average of three separate experiments agreeing with 6% of the mean.

from phosphatidylethanolamine, the product of the enzymatic reaction which has been shown to be inhibitory to binding (Abood *et al.*, 1977). In either case it appears that phosphatidylserine is closely associated with the receptor.

Dunlap *et al.* (1979) have shown that ascorbic acid destroyed stereospecific opiate binding to guinea pig brain irreversibly. Destruction of stereospecific binding by direct chemical reduction was ruled out, since several other reducing agents were found to be without effect. Homogenates incubated with ascorbate in the absence of oxygen or treated with reagents that inhibit ascorbate-catalyzed lipid peroxide formation did not vary from untreated homogenates with respect to stereospecific opiate binding. Phosphatidylserine, when added to homogenates, was the only lipid that partially prevents the destruction of stereospecific opiate binding by ascorbic acid. These data support the contention that a membrane lipid plays a critical role in the structural integrity of the opiate receptor and that this lipid may be phosphatidylserine.

Cerebroside sulfate may also be an integral component of the opiate receptor. Cerebroside sulfate, itself, exhibits many of the structural requirements proposed for an opiate binding site. It exhibits a high affinity for and stereoselective binding to narcotic drugs (Loh *et al.*, 1974). Recently, Craves *et al.* (1980) have shown that administration of antibodies to cerebroside sulfate into the periaqueductal gray region antagonizes the effect of morphine and β-endorphin. A reduction in the availability of brain cerebroside sulfate induced pharmacologically or genetically results in a decrease in the analgesic response to morphine (Law *et al.*, 1978).

3. *Sulfhydryl Reagents*

Studies on agonist opiate binding have shown that *N*-ethylmaleimide (Terenius, 1973b) and other sulfhydryl reagents (Simon *et al.*, 1973; Pasternak *et al.*, 1975) reduced stereospecific agonist binding with little effect on nonspecific binding. The presence of bound opiate, however, provided protection from receptor inactivation by iodoacetamide (Pasternak *et al.*, 1975) or *N*-ethylmaleimide (Simon and Groth, 1975). It was also noted that agonist binding was more readily reduced by sulfhydryl reagents than was antagonist binding (Simon and Groth, 1975; Wilson *et al.*, 1975). Simon and Groth (1975) have further shown that the rate of receptor inactivation by *N*-ethylmaleimide is greatly reduced in the presence of 100 mM NaC1. Data with sulfhydryl reagents appear to suggest that the antagonist conformation of the receptor, which is favored in the presence of Na, is less readily attacked by sulfhydryl reagents (Goldstein and Cox, 1978).

4. *Salt*

As alluded to in the preceding section, salt (particularly Na) does affect the conformation of the receptor. General reviews are available on the effect of salt on opiate binding (Goldstein and Cox, 1978; Snyder *et al.*, 1979), but only work directly applicable to an understanding of the molecular basis of the opiate receptor will be summarized here.

The binding of opiate agonists and antagonists is differentially affected by the presence of Na at concentrations between 1 and 100 mM (Pert *et al.*, 1973; Pert and Snyder, 1974; Simon *et al.*, 1973). The binding of agonists is inhibited by the presence of Na, while antagonist binding is unaffected or increased. This effect is fairly specific for Na, with Li being less active and other monovalent cations being ineffective except at concentrations in excess of 100 mM. This differentiation has led to the postulate that the opiate receptor exists in two forms (Pasternak and Snyder, 1974; Simon *et al.*, 1975b). One form preferentially binds agonists, while the other favors antagonists. Sodium appears to shift the equilibrium toward the antagonist form, promoting antagonist binding at the expense of the agonist.

Pure antagonists, such as naloxone, are unaffected by the presence of Na, binding equally well in its presence or absence. Snyder *et al.* (1979) have divided agonists into two categories. One group contains agonists such as dihydromorphine and fentanyl, whose binding to neural membranes is greatly inhibited by low concentrations of Na. The second group of agonists contains Met-enk, Leu-enk, levorphanol, and etorphine. While susceptible to inhibition by Na, these agonists are two to three times less sensitive to inhibition by Na than the former group of agonists. Snyder *et al.* (1979) reported that agonists which are very sensitive to Na are 20 to 50 times more potent in reducing [^{3}H]dihydromorphine than [^{3}H]enkephalin binding. Agonists less affected by Na are more potent in reducing [^{3}H]enkephalin than [^{3}H]dihydromorphine binding. Although the results indicate the existence of subclasses of receptors, their significance is not understood. Questions remain as to whether there are multiple conformational forms of the receptor or multiple points of attachment for endorphins and enkephalins, in comparison to the alkaloids.

The interaction of Ca with the opiate receptor is not well understood. Tissue Ca in the rat brain was decreased following an *in vivo* administration of a single dose of morphine (Cardenas and Ross, 1975). This effect of morphine was blocked by naloxone and exhibited a high degree of stereospecificity. Iwamoto *et al.* (1978) found that La, as well as morphine, produced an antinociceptive response when injected into the periaqueductal gray region of the rat brain. They also found that morphine or La analgesia was antagonized by a microinjection of Ca into the periaqueductal gray region.

Way *et al.* (1979) have proposed an encompassing Ca hypothesis concerning the action of opiates, postulating that a lowered level of brain Ca is associated with opiate analgesia while a higher level of brain Ca is associated with tolerance and dependence, reflecting a counteradaptive mechanism to the lowered level of acute morphinization.

Pert and Snyder (1973a,b) have demonstrated that at near physiological concentration, Ca inhibited the binding of naloxone to neural membranes. Recently our laboratory undertook a study to ascertain whether

Ca or another multivalent cation was essential for the ability of phosphatidylserine to enhance stereospecific opiate binding (Abood *et al.*, 1980a). Repeated washings with ethylene glycol bis(β-aminoethyl ether) *N,N'*-tetraacetic acid or EDTA did not alter the enhancement in stereospecific binding observed by the addition of phosphatidylserine. The addition of Ca to neural membranes did not affect the stereospecific binding of dihydromorphine.

The section will describe work that our laboratory and others have done on the solubilization of the opiate receptor, the first step in the purification process.

C. Solubilization of the Opiate Receptor

Simon and colleagues (1975a) first reported the solubilization of a stereospecific opiate–macromolecular complex from rat brain. They incubated a synaptosomal fraction from rat brain with [^{3}H]etorphine in the presence of levorphanol or dextrorphan, the inactive enantiomer. This complex was then solubilized with the nonionic detergent Brij 36 T. Since [^{3}H]etorphine binds tightly to the receptor complex, this complex could be characterized. Proteolytic enzymes, *N*-ethylmaleimide, and heating at 50°C all resulted in a loss of the bound radioactivity. By molecular sieve chromatography, the molecular weight of the complex was estimated to be 370,000. The major problem with this preparation was that once solubilized, the opiate receptor could not bind opiates.

Zukin and Kream (1979) covalently bound radioactive Met-enk to the opiate–macromolecular complex by crosslinking the solubilized noncovalent complexes from rat brain. They incubated a synaptosomal preparation with either ^{3}H- or ^{125}I-labeled Met-enk and then solubilized this complex by the same procedure used by Simon *et al.* (1975a). The complex was covalently crosslinked by the addition of dimethyl suberimidate. By molecular sieve chromatography, the covalently bound enkephalin complex appeared to have a molecular weight of 380,000, which closely parallels what Simon *et al.* (1975a) found for the etorphine–macromolecular complex. However, when the ^{125}I-enkephalin complex was electrophoresed on sodium dodecyl sulfate (SDS)–polyacrylamide gels, a major radioactive peak corresponding to a molecular weight of 35,000, along with a number of minor components, were deleted.

Recently our laboratory reported on a solubilized rat neural membrane preparation exhibiting stereospecific opiate binding (Bidlack and Abood, 1980). Neural membranes were treated with 1% Triton X-100, followed by centrifugation at 100,000*g* for 30 min. The supernatant was

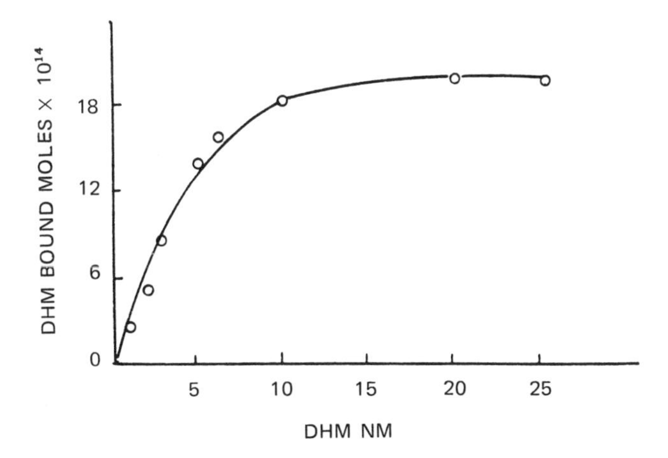

Fig. 1. Saturation curve for the stereospecific binding of [^{3}H]dihydromorphine from equilibrium dialysis.

immediately added to Bio-Beads SM-2, which bind Triton X-100 as described by Holloway (1973). The resulting supernatant was concentrated on Amicon PM-10 membranes. Figure 1 shows the stereospecific binding of [^{3}H]dihydromorphine to the solubilized receptor. The binding is saturable at 1×10^{-8} M, with half-saturation occurring at 3.5×10^{-9} M. Other opiates and naloxone compete for dihydromorphine binding sites in the Triton X-100-solubilized supernatant. The results were generally similar to those reported for neural membranes, with the exception that naloxone was relatively more potent in the solubilized fraction than in neural membranes.

Like the membrane-bound receptor, the solubilized receptor appears to consist of a protein and lipid moiety. Trypsin, chymotrypsin, and phospholipase A_2 destroy stereospecific opiate binding, while the addition of phosphatidylserine to the solubilized preparation enhances binding by 20%. Once stereospecific binding had been destroyed by phospholipase A_2, the addition of phosphatidylserine could not restore binding. Phosphatidylserine decarboxylase inhibited binding almost completely, despite the fact that only 10% of the phosphatidylserine present was converted to phosphatidylethanolamine. Whether phosphatidylserine associated with the receptor was preferentially attacked is not known. The addition of phosphatidylethanolamine to the solubilized receptor did result in a 60% inhibition of stereospecific etorphine binding.

Ruegg *et al.* (1980) have reported on the solubilization of an active opiate receptor from *Bufo marinus*. Treating neural membranes from toad

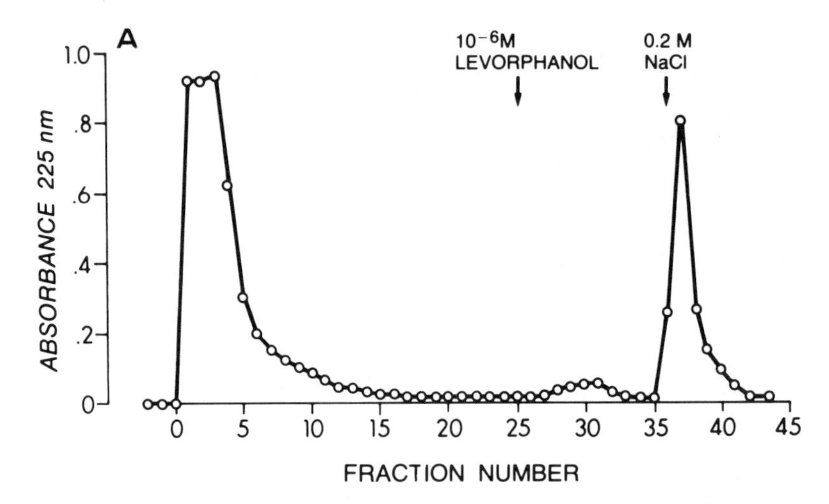

Fig. 2. (A) Elution profile from affinity fraction column. Protein was applied to the column at the onet of fraction 1. The column was eluted with 50 mM Tris, pH 7.5. Buffer was switched where indicated. (B) SDS–polyacrylamide gels of the column fractions from A. Gel A is the solubilized supernatant that was applied to the column. Gel B shows fractions 1–3, the protein that passes rapidly through the column. Gel C shows fractions 20–24. Gel D shows fractions 35–40. Gel F shows another preparation of the opiate receptor, eluted from the affinity column with 10^{-6} M levorphanol.

brain with 1% digitonin resulted in the solubilization of an active opiate receptor. [^{3}H]Diprenorphine bound stereospecifically with a K_d of 4 nM. This solubilized receptor was also inactivated by treatment with trypsin or heat.

Simonds *et al.* (1980) have solubilized the opiate receptor from rat neural membranes using the zwiterionic detergent, 3-[(3-cholamidopropyl) dimethylammonio]-1-propanesulfonate (CHAPS). Optimal solubilization occurred at 10 mM CHAPS, and it was necessary for CHAPS to be present during the binding assays. The binding of [^{3}H]Met-enk to the solubilized receptor showed a K_d of 2.3 nM. Other opiates could displace the binding of [^{3}H]Met-enk. By gel filtration, Klee and colleagues estimated the molecular weight of the opiate receptor to be between 100,000 and 500,000.

D. Purification of the Opiate Receptor

Recently, purification of the opiate receptor has been achieved by affinity chromatography (Bidlack *et al.*, 1980).The affinity gel was prepared by coupling 14-β-bromoacetamidomorphine, a newly synthesized

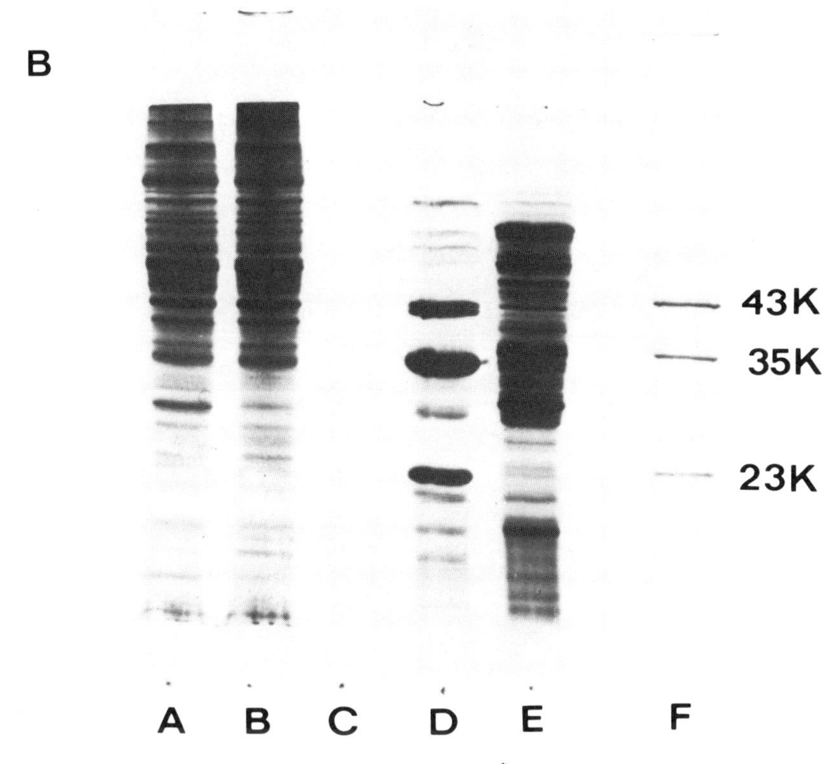

Fig. 2. (*Continued*)

ligand, to ω-aminohexyl-Sepharose. The opiate receptor was eluted from the column with either 10^{-6} M levorphanol or 10^{-6} M etorphine, as shown in Fig. 2A; SDS–polyacrylamide electrophoresis revealed three major proteins associated with the opiate receptor, having molecular weights of 43,000, 35,000, and 23,000 (Fig. 2B). The purified receptor binds 4×10^{-11} mole of dihydromorphine/mg protein with a K_d of 3.8×10^{-9} M. The binding was 100% stereospecific; that is, 10^{-6} M levorphanol completely inhibited all [^{3}H]dihydromorphine binding.

Table II shows the ability of other opiates, naloxone and Met-enk, to inhibit [^{3}H]dihydromorphine binding. As with neural membranes and the solubilized receptors, etorphine was the most potent opiate in inhibiting dihydromorphine binding. Naloxone and Met-enk had IC_{50} values in the low nanomolar range. Levorphanol had only 10% the potency of these latter two compounds. It appears that the purified receptor exhibits similar binding characteristics to the membrane with respect to various opiate alkaloids and peptides as well as in kinetic parameters.

Table II
Relative Potencies of Drugs in Reducing [^{3}H]-Dihydromorphine Binding to the Purified Receptor[a]

Drug	IC_{50} (M)
Etorphine	5×10^{-10}
Naloxone	2×10^{-9}
Met-Enkephalin	3×10^{-9}
Levorphanol	3×10^{-8}
Dextrorphan	1×10^{-4}

[a] The concentration of [^{3}H]dihydromorphine was 4×10^{-9}M. The IC_{50} values were determined from log–probit plots of the data.

E. Opiate Receptors in Blood Cells

1. Erythrocytes

In addition to brain tissue, a stereospecific opiate receptor was identified in erythrocyte membranes (Abood *et al.*, 1976). These membranes bound dihydromorphine in a stereospecific manner, with a K_d of 9 x 10^{-9} M. The binding characteristics of erythrocyte membranes closely resembled those of synaptic membranes. The pharmacological potency of opiates correlated well with that observed for synaptic membranes. Both Na and Ca inhibited stereospecific binding. Inhibition was also observed with phospholipases A_2 and C, as well as trypsin. A comparison of stereospecific opiate binding in controlled human subjects and heroin addicts revealed a 43% increase in the addict group.

2. Leukocytes

Lymphocytes from street opiate addicts have been shown to have a severe reduction in their ability to respond to mitogenic stimulation by phytohemagglutinin, pokeweed mitogen, or concanavalin A (Brown *et al.*, 1974). Wybran *et al.* (1979) have shown that morphine reduced the frequency of rapid-rosetting human T lymphocytes *in vitro*. This effect was reversible by naloxone, suggesting that T lymphocytes may have opiate receptor sites analagous to those observed in neural membranes. McDonough *et al.* (1980) have recently reported that there is a significant decrease in total T lymphocytes *in vivo* in street opiate addicts and a concomitant increase in *null* lymphocytes. Both of these effects *in vitro* were reversed with naloxone.

Recently, it has been shown that leukocytes do in fact bind dihydromorphine in a stereospecific manner (Lopker *et al.*, 1980). Both granulocytes and monocytes exhibited stereospecific opiate binding with an apparent K_d of 10 nM for granulocytes and 8 nM for monocytes. The number of binding sites per cell was 3000 and 4000, respectively.

F. Opioid Receptor in Pathological States

A number of attempts have been made to determine possible alterations in the opiate receptor during addiction and other pathological conditions involving endorphinergic neurons. It has not been possible to detect any change in the number or affinity of opiate receptors in homogenates of whole brain (Klee and Streaty, 1974); however, Davis *et al.* (1975) reported that chronic morphine treatment decreases the affinity for opiates in rat brain slices. There is some suggestion that the number of opiate binding sites is reduced in the brains of mice reared in isolation (Bonnet *et al.*, 1976), while the naloxone-precipitated abstinence syndrome is altered in morphine-dependent rats which are differentially housed (Adler *et al.*, 1975). As described above, there is an indication that the number of opiate binding sites is increased in the erythrocytes of heroin addicts; however, since the funtional role of the opiate receptors on blood cells is unknown, the significance of such a finding cannot be assessed. It is apparent from the few positive findings that changes in membrane recepotrs can occur under altered physiological conditions. The endorphinergic–enkephalinergic system is involved in a variety of physiological roles, such as stress, visceral and affective functions, and sensory activity, any one of which can contribute to changes in membrane receptors.

III. THE BRAIN NICOTINE RECEPTOR

A. Introduction

Although the action of acetylcholine (ACh) in the parasympathetic nervous system can be explained on the basis of its effects at either muscarinic or nicotinic synapses, its mode of action in the mammalian central nervous system is considerably more complex. Within the autonomic ganglia both nicotinic anc cholinergic synapses are interrelatedly involved in the action of ACh. A good example is the superior cervical ganglion where a nicotinic component (responsible for a fast excitatory postsynaptic potential) and a muscarinic one (resulting in a slow postsynaptic potential) regulate the secretion of noradrenalin from a ganglion cell (Libet, 1979). With respect to the brain itself, however, the preponderance of cholinergic synapses appear to be muscariniclike in nature and not interconnected with nicotinic components. Although there is evidence for the existence of nicotinic cholinergic synapses in the spinal cord (e.g., Renshaw cells) and brain (e.g., midbrain reticular formation, septohippocampal pathways), a number of the behavioral effects of the drug nicotine may not be mediated via nicotinic cholinergic receptors.

In an effort to study the characteristics and distribution of nicotinic receptors in brain, a number of radioactive ligands have been employed. These include [^{3}H]nicotine, [^{3}H]tubocurarine, [^{3}H]hexamethonium, and ^{125}I- or ^{3}H-labeled α-bungarotoxin (α-BT). The earliest report of nicotine binding to invertebrate and vertebrate neural tissue involving radiolabeled nicotine dates back to 1967 (Schmiterlaw *et al.*, 1967). Since that time, binding of radiolabeled nicotine has been demonstrated in membrane preparations of the electric organ of eels (Eldefrawi *et al.*, 1971), lobster axons (Denburg *et al.*, 1972; Marquis *et al.*, 1977), and mouse brain (Schleifer and Eldefrawi, 1974). With the technique of equilibrium dialysis the K_d for membranes from mouse brain was found to be 7×10^{-9} M (Schliefer and Eldefrawi, 1974). By means of centrifuge assay and with the use of [^{3}H]nicotine of considerably higher radioactive specific activity, a value of 6×10^{-9} M was obtained for rat brain membranes (Abood *et al.*, 1980c). In the earlier of the two studies, by measuring the competition of various cholinergic ligands with nicotine it was inferred that the nicotine binding sites of mouse brain were essentially nicotinic cholinergic in nature, resembling those of electroplax (Schliefer and Eldefrawi, 1974). A subsequent study by the same investigators of housefly brain suggested that the ACh receptors were somewhat more complex, revealing a competition between nicotine and atropine as well as affinity for such unrelated agents as amphetamine, tyramine, and bretylium (Eldefrawi *et al.*, 1971; Eldefrawi and O'Brien, 1970). The affinity of the housefly brain for both nicotine and ACh was surprisingly low, $K_i = 10^{-5}$ M for [^{3}H]nicotine blockade (Mansour *et al.*, 1977).The receptor from housefly brain, which was partially purified by affinity chromatography, exhibited no binding for α-BT and did not possess any antigenic similarity to the ACh receptors of the *Torpedo* electric organ (Mansour *et al.*, 1977).

A number of studies have described the use of α-BT to characterize the nicotinic cholinergic receptor in mammalian brain. An α-BT-binding component has been purified from rat brain with a $K_d = 5.6 \times 10^{-11}$ M which exhibited an affinity for nicotinic cholinergic drugs (Lowy *et al.*, 1976; Schmidt, 1977). It has been argued that the α-BT binding sites in brain are indicative of nicotinic cholinergic sites, a conclusion based in part on autoradiographic studies demonstrating the presence of α-BT in discrete nicotinic cholinergic pathways, such as the hippocampus (Hunt and Schmidt, 1978). The significance of the data on α-BT binding to vertebrate nervous tissue is difficult to interpret, since α-BT fails to block neurotransmission in a variety of nicotinic cholinergic systems such as Renshaw cells, amphibian ventral–dorsal pathway, and autonomic ganglia (Duggan *et al.*, 1976; Miledi and Szczepaniak, 1975). Another difficulty is that α-BT does not compete with [^{3}H]nicotine binding to rat brain membranes or prevent the prostration syndrome resulting from nicotine administered intraventricularly (i.v.c.). The question arises as to whether mammalian brain contains both nicotinic cholinergic and noncholinergic receptors, both exhibiting differential affinities for nicotine and α-BT.

Recently, with (−)-[^{3}H]nicotine of a high radioactive specific activity, stereospecific, saturable binding of nicotine to rat brain membranes was demonstrated, having a K_d of 6×10^{-9} M and a binding capacity of 2×10^{-14} mole/mg membrane protein (Abood *et al.*, 1980c). The binding was destroyed by exposure to a boiling water bath or organic solvents and trichloracetic acid. Exposure to phospholipase A_2, DNAase, or RNAase was without effect while trypsin and chymotrypsin resulted in about 50% loss of binding. It appears, therefore, that the nicotine binding site is proteinaceous in nature and does not become involved with either nucleic acids or phospholipids. With the use of Triton X-100 an enriched solubilized fraction was obtained with a K_d of 1×10^{-8} and a binding capacity of 3×10^{-13} mole/mg protein.

B. [^{3}H]Nicotine Binding to Rat Brain and *Torpedo* Membranes

A variety of cholinergic and noncholinergic drugs were tested for their ability to compete with (−)-nicotine for *Torpedo* and rat brain membranes as well as the solubilized preparation from brain (Table III). The binding data are expressed as apparent K_i. The binding of nicotine to both *Torpedo* and brain membranes is tereospecific; the ratio of (−)- to (+)-nicotine being about 10 for *Torpedo* and 5 for brain membranes. In producing the prostration syndrome, the (−)-isomer was 45 times more effective than the (+)-isomer. Both benzylnornicotine and benzylpiperidine, which are specific antagonists to the nicotine-induced prostration syndrome (Abood *et al.*, 1980c), exhibited a high affinity to both types

Table III
Competition of Various Drugs for [^{3}H]Nicotine Binding to Intact and Solubilized Membrane Preparations of Rat Brain and Comparison with Pharmocologic Potency[a]

	Torpedo membranes K_i (M)	Rat brain K_i (M)		Prostration ED_{50} (nmole)	
		Membranes	Solubilized membranes	Agonist	Antagonist
(−)-Nicotine	3.1×10^{-7}	6.0×10^{-9}	1×10^{-8}	20	—
(+)-Nicotine	3.4×10^{-6}	3.0×10^{-8}	8×10^{-7}	900	—
(+, −)-Nornicotine	3.2×10^{-7}	2.5×10^{-8}	2×10^{-8}	800	—
+, −)-*N*-Benzylnornicotine	4.0×10^{-7}	6.0×10^{-8}	1×10^{-8}	—	35
N-Benzylpiperidine	5.2×10^{-7}	9.0×10^{-8}	1×10^{-8}	—	50
Mecamylamine	1.5×10^{-6}	5.0×10^{-6}	2×10^{-7}	—	30[b]
Hexamethonium	5.0×10^{-6}	5.0×10^{-6}	2×10^{-7}	—	25
Carbamylcholine	3.0×10^{-7}	1.0×10^{-6}	1×10^{-7}	1000	—
Oxotremorine	2.6×10^{-4}	5.0×10^{-6}	2×10^{-6}	1000	—
(−)-Anabasine	1.3×10^{-7}	5×10^{-6}	2×10^{-6}	500	—
(−)-Cotinine	5.2×10^{-4}	1.0×10^{-6}	2×10^{-6}	1000	—
(QNB)	5.0×10^{-4}	5.0×10^{-6}	—	—	100
α-BT	7.8×10^{-6}	1.0×10^{-6}	—	[c]	

[a] The binding data are given as apparent K_i. The prostration data are expressed as ED_{50}, i.e., the dose required to produce (or antagonize) prostration in all four limbs following intraventricular injection into rats. The dose is expressed as nanomoles in a 5-μl volume. The concentration of [^{3}H]nicotine for the *Torpedo* membranes was 3×10^{-7} and for the brain preparations 2×10^{-8}M. In a typical binding experiment, the difference (in cpm) of [^{3}H]nicotine with and without a 100-fold excess of (−)-nicotine was 6800/mg protein for *Torpedo* membranes. With both preparations, specific binding represented about 10% of the total binding.

[b] Blockade was not complete, i.e., muscle weakness persisted in most animals.

[c] No behavioral effects or antagonism to nicotine was noted at 5 nmol of α-BT.

of membranes and effectively blocked the prostration response from 20 nmole (−)-nicotine i.v.c. Although blockade of prostration occurred with mecamylamine and hexamethonium, the K_i of both substances was three orders of magnitude less than that of the K_d for (−)-nicotine. Both drugs are nicotinic cholinergic antagonists at autonomic ganglia and the neuromuscular junction. Carbamylcholine, a nicotinic cholinergic agonist, and oxotremorine, a muscarinic agonist, exhibited a very low binding affinity to brain membranes and did not produce the prostration syndrome even at very high doses. By contrast, the K_i for carbamylcholine with *Torpedo* membranes was identical to K_d of (−)-nicotine. Anabasine, a close analog of nicotine, also exhibited a much lower affinity to rat brain membranes, while producing prostration at 25 times the dose of (−)-nicotine. Cotinine, the major metabolite of nicotine exhibited only slight binding in all preparations and failed to produce prostration at 50 times the dose of (−)-nicotine. The potent centrally active antimuscarinic drug, 3-quinuclidinylbenzilate (QNB), exhibited low affinity for the nicotine binding site in brain and had no antagonistic action on nicotine-induced prostration. With α-BT, a potent neuromuscular blocking agent, no binding to rat brain was observed at 1×10^{-6} M, while at 5 nmole no antagonism to the prostration response was noted.

Recently, Romano and Goldstein (1980) demonstrated the existence of stereospecific binding sites for nicotine in rat brain membrane using the technique of filtration through glass fiber filters. They were able to overcome nonspecific [^{3}H]nicotine binding to the filters by pretreatment with polyamine. Their K_d for nicotine binding was 28 nM which was almost an order of magnitude greater than that reported by Abood *et al.* (1980c). Although the results with different nicotinic and other drugs were qualitatively similar to those obtained by the centrifuge assay, the IC_{50} values for many agents are at least an order of magnitude greater. Despite their findings that traditional nicotinic cholinergic antagonists exhibited low affinity for the sites, they conclude that the sites are similar to the ganglionic nicotinic receptor.

When at least 20 μmole of (−)-nicotine are administered into the lateral ventricles of rats, an immediate prostration results in all four limbs, usually of 1–2 min duration (Abood *et al.*, 1979). By administering various nicotinic cholinergic as well as an extensive variety of neurotransmitters and psychotropic drugs, it was inferred that the prostration syndrome was not mediated via nicotinic cholinergic synapses. By determining the ability of a number of nicotinic analog and other agents to mimic or prevent the nicotine-induced prostration, a behavioral screening procedure was available that also permitted a comparison of receptor binding ability with the psychopharmacologic efficacy of the agents. A comparison between the binding of drugs to membranes from the *Torpedo* electric organ, which contains a high density of nicotinic cholinergic receptors, and to rat brain membranes was expected to help resolve the question of the existence of a population of noncholinergic receptors in mammalian brain.

On the basis of the comparison of the binding characteristics of various drugs with their ability to mimic or antagonize the behavioral effects of nicotine, it was concluded that the brain sites and mode of action of nicotine cannot be explained merely on the basis of a cholinergic mechanism. With respect to the characteristic prostration syndrome resulting from administering nicotine into the lateral ventricles of the rat brain, the response cannot be mimicked by nicotinic cholinergic agonists or—with the exception of hexamethonium and mecamylamine—blocked by traditional nicotinic antagonists. Furthermore, a number of analogs of nicotine which are effective antagonists to the nicotine-induced prostration syndrome do not appear to compete with cholinergic ligands for the brain cholinergic receptor (Abood *et al.*, 1978b, 1979). Finally, a reasonably good correlation was noted in the ability of a variety of agents with their ability to mimic or antagonize nicotine-induced prostration and [^{3}H]nicotine binding to rat brain but not *Torpedo* membranes (Abood *et al.*, 1980c). At present, no definitive explanation can be offered for the fact that hexamethonium and mecamylamine—effective blockers at nicotinic cholinergic synapses—are able to antagonize the nicotine-induced prostration. There is also the additional discrepancy that hexamethonium and mecamylamine have a considerably lower affinity for the nicotine binding sites in brain membranes than (−)-nicotine. Mecamylamine bears some structural resemblance to nicotine; while hexamethonium, a bis-quaternary with two cationic charges, may assume the appropriate configuration and dimensions to occupy the nicotine receptor in brain. On the other hand, a variety of other nicotinic cholinergic blocks, such as decamethonium, *d*-tubocurarine, and α-BT are ineffective against the nicotine-induced prostration and have a relatively low affinity for the nicotine binding site. The problem regarding the nature of the nicotine receptor in brain and its correlation with the central action of nicotine and related pharmacologic substances is complex.

C. Altered Membranes from Nicotine Use

Although a considerable amount of literature exists on the alterations in erythrocyte and white cell membranes and function resulting from chronic tobacco use (see Larson and Silvette, 1975, for extensive review), few definitive studies have been performed on nicotine receptors. There is a suggestion that the degree of suppression of immunoglobulin M and the immunoglobulin G antibody response of sheep erythrocytes was dependent upon the nicotine content of tobacco smoke (Roszman and Rogers, 1973). Nicotine at concentrations considerably greater than that present in tobacco smoke inhibit endotoxin-induced chemotaxis in leukocytes, the chemotaxis being presumably related to complement fixation (Bridges *et al.*, 1977). Recently Hoss and Davies (unpublished) have observed high-affinity nicotine binding in human neutrophils and monocytes. Conceivably, such receptors could serve as a triggering mechanism for the many functional aspects of such cells.

IV. SUMMARY

The present review has dealt with some of the molecular characteristics of the receptors associated with the action of opiates and nicotine. The binding characteristics of the opiate receptor in brain membrane preparations has been compared with those of a Triton X-100-solubilized extract of membranes as well as a preparation purified from rat brain by affinity chromatography. The purified opiate receptor appears to consist of three proteins; and the major component, which has a molecular weight of 35,000, may be the site of attachment of the opiate. Although phosphatidylserine appears to be associated with the opiate receptor within the membrane, its exact role in receptor binding is not known. The brain receptor for nicotine, which has a K_d in the nanomolar range, appears to have binding characteristics which differ from that of the classical nicotinic cholinergic receptor. This conclusion is corroborated by a characteristic behavioral response elicited by nicotine and not by cholinergic agonists. Finally, a discussion is presented on the presence of opiate receptors in human erythrocytes and leukocytes and the changes occurring in opiate addicts.

V. REFERENCES

Abood, L. G., Atkinson, H. G., and MacNeil, M., 1976, Stereospecific opiate binding in human erythrocyte membranes and changes in heroin addicts, *J. Neurosci. Res.* **2**:427.

Abood, L. G., Salem, N., MacNeil, M., Bloom, L., and Abood, M. E., 1977, Enhancement of opiate binding by various molecular forms of phosphatidylserine and inhibition by other unsaturated lipids, *Biochim. Biophys. Acta* **468**:51.

Abood, L. G., Lowy, K., Tometsko, A., and Booth, H., 1978a, Electrophysiological, behavioral, and chemical evidence for a noncholinergic, stereospecific site for nicotine in rat brain, *J. Neurosci. Res.* **3**:327.

Abood, L. G., Salem, N., MacNeil, M., and Butler, M., 1978b, Phospholipid changes in synaptic membranes by lipolytic enzymes and subsequent restoration of opiate binding with phosphatidylserine, *Biochim. Biophys. Acta* **530**:35.

Abood, L. G., Lowy, K., Tometsko, A., and MacNeil, M., 1979, Evidence for a noncholinergic site for nicotine's action in brain: psychopharmaoclogical, electrophysiological and receptor binding studies, *Arch. Int. Pharmacodyn. Ther.* **237**:212.

Abood, L. G., Butler, M., and Reynolds, D., 1980a, Effect of calcium and physical state of neural membranes on phosphatidylserine requirement for opiate binding, *Mol. Pharmacol.* **17**:290.

Abbood, L. G., Knapp, R. J., and Reynolds, D. T., 1980b, Inhibition of opiate binding to brain membranes by enzymatically inactivated peptidases and tryptic fragments from autolysis, *Substance Alc. Actions/Misuse* **1**:53.

Abood, L. G., Reynolds, D. T., and Bidlack, J. M., 1980c, Stereospecific ^{3}H- nicotine binding to intact and solubilized rat brain membranes and evidence for its noncholinergic nature, *Life Sci.* **27**:1307.

Adler, M. W., Bendotti, C., Ghezzi, D., Samanin, R., and Valzelli, L., 1975, Dependence to morphine in differentially housed rats, *Psychopharmacologia* **41**:15.

Bidlack, J. M., and Abood, L. G., 1980, Solubilization of the opiate receptor *Life Sci.* **27**:331.

Bidlack, J. M., Abood, L. G., Osei-Gyimah, P., and Archer, S., 1980, Purification of the opiate receptor from rat brain, *Proc. Natl. Acad. Sci. USA* **78**:636.

Bridges, P. B., Kraal, J. H., Huang, L. J. T., and Chancellor, M. B., 1977, Effects of cigarette smoke components on in vitro chemotaxis of human polymorphonuclear leukocytes, Infection and Immunity **16**:240.

Bonnet, K. A., Hiller, J. M., and Simon, E. J., 1976, The effects of chronic opiate treatment and social isolation on opiate receptors in the rodent brain, in: "Opiates and Endogenous Opioid Peptides" (H. W. Kosterlitz, ed.), pp. 335, Amsterdam, North Holland.

Brown, S. M., Stimmel, B., Taul, R. N., Kochwa, S., and Rosenfield, R. E., 1974, Immunological dysfunction in heroin addicts, *Arch. Intern. Med.* **134**:1001.

Cardenas, H. L., and Ross, D. H., 1975, Morphine induced calcium depletion in discrete regions of rat brain, *J. Neurochem.* **24**:487.

Craves, F. B., Zalc, B., Leybin, L., Baumann, N., and Loh, H. H., 1980, Antibodies to cerbroside sulfate inhibit the effects of morphine and β-endorphin, *Science* **207**:75.

Davies, B. D., Hoss, W., Lin, J. P., and Lionetti, F., 1982, Evidence for a noncholinergic nicotine receptor on human PhaBgocytic leukocytes, *Molecular and Cellular Biochemistry*. **44**:23.

Davis, M. E., Akera, T., and Brody, M., 1975, Saturable binding of morphine to rat brainstem slices and the effect of chronic treatment, *Res. Commun. Chem. Pathol. Pharm.* **12**:409.

Denburg, J. L., Eldefrawi, M. E., and O'Brien, R. D., 1972, Macromolecules from lobster axon membranes that bind cholinergic ligands and local anesthetics, *Proc. Natl. Acad. Sci. USA* **69**:177.

Domino, E. F., 1973, Neuropsychopharmacology of nicotine and tobacco smoking, *in* "Smoking Behavior, Motives, and Incentives" (W. L. Dunn, ed.), pp. 5–31, Wiley, New York.

Duggan, A. W., Hall, J. G., and Lee, C. Y., 1976, Alpha-bungarotoxin, cobra neurotoxin and excitation of Renshaw cells by acetylcholine, *Brain Res.* **107**:166.

Dunlap, C. E., Leslie, F. M., Rado, M., and Cox, B. M., 1979, Ascorbate destruction of opiate stereospecific binding in guinea pig brain homogenate, *Mol. Pharmacol.* **16**:271.

Eldefrawi, A. T., and O'Brien, R. D., 1970, Binding of muscarone by extracts of housefly brain: Relationship to receptors for acetylocholine, *J. Neurochem.* **17:**1287.

Eldefrawi, M. E., Eldefrawi, A. T., and O'Brien, R. D., 1971, Binding of five cholinergic ligands to housefly brain and *Torpedo* electroplax, *Mol. Pharmacol.* **7:**104.

Faden, A. I., and Holaday, J. W., 1979, Opiate antagonists: A role in the treatment of hypovalemic shock, *Science* **205:**317.

Goldstein, A., and Cox, B. M., 1978, Opiate receptors and their endogenous ligands (endorphins), *Progr. Mol. Subcell. Biol.* **6:**113.

Goldstein, A., Lowney, L. I., and Pal, B. K., 1971, Stereospecific and nonspecific interactions of the morphine congener levorphanol in subcellular fractions of mouse brain, *Proc. Natl. Acad. Sci. USA* **68:**1742.

Hiller, J. M., Pearson, J., and Simon, E. J., 1973, Distribution of stereospecific binding of the potent narcotic analgesic etorphine in the human brain: Predominance in the limbic system, *Res. Commun. Chem. Pathol. Pharmacol.* **6:**1052.

Holloway, P. W., 1973, A simple procedure for removal of Triton X-100 from protein samples, *Anal. Biochem.* **53:**304.

Hughes, J., 1975a, Isolation of an endogenous compound from the brain with pharmacological properties similar to morphine, *Brain Res.* **88:**295.

Hughes, J., 1975b, Search for the endogenous ligand of the opiate receptor, *in* "Opiate Receptor Mechanisms" (S. H. Snyder and S. Matthysse, eds.),pp. 55–58, Neurosciences Research Program Bulletin, Boston.

Hunt, S. P., and Schmidt, J., 1978, Some observations on the binding patterns of α-bungarotoxin in the central nervous system of the rat, *Brain Res.* **157:**213.

Iwamoto, E. T., Harris, R. A., Loh, H. H., and Way, E. L., 1978, Antinociceptive responses after microinjection of morphine or lanthanum in discrete rat brain sites, *J. Pharmacol. Exp. Ther.* **206:**46.

Kawamura, H., and Domino, E. F., 1969, Differential effects of m and n cholinergic agonists on the brainstem activating system, *Int. J. Neuropharmacol.* **8:**105.

Klee, W. A., 1979, Drug actions mediated by changes in adenylate cyclase activity, *in* "Membrane Mechanisms of Drug Abuse" (C. W. Sharp and L. G. Abood, eds.), pp. 219–226, Liss, New York.

Klee, W. A., and Streaty, R. A., 1974, Narcotic receptor sites in morphine-dependent rats, *Nature (London)* **288:**61.

Law, P., Harris, R. A., Loh, H. H., and Way, E. L., 1978, Evidence for the involvement of cerebroside sulfate in opiate receptor binding: Studies with Azure A jimpy mutant mice, *J. Pharmacol. Exp. Ther.* **207:**458.

Libet, B., 1979, Which postsynaptic action of dopamine is mediated by cyclic AMP, *Life Sci.* **24:**1043.

Lin, H. K., and Simon, E. J., 1978, Phospholipase A inhibition of opiate binding can be reversed by albumin, *Nature (London)* **271:**383.

Loh, H. H., Cho, T. M., Wu, Y. C., and Way, E. L., 1974, Stereospecific binding of narcotics to brain cerebrosides, *Life Sci.* **14:**2231.

Lopker, A., Abood, L. G., Hoss, W., and Lionetti, F., 1980, Stereoselective muscarinic acetylcholine and opiate receptors in human phagocytic leukocytes, *Biochem. Pharmacol.* **29:**1361.

Lord, J. A. H., Waterfield, A. A., Hughes, J., and Kosterlitz, H. W., 1977, Endogenous opioid peptides: Multiple agonists and receptors, *Nature (London)* **267:**495.

Lowy, J., MacGregor, J., Rosenstone, J., and Schmidt, J., 1976, Solubilization of an α-bungarotoxin binding component from rat brain, *Biochemistry* **15:**1522.

Mansour, N. A., Eldefrawi, M. E., and Eldefrawi, A. T., 1977, Isolation of putative acetylocholine receptor proteins from housefly brain, *Biochemistry* **16:**4126.

Marquis, J. K., Hilt, D. C., Papadeas, V. A., and Mautner, H. G., 1977, Interaction of cholinergic ligands and local anesthetics with plasma membranes from lobster axons, *Proc. Natl. Acad. Sci. USA* **74:**2278.

Martin, W. R., Eases, C. G., Thompson, J. A., Huppler, R. E., and Gilbert, P. E., 1976, The effects of morphine- and nalorphine-like drugs in the nondependent and morphine dependent chronic spinal dog, *J. Pharmacol. Exp. Ther.* **197:**517.

McDonough, R. J., Madden, J. J., Falek, A., Shafer, D. A., Pline, M., Gordon, D., Bobos, P., Kuenle, J. C., and Mendelson, J., 1980, Alteration of T and null lymphocyte frequencies in the peripheral blood of human opiate addicts: *In vivo* evidence for opiate receptor sites on T lymphocytes *J. Immunol.* **125:**2539.

McDonough, R. J., Madden, J. J. and Falek, A., 1980, Alteration of human lymphocyte-T function by opiates is reversible by naloxone, *Fed. Proc.* **39:**1754.

Miledi. R., and Szczepaniak, A. C., 1975, Effect of *Dendroaspis* neurotoxins on synaptic transmission in the spinal cord of the frog, *Proc. R. Soc. London Ser. B* **190:**267.

Pasternak, G. W., and Snyder, S. H., 1974, Identification of novel high affinity opiate receptor binding in rat brain, *Nature (London)* **253:**563.

Pasternak, G. W., Wilson, H. A., and Snyder, S. H., 1975, Differential effects of protein-modifying reagents on receptor binding of opiate agonists and antagonists, *Mol. Pharmacol.* **11:**340.

Pert, C. B., and Snyder, S. H., 1973a, Opiate receptor: Demonstration in nervous tissue, *Science* **179:**1011.

Pert, C. B., and Snyder, S. H., 1973b, Properties of opiate-receptor binding in rat brain, *Proc. Natl. Acad. Sci. USA* **70:**2243.

Pert, C. B., and Snyder, S. H., 1974, Opiate receptor binding of agonists and antagonists affected differentially by sodium, *Mol. Pharmacol.* **10:**868.

Pert, C. B., Pasternak, G. W., and Snyder, S. H., 1973, Opiate agonists and antagonists discriminated by receptor binding in brain, *Science* **182:**1359.

Pert, C. B., Kuhar, M. J., and Snyder, S. H., 1976, Opiate receptor: Autoradiographic localization in rat brain, *Proc. Natl. Acad. Sci. USA* **73:**3729.

Romano, C., and Goldstein, A., 1980, Stereospecific nicotine receptors in rat brain membranes, *Science* **210:**647.

Roszman, T. L., and Rogers, A. S., 1973, The immunosuppressive potential of products derived from cigarette smoke, *Am. Rev. Resp. Dis.* **108:**1158.

Ruegg, U. T., Hiller, J. M., and Simon, E. J., 1980, Solubilization of an active opiate receptor from *Bufo marinus*, *Eur. J. Pharmacol.* **64:**367.

Schleifer, L. I., and Eldefrawi, M. E., 1974, Identification of the nicotinic and muscarinic acetylcholine receptors in subcellular fractions of mouse brain, *Neuropharmacology* **13:**53.

Schmidt, J., 1977, Drug binding properties of an α-bungarotoxin binding component from rat brain, *Mol. Pharmacol.* **13:**283.

Schmiterlaw, C. G., Hansson, E., Andersson, G., Appelgren, L. E., and Hoffman, P. C., 1967, Distribution of nicotine in the central nervous system, *Ann. N. Y. Acad. Sci.* **142:**2.

Simantov, R., and Snyder, S. H., 1976, Brain-pituitary opiate mechanisms: Pituitary opiate receptor binding, radioimmunoassays for methionine enkephalin and leucine enkephalin and ^{3}H-enkephalin interactions with the opiate receptor, *in* "Endogenous Opioid Peptides" (H. W. Kosterlitz, ed.), pp. 41–48, North-Holland, Amsterdam.

Simantov, R., Childers, S. R., and Snyder, S. H., 1978, The opiate receptor binding interactions of ^{3}H-methionine enkephalin, an opioid peptide, *Eur. J. Pharmacol.* **47:**319.

Simon, E. J., and Groth, J., 1975, Kinetics of opiate receptor inactivation by sulfhydryl reagents: Evidence for conformational change in presence of sodium ions, *Proc. Natl. Acad. Sci. USA* **72:**2404.

Simon, E. J., Hiller, J. M., and Edelman, I., 1973, Stereospecific binding of the potent narcotic analgesic [^{3}H]etorphine to rat brain homogenate, *Proc. Natl. Acad. Sci. USA* **70:**1947.

Simon, E. J., Hiller, J. M., and Edelman, I., 1975a, Solubilization of a stereospecific opiate-macromolecular complex from rat brain, *Science* **190:**389.

Simon, E. J., Hiller, J. M., Groth, J., and Edelman, I., 1975b, Further properties of stereospecific opiate binding sites in rat brain: On the nature of the sodium effect, *J. Pharmacol. Exp. Ther.* **192:**531.

Simonds, W. F., Koski, G., Streaty, R. A., Hjelmeland, L. M., and Klee, W. A., 1980, Solubilization of active opiate receptors, *Proc. Natl. Acad. Sci. USA* **77:**4623.

Smith, A. P., and Loh, H. H., 1976, The sub-cellular localization of stereo-specific opiate binding in mouse brain, *Res. Commun. Chem. Pathol. Pharmocol.* **15:**205.

Snyder, S. H., and Simantov, R., 1977, The opiate receptor and opioid peptides, *J. Neurochem.* **28:**13.

Snyder, S. H., Childers, S. R., and Creese, I., 1979, Molecular actions of opiates: Historical overview and new findings on opiate receptor interactions with enkephalins and guanyl nucleotides, *in* "Advances in Biochemical Psychopharmacology," Vol. 20, "Neurochemical Mechanisms of Opiates and Endorphins" (H. H. Loh and D. H. Ross, eds.), pp. 543–552, Raven Press, New York.

Terenius, L., 1973a, Stereospecific interaction between narcotic analgesics and a synaptic plasma membrane fraction of rat cerebral cortex, *Acta Pharmacol. Toxicol.* **32:**317.

Terenius, L., 1973b, Characteristics of the "receptor" for narcotic analgesics in synaptic plasma membrane fraction from rat brain, *Acta Pharmacol. Toxicol.* **33:**377.

Terenius, L., 1978, Endogenous peptides and analgesia, *Annu. Rev. Pharmacol. Toxicol.* **18:**189.

Terenius, L., and Wahlstrom, A., 1974, Morphine-like ligand for opiate receptors in mammalian brain, *Acta Pharmacol. Toxicol.* **35:**55.

Way, E. L., 1979, Review and overview of four decades of opiate research, *in* "Neurochemical Mechanisms of Opiates and Endorphins" (H. H. Loh and D. H. Ross, eds.), pp. 4–27.

Wilson, H. A., Pasternak, G. W., and Snyder, S. H., 1975, Differentiation of opiate agonists and antagonist receptor binding by protein modifying regents, *Nature (London)* **253:**448.

Wybran, J., Appleboom, T., Famaey, J. P., and Govaerts, A., 1979, Suggestive evidence for receptors for morphine and methionine-enkephalin on normal human blood and lymphocytes, *J. Immunol.* **123:**1068.

Zukin, R. S., and Kream, R. M., 1979, Chemical crosslinking of a solubilized enkephalin macromolecular complex, *Proc. Natl. Acad. Sci. USA* **76:**1593.

Chapter 16

Investigation of Pathological Membranes with Nuclear Magnetic Resonance Spectroscopy

Charles Eric Brown

Department of Biochemistry
The Medical College of Wisconsin
Milwaukee, Wisconsin

I. INTRODUCTION

When a biological tissue is investigated with nuclear magnetic resonance (NMR) spectroscopy, the tissue can be considered to consist of two main fractions. One is the cytosol, and the other is the membrane and particulate components. The cytosol is, for the purpose of this discussion, an aqueous solution that yields relatively well-resolved NMR spectra of the low-molecular-weight components. The recent application of ^{31}P and ^{13}C NMR spectroscopy for detecting and measuring the concentrations of metabolites within the cytosol of whole, functioning organs, cells, and subcellular organelles has provided a new means to study the regulation of metabolism in a nondestructive manner (Burt *et al.*, 1977; Seeley *et al.*, 1977; Shulman *et al.*, 1979; Pettegrew *et al.*, 1979; Ross *et al.*, 1981). In addition, ^{31}P NMR spectroscopy has proven very useful in the detection and characterization of abnormal metabolites that occur in the cytosol in various disease states (Glonek, 1980) and shows promise as a tool for the diagnosis of heart damage (Hollis, 1979). In addition, the determination of water content and of spin–lattice relaxation times (see below) for water with ^{1}H NMR spectroscopy has made it possible to produce cross-sectional images of soft tissues in whole animals (Lauterbur, 1977; Brownell *et al.*, 1982) and may prove valuable for the detection of cancer (Hollis, 1979).

One can expect that NMR spectroscopy also will prove useful for the characterization of changes that occur in membranes during various pathological conditions. This technique is likely to provide information not available with other techniques and, with proper application, may provide a means of investigating components of membranes that are not readily isolated in a stable soluble form. The intent of this chapter is to describe the theoretical and technical considerations that underlie the investigation of biological membranes with NMR spectroscopy and to review the current uses of NMR spectroscopy for characterizing pathological membranes. These shall include descriptions of (i) the various physical parameters that differ between normal and pathological membranes and how these are measured with NMR spectroscopy and (ii) recent technical advances that should expand the capabilities of NMR spectroscopy to characterize pathological membranes.

II. AN INTRODUCTION TO NMR SPECTROSCOPY

I should preface this section by stating that this is not intended to be a complete description of NMR spectroscopy. The material presented is limited to ^{1}H, ^{13}C, and ^{31}P nuclei, which can be detected in biological tissues by NMR spectroscopy without specific isotopic enrichments. These nuclei have a spin quantum number of $\frac{1}{2}$ and thus exhibit only a single transition between two energy levels when the NMR spectrum is recorded. The use of deuterium magnetic resonance for studying lipid membranes was reviewed recently by Seelig (1977) and thus shall not be mentioned further here. The references have been chosen to permit the novice to NMR spectroscopy to learn more at a reasonable pace.

A. The Spectrometer

To begin with, not all elements or isotopes of a given element have atoms whose nuclei exhibit magnetic moments (i.e., whose nuclei point in one direction when placed in a magnetic field). Thus ^{12}C, ^{14}C, and ^{32}P are of no value for these experiments. On the other hand, ^{1}H, ^{13}C, and ^{31}P nuclei do exhibit magnetic moments, and each can be thought of as a small bar magnet with an axis that runs through the north and south poles. This axis is the magnetic moment. When these nuclei are placed in a strong, static external magnetic field, their magnetic moments align with the lines of flux of this magnetic field much like the pointer of a compass does with the earth's magnetic field. Since each nuclear magnetic moment rotates about its axis, each aligned nucleus precesses about the lines of flux of the external field much like a child's spinning top in the earth's gravitational field (Fig. 1A). The frequency of this precession is determined by the strength of the static external magnetic field (B_0), local variations in the value of B_0 (i.e., magnetic shielding) that are produced by the molecule whose spectrum is being recorded, and the identity of the nucleus as follows:

$$\nu_{obs} = \left(\frac{\gamma}{2\pi}\right)(1 - \sigma)B_0 \tag{1}$$

where ν_{obs} is the frequency of precession of the nuclear magnetic moment (usually called the Larmor frequency), σ is the magnetic shielding constant, and γ is the gyromagnetic (or magnetogyric) ratio. The gyromagnetic ratio is a constant that has a different value for each kind of nucleus. Thus, for example, in a 3.5-tesla magnetic field ^1H nuclei resonate at 150.057 MHz, ^{13}C nuclei resonate at 37.735 MHz and ^{31}P nuclei resonate at 60.744 MHz. Spectrometers with different strength magnets must operate at propertionately different frequencies in accordance with Eq. (1). Since the various chemical functional groups produce slightly different local variations in the value of B_0 (i.e., exhibit slightly different values

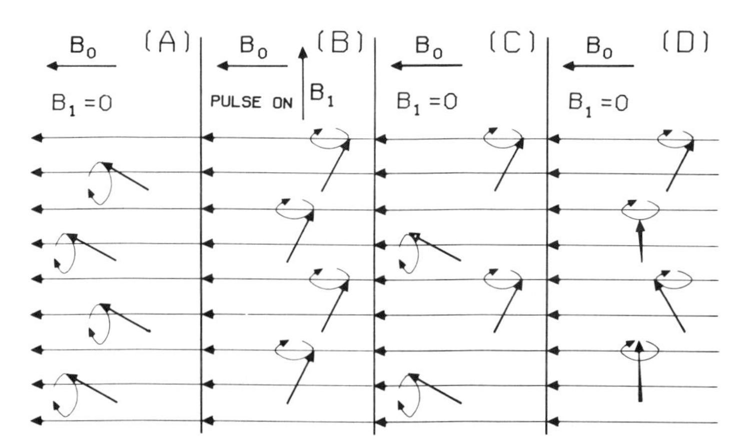

Fig. 1. Representation of precessing magnetic dipole moments in a static external magnetic field, B_0. (A) The orderly precession of the magnetic moments in the direction of B_0 shortly after being placed in the magnetic field. (B) The effect of the pulsed magnetic field, B_1, delivered by the coil around the sample of the pulsed spectrometer in Fig. 2. (C) Return of the nuclear magnetic moments to the lower energy state by spin–lattice relaxation. (D) "Dephasing" of the precession of moments in the high energy state as a result of spin–spin relaxation.

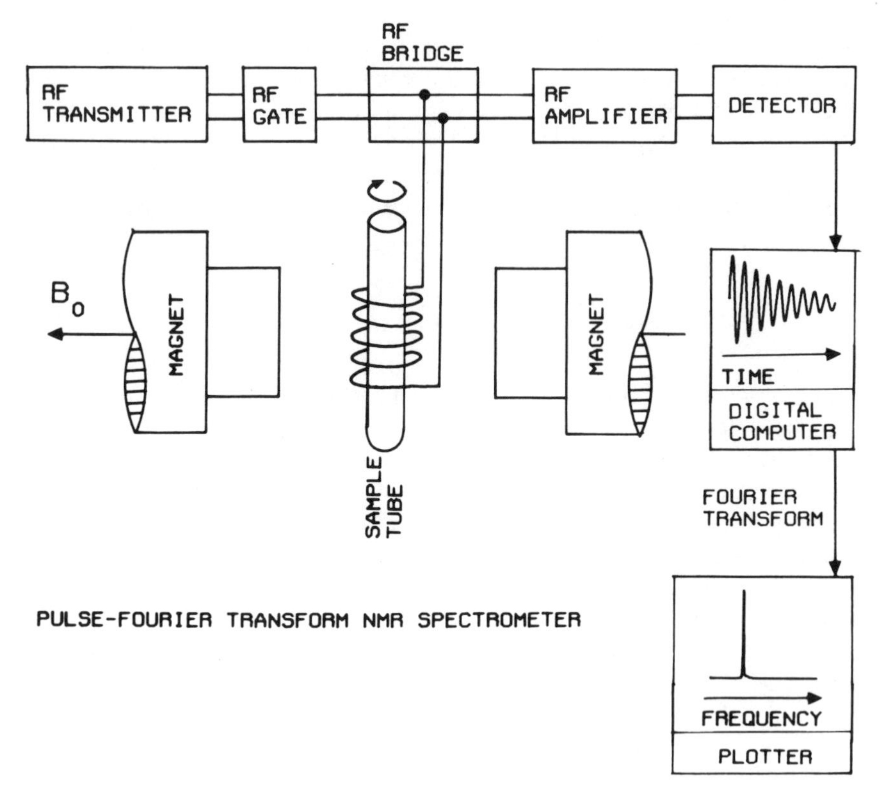

Fig. 2. Block circuitry diagrams of pulse-Fourier transform and continuous-wave NMR spectrometers.

of σ), a molecule with several different chemical functional groups will yield several resonances spread a few kilohertz to either side of the nuclear Larmor frequencies listed above.

When one records an NMR spectrum, one measures the frequency at which the nuclear magnetic moments precess (ν_{obs}) when they are placed in a static external magnetic field of known strength (B_0). This is accomplished with spectrometers of the type in Fig. 2. These spectrometers are composed of a large magnet that places the lines of flux of B_0 through the sample tube horizontally and one or more transmitter/receiver coils around the sample.

The pulse-Fourier transform spectrometer (Fig. 2) has a single transmitter/receiver coil around the sample that can both produce and detect an oscillating magnetic field B_1 perpendicular to B_0 (Fig. 1B). When the experiment begins, the bridge connects the coil to the transmitter and

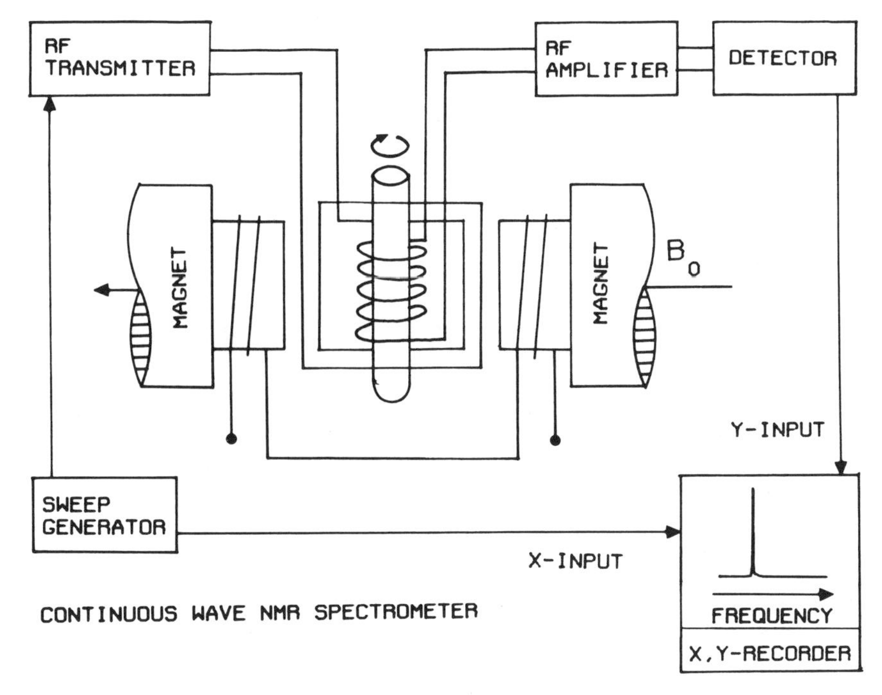

Fig. 2. (*Continued*)

disconnects the detector. The transmitter generates only a single radio frequency (rf) of high intensity and the gate passes the current to the coil for a very short period of time (from 1 to 20 μsec depending on the spectrometer). This "pulse" pumps the nuclei into the higher potential energy state (Fig. 1B) in which the magnetic moments face against B_0. This is analogous to manually turning a compass pointer so that it no longer faces north. Then the bridge disconnects the coil from the transmitter and connects the detector. The nuclei then lose their potential energy with time as an oscillating magnetic field, which induces a current in the receiver coil. The loss of potential energy by the nuclei in the absence of applied rf (called a *free induction decay*) occurs at the Larmor frequency, ν_{obs}, and is exponential with time. (This is analogous to watching the pointer of the compass passing through several dampened oscillations after it has been released until it again faces north.) The rate at which the potential energy is lost is determined by two relaxation mechanisms called spin–lattice and spin–spin relaxation, which are described in Table I and Fig. 1. The free induction decay (FID) is usually measured as an "off-resonance" interference pattern between the detected resonance frequency and a known reference frequency (Fig. 2). By repeated

Table I
NMR Spectral Parameters

1. The chemical shift

The chemical shift, (δ), is a measure of ν_{obs} [Eq. (1)] that has been normalized to unit magnetic field. This normalization is required to permit comparison of the frequency distribution of resonances of spectra recorded on spectrometers with different magnetic fields. Since the frequency differences between functional groups are measured in hertz, whereas the Larmor frequency is measured in megahertz, the chemical shift (δ), expressed in parts per million, from the resonance of a reference compound was defined as follows:

$$\delta = \frac{\nu_{obs} - \nu_{ref}}{\nu_{ref}} \times 10^6$$

where ν_{obs} is the observed resonance frequency of a given nucleus within a particular functional group and ν_{ref} is that of a functional group of an added reference compound. For ^1H and ^{13}C NMR spectroscopy the agreed upon reference signal is from the ^1H and ^{13}C nuclei, respectively, of the methyl groups of tetramethylsilane (TMS). The chemical shift is used to determine the functional groups that comprise the molecule.

2. The areas of the peaks in the spectrum

The relative areas of the various resonance peaks in the spectrum of a molecule provide a measure of the number of each kind of functional group in the molecule. However, extreme care must be exercised in recording the spectrum to avoid introduction of distortion and to permit complete relaxation of the nuclei in all the different functional groups.

3. The number of peaks in multiplets that arise from spin–spin coupling

The magnetic moments of nonequivalent atoms that are connected by a small number of covalent bonds (usually three or less) can couple with each other through these covalent bonds. The equivalent nuclei of each functional group split the resonance of the nonequivalent nucleus to which they are coupled according to the relationship:

$$\text{No. of peaks in multiplet} = 2nI + 1$$

where n is the number of equivalent nuclei that cause the observed splitting, and I is the spin quantum number of these nuclei ($I = \frac{1}{2}$ for ^1H, ^{13}C, and ^{31}P nuclei). [It must be noted that the splitting pattern becomes more complex when the frequency separation between the two coupled resonances approaches the frequency separation between the peaks of the split multiplets—see Bovey (1969) and Pople *et al.* (1959).] The splitting patterns are used to determine the arrangement of the functional groups within the molecule.

4. The value of the spin–spin coupling constant

The spin–spin coupling constant, J, is defined as the amount of separation, measured in hertz, between two adjacent peaks in a split multiplet. The value of J is the same for multiplets arising from nuclei that are coupled and can be correlated with the angle (if joined by two bonds) or dihedral angle (if joined by three bonds) between the coupled nuclei. J is used to determine the preferred conformation of a molecule in solution.

5. The rate of relaxation of excited nuclei

In order to record an NMR spectrum the nuclei that are excited into the higher energy level by the rf field, B_1 (Fig. 1B), must relax back to the lower energy level (Fig. 1A), and two mechanisms exist by which this relaxation can occur. These are spin–lattice relaxation and spin–spin relaxation. Spin–lattice relaxation involves a loss of potential energy by each excited nucleus to the surrounding environment or 'lattice.' Each nucleus must lose its potential energy in a single quantum jump, and thus the potential energy of the population of the nuclei will decrease exponentially with time as the number of nuclei remaining in the

Table I (*Continued*)

higher energy level decreases (Fig. 1C). The rate at which this relaxation process occurs is measured by the spin–lattice relaxation time, T_1 ($1/T$ = rate). Spin–spin relaxation involves interaction through space of the magnetic dipole moments of nuclei in close proximity. This mechanism permits a redistribution of potential energy between individual nuclei without causing a loss of potential energy by the total population of nuclei. This can be visualized as a process in which the energy given up by some nuclei in the population as they return to the lower energy level (Fig. 1A) causes the precession of the remaining nuclei in the higher energy level to become dephased (Fig. 1D). The rate at which this relaxation process occurs is measured by the spin–spin relaxation time, T_2.

T_1 and T_2 generally are relatively long and of equal value for nuclei of small, rapidly tumbling molecules in nonviscous solvents. When the tumbling is slowed as a result of increased viscosity of the solution or high molecular weight, the rate of relaxation by both mechanisms is increased, and the values of T_1 and T_2 become smaller. The rates of these relaxation processes are also increased by the presence of paramagnetic metal ions. When the molecular motion is greatly restricted, as would occur in a solid, the value of T_1 may again be long, but the value of T_2 remains short (Bovey, 1969; Pople *et al.*, 1959). Thus, since

$$\frac{1}{T_2} = \pi(\Delta\nu)$$

where $(\Delta\nu)$ is the "peak width at half-height" expressed in hertz, any molecular interaction that reduces the rate of molecular tumbling will cause the peaks in the NMR spectrum to be broadened.

---

pulsing of the sample and addition of the resulting FIDs the signal-to-noise ratio of the data can be improved. Then a mathematical algorithm called a Fourier transform is performed on the FID that converts the spectrum from intensity vs. time to intensity vs. frequency (Fig. 2).

The continuous-wave spectrometer (Fig. 2) yields the same kind of NMR spectrum as the pulse-Fourier transform spectrometer but has two coils of wire around the sample tube that are wound such that their axes are at right angles to the lines of flux of B_0 and to each other. One of these two coils is attached to an rf transmitter and the other is attached to a detector. The transmitter generates an oscillating current in the coil to which it is attached that in turn produces an oscillating magnetic field, B_1, in the sample at right angles to B_0. Since the axis of the detector coil is orthogonal to the direction of B_1, the detector coil cannot detect this oscillating magnetic field unless the sample changes the apparent direction of B_1. This occurs "at resonance." When the frequency of the rf transmitter is equal to the Larmor frequency, ν_{obs}, of the nucleus at this value of B_0 as defined by Eq. (1), the nuclei absorb energy. The nuclei lose this potential energy with time as an oscillating magnetic field, which can be detected by the detector coil because random molecular motion causes the nucleus to lose the potential energy in random directions. Thus by sweeping the rf transmitter through a range of frequencies (used as the X input of an X, Y-recorder) and using the oscillating frequency generated in the detector coil to drive the pen of an X, Y-recorder in the Y direction, it is possible to record a spectrum in which the pen is deflected from the baseline only when resonance occurs. This method of recording a spectrum is not used often with biological preparations because averaging of multiple scans requires too much time.

B. The Spectrum

1. Liquids and Solutions

The information that can be derived from the high-resolution NMR spectra of liquids and solutions has been the subject of many reviews and texts including the books of Bovey (1969) and Pople *et al.* (1959). Since space does not permit a complete review of the assignments of NMR spectra, I have listed the spectral parameters that are of use in Table I and demonstrate how they are derived and utilized with the example in Fig.3.

Several factors participate in producing the narrow peaks (i.e., linewidths of a few hertz) in high-resolution spectra of liquids and solutions. Rapid thermal tumbling causes both T_1 and T_2 (Table I) to be long. Thus, the contribution of spin–spin relaxation mechanisms to the linewidth [Table I (5)] is small. Since this motion is random and isotropic, the dipolar coupling of magnetic moments through space, which is a vector quantity, averages to zero. In addition, the magnetic shielding constant, σ [Eq. (1)], has a single value that arises from averaging of the three shielding tensors in the directions of the three cartesian axes.

2. Solids

In rigid "solids," molecular motion tends to be slower than in liquids and to be anisotropic. The value of T_2 is shorter than in liquids, so the contribution of spin-spin relaxation mechanisms to the linewidth [Table I (5)] can be substantial. Dipolar coupling through space of protons with the magnetic moments of other nuclei, such as ^{13}C and ^{31}P, is not averaged to zero by molecular motion and adds a large contribution to the apparent linewidth. Furthermore, the shielding tensors in the directions of the three cartesian axes are not averaged (i.e., the sample exhibits chemical shift anisotropy). In a highly ordered single crystal, each nucleus will yield three peaks in the NMR spectrum, one for each shielding tensor. The apparent chemical shifts of these three reasonances depend on the spatial orientation of the crystal relative to the direction of B_0. In samples with randomly oriented molecules, a "power pattern" which is composed of all the overlapping peaks from the three shielding tensors in all random orientations is obtained. Thus, NMR spectra are obtained in which each nucleus gives rise to a reasonance 5000–10,000 Hz wide.

3. Model Phospholipid Bilayers and "Normal" Biological Membranes

A biological membrane is a mixture of relatively "liquid" and "solid" components. Those components with a high degree of freedom of motion yield usable NMR spectra with high-resolution techniques (Section II.A), whereas components whose motion is restricted yield resonances that are very broad, for the reasons described in Section II.B.2. This latter difficulty has led to the use of ligand binding to probe the membrane and the use of techniques that yield well-resolved spectra of solids to detect the less mobile components of membranes.

a. High-Resolution NMR Techniques. Characterization of membranes with the high-resolution pulse and continuous-wave NMR techniques described above has been the subject of numerous reviews (Chapman, 1972; Chapman and Oldfield, 1974; James, 1975; Podo, 1975; McLaughlin *et al.* 1977; Chan *et al.* 1979). Virtually all of this work has been limited to detection of the phospholipid molecules of the bilayer.

The phosphate head groups exhibit the least freedom of motion within the phospholipid molecules of membrane bilayers (Davis and Inesi, 1972). Unsonicated lamellar dispersions of phospholipids in water yield ^{31}P spectra with linewidths that are intermediate between the "powder patterns" (see Section II.B.2) of solid powder preparations of the same phospholipids and the narrow resonances of phospholipids dissolved in organic solvents (Kohler and Klein, 1976, 1977). This indicates that the phosphate head groups undergo slow, anisotropic motion in membrane bilayers. Only at point defects where membrane bilayers fuse does motion of the phospholipid molecules appear to be sufficiently isotropic to give rise to narrow ^{31}P resonances (Hui *et al.*, 1981, DeKruijff *et al.*, 1980). The presence of Ca^{2+} ions has different effects on the freedom of motion of the phosphate groups of various phospholipids, which produce corresponding changes in their ^{31}P NMR spectra (Kohler and Klein, 1977; Cullis and Verkleij, 1979; DeKruijff *et al.*, 1980; Yaari *et al.*, 1982).

Motion increases along the hydrocarbon chains of the fatty acids

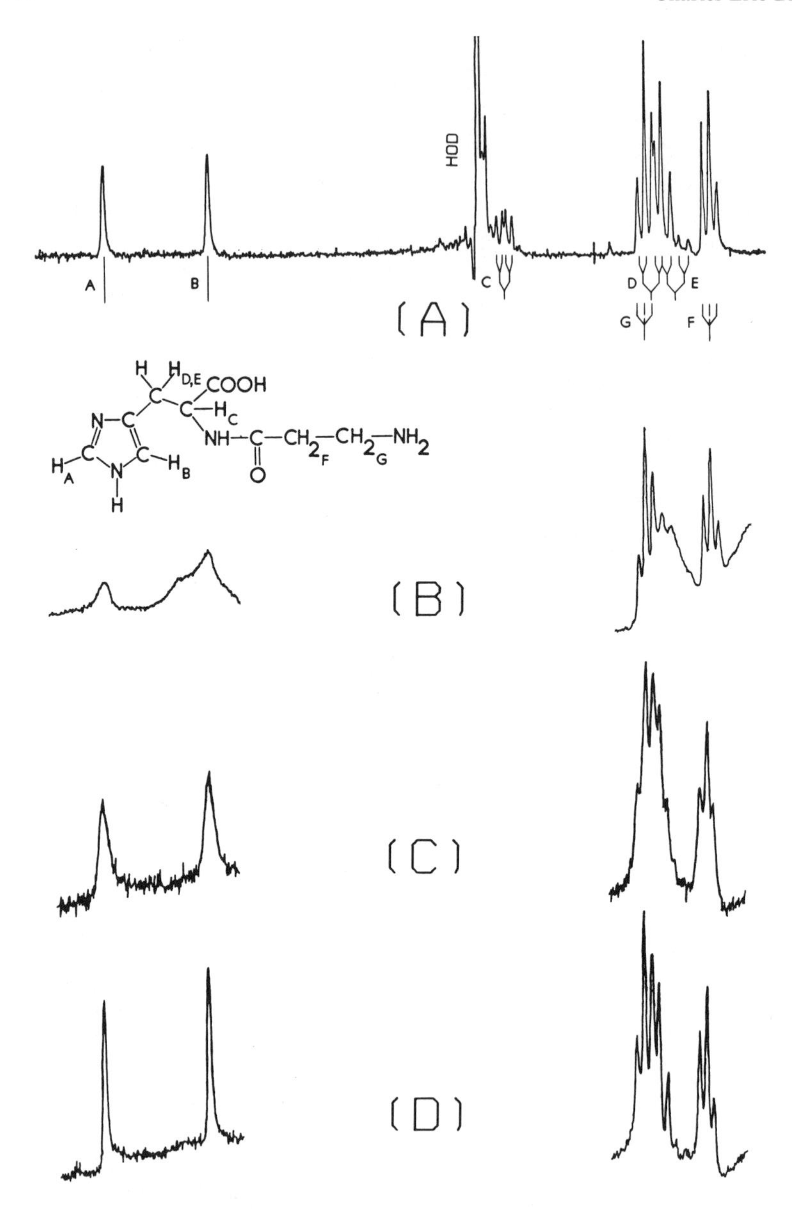

Fig. 3. The ^1H NMR spectra of 10 mM carnosine in D_2O (A) alone, (B) plus 0.8 mM serum albumin, (C) plus the crude particulate fraction of the nasal olfactory epithelium, and (D) plus the crude particulate fraction of the olfactory bulb. All solutions contained 0.15 M NaCl and were adjusted to pH 7.2. The spectral range shown in these spectra is 2.5–8.9 ppm downfield from TMS contained in a coaxial capillary. The assignments of the resonances to the various protons are designated by the letters A–G. The assignments were made in the following manner. The peaks labeled A and B are in the chemical shift range of protons

such that the terminal methyl and methylene groups of model phospholipid vesicles exhibit rather well-resolved resonances with high-resolution techniques (Lee *et al.*, 1973). Similarly, the terminal moieties of flexible head groups that are in contact with water also have sufficient freedom of motion to exhibit detectable resonances in model phospholipid membranes (Lee *et al.*, 1973; James, 1975).

Metal ions that produce paramagnetic line broadening and/or changes in chemical shifts of the head group resonance(s) and that do not diffuse through membranes have been used to determine the distribution of various phospholipids between the inner and outer surfaces of bilayers (Bergelson and Bystrov, 1975; Andrews *et al.*, 1973). This technique has proven useful for studying the phospholipid exchange proteins, which catalyze intermembrane phospholipid exchange (Bergelson and Bystrov, 1975), and for studying transbilayer movement of phosphatidylcholine (DeKruijff and Wirtz, 1977).

The locations of various proteins and cholesterol within both model and biological membranes also have been determined from their effects on the NMR spectra of the phospholipids (James, 1975; Lee *et al.*, 1973; Davis and Inesi, 1972; Chapman *et al.*, 1968; Sheetz and Chan, 1972). These membrane components interact with the head groups and/or alkyl chains of the phospholipids in membrane bilayers and thereby reduce

---

on aromatic rings with a proton two bonds removed from two nitrogen atoms being expected to resonate at lower field (i.e., farther to the left) than a proton two bonds removed from only one nitrogen atom. This distinction between the two protons on the imidazole ring is further supported by the effect of changes of pH on the chemical shifts of these two resonances (Brown *et al.*, 1979) and the ability of the proton on carbon atom-2 of the ring to exchange with deuterons of the solvent. Protons on the α-carbon atoms of amino acids resonate in the chemical shift range of peak C, and peaks D, E, F, and G are in the chemical shift range of methylene groups. The assignments of the resonances to the three methylene groups of carnosine are based on the spin–spin splitting of each resonance. To begin, the resonance of the α-proton of the histidyl residue is split to a doublet of doublets characterized by two spin–spin splitting constants ($J_{H_C-H_D}$ = 5 Hz and $J_{H_C-H_E}$ = 8.5 Hz). Thus, the methylene protons of the histidyl side chain are nonequivalent and are located at two different dihedral angles relative to the α-proton. Since $J_{H_C-H_D} = J_{H_D-H_C}$ and $J_{H_C-H_E} = J_{H_E-H_C}$, one can locate the two multiplets labeled D and E in the methylene region of the spectrum and assign them to the histidyl sidechain. The two nonequivalent protons of this methylene group are coupled to each other with a spin–spin coupling constant $|J_{H_D-H_E}|$ = 15 Hz. On the basis of the measured values of $J_{H_C-H_D}$,$J_{H_C-H_E}$, and $J_{H_D-H_E}$ and of the previous work of Martin and Mathur (1965) and Pachler (1964), there appears to be hindered rotation about the C_α–C_β single bond of the histidyl residue of carnosine with the preferred conformation being that in Fig. 4. The remaining two triplets, labeled F and G, arise from the methylene groups of the β-alanyl residue. To a first approximation the two protons of each methylene group should split the resonance from the other methylene group to a triplet if rotation about the C_α–C_β single bond of this residue is unrestricted. This is indeed observed.

their freedom of motion. Thus these membrane components produce line broadening of the resonances of the functional groups that are involved in the interaction. Variations of linewidths have been used to detect differences in mobility of lipid components of various organs (Daniels *et al.*, 1976).

The primary limitation of investigating membrane structure with high-resolution NMR techniques is that the phospholipid molecules must have a high degree of freedom of motion to yield detectably narrow resonances. The phospholipids of biological membranes tend to be restricted by the many proteins and cholesterol that are present.

b. Ligand Binding as a Probe of the Membrane. This technique recently was reviewed in detail (Brown, 1981a) so I shall mention only a few of the more salient points here. The measurement is based on the fact that a small ligand molecule exhibits long relaxation times when free in solution but short relaxation times when bound to a slowly tumbling macromolecule or membrane. Put another way, the ligand molecule will yield a well-resolved NMR spectrum with narrow peaks when free in solution but will yield resonances too broad to be detected with high-resolution techniques when attached to a rigid membrane.

If the ligand binds to the membrane with a $K_{diss} > \sim 10^{-4}$ M, then the ligand is expected to move between the binding site and the buffer before the excited nuclei relax back to the lower energy state. Thus, the relaxation parameters of the bound state are imparted to the pool of free ligands. This is observed as an apparent broadening of the resonances in the high-resolution spectrum of the *bulk* ligand in solution. The observed linewidth of each NMR spectral peak, $\Delta\nu_{obs}$, is given by

$$\Delta\nu_{obs} = \alpha\Delta\nu_{bound\ state} + (1 - \alpha)\Delta\nu_{free} \qquad (2)$$

where $\Delta\nu_{bound\ state}$ and $\Delta\nu_{free}$ are the linewidths when the ligand is bound to the membrane and free in the buffer, respectively, and α is the fraction of the total ligand molecules that are bound to the membrane at any point in time. The theoretical and technical considerations of this technique are presented in detail elsewhere (Brown, 1981a).

An example of the kind of results that are obtained is given by the binding of carnosine to serum albumin (Fig. 3B) and to a crude particulate preparation of nasal olfactory epithelium (Fig. 3C). Carnosine binds to serum albumin via the imidazole ring and terminal carboxyl group of the histidyl residue, whereas the β-alanyl residue remains free to undergo rapid thermal motion (Fig. 4) (Brown *et al.*, 1979). Thus, the resonances of the histidyl residue of carnosine are broadened, whereas the resonances of the β-alanyl residue are unaffected by addition of serum albumin (Fig. 3B). The binding of carnosine by crude particulate preparations of nasal

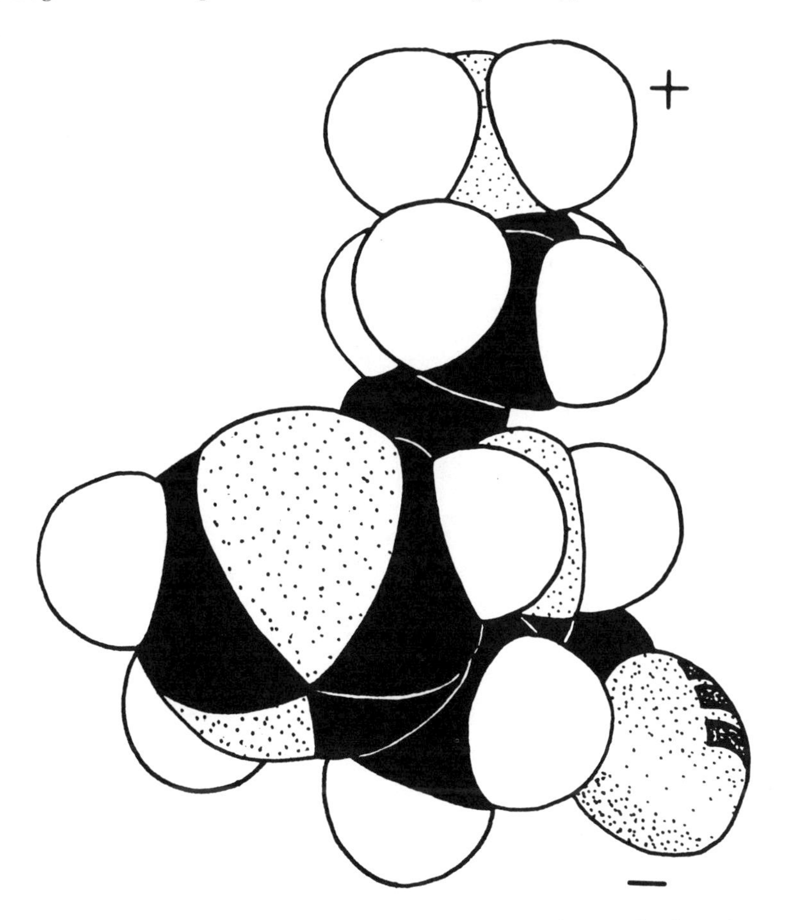

Fig. 4. The rotational isomer of carnosine that is expected to predominate in aqueous solution. When carnosine binds to serum albumin the imidazole ring, adjacent methylene group, and charged carboxyl group are recognized, whereas the β-alanyl residue remains in contact with solvent (Brown *et al.*, 1979). When carnosine interacts with the crude particulate preparation of nasal olfactory epithelium, the positively charged amino group of the β-alanyl group also participates in recognition of the binding site (Brown *et al.*, 1977).

olfactory epithelium (Brown *et al.*, 1977) involves recognition of both the histidyl and β-alanyl residues (Fig. 4), and the resonances of both residues are broadened accordingly (Fig. 3). When carnosine is added to crude particulate preparations that lack this binding site, little spectral line broadening is observed (Fig. 3D). Thus, this technique can be used to characterize the steric requirements of a binding site on a membrane for a specific ligand without the use of pharmacological agents as competitive inhibitors of the binding (Brown, 1981a). The technique has been used

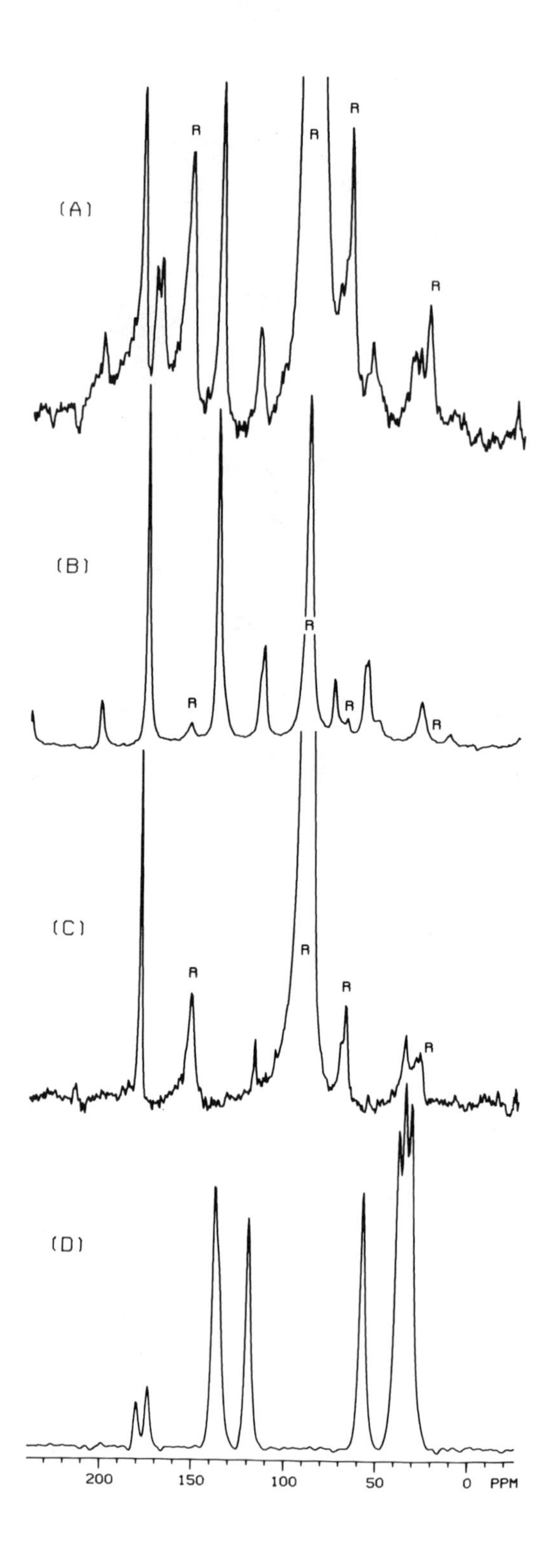

(A)
R
R
R
R
(B)
R
R
R
R
(C)
R
R
R
R
(D)
200
150
100
50
0
PPM

successfully to measure the equilibrium dissociation constant, the enthalpy of binding, the dissociation rate constant, and the activation energy of dissociation for the interaction of choline with the nicotinic receptor of the *Torpedo californica* electroplax (Miller *et al.*, 1979).

A variation of this technique is based on the fact that the relaxation times T_1 and T_2 of atomic nuclei (Table I) also are shortened by close proximity of paramagnetic metal ions such as Mn^{2+}. By adding Mn^{2+} ions to a suspension of cells and then recording ^{31}P NMR spectra of the intracellular metabolites as a function of time, it is possible to measure the rate of transport of the Mn^{2+} ions across the cell membrane from the extracellular space to the intracellular cytosol (Pettegrew and Minshew, 1981, 1982). Similarly, the ability of paramagnetic lanthanide ions to alter the chemical shifts of the phosphate headgroups has been used to measure the transport of the these cations across vesicle bilayer membranes (Ting *et al.*, 1981).

In those cases where the transport of Mn^{2+} ion across the cell membrane is small, the rate of diffusion of water across cell membranes can be measured (Conlon and Outhred, 1972; Pirkle *et al.*, 1979). This is possible because the relaxation times of the water protons become a function of the exchange process whereby water molecules leave and enter the cells. In addition to the rate of diffusion, the activation energy for water exchange and the volume of the cells that are exchanging the water can be determined. This technique was first developed to investigate the state of water in polarized and depolarized nerve cells (Fritz and Swift, 1967).

c. Cross-Polarization, Magic Angle NMR Spectroscopy. Membrane-bound proteins, cholesterol, and low-molecular-weight metabolites and phospholipid molecules bound to them are among the membrane components that yield very broad resonances because of restricted motion. Better-resolved spectra should be obtained from these components with

Fig. 5. Comparison of the cross-polarization, magic angle ^{13}C NMR spectra of (A) solid carnosine (β-alanyl-L-histidine), (B) solid histidine, and (C) solid β-alanine with (D) the high resolution ^{13}C NMR spectrum of carnosine in aqueous solution at pH 7.2. Resonances in the spectra of the solid samples are within ± 1 ppm of their respective chemical shifts in the spectra of the corresponding aqueous solutions. The assignments are as follows: methylene groups of histidyl and β-alanyl residues, 31–35 ppm; α-carbon atom of the histidyl residue, 57 ppm; C-4 of the imidazole ring, 117 ppm; C-2 and C-5 of the imidazole ring, 138 ppm; peptide carbonyl, 172 ppm; and the terminal carboxyl group, 176–180 ppm (Voelter *et al.*, 1971). All spectra were recorded with proton decoupling. A large exponential multiplication factor was applied to the free induction decay of the high-resolution spectrum (D) to produce linewidths equivalent to those in the cross-polarization, magic angle spectra. Peaks arising from the Delrin rotor are labeled with an R. In the absence of magic angle spinning and dipolar decoupling each peak in spectra A–C would be about 100-fold broader.

the techniques that have been developed for recording well resolved NMR spectra of solids. The physical parameters that give rise to the broad resonances (Section II.B.2) are the same for both the restricted membrane components and solid powders.

The "magic angle" sample spinning technique (Andrew, 1971) in which the chemical shift anisotropy (see Section II.B.2) of solid samples is reduced by spinning the sample at a rate of about 2 to 5 KHz with the axis of rotation oriented at an angle at 54°44′ from the direction of B_0 (Fig. 1) has been used successfully to improve the resolution of ^{1}H, ^{13}C, and ^{31}P NMR spectra of phospholipid membrane preparations (Chapman *et al.*, 1972; Haberkorn *et al.*, 1978; Herzfeld *et al.*, 1980). Strong resonant decoupling of dipolar interactions between ^{1}H and both ^{13}C and ^{31}P nuclei has been used successfully to reduce dipolar broadening of ^{13}C and ^{31}P NMR spectra of phospholipid membranes (Kohler and Klein, 1977; Haberkorn *et al.*, 1978). Furthermore, cross polarization (also called *proton-enhanced nuclear induction*) spectroscopy (Hartmann and Hahn, 1962; Pines *et al.*, 1973) has been shown to produce a 5- to 10-fold signal enhancement of the ^{13}C NMR spectrum of phospholipid membranes (Urbina and Waugh, 1974). In addition to making NMR spectra of biological membranes feasible, the cross polarization technique can be used to measure dynamic properties of the phospholipid molecules within the membranes (Urbina and Waugh, 1974). An example of the spectral resolution that can be achieved with combined use of cross polarization, high-power dipolar decoupling and magic angle sample spinning is presented in Fig. 5.

The current applications of cross-polarization and magic angle sample spinning to the characterization of the solid components of biological tissues has been reviewed (Brown and Wilkie, 1982c). ^{31}P NMR spectra of solid materials are rather easy to record (Brown and Wilkie, 1982a,b,c), and it might be possible to probe the headgroups of membrane bilayers with this technique (Brown and Wilkie, 1982b,c).

III. RESULTS WITH PATHOLOGICAL MEMBRANES

The recent advances in understanding the NMR spectra of membrane bilayers has generated interest in determining whether various diseases produce changes in membrane structure and/or function that can be detected with NMR spectroscopy. The work that has been published to date contains good examples of the data that can be obtained.

Viral transformation of cells was studied by direct observation of isolated plasma membranes with ^{31}P and ^{1}H NMR spectroscopy (McLaughlin *et al.*, 1977). The ^{31}P NMR spectra of isolated plasma membrane from normal hamster NIL 8 cells and hamster cells transformed with hamster sarcoma virus were found to exhibit no detectable differences. This is to be expected, however, because the broad ^{31}P resonances obtained from membrane preparations with high-resolution NMR techniques are insensitive to chemical alteration of membrane protein and to increased fluidity in lipid extracts (Davis and Inesi, 1972). In contrast to the insensitivity of ^{31}P spectra, ^{1}H NMR spectra of model membranes are much better resolved and exhibit changes when both the fatty acid chains and head group substituents are immobilized as mentioned in Section II.B.3.a. The ^{1}H NMR spectra of isolated plasma membranes from normal rat embryo cells and rat embryo cells transformed with polyoma virus are very similar with only small differences in the signal intensities from the trimethylammonium group of choline and the mobile methylene groups of the fatty acid residues. However, the ^{1}H NMR spectrum of the transformed plasma membrane exhibits changes after mild treatment such as incubation at 4°C for 4 days, whereas the spectrum of normal plasma membrane remains the same. After incubation at 4°C, the transformed plasma membrane exhibits a large, very narrow peak from the trimethylammonium group of choline and an increase in the signal intensity from the mobile methylene groups. This suggests that the interaction between the lipid head groups and the membrane proteins is more labile in the transformed plasma cell membranes than in the normal cell membranes (McLaughlin *et al.*, 1977). Release of a macromolecule from binding with the choline head group of the membrane would permit increased motion of the head group and thereby give rise to a much narrower choline resonance (Chapman *et al.*, 1968; Davis and Inesi, 1972; Sheetz and Chan, 1972). This increased lability of the interaction between phospholipid head groups and membrane proteins might underlie certain characteristics of transformed cells and neoplastic tumor cells such as loss of contact inhibition.

Direct observation of phospholipid bilayers with NMR spectroscopy also has been applied to the investigation of atherosclerosis (Cullis and Hope, 1980). In the early stages of this disease, there is an increase in the concentration of cholesterol in the membranes of the intimal cells that form the arterial wall. This increase in cholesterol appears to stimulate the synthesis of sphingomyelin or the transfer of sphingomyelin into membranes (Small and Shipley, 1974). With ^{31}P NMR spectroscopy, Cullis and Hope (1980) demonstrated that elevated concentrations of cholesterol destabilize the bilayer structure of phospholipid membranes. However, sphingomyelin was observed to stabilize the bilayer structure, especially in the presence of cholesterol. Thus, the elevated level of sphingomyelin that accompanies the early stages of atherosclerosis was suggested to be a compensatory response to overcome the destabilizing effect of increased concentrations of cholesterol on cell membranes.

In addition to direct observation of the membrane preparation, various ligand binding assays also have been used to characterize the properties of pathologial membranes. The application of such NMR binding assays to cancer research was suggested at least as early as 1973 (Höglund *et al.*, 1973).

The measurement of carnosine binding by serum albumin and crude particulate preparations (Brown *et al.*, 1977, 1979), which was used for purposes of demonstration in Sections II.B.1 and II.B.3.b. has provided new insight into the possible role(s) of the dipeptide carnosine in regulating the metabolism of skeletal muscle and the orderly turnover of the primary olfactory neuron (Brown, 1981b). In addition, the combined results of the above assays of carnosine binding by crude particulate preparations and of NMR and electron spin resonance investigations into the formation of copper–carnosine chelates (Brown and Antholine, 1979; Brown *et al.*, 1980) have suggested a possible mechanism whereby the neurological damage in Wilson's disease (hepatolenticular degeneration) might occur (Brown and Antholine, 1980). Carnosine is suggested to function as a carrier that, under the abnormal conditions present in Wilson's disease, transports toxic amounts of copper across membranes into the basal ganglia.

Transport of water across the erythrocyte membranes of patients with a number of pathological conditions also has been measured with NMR spectroscopy (Morariu and Benga, 1977; Ashley and Goldstein, 1981). Permeability to water was found to be reduced from that of normals in Gaucher's disease, essential hyperlipemia, obstructive jaundice, chronic hepatitis, nephrotic syndrome, and Duchenne muscular dystrophy. The decreased permeability in Gaucher's disease and liver diseases was suggested to arise from increased levels of cholesterol within the erythrocyte membrane (Morariu and Benga, 1977). The decreased permeability in Duchenne muscular dystrophy was attributed to a reduced amount of Band III protein in the membrane, which in turn might reflect a change in the lipid environment of the erythrocyte membrane (Ashley and Goldstein, 1981).

The effect of paramagnetic metal ions on the NMR spectrum of intracellular metabolites has been used to investigate the structural integrity of cellular membranes during hepatic encephalopathy, including Reye syndrome (Pettegrew and Minshew, 1981, 1982). Elevated concentrations of short-chain fatty acids have been implicated in the pathogenesis of this disease. Infusions of short-chain fatty acids into animals produce coma, presumably by a direct effect on neuronal membranes. Fluorescence polarization spectroscopy has provided evidence for significant alterations in the physical properties of cellular membranes, particularly of the hydrocarbon core, following brief exposure to sodium octanoate at concentrations that are observed in patients with hepatic encephalopathy (Minshew *et al.*, 1981). By measuring the concentrations of intracellular metabolites with ^{31}P NMR spectroscopy, it was shown that the intermediary metabolism of intact erythrocytes is not altered by sodium octanoate at concentrations observed clinically. However, sodium octanoate does increase the permeability of the cell membrane to Mn^{2+} ion. Line broadening of the resonances from the intracellular metabolites upon addition of Mn^{2+} ions to the extracellular medium is observed in the presence of sodium octanoate but not in its absence (Pettegrew and Minshew, 1981, 1982).

IV. CONCLUSIONS

The use of NMR spectroscopy for the characterization of biological membranes is well established, and experiments with pathological membranes are beginning to appear in the literature with greater frequency. The results that have been published to date indicate that the technique holds great promise for characterizing changes in membrane structure that occur in a number of pathological conditions.

The measurement of ligand binding, water exchange, and permeability to metal ions with NMR spectroscopy experienced its first application to the investigation of biological membranes in the area of neuroscience (Brown, 1981a; Fritz and Swift, 1967). Several applications of these measurements to characterize pathological membranes have been reported, and more can be expected. The high-resolution NMR spectrometers that are used to make these measurements are of the type commonly found on academic campuses, and the information provided often cannot be obtained readily with other techniques.

Direct observation of the membrane with high-resolution NMR spectrometers is valuable with model systems but experiences difficulties with biological membranes. ^{1}H and ^{13}C NMR spectra of model phospholipid bilayers exhibit resolved resonances that arise from functional groups with high degrees of freedom of motion in the lipid core and head group regions. These resonances become less resolved when cholesterol and membrane proteins restrict the motion of the phospholipid molecules.

Thus, many biological membranes, which contain cholesterol and several types of proteins, tend not to yield highly informative ^{1}H or ^{13}C NMR spectra with high-resolution spectrometers. Since the phosphate groups are very restricted, in both model and biological membrane bilayers, the ^{31}P NMR spectra obtained with high-resolution spectrometers also tend to be poorly resolved and to be relatively insensitive to changes in membrane structure unless they involve disruption of the bilayer. In spite of these difficulties, high-resolution NMR spectroscopy has provided evidence for changes in membrane structure in cell transformation and atherosclerosis.

Cross-polarization, high-power dipolar decoupling, and magic angle sample spinning techniques, which were developed for recording NMR spectra of solids, overcome some of the difficulties in recording NMR spectra of membranes and other biological preparations. These techniques have been demonstrated to yield well-resolved ^{13}C and ^{31}P NMR spectra of model membranes and already have been used to investigate the structures of proteins (Jardetzky and Wade-Jardetzky, 1980), dental enamel (Rothwell *et al.*, 1980), and bones (Herzfeld *et al.*, 1980), and the utilization by soybeans of nitrogen from fertilizer (Schaefer *et al.*, 1979). To the best of my knowledge, work has not yet been published in which these techniques have been applied to pathological membranes. It will, however, be interesting to see what applications are found for these techniques in the future. Better-resolved ^{13}C and ^{31}P NMR spectra of the phospholipids may afford increased sensitivity to structural differences between normal and pathological membranes. In addition, it should become feasible to observe highly immobile components of biological membranes such as cholesterol and membrane-bound proteins.

ACKNOWLEDGMENTS. I thank the Roche Institute of Molecular Biology, The Chemical Research Department of Hoffman–LaRoche, Inc., The Medical College of Wisconsin, and the Nicolet Instrument Corporation for their encouragement and technical support. This work was supported by a Biomedical Research Support Grant from The Medical College of Wisconsin and by a Cottrell Research Grant from the Research Corporation.

V. REFERENCES

Andrew, E. R., 1971, Narrowing of NMR spectra of solids by high-speed specimen rotation and the resolution of chemical shift and spin multiplet structures for solids, *Prog. Nucl. Magn. Reson. Spectrosc.* **8**:1.

Andrews, S. B., Faller, J. W., Gilliam, J. M., and Barrnett, R. J., 1973, Lanthanide ion-induced isotropic shifts and broadening for nuclear magnetic resonance structural analysis of model membranes, *Proc. Natl. Acad. Sci. USA* **70:**1814.

Ashley, D. L., and Goldstein, J. H., 1981, Nuclear magnetic resonance evidence for abnormal water transport in Duchenne muscular dystrophy erythrocytes, *Biochem. Biophys. Res. Commun.* **100:**364.

Bergelson, L. D., and Bystrov, V. F., 1975, Use of shift and broadening reagents in the NMR investigation of membranes, *in* "Biomembranes: Structure and Function" (G. Gardos and I. Szasz, eds.), pp. 33–46, North-Holland, Amsterdam.

Bovey, F. A., 1969, "Nuclear Magnetic Resonance Spectroscopy," Academic Press, New York.

Brown, C. E., 1981a, Measurement of ligand binding with nuclear magnetic resonance spectroscopy, *J. Neurosci. Methods* **3:**339.

Brown, C. E., 1981b, Interactions among carnosine, anserine, ophidine and copper in biochemical adaptation, *J. Theor. Biol.* **88:**245.

Brown, C. E., and Antholine, W. E., 1979, Chelation chemistry of carnosine. Evidence that mixed complexes may occur in vivo, *J. Phys. Chem.* **83:**3314.

Brown, C. E., and Antholine, W. E., 1980, Evidence that carnosine and anserine may participate in Wilson's disease, *Biochem. Biophys. Res. Commun.* **92:**470.

Brown, C. E., and Wilkie, C. A., 1982a, Phosphorus-31 NMR spectra of solid materials, *in* "Relaxation Times," Vol. 3, No. 1 (D. Dalrymple, ed.) Nicolet Instrument Corporation, Mountain View.

Brown, C. E., and Wilkie, C. A., 1982b, Approaches for recording ^{31}P NMR spectra of the 'Solid' components of biological tissues—nuclear polarization, *in* "Program and Abstracts, 23rd Experimental NMR Conference," Madison, Wisconsin, Abs. A-11.

Brown, C. E., and Wilkie, C. A., 1982c, Characterization of the solid components of biological tissues by cross polarization, magic angle NMR spectroscopy, *CRC Critical Reviews in Biomedical Engineering*, in press.

Brown, C. E., Margolis, F. L., Williams, T. H., Pitcher, R. G., and Elgar, G., 1977, Carnosine in olfaction: Proton magnetic resonance spectral evidence for tissue-specific carnosine binding sites, *Neurochem. Res.* **2:**555.

Brown, C. E., Margolis, F. L., Williams, T. H., Pitcher, R. G., and Elgar, G. J., 1979, Carnosine binding: Characterization of steric and charge requirements for ligand recognition, *Arch. Biochem. Biophys.* **193:**529.

Brown, C. E., Antholine, W. E., and Froncisz, W., 1980, Multiple forms of the Copper(II)-Carnosine Complex, *J. Chem. Soc. Dalton Trans.*, **1980:**590.

Brownell, G. L., Budinger, T. F., Lauterbur, P. C. and McGeer, P. L., 1982, Positron tomography and nuclear magnetic resonance imaging, *Science* **215:**619.

Burt, C. T., Glonek, T. and Bárány, M., 1977, Analysis of living tissue by phosphorus-31 magnetic resonance, *Science,* **195:**145.

Chan, S. I., Lindsey, H., Eigenberg, K. E., Croasmun, W. R., and Campbell, G. W., 1979, The motional states of two unusual bilayer membrane systems: Diphytanoylphosphatidylcholine and Chlorophyll α-Distearoylphosphatidylcholine, *in* "NMR and Biochemistry, A Symposium Honoring Mildred Cohn," (S. J. Opella and P. Lu, eds.), pp. 249–268, Marcell Dekker, New York.

Chapman, D., 1972, Recent studies using nuclear magnetic resonance spectroscopy of lipids and biological membranes, *Biomembranes.* **3:**281.

Chapman, D. and Oldfield, E., 1974, Nuclear magnetic resonance studies of biological and model membrane systems, *in* "Methods in Enzymology, Vol. XXXII, Part B" (L. Packer and S. Fleischer, eds.), pp. 198–211, Academic Press, New York.

Chapman, D., Kamat, V. B., de Gier, J., and Penkett, S. A., 1968, Nuclear magnetic resonance studies of erythrocyte membranes, *J. Mol. Biol.* **31:**101.

Chapman, D., Oldfield, E., Doskočilová, D., and Schneider, B., 1972, NMR of gel and liquid crystalline phospholipids spinning at the 'Magic Angle,' *FEBS Lett.* **25**:261.

Conlon, T., and Outhred, R., 1972, Water diffusion permeability of erythrocytes using an NMR technique, *Biochim. Biophys. Acta* **288**:354.

Cullis, P. R., and Hope, M. J., 1980, The bilayer stabilizing role of sphingomyelin in the presence of cholesterol, a ^{31}P NMR study, *Biochim. Biophys. Acta* **597**:533.

Cullis, P. R., and Verkleij, A. J., 1979, Modulation of membrane structure by Ca^{2+} and dibucaine as detected by ^{31}P NMR, *Biochim. Biophys. Acta* **552**:546.

Daniels, A., Williams, R. J. P., and Wright, P. E., 1976, Nuclear magnetic resonance studies of the adrenal gland and some other organs, *Nature (London)* **261**:321.

Davis, D. G., and Inesi, G., 1972, Phosphorus and proton nuclear magnetic resonance studies in sarcoplasmic reticulum membranes and lipids, a comparison of phosphate and proton group mobilities in membranes and lipid bilayers, *Biochim. Biophys. Acta* **282**:180.

DeKruijff, B., and Wirtz, K. W. A., 1977, Induction of a relatively fast transbilayer movement of phosphatidylcholine in vesicles, a ^{13}C NMR study, *Biochim. Biophys. Acta* **468**:318.

DeKruijff, B., Cullis, P. R., and Verkleij, A. J., 1980, Non-bilayer lipid structures in model and biological membranes, *Trends Biochem. Sci.* **5**:79.

Fritz, O. G., and Swift, T. J., 1967, The state of water in polarized and depolarized frog nerves, a proton magnetic resonance study, *Biophys. J.* **7**:675.

Glonek, T., 1980, Applications of ^{31}P NMR to biological systems with emphasis on intact tissue determinations, *in* "Phosphorus Chemistry Directed toward Biology" (W. J. Stec, ed.), pp. 157–174, Pergammon, New York.

Haberkorn, R. A., Herzfeld, J., and Griffin, R. G., 1978, High resolution ^{31}P and ^{13}C nuclear magnetic resonance spectra of unsonicated model membranes, *J. Am. Chem. Soc.* **100**:1296.

Hartmann, S. R., and Hahn, E. L., 1962, Nuclear double resonance in the rotating frame, *Phys. Rev.* **128**:2042.

Herzfeld, J., Roufosse, A., Haberkorn, R. A., Griffen, R. G., and Glimcher, M. J., 1980, Magic angle sample spinning in inhomogeneously broadened biological systems, *Phil. Trans. R. Soc. London. B* **289**:459.

Högland, A., Joelsson, I., Ingelman-Sundberg, A., and Odeblad, E., 1973, Acridine orange in gynecological cancer II. The effect of stain receptors on some proton magnetic resonance parameters, *Acta Obstet. Gynec. Scand.* **52**:1.

Hollis, D. P., 1979, Nuclear magnetic resonance studies of cancer and heart disease, *Bulletin Magnetic Resonance* **1**:27.

Hui, S. W., Stewart, T. P., and Boni, L. T., 1981, Membrane fusion through point defects in bilayers, *Science* **212**:921.

James, T. L., 1975, "Nuclear Magnetic Resonance in Biochemistry, Principles and Applications," pp. 298–388, Academic Press, New York.

Jardetzky, O., and Wade-Jardetzky, N. G., 1980, Comparison of protein structures by high resolution solid state and solution NMR, *FEBS Lett.* **110**:133.

Kohler, S. J., and Klein, M. P., 1976, ^{31}P nuclear magnetic resonance chemical shielding tensors of phosphorylethanolamine, lecithin, and related compounds: Applications to head-group motion in model membranes, *Biochemistry* **15**:967.

Kohler, S. J., and Klein, M. P., 1977, Orientation and dynamics of phospholipid head groups in bilayers and membranes determined from ^{31}P nuclear magnetic resonance chemical shielding tensors, *Biochemistry* **16**:519.

Lauterbur, P. C., 1977, Spatially-resolved studies of whole tissues, organs and organisms by NMR zeugmatography, *in* "NMR in Biology" (R. A. Dwek, I. D. Campbell, R. E. Richards, and R. J. P. Williams, eds.), pp. 323–335, Academic Press, New York.

Lee, A. G., Birdsall, N. J. M., and Metcalfe, J. C., 1973, NMR studies of biological membranes, *Chem. Br.* **9:**116.

Martin, R. B., and Mathur, R., 1965, Analysis of proton magnetic resonance spectra of cysteine and histidine and derivatives, conformational equilibria, *J. Am. Chem. Soc.***87:**1065.

McLaughlin, A. C., Cullis, P. R., Hemminga, M., Brown, F. F., and Brocklehurst, J., 1977, Magnetic resonance studies of model and biological membranes, *in* "NMR in Biology" (R. A. Dwek, I. D. Campbell, R. E. Richards, and R. J. P. Williams, eds.), pp. 231–246, Academic Press, New York.

Miller, J., Witzemann, V., Quast, U., and Raftery, M. A., 1979, Proton magnetic resonance studies of cholinergic ligand binding to the acetylcholine receptor in its membrane environment, *Proc. Natl. Acad. Sci. USA* **76:**3580.

Minshew, N. J., Henderson, A. C., and Pettegrew, J. W., 1981, Hepatic encephalopathy: Effect of short-chain fatty acids on cellular membranes, *Neurology* **31**(2):142.

Morariu, V. V., and Benga, G., 1977, Evaluation of a nuclear magnetic resonance technique for the study of water exchange through erythrocyte membranes in normal and pathological subjects, *Biochim. Biophys. Acta* **469:**301.

Pachler, K. G. R., 1964, Nuclear magnetic resonance study of some amino acids. II. Rotational isomerism, *Spectrochim. Acta* **20:**581.

Pettegrew, J. W., and Minshew, N. J., 1981, Effects of short-chain fatty acids on cellular membranes and energy metabolism: A nuclear magnetic resonance study, *Neurology* **31**(2):143.

Pettegrew, J. W., and Minshew, N. J., 1982, Effects of short chain fatty acids on cellular membranes and energy metabolism: A nuclear magnetic resonance study, in preparation.

Pettegrew, J. W., Glonek, T., Baskin, F., and Rosenberg, R. N., 1979, Phosphorus-31 NMR of neuroblastoma clonal lines, effect of cell confluency state and dibutyryl cyclic AMP, *Neurochem. Res* **4:**795.

Pines, A., Gibby, M. G., and Waugh, J. S., 1973, Proton-enhanced NMR of dilute spins in solids, *J. Chem. Phys.* **59:**569.

Pirkle, J. L., Ashley, D. L., and Goldstein, J. H., 1979, Pulse nuclear magnetic resonance measurements of water exchange across the erythrocyte membrane employing a low Mn concentration, *Biophys. J.* **25:**389.

Podo, F., 1975, The application of nuclear magnetic resonance spectroscopy to the study of natural and model membranes, *Biochimie* **57:**461.

Pople, J. A., Schneider, W. G., and Bernstein, H. J., 1959, "High-Resolution Nuclear Magnetic Resonance," McGraw–Hill, New York.

Ross, B. D., Radda, G. K., Gadian, D. G., Rocker, G., Esiri, M., and Falconer-Smith, J., 1981, Examination of a case of suspected McArdle's syndrome by ^{31}P nuclear magnetic resonance, *N. Engl. J. Med.* **304:**1338.

Rothwell, W. P., Waugh, J. S., and Yesinowski, J. P., 1980, High-resolution variable-temperature ^{31}P NMR of solid calcium phosphates, *J. Am. Chem. Soc.* **102:**2637.

Schaefer, J., Stejskal, E. O., and McKay, R. A., 1979, Cross-polarization NMR of N-15 labeled soybeans, *Biochem. Biophys. Res. Commun.* **88:**274.

Seeley, P. J., Sehr, P. A., Gadian, D. G., Garlick, P. B., and Radda, G. K., 1977, Phosphorus NMR in living tissue, *in* "NMR in Biology" (R. A. Dwek, I. D. Campbell, R. E. Richards, and R. J. P. Williams, eds.), pp. 247–275, Academic Press, New York.

Seelig, J., 1977, Deuterium magnetic resonance: Theory and application to lipid membranes, *Q. Rev. Biophys.* **10:**353.

Sheetz, M. P., and Chan, S. I., 1972, Proton magnetic resonance studies of whole human erythrocyte membranes, *Biochemistry* **11:**548.

Shulman, R. G., Brown, T. R., Ugurbil, K., Ogawa, S., Cohen, S. M., and den Hollander, J. A., 1979, Cellular applications of ^{31}P and ^{13}C nuclear magnetic resonance, *Science* **205:**160.

Small, D. M., and Shipley, G. G., 1974, Physical-chemical basis of lipid deposition in atherosclerosis, *Science* **185:**222.

Ting, D. Z., Hagan, P. S., Chan, S. I., Doll, J. D., and Springer, Jr., C. S., 1981, Nuclear magnetic resonance studies of cation transport across vesicle bilayer membranes, *Biophys. J.,* **34:**189.

Urbina, J., and Waugh, J. S., 1974, Proton-enhanced ^{13}C nuclear magnetic resonance of lipids and biomembranes, *Proc. Nat. Acad. Sci. USA,* **71:**5062.

Voelter, W., Jung, G., Breitmaier, E., and Bayer, E., 1971, ^{13}C-chemische verscheibungen von aminosauren und peptiden, *Z. Naturforsch.* **26b:**213.

Yaari, A. M., Shapiro, I. M. and Brown, C. E., 1982, Evidence that phosphatidylserine and inorganic phosphate may mediate calcium transport during calcification, *Biochem. Biophys. Res. Commun.* **105:**778.

Index